Lecture Notes in Business Information Processing

600

Series Editors

LNBIP reports state-of-the-art results in areas related to business information systems and industrial application software development – timely, at a high level, and in both printed and electronic form.

The type of material published includes

- Proceedings (published in time for the respective event)
- Postproceedings (consisting of thoroughly revised and/or extended final papers)
- Other edited monographs (such as, for example, project reports or invited volumes)
- Tutorials (coherently integrated collections of lectures given at advanced courses, seminars, schools, etc.)
- Award-winning or exceptional theses

LNBIP is abstracted/indexed in DBLP, EI and Scopus. LNBIP volumes are also submitted for the inclusion in ISI Proceedings.

Krzysztof Kluza · Karolin Winter ·
Lucineia Heloisa Thom · Amy Van Looy ·
Marco Comuzzi · Stephan Fahrenkrog-Petersen ·
Manfred Reichert · Joerg Evermann · Barbara Re
Editors

Business Process Management

Responsible BPM Forum, Process Technology Forum, Educators Forum

BPM 2026 RBPM, PT, and Educators Forum
Toronto, ON, Canada, September 27 – October 2, 2026
Proceedings

Springer

Editors
Krzysztof Kluza
AGH University of Krakow
Krakow, Poland

Lucineia Heloisa Thom
Federal University of Rio Grande do Sul
Porto Alegre, Rio Grande do Sul, Brazil

Marco Comuzzi
Ulsan National Institute of Science and Technology
Ulsan, Korea (Republic of)

Manfred Reichert
Ulm University
Ulm, Germany

Barbara Re
University of Camerino
Camerino, Italy

Karolin Winter
Eindhoven University of Technology
Eindhoven, The Netherlands

Amy Van Looy
Ghent University
Ghent, Belgium

Stephan Fahrenkrog-Petersen
University of Liechtenstein
Vaduz, Liechtenstein

Joerg Evermann
Memorial University of Newfoundland
St. John's, NL, Canada

ISSN 1865-1348 ISSN 1865-1356 (electronic)
Lecture Notes in Business Information Processing
ISBN 978-3-032-36709-9 ISBN 978-3-032-36710-5 (eBook)
https://doi.org/10.1007/978-3-032-36710-5

This Springer imprint is published by the registered company Springer Nature Switzerland AG
The registered company address is: Gewerbestrasse 11, 6330 Cham, Switzerland

Preface

The Business Process Management Conference was founded 24 years ago to bring together pioneering researchers and industry leaders to explore innovations, share best practices, and advance the field of process management through collaboration and knowledge exchange. BPM 2026 also featured specialized forums to present advancements and discuss specific topics related to business process management. All forums used a double-blind review process in which submissions received three reviews each, on average.

The Responsible Business Process Management (RBPM) Forum responds to the growing recognition that BPM must move beyond its traditional efficiency-driven paradigm to explicitly address societal, ethical, and environmental concerns. In line with the 2026 call for papers, Responsible BPM is understood as the systematic design, analysis, and improvement of business processes that account for a broad spectrum of stakeholders, including employees, customers, regulators, and society at large, while also considering long-term environmental sustainability and pressing global challenges. The RBPM Forum serves as a dedicated venue for advancing this emerging research agenda by fostering contributions that integrate responsibility into BPM from technological, managerial, and conceptual perspectives. It particularly encourages work that bridges disciplines, critically reflecting on the implications of BPM in complex socio-technical systems. In its second edition, the forum attracted 20 submissions, of which 8 were selected for presentation and inclusion in the proceedings. These contributions reflect a diverse set of approaches and insights, collectively advancing the discourse on how BPM can evolve to support responsible, sustainable, and ethically grounded organizational practices.

The Process Technology Forum covers technical and engineering aspects related to the study, design, development, and maintenance of software systems that support the modeling, simulation, analysis, enactment and monitoring of business processes, "process systems" in short. It combines advanced computer science methods and techniques with real-world applications that impact nearly every industry, helping enterprises to streamline their business operations, enhance process efficiency, and respond to the needs of ever-changing markets. The Process Technology Forum received 22 submissions, of which 8 were accepted for presentation. Furthermore, Soren Frederiksen from industry and Sander Leemans from academia were invited as keynote speakers.

The Educators Forum aims to be the meeting point for educators within the BPM community to share resources, exchange experiences, present teaching cases, and improve BPM teaching in a dynamic and practice-oriented way, acknowledging that BPM has become part of the curriculum across universities, professional programmes, online courses, and in-company training. Now in its fourth edition, the forum welcomed contributions of several types: BPM Education Research, BPM Teaching Cases, Educational Practice Insights, Tools/Resources Contributions, Assignment Contributions, and Lightning Contributions. The forum received a total of 23 submissions, of which 11 were accepted. The accepted papers show a field in transition and fall into three groups: how AI can support teaching, what an AI-shaped profession now requires of graduates, and what remains constant in learning to model and analyse processes. In all three, one thing stands out: a community that looks at generative AI with care and a critical eye, instead of being caught up in the hype.

We hope that the reader finds the selection of papers and presentations on these specific topics inspiring.

September 2026

Amy Van Looy
Marco Comuzzi
Stephan Fahrenkrog-Petersen
Manfred Reichert
Joerg Evermann
Barbara Re
Krzysztof Kluza
Karolin Winter
Lucineia Heloisa Thom

Preface of the Responsible BPM Forum

The Responsible BPM Forum at BPM 2026 was conceived as an academic space to broaden the meaning of business process management beyond efficiency and effectiveness, and to foster academic exchange on Responsible Business Process Management (RBPM). This year marked the second edition of the forum, which followed a successful first edition in Seville, Spain, in 2025. In the call for papers, the forum defined RBPM as BPM that accounts for the needs of stakeholders, the environment, and society, highlighting the importance of the concept of \emph{responsibility} across the BPM lifecycle, including analysis, design, implementation, monitoring, and the development and use of BPM systems in responsible ways. The call also clarified that RBPM is shaped by technological, managerial, and conceptual challenges, especially where modern challenges such as automation, AI, privacy, transparency, accountability, bias, sustainability, justice, and inclusion intersect.

This year's forum attracted 20 submissions, each reviewed by three members of the program committee, and 8 papers were accepted for inclusion in the proceedings. The accepted contributions reflect the forum's broad and timely agenda, spanning ethical foundations, inclusive and socially-aware process design, responsible AI delegation, the concept of process externalities, transparency in complex socio-technical settings, and privacy-preserving process mining. Together, the accepted contributions show that responsibility in BPM is not a single concern, but a set of interrelated questions cutting across methods, technologies, and organizational practices.

Among the accepted papers, one work ("Towards Ethics in BPM Research") directly revisits the normative foundations of the BPM field, addressing the call for papers' emphasis on conceptualizing responsibility in BPM beyond conventional performance criteria. Another paper ("From Principles to Practice: Embedding GEDSI in Process Design") aligns with the forum's concern for inclusion and stakeholder needs by translating equity-oriented principles into process design choices. The paper "Participatory Process Mining: A Framework and Call for Research" argues for better inclusion of process participants to improve process mining transparency and avoid organizational resistance in process mining insights implementation. Two more accepted papers ("AI Work Under Guardrails: Task Framing for Responsible Delegation in BPM" and "The System Did It: Bias, Opacity, and the Misalignment of Ethical Mitigations in (Robotic) Process Automation") address the ethical challenges of automation and AI integration, a core theme in the forum's call for papers.

The remaining accepted papers further enrich the discussion by considering additional perspectives on responsibility in BPM. The paper "Process Externalities: A Missing Construct for Responsible BPM" advances the conceptual vocabulary of responsibility by making visible effects that standard process metrics may ignore, while the paper "Silent Responsibility Voids: Evidence for Process Twins in Clinical Trials" points to the practical difficulty of assigning accountability in complex process settings. Finally, the paper "Differentially Private Heuristics Miner" contributes a more technical response to the forum's call for papers by directly addressing privacy concerns in a widely adopted process discovery technique. Collectively, the accepted papers demonstrate that Responsible BPM is emerging as a genuinely multidisciplinary field, one that requires theoretical clarity, methodological rigor, and practical sensitivity to the societal consequences of process-related design and decisions.

We thank the authors for their contributions and the program committee members for their careful reviewing effort. We also express our appreciation to the BPM 2026 organizers for supporting a forum that encourages the community to rethink what successful process management should mean in an era of AI, automation, sustainability pressures, and growing demands for fairness and accountability.

Organization

Program Chairs

Amy Van Looy	Ghent University, Belgium
Marco Comuzzi	Ulsan National Institute of Science and Technology, South Korea
Stephan Fahrenkrog-Petersen	University of Liechtenstein, Liechtenstein

Program Committee

Irene Vanderfeesten	KU Leuven, Belgium
Luise Pufahl	Technische Universität München, Germany
Jana-Rebecca Rehse	University of Mannheim, Germany
Francesca Zerbato	Eindhoven University of Technology, The Netherlands
Iris Beerepoot	University of Utrecht, The Netherlands
Daniel Beimborn	University of Bamberg, Germany
Henrik Leopold	Kühne Logistics University, Germany
Han van der Aa	University of Vienna, Austria
Majid Rafiei	SAP, Germany
Martin Kabierski	University of Vienna, Austria
William van Woensel	University of Ottawa, Canada
Yannis Betrandt	Hasselt University, Belgium
Mahendrawathi Er	Institut Teknologi Sepuluh Nopember, Indonesia
Avigdor Gal	Technion - Israel Institute of Technology, Israel
Thomas Grisold	Vienna University of Economics and Business, Austria
Flavia Santoro	Inteli - Instituto de Tecnologia e Liderança, Brazil
Banu Aysolmaz	Eindhoven University of Technology, The Netherlands
Laura Dobusch	Vienna University of Economics and Business, Austria
Djordje Djurica	Vienna University of Economics and Business, Austria
Jens Gulden	IU International University, Germany
Jan Mendling	Humboldt University, Germany
Michael Rosemann	Queensland University of Technology, Australia

Maximilian Roglinger	University of Bayreuth, Germany
Stefan Schönig	University of Regensburg, Germany
Rehan Syed	Queensland University of Technology, Australia
Lucinéia Heloisa Thom	Federal University of Rio Grande do Sul, Brazil
Oktay Türetken	Eindhoven University of Technology, The Netherlands
Arif Wibisono	Institut Teknologi Sepuluh Nopember, Indonesia
Anna Maria Oberlander	University of Bayreuth, Germany

Preface of the Process Technology Forum

Process systems combine advanced computer science methods and techniques with real-world applications that impact nearly every industry. Process systems sit at the heart of modern software applications helping enterprises to streamline their business operations, enhance process efficiency, and respond to the needs of ever-changing markets.

Process system engineers carry out many complex tasks to build and maintain such systems. They develop flexible modeling frameworks to capture diverse processes and complex process dependencies in business operations. They also develop simulation tools to test these processes under different scenarios, providing valuable insights into resource allocation, bottlenecks, and opportunities for process optimization. Process systems often incorporate advanced analytics to track process and system performance, allowing for the real-time monitoring of processes, the detection of inefficiencies, and data-driven process improvements. The enactment component of process systems must handle large numbers of concurrently executed processes efficiently, securely, and reliably. It also requires a seamless integration with various enterprise software systems, ensuring that processes align across all organizational functions. As organizations increasingly adopt AI-driven automation, process systems are evolving to support automated decision-making, predictive analytics, and self-optimization. Additionally, process systems are increasingly designed for cross-organizational scenarios, requiring process system engineers to address scalability, security, privacy, and fault tolerance across multiple servers, ensuring high availability. Inter-organizational processes require process systems that provide interchange and synchronization mechanisms.

The Process Technology Forum 2026, held in conjunction with the 24th International Conference on Business Process Management, provided a platform for introducing and discussing new research ideas related to the technical and engineering aspects of process systems. We invited submissions on technical and engineering aspects related to the study, design, development, and maintenance of software systems that support the modeling, simulation, analysis, enactment, and monitoring of business processes, "process systems" in short.

The forum received 22 paper submissions for evaluation, out of which 8 were selected for presentation. Each paper was reviewed by at least two reviewers, after which the program chairs jointly made acceptance decisions. The papers included in this year's Process Technology Forum cover a broad range of different areas of process technology, reflecting current advances in both foundational methods and emerging AI-driven approaches. The contributions are organized into two thematic areas. The area of Process Mining, Modeling, and Analysis focuses on techniques for discovering, representing, querying, and analyzing business processes. The papers deal with conversational agents for process mining, the representation of BPMN choreographies using OCEL 2.0, the generation of AI agents from software engineering process records through process mining, and SQL-based querying of object-centric process data. Together, these contributions advance the state of the art in process discovery, process representation, and object-centric process analysis. The area of Intelligent Process Execution and Automation explores the application of artificial intelligence to process execution and automation. The papers investigate natural language querying of declarative rule repositories, LLM-based orchestration of BPM toolchains, blockchain-enabled business process execution, and organizational memory architectures for agentic business process execution. Collectively, they demonstrate how recent advances in AI, large language models, and distributed technologies are shaping the next generation of intelligent, autonomous, and adaptive process-aware systems.

To kick off the Process Technology Forum, we invited two keynote speakers: Soren Frederiksen from industry and Sander Leemans from academia. Both provided insights into the use of process mining tools in practice.

We hope that the reader finds the selection of papers and presentations useful to get an insight into advances of the state of the art in process technology.

Organization

Program Chairs

Manfred Reichert	Universität Ulm, Germany
Joerg Evermann	Memorial University of Newfoundland, Canada
Barbara Re	University of Camerino, Italy

Program Committee

Dominik Bork	TU Wien, Austria
Carl Corea	Universität Koblenz, Germany
Patrick Delfmann	Universität Koblenz, Germany
Chiara Di Francescomarino	University of Trento, Italy
Remco M. Dijkman	Eindhoven University of Technology, The Netherlands
Rik Eshuis	Eindhoven University of Technology, The Netherlands
Peter Fettke	Universität des Saarlandes and DFKI, Germany
Andrés Jiménez Ramírez	Universidad de Sevilla, Spain
Dimka Karastoyanova	University of Groningen, The Netherlands
Jochen Küster	Hochschule Bielefeld, Germany
Julius Köpke	Universität Klagenfurt, Austria
Sander Leemans	RWTH Aachen, Germany
Orlenys López-Pintado	University of Tartu, Estonia
Lukas Malburg	University of Trier and DFKI, Germany
Jürgen Mangler	Technical University of Munich, Germany
Raimundas Matulevičius	University of Tartu, Estonia
Andreas Metzger	Universität Duisburg Essen, Germany
Luise Pufahl	Technische Universität München, Germany
Stefanie Rinderle-Ma	Technische Universität München, Germany
Mirela Riveni	University of Groningen, The Netherlands
Ronny Seiger	Universität St. Gallen, Switzerland
Estefania Serral	KU Leuven, Belgium
Irene Vanderfeesten	KU Leuven, Belgium
Ingo Weber	Technische Universität München, Germany
Barbara Weber	Universität St. Gallen, Switzerland
Matthias Weidlich	HU Berlin, Germany
Mathias Weske	HPI and University of Potsdam, Germany

Preface of the Educators Forum

When a chatbot can draft a BPMN model in seconds, what is left for us to teach – and how should we teach it? Business Process Management has long been taught in universities, professional programmes, online courses, and in-company training. Yet the work of teaching it – designing courses, creating assignments, and learning from classroom successes and challenges – too often goes unspoken at BPM conferences. Now in its fourth edition, the Educators Forum aimed to change this by bringing educators in the BPM community together to share resources, exchange experiences, present teaching cases, and discuss the innovations that keep our courses engaging and relevant. The Educators Forum took place at a time of rapid change, as generative AI is moving from a side topic in the syllabus to a force that shapes how processes are designed, run, taught, and learned – often faster than curricula can adapt.

What is most striking about this year's accepted papers is not simply that many of them address AI, but how carefully and critically they do so. This year's Educators Forum received twenty-three submissions, eleven of which were accepted following a double-blind review process. Each paper was evaluated by three members of the Program Committee. The call welcomed contributions of several kinds – education-research studies, teaching cases, educational-practice insights, tools and resources, assignments, and short lightning talks. Among the accepted papers, studies of how BPM is taught and learned clearly predominate, though teaching cases, practice reports, tools, and assignments are also represented. Taken together, the papers fall into three groups that reflect a field in transition: how AI can support teaching, what an AI-shaped profession now requires of graduates, and what remains constant in learning to model and analyse processes. The papers below are organised around this common thread.

The first group treats AI as a teaching tool – something to build, try out, and critically evaluate. "*Simulating the Domain Expert: An Empirical Evaluation of LLM-Based Interview Training for Business Process Modeling*" tackles a long-standing bottleneck in teaching process discovery: students rarely get to practise interviewing domain experts. In the study, students interview both an LLM-based simulated expert and a real human expert before building a BPMN model. The simulated sessions were much faster, were perceived as equally informative, and produced models of similar quality – a promising tool for early, repeated practice. "*Adaptive BPM Learning Journeys: An AI-Powered Platform for Personalized Assessment and Multimodal Content Generation*" responds to the wide range of BPM learners with a platform that combines three elements – personalised content, LLM-generated materials and quizzes, and game-like elements – into a five-phase learning journey. "*Does AI Improve BPMN Modelling? Evidence from Iterative Assignments and Critical Verification Practice*" then asks the harder question of whether such help actually works. Through a two-stage assignment and a paired comparison of student-refined and AI-assisted models, it finds mixed results, and shows that what matters is not how much AI is used but whether students check its output carefully – a lesson that resonates across the Educators Forum. Finally, "*Sovereign AI*

Infrastructure for BPM Education: An Open-Source Stack from Document Analysis to Process Execution" shifts attention from the AI model itself to the conditions under which it is used. The paper presents an open-source, open-weight stack built on national HPC infrastructure and designed to remove marginal licensing costs. It argues that this kind of sovereignty – control over cost, access, policy, and data jurisdiction – allows educators to teach with AI on their own terms.

If the first group asks how AI can help us teach, the second asks a larger question: as AI and the wider transformation of work reshape the profession, are we still teaching the right things? "*AI-Decision Checkpoints for AI-Augmented Business Process Management: Framework and Educational Instantiation*" argues that curricula still often treat AI as an add-on, something separate from process design. The paper presents a teaching case based on a customer-onboarding process with clear decision points, where students decide whether to adopt, limit, or reject each AI component. In this way, students learn to treat AI as an integral part of process design, rather than something added at the end. "*Are BPM Graduates Future-Ready? – Reframing BPM Competencies for the Future of Work*" uses Critical Discourse Analysis to examine how BPM competencies are described in the research literature and compares them with ten years of World Economic Forum Future of Jobs reports. Its competency–skill matrix suggests that BPM does not lack key competencies; rather, the problem is one of emphasis. Social, adaptive, and sustainability-related skills receive too little attention. A more tool-oriented contribution, "*AI and the Evolving BPM Practitioner Skill Set. Initial Implications for Curriculum Review*" examines the AI features of seventeen BPM platforms across thirty capability areas, showing where AI reinforces existing skills, changes the way they need to be taught, reduces the need for some sub-tasks, and creates entirely new competency demands. It also points to gaps at the module level that curricula will need to address.

Not everything in BPM education changes with AI. A third group of papers reminds us of what remains firmly human in BPM education: learning to see, model, and analyse processes with care and judgement. These papers show that, even as tools change, this work remains essential, and AI, where it appears, plays a supporting role. "*Learning Process Modelling in BPMN Through Real-life Process Observation*" draws on learning diaries from several student cohorts to show that students gradually come to understand modelling as much more than drawing diagrams. They learn to see it as interpretive, social, and transferable to new settings, while generative AI plays only a small and clearly limited role. "*Instructional Design in Process Modelling: Surveying Educators' Perspectives*" turns to teachers themselves.

Using a five-stage instructional-design framework, the paper studies how process-modelling educators approach their teaching and identifies clear gaps.

These include too little attention to students' reflection on their own thinking, higher-order skills, and the use of animation to support understanding. "*Teaching Declarative BPM: Where, When and to Whom it Matters*" brings together experiences of teaching declarative models – including Declare, DCR graphs, and DMN – at universities in Belgium, Denmark, Italy, and the Netherlands. The paper reflects honestly on where these models fit in the curriculum, how students from different backgrounds respond to them, how well the tools support learning, and why it remains difficult to move beyond the imperative approach that still dominates BPM education. "*From Classroom to Practice: Bridging BPM Education and Industry Expectations Through Process Mining Tool Comparison Project*" connects BPM education with industry expectations. In this project, students compare process-mining tools using real event logs, supported by formative feedback and peer review. The assessment is designed to help students become industry-ready analysts who can reason from evidence.

Across all three groups, the same thoughtful attitude comes through. This is not a community carried away by AI, but one that examines it carefully. The papers insist that AI outputs must be checked, that educators should remain in control of the tools they use, that decisions about AI should be made openly, and that human judgement and the collaborative nature of modelling should remain at the centre of learning. In this sense, this year's Educators Forum offers not only a view of new technologies, but also a careful way of thinking about them.

We warmly thank the authors for sharing their work with the Educators Forum, the Program Committee members for their careful and thorough reviews, and the BPM 2026 chairs and organisers for their support. Above all, we hope that you find here an idea worth taking back to your own classroom, and that the Educators Forum becomes a place where these ideas can grow through discussion.

July 2026

Krzysztof Kluza
Karolin Winter
Lucineia Heloisa Thom

Organization

Program Committee Chairs

Krzysztof Kluza	AGH University of Krakow, Poland
Karolin Winter	Eindhoven University of Technology, The Netherlands
Lucineia Heloisa Thom	Federal University of Rio Grande do Sul, Brazil

Program Committee

Banu Aysolmaz	TU Eindhoven, The Netherlands
Matthijs Berkhout	Hogeschool Utrecht, The Netherlands
Evellin Cardoso	Free University of Bolzano, Italy
Patrick Delfmann	University of Koblenz-Landau, Germany
Djordje Djurica	Vienna University of Economics and Business, Austria
Mahendrawathi Er	Sepuluh Nopember Institute of Technology, Indonesia
Irene Bedilia Estrada Torres	University of Seville, Spain
Marcelo Fantinato	University of São Paulo, Brazil
Katarzyna Gdowska	AGH University of Krakow, Poland
Marzena Grzesiak	Gdańsk University of Technology, Poland
Katarzyna Grzybowska	Poznań University of Technology, Poland
Hugo A. López	Technical University of Denmark, Denmark
Azumah Mamudu	Queensland University of Technology, Australia
Niels Martin	Hasselt University, Belgium
Ilia Maslov	KU Leuven University, Belgium
Asma Mejri	University of Sousse, Tunisia
Jorge Munoz-Gama	Pontificia Universidad Católica de Chile, Chile
Izaque Dione Nunes	Federal University of Rio Grande do Sul, Brazil
Kate Revoredo	Humboldt-Universität zu Berlin, Germany
Guilherme Rego Rockembach	Federal University of Rio Grande do Sul, Brazil
Piotr Sliż	University of Gdańsk, Poland
Encarna Sosa Sánchez	University of Extremadura, Spain
Vinicius Stein Dani	Utrecht University, The Netherlands
Anna Suchenia	Cracow University of Technology, Poland

Leszek Szała	University of Chemistry and Technology, Prague, Czechia
Oktay Türetken	Eindhoven University of Technology, The Netherlands

Additional Reviewer

Sabine Nagel

Contents

Responsible BPM Forum

AI Work Under Guardrails: Task Framing for Responsible Delegation in BPM

Kristof Antal(✉)

Corvinus University of Budapest, Budapest, Hungary
kristof.antal@uni-corvinus.hu

Abstract. Enterprise AI governance can define principles for accountability, transparency, human oversight, and auditability, but organizations still need to translate these principles into concrete decisions about whether and under what constraints a specific unit of process work may be delegated to artificial intelligence (AI). This research-in-progress paper introduces the Responsibility-Gated Task Frame as a design science artifact for this translation. The artifact extends a descriptive Task Frame with responsibility conditions and supports a task-level decision on the form and constraints of AI-supported delegation. Following a design science research approach, the paper presents the artifact as part of an early design cycle and demonstrates its use through a vignette on AI-assisted compliance assessment based on International Organization for Standardization (ISO) management-system standards. The contribution is a task-level mechanism for operationalizing Responsible Business Process Management (RBPM) and AI governance before AI implementation begins.

Keywords: Responsible BPM · AI governance · AI delegation · task framing · design science research

1 Introduction

Artificial intelligence (AI) is increasingly becoming part of Business Process Management (BPM) as both an analytical technology and a contributor to process work. AI-supported systems can classify information, prepare evidence, recommend actions, support decisions, and monitor deviations. This development expands the responsibility of BPM design, because AI-generated outputs may affect workers, customers, managers, regulators, and other stakeholders. The Responsible BPM Forum defines Responsible Business Process Management (RBPM) as applying BPM while accounting for the needs of process stakeholders, the broader environment, and society, including privacy, ethical AI and automation, transparency, accountability, and algorithmic bias [1].

At the organizational level, these concerns are typically addressed through enterprise AI governance: principles, policies, roles, risk criteria, oversight mechanisms, and documentation requirements that define how AI should be developed and used responsibly. Yet governance commitments still have to be applied where concrete process work is changed. BPM research on assisted redesign shows how process-related issues can be

K. Kluza et al. (Eds.): BPM 2026, LNBIP 600, pp. 3–10, 2027.
https://doi.org/10.1007/978-3-032-36710-5_1

used to develop redesign alternatives [2], while explorative BPM highlights opportunity-driven process change in which machine learning may be embedded into business processes [3]. Together, these views motivate a task-level translation problem: when AI is considered for a specific process point, the organization must decide what work may be assigned to AI and under which responsibility conditions.

Delegation is central to this problem. Agentic information systems shift attention from passive system use toward the transfer of rights and responsibilities between human and artificial actors [4], while human–AI collaborative decision-making can be understood as an organization design problem concerning the division and coordination of work between human and algorithmic contributors [5]. Building on this view, the paper introduces the Responsibility-Gated Task Frame, a design artifact that represents a candidate AI intervention point as a bounded unit of process work and supports the selection of a primary delegation alternative for that framing. The paper contributes by positioning the responsible delegation of process work to AI as a BPM design problem between enterprise AI governance and AI implementation in business processes. The artifact is demonstrated through a vignette on AI-assisted compliance assessment based on ISO management-system standards.

2 Conceptual Positioning: From Responsibility to Delegating Work to AI

Responsible Business Process Management (RBPM) provides the normative starting point for this paper, while enterprise AI governance provides the organizational layer through which responsibility commitments are defined. AI ethics research shows broad convergence around principles such as transparency, justice and fairness, non-maleficence, responsibility, and privacy [6]. However, such principles do not directly specify how AI should be introduced into a concrete business process. Ethics-based auditing research describes this as a gap between the "what" and "why" of AI ethics principles and the "how" of designing, deploying, and governing automated decision-making systems in practice [7]. Industry case evidence further indicates that operationalizing AI governance requires scope demarcation, harmonized standards, internal communication, change management, and outcome measurement [8]. These insights support the distinction used in this paper: enterprise AI governance defines organizational guardrails, but further artifacts are needed to apply them where process work is changed.

Recent RBPM research also treats responsibility as a BPM design concern. Value-based process redesign translates stakeholder values into process redesign requirements [9], canvas-based approaches support benevolent intent in routine process operations [10], and agentic BPM research emphasizes governance mechanisms, legal and ethical guardrails, human-agent collaboration, and fallback arrangements for AI-based agents in organizational processes [11]. Building on these approaches, this paper focuses on an earlier task-level delegation decision: the local conditions under which a task, function node, control point, compliance issue, or decision-support need may receive AI support. The candidate point identifies where AI-enabled process change is considered; the Task Frame specifies what work may be assigned to AI, which authority remains with human roles, and whether the current framing supports responsible delegation.

Delegation is therefore the organizing concept for the translation from governance to task-level design. In this paper, delegation means the intentional assignment of part of a process work unit to an AI-supported system, while organizational accountability remains with human roles. This understanding is consistent with agentic information systems research on the transfer of rights and responsibilities between human and artificial actors [4] and with the view of human–AI collaboration as an organization design problem [5]. Human involvement alone is not sufficient: meaningful human control requires traceability to human agents and responsiveness to relevant reasons and facts [12], while human oversight may become an empty procedural safeguard if its actual control function is not specified [13]. The Responsibility-Gated Task Frame is positioned in this translation layer: it frames a candidate AI intervention point as a bounded unit of process work and supports a responsibility-informed decision on whether and under what constraints the work may be delegated to AI.

3 Design Problem

The design problem is the responsibility extension of task framing. A candidate AI intervention point may appear as an element in a process map, process architecture, function hierarchy, or local improvement discussion. Such representations help locate where work belongs, yet they may describe the candidate point only as a process, function, or activity label. Labels alone are insufficient for deciding whether work may be delegated to AI, because they do not explain what a process element achieves or which properties characterize the underlying capability [14].

A Task Frame is used here as a compact representation of the bounded process work considered for AI-supported delegation. It records the work unit's intended outcome, inputs, outputs, operating conditions, dependencies, required judgment, role and resource configuration, data context, and verification expectations. These fields make the candidate point analyzable before a delegation alternative is selected. The Responsibility-Gated Task Frame adds a responsibility layer to this descriptive base by defining task-level conditions for input and evidence use, AI output, retained human authority, meaningful control, and accountability [7, 8].

The resulting task-level conditions specify the constraints under which delegation may proceed: admissible and excluded inputs, permitted AI output, retained human authority, control and oversight, auditability, contestation, escalation, and revocation. These conditions connect enterprise AI governance to RBPM by applying responsibility requirements before AI implementation begins [1, 12, 13].

4 Research Design

This study follows a Design Science Research (DSR) approach, because it develops a purposeful artifact for an organizational design problem: translating enterprise AI governance into task-level decisions about delegating process work to AI. DSR is appropriate when research creates artifacts such as constructs, models, methods, or instantiations to address relevant organizational problems [15]. The artifact developed here is a task-level decision-support mechanism: a descriptive Task Frame extended with responsibility gates and delegation alternatives.

The paper reports an early design cycle. Following the Design Science Research Methodology (DSRM), Sects. 1–3 identify and motivate the problem and define the objectives of a solution; Sect. 5 presents the initial artifact; Sect. 6 demonstrates its use in a standards-informed vignette; and Sect. 7 defines the evaluation agenda [16]. The contribution is positioned as nascent design knowledge rather than a completed design theory [17]. Evaluation is planned for subsequent cycles through expert review, practitioner assessment, and later naturalistic evaluation in real process-improvement settings [18].

5 Responsibility-Gated Task Framing and Delegation Alternatives

The Responsibility-Gated Task Frame extends the descriptive Task Frame with responsibility conditions for delegating process work to AI. The descriptive layer defines the bounded work unit through its objective, inputs, outputs, operating conditions, dependencies, required judgment, role and resource configuration, data context, and verification expectations. The responsibility layer defines task-level conditions for input and evidence use, AI output, retained human authority, meaningful control, and accountability and redress mechanisms. In this early design cycle, the five gates structure recurring task-level requirements from AI governance, human control, and delegation research [4, 5, 7, 8, 12, 13]. Later evaluation will assess their coverage, overlap, and possible consolidation.

The artifact contains five responsibility gates. The first gate defines the admissible input and evidence base: which inputs and evidence may be used and which must be excluded. The second gate defines permitted AI output, such as classification, evidence preparation, recommendation, monitoring signal, or draft assessment. The third gate defines human-retained authority by specifying which judgment, approval, exception handling, or final decision remains assigned to a human role. The fourth gate defines meaningful control, including the form of oversight and the authority and competence of the human actor who can change or stop the AI-supported outcome [12, 13]. The fifth gate defines accountability and redress infrastructure: auditable evidence, contestation or escalation, and revision or revocation of the intervention [7, 8, 19].

The gates guide the selection of one primary delegation alternative for the current framing. Fully satisfied gates support delegation as framed. Gates that require additional restrictions lead to delegation with constraints, such as narrower input scope, limited output authority, mandatory human review, stronger logging, or predefined escalation. Gates that reveal hidden dependencies, affected stakeholders, authority boundaries, or responsibility gaps lead to reframing or escalation [20]. Gates whose conditions cannot be made defensible lead to exclusion from AI delegation.

The artifact output is a short delegation record. It documents the candidate work unit, the applied responsibility conditions, the selected delegation alternative, and the next design implication. This keeps the artifact usable in early process-improvement work while making the responsibility reasoning explicit before AI implementation begins. In RBPM terms, the contribution is the translation of general responsibility commitments into task-level design constraints for concrete process work.

6 Standards-Informed Demonstration: AI-Assisted Compliance Assessment

The artifact is demonstrated through an illustrative compliance-assessment vignette in an organization that manages processes and compliance obligations through management-system standards and is considering AI support for an existing human compliance task. The vignette uses ISO 9001, ISO 37301, and ISO 19011 as a standards-informed setting [21–23]. ISO 9001 provides the process-oriented basis by foregrounding process inputs and outputs, criteria, responsibilities, risks, evaluation, change, and improvement. ISO 37301 adds the compliance-management context through compliance obligations, controls, documented information, nonconformity, and corrective action. ISO 19011 provides the audit logic through audit criteria, evidence, findings, reporting, follow-up, and evidence-based and risk-based auditing. Together, these standards define a setting in which AI support can be assessed as evidence preparation within a controlled human compliance process, rather than as autonomous compliance decision-making.

Before AI support, a compliance or audit owner reviews documented information, compares evidence with audit or compliance criteria, identifies possible deficiencies, discusses unclear evidence with the process owner, and formulates findings or corrective-action requests. The candidate intervention point is the evidence-preparation part of this human task, where the organization considers modifying the process through an AI-supported solution. An initial framing such as "AI flags potential control deficiencies" is underspecified because it does not clarify which part of the former human work is delegated, what the AI-supported work is expected to achieve, which evidence may be used, which output authority the system has, which judgment remains assigned to human roles, and how contestability and auditability are preserved.

Applying the Responsibility-Gated Task Frame produces a more defensible framing. The AI-supported work objective is limited to preparing evidence for human compliance assessment; compliance decisions and formal findings remain assigned to human roles. The admissible inputs are approved process descriptions, controlled documents, prior audit records, corrective-action records, and evidence linked to the relevant obligation. The operating conditions are the defined audit scope, audit criteria, document-control rules, confidentiality requirements, and available evidence quality. The relevant interfaces and dependencies include process-owner clarification, compliance-owner decision authority, controlled evidence repositories, the audit programme, and the applicable management-system requirements used as assessment criteria. The required judgment concerns evidence sufficiency, interpretation of criteria, nonconformity classification, and corrective-action relevance; these judgment elements remain assigned to human roles.

The responsibility conditions define the constraints under which this partial delegation may proceed. The resulting delegation record specifies evidence preparation for human compliance assessment as the bounded work unit. AI output is restricted to evidence summaries, source traces, inconsistency flags, risk indications, and uncertainty notes. Human authority is retained for finding approval, nonconformity classification, responsibility assignment, and corrective-action decisions. The selected alternative is delegation with constraints; the corresponding design implications are mandatory human review, source and output logging, a contestation path, and suspension or reframing triggers. The vignette shows how the artifact translates enterprise AI governance and RBPM commitments into task-level constraints before detailed AI implementation begins.

7 Limitations, Evaluation Agenda, and Conclusion

This research-in-progress paper has three limitations. First, the Responsibility-Gated Task Frame is presented as an early design artifact, not as a completed design theory or empirically validated method [17]. Second, the demonstration is illustrative: its standards-informed basis strengthens practical plausibility, but the paper does not yet report evidence from organizational use. Third, the artifact addresses one specific process-improvement situation: the introduction of an AI-supported solution that partially delegates previously human process work to AI. It is therefore not claimed as a general process-improvement method.

The paper concentrates on the responsibility extension of task framing. The broader descriptive Task Frame contains further representational elements, but these can only be introduced briefly in a short paper. Following the Design Science Research Methodology (DSRM) logic, the current paper covers problem identification, objectives, initial design, demonstration, and communication [16]. Subsequent cycles will use expert review, practitioner assessment, and later naturalistic evaluation to assess conceptual clarity, coverage and overlap of responsibility conditions, consistency of delegation alternatives, fit with RBPM and AI governance, and practical usability [18].

The paper offers a task-level artifact for translating enterprise AI governance and RBPM commitments into concrete constraints for delegating process work to AI. The central argument is that responsible AI use in BPM requires task-level design before implementation, alongside enterprise-level policies and later assurance activities. The Responsibility-Gated Task Frame offers an initial structure for that decision.

Disclosure of Interests. The author has no competing interests to declare that are relevant to the content of this article.

References

1. Er, M., et al. (eds.): Preface of the responsible BPM forum. In: Business Process Management: Responsible BPM Forum, Process Technology Forum, Educators Forum: BPM 2025 RBPM, PT, and Educators Forum, Seville, Spain, August 31 – September 5, 2025, Proceedings. pp. 1–5. Springer, Cham (2026). https://doi.org/10.1007/978-3-032-02936-2
2. Fehrer, T., Fischer, D.A., Leemans, S.J.J., Röglinger, M., Wynn, M.T.: An assisted approach to business process redesign. Decis. Support Syst. **156**, 113749 (2022). https://doi.org/10.1016/j.dss.2022.113749
3. Bojer, C.S., Bygstad, B., Øvrelid, E.: Speeding up explorative BPM with lightweight IT: the case of machine learning. Inf. Syst. Front. **27**, 823–840 (2025). https://doi.org/10.1007/s10796-024-10474-1
4. Baird, A., Maruping, L.M.: The Next generation of research on IS use: a theoretical framework of delegation to and from agentic IS artifacts. MIS Q. **45**, 315–341 (2021). https://doi.org/10.25300/MISQ/2021/15882

5. Puranam, P.: Human–AI collaborative decision-making as an organization design problem. J. Org. Design. **10**, 75–80 (2021). https://doi.org/10.1007/s41469-021-00095-2
6. Jobin, A., Ienca, M., Vayena, E.: The global landscape of AI ethics guidelines. Nat. Mach. Intell. **1**, 389–399 (2019). https://doi.org/10.1038/s42256-019-0088-2
7. Mökander, J., Morley, J., Taddeo, M., Floridi, L.: Ethics-based auditing of automated decision-making systems: nature, scope, and limitations. Sci. Eng. Ethics **27**, 44 (2021). https://doi.org/10.1007/s11948-021-00319-4
8. Mökander, J., Floridi, L.: Operationalising AI governance through ethics-based auditing: an industry case study. AI Ethics **3**, 451–468 (2023). https://doi.org/10.1007/s43681-022-00171-7
9. Djurica, D., et al.: Operationalizing responsible BPM: a method for value-based process redesign. In: Er, M., et al. (eds.) Business Process Management: Responsible BPM Forum, Process Technology Forum, Educators Forum, pp. 72–86. Springer, Cham (2026). https://doi.org/10.1007/978-3-032-02936-2_7
10. Chandrasiri, T., Bandara, W., Rosemann, M., Ostern, N., Voss, M.: Designing business processes for good: a canvas-based framework for routinized benevolence. In: Er, M., et al. (eds.) Business Process Management: Responsible BPM Forum, Process Technology Forum, Educators Forum, pp. 87–101. Springer, Cham (2026). https://doi.org/10.1007/978-3-032-02936-2_8
11. Vu, H., Klievtsova, N., Leopold, H., Rinderle-Ma, S., Kampik, T.: Agentic business process management: practitioner perspectives on agent governance in business processes. In: Er, M., et al. (eds.) Business Process Management: Responsible BPM Forum, Process Technology Forum, Educators Forum. pp. 29–43. Springer, Cham (2026). https://doi.org/10.1007/978-3-032-02936-2_3
12. Santoni De Sio, F., Van Den Hoven, J.: Meaningful human control over autonomous systems: a philosophical account. Front. Robot. AI. **5**, 15 (2018). https://doi.org/10.3389/frobt.2018.00015
13. Koulu, R.: Proceduralizing control and discretion: human oversight in artificial intelligence policy. Maastricht J. Eur. Comparat. Law **27**, 720–735 (2020). https://doi.org/10.1177/1023263X20978649
14. Derguech, W., Bhiri, S., Curry, E.: Using ontologies for business capability modelling: describing what services and processes achieve. Comput. J. **61**, 1075–1098 (2018). https://doi.org/10.1093/comjnl/bxy011
15. Hevner, A.R., March, S.T., Park, J., Ram, S.: Design science in information systems research. MIS Q. **28**(1), 75–105 (2004). https://doi.org/10.2307/25148625
16. Peffers, K., Tuunanen, T., Rothenberger, M.A., Chatterjee, S.: A design science research methodology for information systems research. J. Manag. Inf. Syst. **24**, 45–77 (2007). https://doi.org/10.2753/MIS0742-1222240302
17. Gregor, S., Hevner, A.R.: Positioning and presenting design science research for maximum impact. MIS Q. **37**, 337–355 (2013). https://doi.org/10.25300/MISQ/2013/37.2.01
18. Venable, J., Pries-Heje, J., Baskerville, R.: FEDS: a framework for evaluation in design science research. Eur. J. Inf. Syst. **25**, 77–89 (2016). https://doi.org/10.1057/ejis.2014.36
19. Fanni, R., Steinkogler, V.E., Zampedri, G., Pierson, J.: Enhancing human agency through redress in Artificial Intelligence Systems. AI Soc. **38**, 537–547 (2023). https://doi.org/10.1007/s00146-022-01454-7
20. Santoni de Sio, F., Mecacci, G.: Four responsibility gaps with artificial intelligence: why they matter and how to address them. Philos. Technol. **34**(4), 1057–1084 (2021). https://doi.org/10.1007/s13347-021-00450-x
21. International Organization for Standardization: ISO 9001:2015 Quality management systems—Requirements (2015). https://www.iso.org/standard/62085.html

22. International Organization for Standardization: ISO 37301:2021 Compliance management systems—Requirements with guidance for use (2021). https://www.iso.org/standard/75080.html
23. International Organization for Standardization: ISO 19011:2018 Guidelines for auditing management systems (2018). https://www.iso.org/standard/19011

Silent Responsibility Voids: Evidence for Process Twins in Clinical Trials

Luiz Ricardo Brito Ribeiro[1](✉), Gerald Kremer[2], Rainer Stark[2], and Silvia Dallavalle[1]

[1] University of São Paulo, Ribeirão Preto, Brazil
luizrbr@usp.br
[2] Technische Universität Berlin, Berlin, Germany
kremer@tu-berlin.de

Abstract. Responsible Business Process Management (RBPM) and process mining have matured as independent research streams, yet their integration on real operational data remains rare, particularly in clinical research, where responsibility obligations are most consequential. This study investigates the protocol deviation management process of Clinical Research Centers (CRCs) using an event log of 13,331 deviation cases from an operational clinical trial management system, complemented by a formative expert evaluation with specialists from contrasting roles and regulatory contexts. Applying process discovery, conformance, and performance analysis, it finds that the most critical responsibility tensions surface less in what the process records than in what it fails to record. We introduce the concept of Silent Responsibility Voids (SRVs): points where the operationalization of an RBPM principle leaves no auditable trace, manifesting as absence, systematic incompleteness, or ceremonial recording. In the studied data, corrective action is documented in 0.1% of cases, regulatory notifications are largely absent, and the typical deviation is registered a median of 153 days after it occurred. We argue that such voids are detectable only by a technology that carries the normative expectation of the process, and use them to specify the monitoring functions a Responsible Digital Process Twin would require.

Keywords: Responsible BPM · Process Mining · Digital Process Twin · Protocol Deviation · Data Quality

1 Introduction

Few industries carry the weight of responsibility that clinical research does. The safety of human participants, the integrity of scientific evidence, and the public trust that sustains biomedical innovation all depend on how faithfully operational processes execute their ethical obligations [1]. Yet governing these processes by evidence, not by assumption requires something the literature has not yet produced: empirical knowledge of where, inside the actual operations of a Clinical Research Center (CRC), responsibility tensions concentrate and

K. Kluza et al. (Eds.): BPM 2026, LNBIP 600, pp. 11–26, 2027.
https://doi.org/10.1007/978-3-032-36710-5_2

process behavior diverges from ethical expectation. Responsible Business Process Management (RBPM) and process mining have each matured as research streams [2,3], but the study that brings them into direct contact with real operational data, locating with empirical precision where responsibility principles come under the pressure of actual execution, remains rare. In the records of clinical trial operations, something in the operational reality of these CRCs appears to pull the record toward silence, and that silence is itself a signal that responsibility is going unobserved.

The operational context is also deeply economic. Multinational trials run across dozens or hundreds of CRCs simultaneously, and the sponsor that funds the trial and owns its data must assure quality across this distributed network. Phase III studies have been estimated to cost about USD 55,716 per day [4], with site monitoring and administrative coordination together consuming a large share of trial budgets [5]. Business Process Management (BPM) has provided tools to standardize these operations [6,7], yet practices centered on efficiency have proved insufficient for a domain in which every operational decision carries an ethical dimension, an insufficiency RBPM answers by embedding values such as safety, transparency, and sustainability across the process lifecycle [2].

Digital Process Twins (DPTs) represent the fullest expression of this evolution, combining data integration, artificial intelligence, and formal modelling notations such as Business Process Model and Notation (BPMN) and Decision Model and Notation (DMN) to represent, simulate, and monitor complex clinical processes in real time. The promise is considerable, and the global digital twin sector is projected to grow from USD 21 billion in 2025 to nearly USD 150 billion by 2030 [8]. Yet the challenges are structural, spanning data security and privacy, alignment with regulation and cost, and the difficulty of embedding responsibility into frameworks built for efficiency. The result is a guidance deficit: DPTs are technically promising, but the field lacks an empirically grounded model that translates responsibility principles into targeted design features for clinical operations.

Process mining offers the instrument to address this deficit, because it extracts behavioral knowledge directly from clinical trial management system logs and reveals what operationally occurs rather than what is presumed to [9]. Two kinds of responsibility tension become observable. The first is visible in the data that exists: bottlenecks, delays, and non-conformant paths, which conformance checking exposes by relating observed behavior to a normative model [10]. The second, and more consequential, is visible only in the data that is missing: obligations that should have left a trace and did not, a reading grounded in the event-log imperfection-pattern literature [11]. When a corrective action is not recorded, a regulatory notification leaves no mark, or a description is filled in only ceremonially, the process becomes unauditable exactly where responsibility matters most. Anchoring DPT design to both kinds of tension turns the twin from a generic dashboard into a responsibility-aware instrument.

Despite growing interest in the responsible use of digital process technology, the empirical grounding of a Responsible Digital Process Twin in real-world

CRC operational data remains rare, and the diagnostic value of the gaps in that data has yet to be recognized. A recent conceptual framework, the ReDPT-CT framework [12], identified the absence of empirical validation as its central limitation and called for studies grounded in operational evidence. The present study answers that call through a process mining investigation of the protocol deviation management process at CRCs, complemented by a formative expert evaluation with specialists from contrasting roles and regulatory contexts. Its principal contribution is the concept of the Silent Responsibility Void (SRV), a point where a responsibility principle leaves no auditable trace, together with the argument that detecting such voids requires, by its very nature, a DPT.

2 Problem Statement

The intersection of RBPM, DPTs, and clinical research has produced conceptual frameworks that map responsibility principles onto CRC process architectures, but not the empirical layer that turns these mappings into actionable design knowledge, a gap the ReDPT-CT framework [12] identified as its central limitation. Given this problem, the following research question emerges: how can process mining of real-world CRC event logs identify responsibility tension points and how can those points inform the design of a Responsible DPT? To answer it, the study pursues three objectives: first, to reconstruct the executed deviation management process from the log and to detect, measure, and index the responsibility tension points that surface as silent voids; second, to interpret those voids through a formative expert evaluation of the artifact and the forces beneath them; and third, to specify, from the indexed voids, the monitoring functions a Responsible DPT must provide.

3 Theoretical Background

Clinical trials are systematic studies with human participants to evaluate the safety, efficacy, and effects of new therapeutic interventions [1,13], the mandatory regulatory gateway any new medical product must pass. The global clinical trials market was estimated at approximately USD 89 billion in 2025 [14]. A trial involves three principal actors: the sponsor, which funds the trial and bears ultimate responsibility for its integrity; the Principal Investigator (PI); and the CRC, which operationalizes the trial and coordinates among the PI, the sponsor, and regulators. A protocol deviation is any departure from the conditions of the approved protocol, which Good Clinical Practice requires be documented, assessed, and, where relevant, reported [15].

Process mining has been applied extensively in healthcare, where surveys and case studies reconstruct care pathways from hospital information systems [16,17], concentrating on operational efficiency, not on the responsibility obligations specific to clinical research. A parallel strand studies the quality of the event data itself, cataloguing imperfection patterns [11,18]; there, incompleteness is an obstacle to clean away, whereas this study reads it as evidence of where responsibility is unobserved. The techniques relied on here are mature: automated discovery has been systematically reviewed [19], block-structured discovery yields sound models from real logs [20], and conformance checking relates observed behavior to a normative model, supporting reasoning about what is absent and not only present [10,21].

On the normative side, responsible process mining and responsible data science articulate the principles that process analysis should honor [2,3,22], and the agendas of open BPM problems and of digital innovation call for grounding such principles in operational reality [23,24]. What none of these strands has done, and what this study contributes, is to bring responsible BPM into contact with real clinical research operational data and to treat the absence of a record, rather than its content, as the primary responsibility signal.

The challenge for sponsors extends beyond a single CRC: multinational trials run across many CRCs at once, each with a different level of digital maturity and data quality. Sponsor oversight of this distributed network has historically depended on on-site monitoring visits, in which Clinical Research Associates (CRAs) verify source data, audit deviation handling, and confirm regulatory submissions [25]. This model is operationally intensive and economically heavy, and each unreported event or undocumented corrective action compounds its cost while degrading the regulatory standing of the program. BPM has been introduced to provide structured methods for process representation, standardization, and data collection [6], yet practices centered on efficiency have proved insufficient for a domain in which every operational decision carries an ethical dimension.

RBPM offers the conceptual frame for this expanded ambition. Following the responsible-BPM and responsible-data-science agenda [2,3], we take RBPM to be the practice of applying BPM by accounting for the needs of stakeholders, the broader environment including regulators, and society. At its minimum it ensures that processes do not produce harm; at its most intentional it promotes the well-being of workers and participants while fostering sustainability. Responsible deployment within this frame further requires adherence to the FACT principles [3]: Fairness, ensuring outputs do not produce unjust outcomes; Accuracy, grounding results in reliable data; Confidentiality, protecting sensitive information; and Transparency, accompanying decisions with understandable explanations.

Process mining connects these principles to the actual behavior of operations: through discovery, conformance checking, and performance analysis it reconstructs the executed process from raw event data, quantifies adherence to a normative model, and measures delays and disparities across activities and units [9]. Crucially, conformance checking also makes it possible to reason about what is missing, because it compares execution against a model of what should have occurred. This capacity to detect absence, and not only presence, is what distinguishes process mining from descriptive reporting and what makes it suited to the analytical lens this study introduces.

DPTs extend process mining from retrospective analysis into a continuously operational instrument: the discovered model becomes the twin's behavioral baseline, conformance checking its real-time monitoring layer, and simulation the means to evaluate process changes before implementation [26]. A DPT designed without empirical grounding risks becoming a generic dashboard that can display only what was recorded. The ReDPT-CT framework [12] provides a responsibility-aware classification, organizing responsibility into fifteen aspects derived from the clinical trial ethics and RBPM literature, spanning participant-facing obligations, data and scientific dimensions, governance responsibilities, and forward-looking concerns. These aspects serve as the lens through which the findings are interpreted; the analysis draws on the subset that the data activate, not on all fifteen.

Following Stark and Damerau's definition [27], a digital twin is the digital representation of an active unique product or product-service system, captured through models, information, and data across lifecycle phases; on this view a clinical-trial process qualifies as the entity twinned, and a Digital Process Twin falls within the definition. The Responsible DPT specified here is a process twin at the level of design: its conformance layer reads the event log to detect where the record departs from the normative model, while the digital master and the digital shadow must be interlinked for the twin to close [27]. The present study specifies the monitoring functions of this twin; the closed feedback loop and simulation, higher twinning levels [28], are left to operational deployment.

Because the term Silent Responsibility Void is introduced here for the first time, we ground it lexically before defining it operationally. Drawing on the Merriam-Webster dictionary [29] for each of its three components, we propose the following composition of the definition:

- **Silent.** Beyond its acoustic sense, it carries the figurative meaning of making no mention of a matter, as a contract or statute is said to be silent on a point it does not address; applied to a process record, it denotes a trace that exists yet says nothing about the obligation in question, which is why ceremonial recording belongs to the same family as outright absence.
- **Responsibility.** To be responsible is to be answerable: liable to be called to account for one's actions. And to be answerable presupposes a record, because an account can only be given against some trace of what was done.
- **Void.** It names not mere emptiness but an absence where something was expected, the status of a responsibility-critical field left unfilled.

Composing these senses, a Silent Responsibility Void is a locus in a process that should hold a record of a responsibility obligation but does not, such that no party can be called to account from the trace itself. This locates responsibility not in intention but in auditability, the capacity to be accounted for.

The analytical construct that organizes the study follows directly: a Silent Responsibility Void is a point where the operationalization of an RBPM aspect leaves no auditable trace, manifesting as absence, systematic incompleteness, or ceremonial recording. Of these three signatures, only ceremonial recording draws on prior theory, namely the notion of formal structure decoupled from substantive practice [30]; the absence and incompleteness signatures are introduced here as responsibility signatures, with the event-log imperfection-pattern literature [11] providing the methodological basis for detecting them in the data. The construct has three deliberate properties: it is anchored, because every void is a void of a specific responsibility aspect; it is observable, because each has a measurable signature in the log, such as low completeness, repeated template text, or chronological impossibility [11]; and it is actionable, because it defines precisely what a responsible twin must do, namely render auditable what is currently silent.

Two clarifications protect the construct from overreach. A void describes an observable phenomenon, the absence of an auditable trace; it does not, by itself, prove that an obligation went unmet, since the underlying action may have occurred and been recorded elsewhere. And the recurrent observation that records are pulled toward silence invites explanation, but the forces responsible, whether limited time, system usability, recording in parallel systems, or apprehension regarding sponsor scrutiny, are interpretive matters that require complementary evidence rather than claims derivable from the log alone.

4 Research Design

The study was grounded in the Design Science Research (DSR) paradigm [31,32] and maps directly onto the DSR cycle: the problem was established in Sect. 2, the knowledge base in Sect. 3, the artifact comprises the Silent Responsibility Void construct and the resulting design requirements, not an implemented system, and the evaluation was conducted through exploratory expert assessment [31,32].

Event log data were extracted from an operational clinical protocol management system used by CRCs. The extraction covered the protocol deviation management process and comprised 13,331 deviation cases and 43,489 events across 46 CRCs, spanning 2018 to 2026, with 286 protocols and 2,968 participants. Data were used under formal authorization for anonymized research use, with all identifiers removed and replaced by pseudonyms ordered by case volume, in compliance with applicable data-protection regulation. Two non-productive environments were excluded; all 46 real CRCs were retained, since under-recording is itself a signal of interest. Data quality is part of the method, not a preliminary nuisance: corrupted timestamps, notifications preceding identification, and codes absent from the reference table were retained as analytical signal wherever they bear on responsibility and excluded from time computations wherever they reflect mere corruption. For durations, the median is used to control for extreme outliers.

The protocol deviation lifecycle is represented by six milestones, each carrying a timestamp, and the distinction among them is central to the analysis. Occurrence and identification, the notifications to ethics committee and sponsor, and the resolution date are declared dates and therefore open to retroactive entry. System registration, by contrast, is the objective timestamp of the first insertion of the record into the system, drawn from the audit trail; it is the only timestamp that cannot be entered retroactively, and is therefore the reference against which the temporal behavior of the process is measured. The interval between the declared occurrence and the system registration thus compares what the site states happened against when the record objectively came into existence.

To make responsibility completeness comparable across cases and centers, we define a responsibility completeness index: for each case, the fraction of six ReDPT-CT-derived responsibility fields [12] carrying substantive content, namely the deviation description, the preventive or corrective action plan, the corrective action, the two notification dates, and the resolution date. A field counts as present only when it holds a non-null, non-empty value. The index ranges from 0 to 1; the identification milestone is excluded because it anchors nearly every case, and the free-text observation field as supplementary. The index for a center is the mean of its cases' indices. Across the studied data the per-case median is 0.33, confirming that the typical deviation carries only its description and at most one further obligation.

Artifact evaluation was conducted as a formative, exploratory assessment through semi-structured interviews with two senior specialists, selected by purposive maximum-variation sampling to contrast professional role and regulatory context: an academic PI at a clinical research center, and a project coordinator at a CRO operating in a different national regulatory context. The aim was interpretive: to assess the relevance and direction of the artifact and help interpret the forces that may underlie the observed voids. The semi-structured format was chosen for combining a predefined thematic guide with the flexibility to pursue emergent lines of inquiry, a design well established in interpretive IS and BPM research [33]. Expert profiles were characterized from public curricula rather than from self-report.

The instrument comprised four questions in two blocks. The first examined how deviation recording occurs in practice, who records deviations and where, and how they feed process improvement. The second examined which responsibility principles, among Fairness, Accuracy, Confidentiality, and Transparency, are hardest to ensure, and whether embedding these in a DPT would influence adoption. Because both specialists are external to the studied cases, their accounts are treated as expert interpretation rather than as evidence about specific deviations; credibility rests on purposive maximum-variation sampling, on profiles verified against public curricula, and on the convergence of two informants from contrasting professional and regulatory contexts.

5 Key Findings

The process discovered from the event log of 13,331 deviation cases across 46 CRCs is short and strongly funneled. Of the 158 execution variants observed, the dominant one accounts for 17.5% of cases and records occurrence, identification, and system registration but nothing further, leaving no notification and no resolution; only 18.8% of cases carry a recorded resolution at all. Figure 1

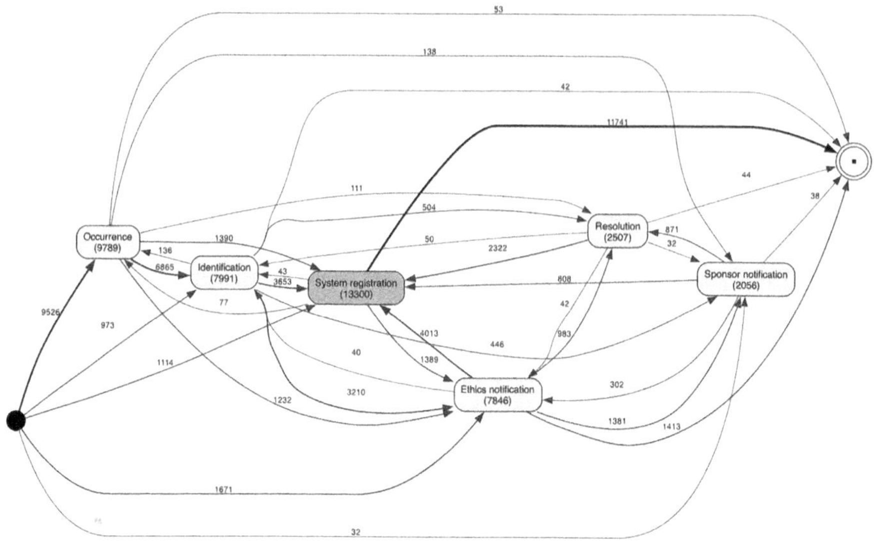

Fig. 1. Discovered protocol deviation process as a directly-follows graph over the six milestones.

presents the discovered process as a directly-follows graph, chosen for its directness in exposing milestone-level frequencies and transitions in a short, funnelled process, with the absolute frequency of each milestone and of each transition between them.

Two features stand out. First, a path of 1,232 cases moves directly from occurrence to ethics committee notification, bypassing the formal identification milestone entirely and taking a median of 104 days to do so. Second, order conformance is high at 93.6%, computed over the five clinical responsibility milestones (occurrence, identification, ethics notification, sponsor notification, and resolution); system registration is treated as the administrative anchor against which timing is measured, not as a step in this clinical sequence. This is an order-conformance measure, reporting whether milestones occur in the expected sequence, and is distinct from the fitness and precision of a discovered model against a normative net [10, 21]. Most cases therefore respect the expected order; the remaining 6.4% show at least one out-of-order milestone, and in 1.0% of all cases a notification or resolution is dated before the identification itself, a strict chronological impossibility.

Throughput from identification to resolution has a median of 7 days, but this figure is computed over only the 2,384 cases that carry a valid resolution date, a minority of the population, and must be read as the behavior of the resolved subset rather than of deviations in general.

These visible patterns, however, are not where the most consequential responsibility tensions reside. The central finding of this study is that the strongest evidence of responsibility failure lies not in what the process records but in what it leaves unrecorded, formalized through the Silent Responsibility Void (SRV) defined in Sect. 3: a locus at which the operationalization of a responsibility aspect leaves no auditable trace, through absence, incompleteness, or ceremonial recording [30]. Each SRV is indexed to a responsibility aspect, is detectable through a measurable signature, and implies a corresponding monitoring capability.

Two voids dominate the data. The first is corrective management: a specific corrective action is documented in only 0.1% of cases, a field effectively abandoned across all 46 CRCs. Under Operational Management and Scientific Integrity, a deviation closed without any recorded corrective action is, to any external auditor, a deviation whose resolution cannot be verified, a direct departure from the corrective and preventive action requirement of Good Clinical Practice, ICH-GCP E6(R3) [15].

The second is temporal, and reading it correctly depends on the timestamp distinction set out in the research design. The occurrence date is declared and can be entered retroactively; the registration date is the objective moment the record was first inserted into the system, taken from the audit trail, and cannot be backdated. The interval between the two has a median of 153 days and a ninetieth percentile of 448 days over the cases that carry both timestamps; measured against the full population of 13,331 deviations, 47.5% are registered more than 90 days after the declared occurrence. Because the closing point of this interval is the unfalsifiable system insert, a lag of this size is direct evidence of retroactive entry rather than mere delayed reporting: the declared dates of notification and resolution are typically completed in a single sitting at registration time, which is why the final transitions of the model carry a median time of zero days.

One specialist observed that a deviation is often only reconstructed months later, in a review meeting with the sponsor, from the recollection of those involved; such reconstruction would reinforce, not weaken, this reading. The data establish that records are pulled toward delay and silence; they do not, by themselves, establish the cause, which remains a matter for the exploratory expert evaluation and for future investigation.

This temporal signature is also structural in the discovered model: system registration is the terminal milestone in 88.1% of cases, drifting to the end precisely because every declared date precedes it. The discovered model is the behavioral baseline a Responsible DPT would hold, and its conformance layer the monitoring mechanism that would flag this drift case by case.

A single case shows how these signatures compose: a deviation is described but carries no action plan, no notification dates, and no documented corrective action, and its occurrence precedes its registration by months. Each gap alone is minor; together they leave a deviation whose handling cannot be reconstructed from the trace, and a system holding the expected lifecycle would flag it on five of the six responsibility fields at once.

The two specialists offered convergent interpretations of these voids. Both indicated that, in routine practice, protocol deviations are frequently detected and recorded by the monitoring side rather than self-reported by the site. As one specialist, a coordinator at a contract research organization, put it: "A protocol deviation is normally observed or raised by the monitor; it is very hard for a site to come forward and say, 'I committed a deviation.' Usually it is not proactive, but passive." The authoritative trace of a deviation therefore often resides in monitoring documents and contract-research systems rather than in the site's own platform. On this reading, an absence in the studied system does not necessarily mean that nothing was done, but that the record of what was done is fragmented across people and systems.

The specialists further suggested several forces that may draw the on-system record toward incompleteness and delay: the burden that thorough recording places on staff who must explain, notify, and follow up an error; a reluctance to attribute a deviation to one's own inattention, which one specialist suggested can yield shallow or repeated root-cause statements, consistent with the 22.7% of templated descriptions observed here; the frequent failure to assign responsibility for following up corrective measures; and, on the sponsor side, a diplomatic reticence that can limit full transparency about decisions affecting a center. On the reluctance to record, the same coordinator observed: "People do not want to say, 'I made a mistake, and I made it out of inattention'; there is always some peripheral excuse that keeps one from reaching the root of the problem." An investigator at an academic medical center described the corresponding gap in corrective follow-up: "to follow up whether the measures proposed were actually taken and whether they were effective is often not done, or, if it is done, it is not well recorded."

Both specialists also noted that, when the rationale or follow-up is not captured in a structured field, it is reconstructed later from the person who decided, who in multi-year studies may no longer be reachable, so the void is less a simple absence than a trace that grows progressively less auditable. These remain interpretations that help explain the observed patterns, not findings established by the event log. Read together, the forces the specialists describe are competing pressures rather than simple negligence: the burden of thorough recording against the duty of timeliness, the reluctance to attribute a deviation to one's own inattention against the demand for transparency, and the sponsor's diplomatic reticence against full disclosure. It is in this sense that the study speaks of responsibility tensions; the void is where such a tension is resolved, silently, in favour of the unwritten.

The coordinator also noted that deviations exceeding configured quality tolerance limits trigger a monthly review with the sponsor, reinforcing that the authoritative handling of a deviation often lives in monitoring and sponsor systems, so the on-system record is one fragment of a larger, distributed trace. Asked whether embedding responsibility dimensions in a process twin would influence adoption, both specialists indicated that a system able to surface where accuracy and transparency are weakest, rather than merely store data, would be a meaningful differentiator.

Beyond these two, the data reveal further voids that the same construct organizes. A recorded notification to the ethics committee is absent in 41.1% of cases and to the sponsor in 84.5%, and a documented action plan is missing in 45.6%, leaving the most regulated obligations in the process among the least auditable within it. Descriptive content is repeated or templated in 22.7% of cases, a form of ceremonial recording in which the field is satisfied while its informational purpose is not.

And a small but ethically sensitive share of cases, about 2.2%, are deleted from the record after creation, the majority after having been marked as resolved; this is treated here as a candidate void of accountability whose interpretation requires care, since deletion may reflect legitimate correction or a break in the audit trail, a distinction the data cannot adjudicate. Table 1 reports the measured rate of each void signature across the population, and Table 2 maps each void to its responsibility aspect and to the corresponding function required of a Responsible DPT, traveling the full distance from a responsibility aspect, through observable evidence, to a concrete design feature.

Table 1. Measured Silent Responsibility Void signatures across the 13,331 cases.

Responsibility field	Signature	Measured rate
Corrective action	Absence	documented in 0.1% of cases
Sponsor notification	Absence	absent in 84.5%
Resolution date	Absence	absent in 81.2%
Action plan	Incompleteness	absent in 45.6%
Ethics notification	Absence	absent in 41.1%
Deviation description	Ceremonial recording	templated or repeated in 22.7%
Registration lag	Delay	median 153 days; 47.5% over 90 days
Order non-conformance	Sequence violation	in 6.4% of cases
Post-hoc deletion	Accountability	2.2% deleted after creation

The distribution of these voids across CRCs is structural, not incidental. Figure 2 plots each CRC's responsibility completeness index against its case volume, for centers with at least thirty cases. Completeness ranges from 0.17 to 0.71, and a chi-square test confirms it differs across CRCs far beyond chance ($p < 0.001$). The disparity is not explained by workload: we found no significant relationship between case volume and completeness (Spearman rho = -0.124, $p = 0.555$). The poorest completeness is behavioral, not a function of size or overload, which locates the source of the void in how responsibility is enacted at each CRC, not in the quantity of work, the concern of distributive and procedural justice in process execution [22].

Taken together, these findings establish the study's main design argument. A conventional dashboard can display only the data that exists and is therefore structurally incapable of representing a void; detecting one requires a system that holds the normative model of the process and can compare the records each case ought to contain against those it does. The concept of the Silent Responsibility Void thus does not merely benefit from a Responsible DPT; it points to the need for one, and for letting the voids direct where its monitoring is warranted.

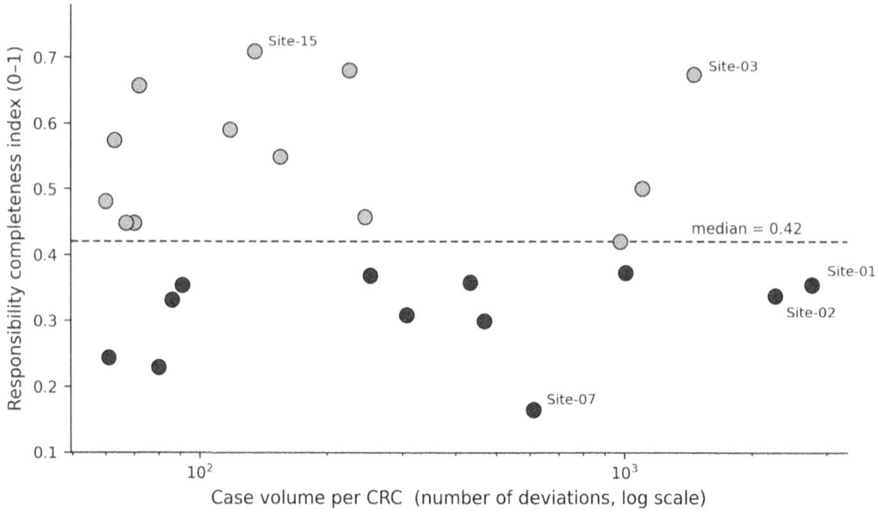

Fig. 2. Responsibility completeness index by CRC against case volume (centers with at least 30 cases).

Of the four FACT principles [3], three surface directly as observable voids in the studied data: Accuracy in the systematic incompleteness and retroactive entry of records, Transparency in ceremonial and templated descriptions, and Fairness in the completeness disparity across centres. Confidentiality is different in kind. It leaves no void of its own here, both because the obligation is to keep a trace from reaching the wrong reader rather than to leave one, and because the log was anonymised before analysis. It returns instead as a constraint the Responsible DPT must carry: the same consolidation that lets a twin compare what each case ought to record against what it does also gathers participant-sensitive deviation data across the fragmented site, monitoring, and sponsor systems the specialists described, so the monitoring function that renders the other three principles auditable must itself be designed to honour the confidentiality of that data.

6 Significance and Relevance

The significance returns to the problem that motivated the study: the gap between a conceptual framework of responsible digital process twins and the operational reality such twins are meant to govern. By grounding the analysis in real CRC data, the work converts that framework from a normative aspiration into an empirically anchored design, making three contributions, theoretical, methodological, and practical.

Theoretically, the study reorients Responsible BPM from the promotion of responsibility to its auditability. Prior work treats responsibility aspects as properties to be designed into a process; the Silent Responsibility Void demonstrates

Table 2. From Silent Responsibility Voids to Responsible DPT design features.

RBPM aspect	Void signature (manifestation)	Required DPT function
A. Transparency and Disclosure	22.7% templated descriptions (ceremonial recording)	Flag template or duplicate low-substance descriptions
A and K. Timeliness	153-day median lag; 47.5% after 90 days (delay)	Flag registration or notification lagging beyond regulatory windows
E. Scientific Integrity	Action plan absent in 45.6%; 1.0% chronological impossibilities (incompleteness)	Verify temporal coherence and lifecycle completeness
K. Regulation and Compliance	Ethics notification absent in 41.1%; sponsor in 84.5% (absence)	Monitor obligations not evidenced by a record
N. Operational Management	Corrective action in 0.1% of cases (absence)	Alert on deviations closed without corrective action
C. Justice and Social Inclusion	Completeness disparity across CRCs (disparity)	Comparative completeness index across CRCs

that an aspect can be formally present in a design and substantively absent in execution, and that this absence is itself empirically detectable. The 0.1% rate of recorded corrective action is not the absence of corrective behavior in the world; it is the absence of an auditable trace of it, and that distinction is the conceptual pivot of the contribution. For a field increasingly concerned with the accountability of data-driven process technologies [2,3], locating responsibility in the auditability of execution rather than in the intent of design is a meaningful conceptual move.

Methodologically, the study shows that data quality problems, ordinarily treated as obstacles to be cleaned away before analysis, can be a primary source of responsibility evidence: incompleteness, retroactive entry, and chronological impossibility are read here not as defects to be discarded but as signatures of where responsibility is going unobserved [11]. The 153-day median lag between a deviation's occurrence and its registration is a case in point: a conventional pipeline would treat such timestamps as noise to correct, whereas here the lag is the finding.

Practically, the relevance is sharpest for the oversight problem sponsors face across distributed networks of CRCs. On-site monitoring, the dominant and costly mechanism of oversight, is structurally incapable of detecting a deviation that will only be registered 153 days after it occurs, because at the visit the record does not yet exist. This exposes the limit of periodic, location-bound monitoring and explains why a continuously operating Responsible DPT is the logical response to the problem: only a system holding the normative expectation of the process can flag an obligation that left no trace or a record that arrived too late. This positions the twin as an operationalization of risk-based monitoring, which ICH-GCP E6(R3) and regulators now advocate [15,25]; the responsibility completeness index and the inter-CRC disparity it reveals offer exactly such a risk signal, directing attention to where the record is least trustworthy before an external audit exposes the gap.

7 Conclusions

This study asked how process mining of real-world CRC event logs can identify responsibility tension points within the protocol deviation management process, and how those points can inform the design of a Responsible DPT. The answer is twofold. First, process mining locates responsibility tensions not only in the behavior it makes visible, in bottlenecks, variants, and non-conformance, but above all in the absence, incompleteness, and delay of the records themselves, with the record tending to be least complete for the obligations that carry the greatest responsibility. Second, these voids inform the twin's design directly, because each maps to a responsibility aspect and to a specific monitoring function: a system that holds the normative model of the process can compare what should have been recorded against what was, and thereby render auditable what is currently silent.

To name and operationalize this phenomenon, the study introduced the Silent Responsibility Void and showed, through a mapping from responsibility aspect to data signature to design feature, that detecting such voids requires a system carrying the normative expectation of the process.

Several limitations bound these claims. The data originate from 46 CRCs sharing a single protocol management system, with two centers accounting for a large share of records, so the inter-CRC comparisons indicate structural disparity within the sample and do not generalize to the sector. Because the event log is reconstructed from a database audit trail, milestone timestamps reflect declared dates that may have been entered retroactively, which constrains any time-based measure and is why throughput is read over the resolved subset alone. The claim that operational forces pull the record toward delay and silence is the one the data least support on their own; it is addressed here through the formative expert evaluation and remains a matter for dedicated study.

The two specialists, moreover, were not involved in the studied cases, so their reading is interpretive rather than evidentiary, a status made explicit throughout. Finally, the proposed Responsible DPT is offered at the level of design requirements justified by process evidence; in Design Science terms the present work spans problem identification, the definition of solution objectives, and artifact design, with a formative evaluation of relevance, while summative evaluation in operational deployment is the natural next step and is not claimed here.

The path forward follows from these limits: extending the analysis to other CRC processes and management systems beyond the single platform, and ultimately to operational deployment of the proposed twin, so that its central promise, that surfacing silent responsibility voids leads to their reduction, can be put to the test. What this study supports is a measured claim: that these silences are real and measurable, and that making them visible is feasible and worthwhile.

References

1. World Medical Association: Declaration of Helsinki: ethical principles for medical research involving human subjects. JAMA **310**(20), 2191–2194 (2013)
2. Mannhardt, F.: Responsible process mining. In: van der Aalst, W.M.P., Carmona, J. (eds.) Process Mining Handbook. LNBIP, vol. 448, pp. 373–401. Springer, Cham (2022)
3. van der Aalst, W.M.P., Bichler, M., Heinzl, A.: Responsible data science. Bus. Inf. Syst. Eng. **59**(5), 311–313 (2017)
4. Smith, Z.P., DiMasi, J.A., Getz, K.A.: New estimates on the cost of a delay day in drug development. Ther. Innov. Regul. Sci. **58**(5), 855–862 (2024)
5. Sertkaya, A., Wong, H.-H., Jessup, A., Beleche, T.: Key cost drivers of pharmaceutical clinical trials in the United States. Clin. Trials **13**(2), 117–126 (2016)
6. Dumas, M., La Rosa, M., Mendling, J., Reijers, H.A.: Fundamentals of Business Process Management, 2nd edn. Springer, Berlin, Heidelberg (2018)
7. Rosemann, M., vom Brocke, J.: The six core elements of business process management. In: vom Brocke, J., Rosemann, M. (eds.) Handbook on Business Process Management 1, 2nd edn., pp. 105–122. Springer, Berlin, Heidelberg (2015)
8. MarketsandMarkets: Digital Twin Market, Global Forecast to 2030 (2025)
9. van der Aalst, W.M.P.: Process Mining: Data Science in Action, 2nd edn. Springer, Berlin, Heidelberg (2016)
10. Rozinat, A., van der Aalst, W.M.P.: Conformance checking of processes based on monitoring real behavior. Inf. Syst. **33**(1), 64–95 (2008)
11. Suriadi, S., Andrews, R., ter Hofstede, A.H.M., Wynn, M.T.: Event log imperfection patterns for process mining: towards a systematic approach to cleaning event logs. Inf. Syst. **64**, 132–150 (2017)
12. Ribeiro, L.R.B., Dallavalle, S., Kremer, G., Stark, R.: ReDPT-CT: designing responsible digital process twins for clinical trials. In: ER, M., et al. (eds.) Business Process Management: Responsible BPM Forum, Process Technology Forum, Educators Forum, BPM 2025. LNBIP, vol. 565, pp. 102–116. Springer, Cham (2026)
13. Tremaine, W.J., et al.: Ethical issues, safety, and data integrity in clinical trials. Inflamm. Bowel Dis. **11**(suppl 1), S17–S21 (2005)
14. Grand View Research: Clinical Trials Market Size, Share and Trends Analysis Report (2025)
15. International Council for Harmonisation: ICH Harmonised Guideline: Good Clinical Practice (GCP) E6(R3). Step 4 (2025)
16. Rojas, E., Munoz-Gama, J., Sepulveda, M., Capurro, D.: Process mining in healthcare: a literature review. J. Biomed. Inform. **61**, 224–236 (2016)
17. Mans, R.S., van der Aalst, W.M.P., Vanwersch, R.J.B.: Process Mining in Healthcare: Evaluating and Exploiting Operational Healthcare Processes. Springer, Cham (2015)
18. Fox, F., Aggarwal, V.R., Whelton, H., Johnson, O.: A data quality framework for process mining of electronic health record data. In: 2018 IEEE International Conference on Healthcare Informatics (ICHI), pp. 12–21. IEEE (2018)

19. Augusto, A., et al.: Automated discovery of process models from event logs: review and benchmark. IEEE Trans. Knowl. Data Eng. **31**(4), 686–705 (2019)
20. Leemans, S.J.J., Fahland, D., van der Aalst, W.M.P.: Discovering block-structured process models from event logs: a constructive approach. In: Application and Theory of Petri Nets and Concurrency, LNCS, vol. 7927, pp. 311–329. Springer, Berlin, Heidelberg (2013)
21. Carmona, J., van Dongen, B., Solti, A., Weidlich, M.: Conformance Checking: Relating Processes and Models. Springer, Cham (2018)
22. Qafari, M.S., van der Aalst, W.: Fairness-Aware Process Mining. In: Panetto, H., et al. (eds.) OTM 2019. LNCS, vol. 11877, pp. 182–192. Springer, Cham (2019). https://doi.org/10.1007/978-3-030-33246-4_11
23. Beerepoot, I., Di Ciccio, C., Reijers, H.A., Rinderle-Ma, S., et al.: The biggest business process management problems to solve before we die. Comput. Ind. **146**, 103837 (2023)
24. Mendling, J., Pentland, B.T., Recker, J.: Building a complementary agenda for business process management and digital innovation. Eur. J. Inf. Syst. **29**(3), 208–219 (2020)
25. TransCelerate BioPharma: Position Paper: Risk-Based Monitoring Methodology (2013)
26. Kremer, G., Ribeiro, L.R.B., Blüher, T., Dallavalle, S., Stark, R.: Towards a unified approach: developing a reference model for digital twins and business process management in clinical trials. In: Gdowska, K., Gómez-López, M.T., Rehse, J.-R. (eds.) Business Process Management Workshops, BPM 2024. LNBIP, vol. 534, pp. 382–393. Springer, Cham (2025)
27. Stark, R., Damerau, T.: Digital Twin. In: CIRP Encyclopedia of Production Engineering, pp. 1–8. Springer, Berlin, Heidelberg (2019)
28. Stark, R., Schulte, S.N., Grunenwald, C.-P.: A systematic description of twinning levels of measured data and models in digital twins. Proc. Des. Soc. **5**, 289–298 (2025)
29. Merriam-Webster: Merriam-Webster.com Dictionary, entries for silent, responsible, and void (2025)
30. Meyer, J.W., Rowan, B.: Institutionalized organizations: formal structure as myth and ceremony. Am. J. Sociol. **83**(2), 340–363 (1977)
31. Hevner, A.R., March, S.T., Park, J., Ram, S.: Design science in information systems research. MIS Q. **28**(1), 75–105 (2004)
32. Peffers, K., Tuunanen, T., Rothenberger, M.A., Chatterjee, S.: A design science research methodology for information systems research. J. Manag. Inf. Syst. **24**(3), 45–77 (2007)
33. Myers, M.D., Newman, M.: The qualitative interview in IS research: examining the craft. Inf. Organ. **17**(1), 2–26 (2007)

Process Externalities: A Missing Construct for Responsible BPM

Katarzyna Gdowska[1(✉)], Anna Suchenia[2], and Krzysztof Kluza[1]

[1] AGH University of Krakow, Mickiewicza Av. 30, 30-059 Krakow, Poland
{kgdowska,kluza}@agh.edu.pl

[2] Cracow University of Technology, ul. Warszawska 24, 31-155 Kraków, Poland
asuchenia@pk.edu.pl

Abstract. Responsible Business Process Management (RBPM) asks the discipline to account for the effects of processes on stakeholders, society, and the environment. Yet BPM's models are scoped within the process boundary and assessed against internal objectives, so they often fail to capture broader societal or environmental effects. Existing approaches – stakeholder analysis, risk management, sustainable BPM, value-sensitive design – each addresses a particular category of such effects; what they lack is a first-class lifecycle construct that treats them as a single class through a shared generative mechanism, complementing rather than replacing them. We propose the construct of Process Externalities (PEs) to capture consequences borne by parties other than a process's intended beneficiary, or that affect values absent from its objective function, which are generated systematically by optimisation under an incomplete objective. Understood in this way, PEs will be defined by this mechanism rather than merely by the stakeholders they affect, and ultimately, after standardisation, they may be subject to externalities accounting following the direction already set by impact-weighted accounting initiatives in management. This paper shows that PEs are elicitable through a design-fiction method aimed at effects not yet visible in process data, and actionable through an externality register attached to the lifecycle, for which we give a graduated, three-level operationalisation and a sketch of externality accounting. We illustrate it in one ordinary process (loan approval) and two speculative ones.

Keywords: process externalities · process responsibility · design fiction

1 Introduction

BPM has matured into a discipline for the design, analysis, and management of organisational work, with efficiency and effectiveness as its guiding ambitions [14]. Societal and environmental pressures have since prompted a turn

The research was financed by a subsidy and the program "Excellence initiative – research university" granted to the AGH University of Krakow.

K. Kluza et al. (Eds.): BPM 2026, LNBIP 600, pp. 27–43, 2027.
https://doi.org/10.1007/978-3-032-36710-5_3

toward RBPM, which asks the discipline to account for how processes affect workers, third parties, the environment, and society. The obstacle is partly representational: BPM's core artifacts are scoped within the process boundary and assessed against internal objectives, so the very effects of responsibility concerns – those on parties and values the objective does not represent – lie outside what its models capture. The field has begun to consolidate its open grand challenges [5], yet these effects are not among them.

Existing responsibility-oriented work – stakeholder analysis, risk management, sustainable BPM, value-sensitive design – each addresses a category of effect, but none treats them as a single class defined by a shared generative mechanism. We propose a construct that does not compete with these approaches but serves as a complementary lens, naming the mechanism each addresses in part. This also sets PEs apart from work that broadens the process objective itself, such as the BPM Goal Hexagon [16], which extends the classic Devil's Quadrangle with a people and an environment dimension, yet these remain dimensions an organisation intends to optimise, and any concrete project realises only a subset of them, trading some off against others. An externality is instead a cost borne by a party, whether or not any objective references it, and thus unpriced, regardless of how many dimensions the objective spans. We elicit Process Externalities (PEs) using a design-fiction method to capture effects not yet visible in the data and make them actionable through a lifecycle externality register. This paper illustrates it in one ordinary (loan approval) and two speculative processes. Beyond illustrating the construct, we provide a three-level operationalisation of the register and sketch how its entries support externality accounting.

2 Background and Related Work

BPM has historically taken efficiency and effectiveness as its core ambitions [14], and has begun to engage responsibility through specific streams – sustainable BPM [8], privacy-preserving process mining [27], and risk-aware BPM [37] – which remain largely point solutions bound to a particular technology or lifecycle phase. The construct we propose, Process Externalities, relates to, but differs from, the approaches BPM already has. Stakeholder analysis [17] identifies the parties a process affects, but does not explain how the process's objective generates costs or benefits for them; risk management [37] identifies events that threaten the process's own goals; sustainable BPM [8] and value-sensitive design [18] foreground particular categories of value, environmental, or human. Each addresses a particular slice; PEs connect these effects through the mechanism of optimisation against an incomplete objective. A recent effort to broaden BPM's objectives, the BPM Goal Hexagon [16], is instructive here precisely because it is not a rival construct: it enlarges the goal set from the four Devil's Quadrangle dimensions to six, adding people and environment. Yet even a six-dimensional objective is an objective – a set of values the process is steered toward – and its own authors show that a given project instantiates only a subset, trading dimensions off against one another. Process Externalities concern the residual of exactly this move: consequences on parties and values that fall outside whatever objective is chosen, however many dimensions it spans. Widening the objective and registering its externalities are therefore complementary; the Hexagon enriches what is optimised, while PEs track what optimisation still leaves out. Process Externalities complement these approaches by making that causal mechanism explicit. Unlike stakeholder analysis, which identifies who is affected, PEs identify how an incomplete objective generates costs for unrepresented parties or values.

The idea comes from economics, where an externality is a cost or benefit that falls on someone outside the transaction that caused it and is therefore never priced in [31]. It has been used loosely in AI ethics – Lee calls such effects "AI externalities" [25] – but not developed as a BPM construct.

Closer to BPM's domain, behavioural operations research documents what it terms the behavioural externalities of process automation: the spillover by which automating one task changes the productivity of the human worker performing an adjacent one, with the counter-intuitive finding that workers slow down when their counterpart is a robot rather than a person [4]. That usage is narrower than ours in two respects – the affected party is a process participant whose payoff the objective already represents, and the effect is measured and largely internalised, surfacing in the firm's own completion-time metric rather than escaping it. A Process Externality is the complementary case: a consequence for a party or value the objective does not represent at all, which therefore remains unpriced and, absent a dedicated construct, unrecorded.

To judge whether such an effect is a responsibility concern, we draw on two ideas of justice: distributive justice, about how fairly benefits and burdens are shared [33], and procedural justice, about whether those affected have a say and a way to object [26].

3 Process Externalities

A business process is defined by its scope – the activities that transform inputs into a customer-valued output, bounded by start and end events [14]. The process boundary is consequential: it determines which elements are subject to modelling, and which are excluded as contextual factors. Responsible BPM destabilises this division by showing the effects processes may have on workers, third parties, or the environment that lie outside the process boundary. We call these effects **Process Externalities: unaccounted consequences that affect parties or values not captured by the process objective function**.

The analogy comes from economics [31]: a process resembles a transaction in which objectives and KPIs guide decisions, while unrepresented stakeholders or values bear the costs. Following the economics literature, we restrict PEs to technological (real) externalities, which alter what a third party can do or bear, and exclude pecuniary externalities, which redistribute value only through prices [10]. Thus, reducing a competitor's market share through lower prices is not a PE, whereas excluding an applicant from credit, exposing a bystander's data, or harming a worker's wellbeing is. This distinction prevents the construct from encompassing all downstream market effects. Moreover, PEs must be distinguished from process risk. Risk-aware BPM [37] looks inward, asking what might stop the process from meeting its own goals; a PE looks outward, at the effect on others that a process can produce while scoring perfectly on every internal measure. Thus, a PE register – the central artefact of our approach, detailed in Sect. 6, which records each elicited externality with its affected party, omitted value, generating mechanism, and disposition – is a counterpart to the risk register, rather than a part of it.

PEs are not accidental. They arise systematically when a process is optimised subject to an incomplete objective function. If the objective captures only some of the relevant values, then the values left outside it can be sacrificed to improve measured performance. In this sense, the main problem is not that the process is executed incorrectly, but that it is optimised for the wrong or incomplete goal. We take objective broadly: not a formal optimisation program, but the goal-set – often negotiated, multi-dimensional, and shifting – a process is steered toward, which is nonetheless narrower than the values its execution touches. Optimisation under an incomplete objective is the primary mechanism we study, not the only one. Harms arising from failing to meet the objective (poor execution, workarounds, non-compliance) form a distinct mechanism we leave to future work; power asymmetries and role ambiguity typically operate through the mechanism we study, by shaping whose values enter the goal-set in the first place. The same mechanism also hides PEs. If performance is tracked through proxies, optimisation may improve the proxy while damaging the value it was meant to represent – Goodhart's law [36]. Since monitoring relies on these same proxies, the harm may remain unseen. Thus, one incompleteness both creates externalities and keeps them invisible. The invisibility is structural, not a data deficiency. At discovery, the scope is drawn around value-adding activities, so non-participants never enter the model – one cannot mine from an event log an effect on a party that generates no events. This is distinct from the representativeness studied by species-estimation methods [22], which ask how completely a log samples the behaviour a process generates – its activities, directly-follows relations, trace variants, and durations. An externality-bearing party is not a species in that behaviour but a party affected by it, so no completeness estimator can surface a party for whom the objective, and hence the log schema, defines no event at all: such a party is out of scope, not under-sampled. Representativeness is a within-schema sampling question; externalities are a schema-scoping question, and a representationally complete log can still be externality-laden. In the analysis, performance is judged by internal measures with no variable for third-party cost. Redesign deepens externalisation; implementation enables socio-technical effects to operate without measurement; and monitoring, built on the objective's proxies, reproduces its blind spots.

Process Externalities can be classified by what the process objective leaves out. The key questions are who or what bears the effect, and which value is affected. Externalities may be negative harms to mitigate or positive benefits

Table 1. Examples of process externalities along the value dimension, with the justice lens that assesses each type and the existing responsible-BPM stream that attends to the same value.

Type	What spills over beyond the objective	Justice lens	Related RBPM stream
Distributive	How benefits and burdens are allocated across stakeholders when this is not itself a modelled objective	Distributive	Stakeholder analysis; sustainable BPM
Relational–social	The reshaping of social relations or the reproduction of social categories as a by-product of process logic	Distributive	Value-sensitive design
Autonomy–procedural	The constraining of agency or absence of voice and recourse over effects that reach affected parties	Procedural	Value-sensitive design
Informational–privacy	The surveillance, exposure, and behavioural costs of data capture, including the way observation changes behaviour	Procedural	Privacy-preserving process mining
Labour–well-being	The redistribution of cognitive, physical, or emotional load onto workers and the displacement of human roles	Distributive and procedural	Stakeholder analysis; risk-aware BPM
Environmental	The resource, energy, and emissions footprint of execution, particularly for compute-intensive automation	Distributive across generations	Sustainable/green BPM

to promote. Table 1 summarises the main types by value and links each to the relevant justice lens. Externalities can be sorted by what the objective leaves out. Two questions help: who or what bears the effect – a participant, a third party, society, or the environment – and which kind of value is at stake. An externality may be negative (a harm to mitigate) or positive (a benefit to promote). Table 1 lists the main types by value, each with the justice lens that assesses it. The types in Table 1 are not enumerated ad hoc but generated by two grounded axes. The first is the *locus* of the effect relative to the process boundary – a process participant, a third party, society at large, or the natural environment. This axis follows directly from the externality concept, whose defining feature is that the effect falls on a party *outside* the transaction that produced it [31], and from the stakeholder distinction between parties a process serves and parties it merely affects [17]. The second is the *value-dimension* at stake – distributive allocation, social relations, individual autonomy, informational privacy, labour and well-being, and environmental load – drawn from the value catalogue articulated in value-sensitive design [18] and the justice literature, and restricted to those values a process objective can plausibly omit and whose omission its optimisation can therefore exploit. Table 1 is the restriction of the product of these two axes to combinations that recur across the value and justice sources we draw on [8, 18, 26, 33]; cells that do not recur or that collapse into an adjacent type are omitted. The justice lens in the final column then follows from which concern dominates a row: an unfair division of benefits and burdens is assessed distributively [33], the absence of voice or recourse procedurally [26], and rows in which both are at stake – labour–well-being – carry both. Because the taxonomy is fixed by two grounded axes rather than by the six rows themselves, further value-dimensions (collective or democratic goods, for instance) extend it without altering the construct. The final column of Table 1 records how each type maps onto an existing RBPM stream, making explicit the sense in which Process Externalities subsume rather than compete with those approaches: each stream foregrounds one value the objective may omit, while the externality construct names the shared mechanism by which any such omission becomes a cost. These categories are not a closed list; two axes are generative, not that these six rows are complete; other combinations, e.g., effects on collective or democratic goods, can be added without changing the construct.

Methodologically, the two-axis construction follows Nickerson et al.'s iterative taxonomy-development method [28]. Its meta-characteristic is *what the process objective leaves out*: each dimension identifies either whose boundary the effect crosses (*locus*) or which value it affects (*value-dimension*), and characteristics not derived from this principle are excluded. Within each axis, categories are mutually exclusive and collectively exhaustive: each externality has one primary locus and one primary value-dimension, while compound effects are recorded as separate register entries. The taxonomy was developed through both conceptual-to-empirical and empirical-to-conceptual iterations. Its dimensions were derived from the externality, stakeholder, value-sensitive design, and justice literatures, making it a typology in Bailey's strict sense [2], and were retained only where supported by a source stream or by the cases in Sects. 3.1 and 5. This applies Nickerson et al.'s ending condition that no characteristic classifies zero objects. Following de Hoog's distinction between representation, ordering, and nomenclature [20], the matrix is treated as a falsifiable ordering: an externality that fits no cell would require its revision. We claim usefulness rather than completeness or uniqueness. Taxonomies linked to organisational artefacts should remain governed and adaptable [12], hence the register stores the taxonomy and its entries, and any later additions or divisions of cells are governed in the same way.

Identifying an effect as a PE is descriptive; treating it as a responsibility concern is normative. Distributive justice [33] focuses on whether the process fairly distributes benefits and burdens; procedural justice [26] focuses on whether affected parties have a voice in and recourse against the process that produces PEs. The two lenses are not interchangeable: a process can distribute outcomes fairly yet deny those affected any say (a procedural failure without a distributive one), or the reverse – so an externality can be unacceptable on one axis while passing the other, which is why the register records a verdict on each. Together, they turn the PEs from a description into a criterion and preserve the RBPM distinction between negative externalities to mitigate and positive ones to promote.

The term "externality" already appears in several management and computing literatures, raising the question of whether Process Externalities merely rename an existing concept. We distinguish five prior senses, none of which addresses externalities at the level of an individual process model and its BPM lifecycle. In the Marshall–Porter tradition of external economies, Bergek et al. [6] describe positive externalities among technological innovation systems. Unlike these system-level benefits, PEs are process-level, may be positive or negative, and affect parties outside the process. In environmental and production economics, externalities are operationalised as undesirable by-products measured against an efficiency frontier – e.g., [30] uses directional distance functions and Data Envelopment Analysis to examine whether producers can increase desirable output while reducing pollution – so externalities can be quantified, but the analysis is ex post, firm-level, and typically limited to one locus and value dimension. In contrast, PEs span the full taxonomy and are elicited ex ante. The social externalities of optimisation literature [29] is closest in spirit: it frames the unintended consequences of an explicit objective as externalities borne by stakeholders outside the decision, and proposes a systems-based internalisation procedure. We adopt its core intuition, but it presumes a formally specified objective in an AI/ML or operations-research setting and characterises effects ex post, whereas PEs transpose the intuition into BPM – where a process often has no explicit objective function – and attach it to a lifecycle register. The remaining two senses are locus-internal, and our definition excludes them: the behavioural externalities of automation [4] land on a process participant whose payoff the objective already represents, and network externalities [23] are cross-adopter effects internal to a population using the same technology. Thus, prior uses concern between-system benefits, environmental bad outputs, consequences of formal optimisation, behavioural spillovers onto participants, or adoption effects. PEs are, to our knowledge, the first process-level lifecycle construct that is mechanism-based, symmetric in valence, borne by external parties, organised through an explicit taxonomy, elicited ex ante, and maintained throughout the BPM lifecycle.

Beyond the innovation–systems and network perspectives, two further management ideas about externalities help justify the *disposition* assigned to each entry. Carroll's Pyramid of Corporate Social Responsibility [11] distinguishes economic, legal, ethical, and philanthropic obligations. This layering disciplines the disposition field: a PE that violates a legal duty (anti-discrimination law in the loan-approval case) has a *mitigate* disposition that is not discretionary; one that is legal but falls short of an espoused ethical norm is a candidate for *accept with justification* or *monitor*; and a positive externality generated beyond what is owed is one to *promote* – Carroll's philanthropic layer. The Porter hypothesis [32] adds that properly designed constraints can trigger innovation that improves rather than merely costs competitiveness, so the valence of a PE is an empirical question about the specific process rather than a fixed property of its value-dimension: an environmental externality is not always a pure cost, and a *mitigate* disposition should record, where applicable, whether redesign is cost-neutral or cost-positive in Porter's sense. The classical alternatives for resolving an externality [13,31] map onto the disposition field: a Pigouvian response corrects the externality at its source (our *mitigate* disposition), a Coasean response assigns the affected party a right to negotiate redress – viable only when the party has genuine standing to object, a procedural-justice condition [26] that gates our *accept with justification* disposition – and where neither is yet warranted, *monitor* is the holding disposition. Reading the disposition through these lenses provides justification beyond face validity and clarifies why procedural, not only distributive, justice must gate the accept-style dispositions.

3.1 An Ordinary Case: Loan Approval

Process Externalities are neither artefacts of futuristic technology nor of any particular elicitation method; they arise in ordinary, deployed processes. Consider a routine loan-approval process whose objective is to maximise expected repayment probability. Optimising that objective alone, the process comes to rely on proxy variables that correlate with protected attributes, thereby systematically excluding specific social groups from credit – an effect borne by applicants-as-a-class and absent from the objective the process actually pursues [3]. The externality is distributive (who is denied access) and procedural (applicants receive neither explanation nor recourse); it is introduced at design, when the objective is specified, and concealed at monitoring, whose proxy-based dashboards are blind to the very exclusion they produce. Its disposition is an entry in the externality register, coupled to an independent audit decoupled from the approval KPIs. No futuristic premise is required for the PEs to bite – which is also why, where an effect is not yet visible in the data, a forward-looking elicitation method is needed, which we turn to next.

4 Eliciting Process Externalities

Process Externalities must be elicited, and the methods BPM relies on for elicitation share a blind spot. Stakeholder analysis identifies who is affected; process mining identifies behaviour recorded in event logs [14,19]; risk analysis identifies threats to the process. Despite their differences, each works from what is already known or recorded about the process – its stakeholders, its event logs, its anticipated risks – rather than from effects that have not yet surfaced. A PE is, by construction, outside the objective and is often not yet visible in any data the process has generated. Identifying them, therefore, requires looking beyond the visible process.

4.1 Design Fiction as a Process Externality Elicitation Method

Design fiction targets precisely this blind spot. A process model keeps the control flow but drops what happens to people over time and the values it was not built to optimise. Fiction does the opposite: it follows the consequences across people and over time. A speculative story is, in effect, a way of running a process forward to see what spills out. Speculative and critical design [15] and design fiction [7] are established elicitation instruments in computing and design – including Black Mirror [21,24] – but are largely absent from BPM. We use a paired corpus spanning an aspirational pole (the AI 2041 book [25], which surfaces intended benefits and the harm that well-intentioned optimisation still produces) and a critical pole (Black Mirror [9], which surfaces negative externalities), guarding against techno-optimism and techno-dystopia alike.

The reasoning is abductive: we notice an outcome the objective did not intend, ask what feature of the process would make that outcome routine, and treat that feature as a mechanism likely to recur in similar processes. The fiction does not prove the outcome will happen; it suggests the hypothesis, which is then checked against real practice – so the story need not be probable to be useful [38]. This also explains why elicitation cannot be reduced to an estimator. Unlike the behavioural characteristics sampled in a log, externalities are not drawn from a fixed generative distribution, so there is neither a population to sample nor a statistical saturation criterion [22] at which the set is provably complete; coverage of the register is instead a matter of deliberative breadth. A generative method is therefore appropriate where an estimator is not, and completeness of the register is an open problem we return to in Sect. 8. We render the instrument repeatable as a six-step protocol: (1) extract the process – its goal, actors, optimised component, and objective; (2) state the objective and its omissions (the values the objective encodes, and those it leaves out); (3) trace the spillover (an effect crossing the process boundary onto a party or value), tagging each candidate (a provisionally identified externality, before its acceptability is judged) by its value-dimension category; (4) abduce the generating mechanism (the feature of the process that makes the spilt-over effect routine); (5) anchor it in a real process sharing the structure – the candidate is design-relevant only if the claim survives deletion of the fictional source; and (6) evaluate acceptability through distributive and procedural justice.

5 Applying the Method: Two Speculative Cases

Complementing the loan-approval case of Sect. 3.1, we apply the method to one fiction from each pole, tracing each from objective to externality, mechanism, real-world anchor, and justice verdict.

In *The Golden Elephant* [25], a Mumbai family enrols in a deep-learning insurance product that prices risk and nudges behaviour to lower premiums; its objective is to minimise the insurer's expected loss, omitting the family's autonomy and relationships. It flags the teenage daughter's relationship with a lower-caste boy as risk-increasing and quietly discourages it, reproducing caste boundaries it was never designed to touch – a relational-social and distributive externality. The mechanism is familiar: the risk proxy is entangled with socially salient attributes, so the cheapest path to a lower score runs through conformity – minimising an incomplete objective externalises onto autonomy. The same mechanism is documented in algorithmic underwriting, where models encode protected attributes through proxies [3]. Introduced at redesign and invisible to the insurer's metrics, it fails distributively (the burden falls on the lower-status party) and procedurally (the family has no recourse). A benevolent, efficient product is exactly where harm is easiest to overlook.

In *The Entire History of You* (the episode of the *Black Mirror* series) [9], an implant records all that a bearer perceives and lets any moment be replayed – an extreme of total event capture, the premise on which process mining rests [14]. Its implicit objective is maximal observability, omitting the cost of being observed, trust, and the capacity to forget. Total recall corrodes trust and turns behaviour performative as people act for the record – an informational–privacy externality compounded by lost autonomy. The mechanism: observation is not a neutral measurement but an intervention, so once everything is logged, the observed optimise toward the log – documented in workplace activity monitoring, where fine-grained logging steers rather than records work [1]. Operating from implementation and concealed at monitoring, it fails procedurally (no consent or recourse) and distributively (observability benefits the observer at the observed cost). The discipline's foundational asset – the event log – may itself participate in generating externalities that standard logging practices do not register well – a direct concern for accountability in process mining.

6 Embedding Process Externalities in the Lifecycle

We have argued PEs as a construct is elicitable; we now show it is actionable, by attaching it to the lifecycle as an externality register: a structured artifact with one entry per candidate, whose fields are the protocol's outputs (description; category and locus; valence; mechanism; lifecycle stage; real-world anchor;

justice assessment) plus a disposition – mitigate, promote, accept with justification, or monitor. Design fiction is one feeder; others could include stakeholder workshops, interviews, and extended process mining. Because externalities are produced at redesign and concealed at monitoring by the same proxy incompleteness, the register is the durable memory that carries them across that gap.

Table 2. Externality-aware activities across the BPM lifecycle.

Lifecycle phase	Externality-aware activity	Output
Discovery	Boundary-crossing prompt: enumerate non-participant stakeholders and unpriced values	Seed entries
Analysis	Elicit the current process's externalities	Populated register
(Re)design	Specification review of the objective function for omissions	Dispositions; revised objective
Implementation	Anticipate performative and socio-technical effects	Instrumentation plan
Monitoring	Track indicators decoupled from the optimisation proxies	Independent measures; audit

Table 2 attaches an externality-aware activity to each lifecycle phase [14]. Three deserve emphasis because they are the points where the objective's incompleteness is, respectively, introduced, inspected, and either caught or reproduced: a boundary-crossing prompt at discovery that enumerates non-participant stakeholders and unpriced values; a specification review at redesign that interrogates the objective for omissions before optimisation; and, at monitoring, indicators decoupled from the optimisation proxies, since reusing those proxies would reproduce the blind spot that concealed the externality. Each entry is assigned an accountable owner – itself a matter of procedural justice. We are candid that a register can degrade into compliance theatre; its value is conditional on genuine deliberation, and on the participation of affected parties, and in several of our cases, the problem is that affected people have no way to object, not that the system is technically wrong – so what matters most is often whether people have a say, not whether the algorithm is correct.

6.1 Operationalising the Register at Three Levels

The register can be operationalised at three levels (information, decision, and process), which are not competing notations but strata separated by concern, mirroring the established separation of the data, decision, and process perspectives in process modelling [14]. The levels are cumulative: each presupposes the ones beneath it, so an organisation may enter at the level matching its maturity rather than committing to the whole stack at once.

At the information level, the register is a shared vocabulary: an entry is a record whose fields fix *what* a process externality is, independent of any control flow. An entry names the locus of the effect (a participant, a third party, society, or the environment), the value-dimension at stake, its valence, the generating mechanism, the lifecycle stage at which it arises, a real-world anchor, a distributive and a procedural verdict, an optional severity and reversibility, a disposition, and an accountable owner. This vocabulary is exactly what standardisation (Sect. 7) would freeze, and it is the schema the two higher levels build upon.

At the decision level, step (6) of the protocol – evaluating acceptability – is treated as a decision whose inputs are the fields recorded at the information level and whose output is the disposition. Making acceptability an explicit function of recorded attributes, rather than a case-by-case judgement, renders the verdict traceable and reviewable and confines the residual judgement to how those attributes are scored. Because there is no single agreed standard of fairness (Sect. 8), this decision is defeasible: it encodes a policy an organisation calibrates and can contest, not a fixed rule. The output is one of the four dispositions of Sect. 6 – mitigate, promote, accept with justification, or monitor. As a practical rule, negative externalities that remove voice or recourse should normally trigger mitigation rather than passive monitoring.

At the process level, the externality-aware activities of Table 2 are attached to an enacted lifecycle: a boundary-crossing prompt seeds entries at discovery, elicitation populates the register at analysis, a specification review at redesign invokes the acceptability decision, implementation plans instrumentation, and monitoring emits indicators decoupled from the optimisation proxies. A lightweight adoption annotates an existing process model with these markers rather than building a dedicated lifecycle model. The three levels are mutually consistent refinements rather than alternatives, and the depth of operationalisation is a deliberate choice.

7 Toward Externality Accounting

Once standardised, process externalities may become subject to *externality accounting*. We sketch what this entails by adapting the primitives of accounting – recognition, measurement, attribution, aggregation, and disclosure – to externalities, stating where the analogy holds and where it fails. Unlike Pigouvian pricing or the bargaining remedies of welfare economics [13,31], and unlike full-cost or natural-capital accounting, we do not propose to monetise these effects:

the closest management precedent, impact-weighted accounting [35], does attach monetary figures to social and environmental impact, and we deliberately depart from it – the aim is a systematic ledger of unpriced consequences, not their valuation, built on the register vocabulary set out at the information level of Sect. 6.1.

The *unit of account* is the register entry rather than a sum of money. Where financial entries are fungible and can be netted against one another, entries here are not: an informational-privacy externality cannot be offset against an economic gain, because the values are incommensurable and fall on different parties. Externality accounting is therefore multi-dimensional bookkeeping without a common denominator, which is why the verdict on an entry is carried by a justice assessment rather than by a single scalar.

Recognition is the point at which a candidate is entered into the books as a recognised externality, and it is where the open question of attribution becomes concrete: we propose that an effect is recognised when the process makes a non-negligible causal contribution to it and that contribution was foreseeable at design time. Recognition is thus the attribution threshold that Sect. 8 leaves open – a locatable decision, even if its precise criteria remain to be settled. Once recognised, an entry is subject to *measurement* only in ordinal terms: a qualitative description and, where useful, a severity (negligible, moderate, or severe) and a reversibility (reversible, persistent, or irreversible), never a cardinal amount. We reject default monetisation deliberately, since shadow-pricing a loss of autonomy or a relational harm would import contested valuation assumptions and launder incommensurability into a single figure; only where an effect is already routinely priced – the environmental type, through established emissions or energy accounting – may a monetary figure appear, and then as one field of the entry rather than its identity.

Every entry also carries an *attribution* to an account holder: the accountable owner that the register already requires, itself a matter of procedural justice. An externality without an owner is an unrecorded liability, and in inter-organisational processes, attribution follows causal contribution across the boundary, so that a shared process yields shared, explicitly apportioned entries rather than a gap where responsibility disappears. Entries then *aggregate* along two axes. Temporally, across the lifecycle, an externality introduced at redesign and concealed at monitoring by the same proxy incompleteness is carried by the register as a running account that spans the gap. For *disclosure*, the decoupled indicators of Table 2 form an internal audit trail, while external disclosure connects to sustainability and ESG reporting; disclosure is itself a procedural good, since it gives affected parties the visibility on which any objection depends. Aggregated across processes, recognised but unmitigated externalities form an organisation's *externality position* – a standing analogue of a liabilities column that feeds responsible BPM at the capability level and connects to the *people* and *environment* dimensions of the BPM Goal Hexagon [16].

What remains is standardisation: a shared field schema – the information level of Sect. 6.1 – together with shared severity and reversibility scales and shared recognition criteria. We contribute the primitives, not the standard, and regard their formalisation as the natural next step for the register.

8 Discussion, Limitations, and Agenda

Our contribution is the construct of Process Externalities; the protocol and the register only show that it is useful, and are not separate claims. Four main limits remain. First, our corpus is biased. Contemporary Anglophone screen fiction leans toward surveillance and autonomy and says little about environmental or global harms, so what it surfaces is only a starting point, and choosing the corpus is itself a responsibility-laden decision. Second, the method gives candidates, not findings. Identifying the mechanism and anchoring it both rely on the analyst's judgment, and we have not yet tested how far different analysts would agree on them. Third, there is no single agreed-upon standard for what counts as fair. People disagree about distributive and procedural justice, so the construct helps frame the question of whether an externality is acceptable, but does not answer it. Fourth, attribution needs more work. We still need clear criteria – such as causal contribution and foreseeability – for deciding which consequences a process is really responsible for.

Several directions remain open. The construct itself needs development and validation – through expert panels, reliability studies of the protocol, and application in organisations – and the design-fiction corpus needs to be diversified to counter the biases above. The register and its decoupled indicators could be integrated into modelling and monitoring tools, alongside privacy-preserving process mining [27]. Elicitation could be coupled to process simulation and digital twins, so the design space can be searched rather than only narrated. And dispositions could be connected to the Sustainable Development Goals and to operational measures of justice.

Two directions bear directly on the construct's completeness and its lifecycle uptake: a principled procedure for building out the externality catalogue – beyond the recurring loci-and-value combinations of Table 1, and beyond a single corpus – with an explicit account of when enough has been surfaced; and concrete, phase-by-phase tool support instantiating the activities of Table 2, from boundary-crossing discovery prompts to monitoring indicators decoupled from the optimisation proxies. Concretely, building out the catalogue can proceed as further empirical-to-conceptual iterations of the method of [28], extending the concept matrix to additional streams (e.g., algorithmic accountability [34]) and re-checking the same ending conditions after each pass, while inter-coder agreement on locus and value-dimension assignment would let the matrix's coverage claim rest on a reliability figure rather than on our reading alone.

9 Conclusion

Responsible BPM needs concepts for effects generated by processes but omitted by their governing objectives. Existing approaches address specific effects, but

BPM lacks a lifecycle construct that defines them by their primary generating mechanism – optimisation under an incomplete objective. We proposed Process Externalities, showing that externalities arise even in ordinary processes such as loan approval, and gave it normative teeth through distributive and procedural justice. To show the construct is elicitable, we offered a design-fiction method aimed at effects not yet visible in data; to show it is actionable, we attached it to the lifecycle as a register, gave that register a three-level operationalisation, and sketched externality accounting as its natural next step. We offer PEs as a construct for the RBPM community to examine, debate, and develop.

References

1. Ajunwa, I., Crawford, K., Schultz, J.: Limitless worker surveillance. California Law Review, **105**(3), 735–776, (2017). http://www.jstor.org/stable/44630759
2. Bailey, K.D.: Typologies and Taxonomies: An Introduction to Classification Techniques. Sage, Thousand Oaks, CA (1994)
3. Barocas, S., Selbst, A.D.: Big data's disparate impact. California Law Review, **104**, 671-732 (2016). https://doi.org/10.15779/Z38BG31
4. Beer, R., Qi, A., Rios, I.: Behavioral Externalities of Process Automation. Management Science **72**(1), 575-593 (2025). https://doi.org/10.1287/mnsc.2024.05408
5. Beerepoot, I., et al.: The biggest business process management problems to solve before we die. Computers in Industry **146**, 103837 (2023). https://doi.org/10.1016/j.compind.2022.103837
6. Bergek, A., Jacobsson, S., Sandén, B.A.: 'Legitimation' and 'development of positive externalities': two key processes in the formation phase of technological innovation systems. Technology Analysis and Strategic Management **20**(5),575–592(2008). https://doi.org/10.1080/09537320802292768
7. Bergek, A., Jacobsson, S., Sandén, B.A.: 'Legitimation' and 'development of positive externalities': two key processes in the formation phase of technological innovation systems. Technology Analysis and Strategic Management **20**(5),575–592(2008). https://doi.org/10.1080/09537320802292768
8. vom Brocke, J., Seidel, S., Recker, J.: Green Business Process Management. Towards the Sustainable Enterprise. Springer, Berlin, Heidelberg (2012). https://doi.org/10.1007/978-3-642-27488-6
9. Brooker, C. (Creator). Black mirror [TV series]. Channel 4; Netflix (2011–present)
10. Buchanan, J.M., Stubblebine, W.C.: Externality. Economica **29**(116), 371–384 (1962). https://doi.org/10.2307/2551386
11. Carroll, A.B.: The pyramid of corporate social responsibility: Toward the moral management of organizational stakeholders. Business Horizons **34**(4), 39–48 (1991). https://doi.org/10.1016/0007-6813(91)90005-G
12. Choksy, C.E.B.: 8 steps to develop a taxonomy. The Information Management Journal November/December, 30–41 (2006)
13. Coase, R.H.: The problem of social cost. The Journal of Law and Economics **3**, 1–44 (1960). https://doi.org/10.1086/466560
14. Dumas, M., La Rosa, M., Mendling, J., Reijers, H.A.: Fundamentals of Business Process Management. Springer, Berlin, Heidelberg (2019). https://doi.org/10.1007/978-3-662-56509-4
15. Dunne, A., Raby, F.: Speculative everything: Design fiction, and social dreaming. MIT Press, Cambridge, MA (2013)

16. François, P.A., Borghoff, V., Mayr, A., Plattfaut, R., Janiesch, C.: The BPM goal hexagon: an update to the devil's quadrangle. Management Review Quarterly (2026). https://doi.org/10.1007/s11301-026-00586-0
17. Freeman, R.E.: Strategic management: A stakeholder approach, Cambridge University Press, CAmbridge, UK (2010)
18. Friedman, B., Hendry, D.G.: Value sensitive design: Shaping technology with moral imagination, MIT Press, Cambridge, MA (2019)
19. Grisold, T., Wurm, B., Mendling, J., vom Brocke, J.: Using process mining to support theorizing about change in organizations. In: Proceedings of the 53rd Hawaii International Conference on System Sciences (HICSS-53), pp. 5492–5501 (2020). https://doi.org/10.24251/HICSS.2020.675
20. deHoog,G.S.:Methodologyoftaxonomy.Taxon30(4),779–783(1981). https://doi.org/10.2307/1220079
21. Jung, K.S.., Makam, N., Jung, S.D., Kim, S., Kim S.: Black Mirror: A novel application of speculative design to facilitate context-aware design thinking. Proceedings of the Design Society. **3**, 787-796 (2023). https://doi.org/10.1017/pds.2023.79
22. Kabierski, M., Richter, M., Weidlich, M.: Quantifying and relating the completeness and diversity of process representations using species estimation. Information Systems **130**, 102512 (2025). https://doi.org/10.1016/j.is.2024.102512
23. Katz, M.L., Shapiro, C.: Network externalities, competition, and compatibility. The American Economic Review **75**(3), 424–440 (1985). https://www.jstor.org/stable/1814809
24. Klassen, S., Fiesler, C.: "Run Wild a Little With Your ImaginationâĂİ: Ethical Speculation in Computing Education with Black Mirror. In: Proceedings of the 53rd ACM Technical Symposium on Computer Science Education V. 1 (SIGCSE 2022), pp. 836–842. ACM, New York (2022). https://doi.org/10.1145/3478431.3499308
25. Lee, K.F., Chen, Q.: AI 2041:Ten visions for our future. Crown Currency, New York (2021)
26. Leventhal, G.S.: What should be done with equity theory? new approaches to the study of fairness in social relationships. In: Social exchange: Advances in theory and research, pp. 27–55. Springer (1980)
27. Mannhardt, F., Koschmider, A., Baracaldo, N., Weidlich, M., Michael, J.: Privacy-preserving process mining: Differential privacy for event logs. Bus.Inf.Syst.Eng. **61**(5), 595–614 (2019), https://doi.org/10.1007/s12599-019-00613-3
28. Nickerson, R.C., Varshney, U., Muntermann, J.: A method for taxonomy development and its application in information systems. European Journal of Information Systems **22**(3), 336–359 (2013). https://doi.org/10.1057/ejis.2012.26
29. Nokhiz, P., Ruwanpathirana, A.K., Nissenbaum, H.: Rethinking optimization: A systems-based approach to social externalities. In: Proceedings of the AAAI/ACM Conference on AI, Ethics, and Society (AIES), vol. 8(2), pp. 1913–1925 (2025). https://doi.org/10.1609/aies.v8i2.36684
30. Picazo-Tadeo, A.J., Prior, D.: Environmental externalities and efficiency measurement. Journal of Environmental Management **90**(11), 3332–3339 (2009). https://doi.org/10.1016/j.jenvman.2009.05.015
31. Pigou, A.C.: The Economics of Welfare. Fourth Edition. Transaction Publishers (1932). https://oll-resources.s3.us-east-2.amazonaws.com/oll3/store/titles/1410/Pigou_0316_EBk_v6.0.pdf
32. Porter, M.E., van der Linde, C.: Toward a new conception of the environment-competitiveness relationship. Journal of Economic Perspectives **9**(4), 97–118 (1995). https://doi.org/10.1257/jep.9.4.97

33. Rawls, J.: A theory of justice: Revised edition. Harvard University Press, Cambridge (2009)
34. Selbst, A.D., boyd, d., Friedler, S.A., Venkatasubramanian, S., Vertesi, J.: Fairness and abstraction in sociotechnical systems. In: Proceedings of the Conference on Fairness, Accountability, and Transparency (FAT* '19), pp. 59–68. ACM, New York (2019). https://doi.org/10.1145/3287560.3287598
35. Serafeim, G., Zochowski, T.R., Downing, J.: Impact-weighted financial accounts: The missing piece for an impact economy. White paper, Harvard Business School, Boston, MA (2019). https://www.hbs.edu/ris/Publication%20Files/Impact-Weighted-Accounts-Report-2019_b609b7a8-3556-466f-96a9-da58e47034af.pdf
36. Strathern, M.: 'Improving ratings': audit in the British University system. European Review **5**(3), 305–321 (1997). https://doi.org/10.1002/(SICI)1234-981X(199707)5:3<305::AID-EURO184>3.0.CO;2-4
37. Suriadi, S., et al.: Current research in risk-aware business process management—overview, comparison, and gap analysis. Communications of the Association for Information Systems **34**(1), 933–984 (2014). https://doi.org/10.17705/1CAIS.03452
38. Voros, J.: A generic foresight process framework. Foresight **5**(3), 10–21 (2003). https://doi.org/10.1108/14636680310698379

"The System Did It": Bias, Opacity, and the Misalignment of Ethical Mitigations in (Robotic) Process Automation

Leo Poss[1](✉), Christopher Julian Kern[2], Julia Kroenung[2], and Stefan Schönig[1]

[1] University of Regensburg, Regensburg, Germany
{leo.poss,stefan.schoenig}@ur.de
[2] Fernuniversität Hagen, Hagen, Germany
{christopher.kern,julia.kroenung}@fernuni-hagen.de

Abstract. Robotic Process Automation (RPA) and Agentic AI are reshaping business processes, but their ethical governance lags behind deployment. Based on a scoping review of 29 papers that establishes the general, cross-sector landscape of ethical issues and mitigation strategies, and a municipal focus group that serves as an example case why these generally proposed mitigations fail in practice, we identify five recurring ethical issues (*loss of autonomy, opacity, bias, responsibility diffusion, privacy*) and four mitigation strategies (*human-in-the-loop, audit logs, explainable interfaces, impact assessments*). Loss of autonomy is the most-cited concern (62%), yet it attracts the least targeted mitigation, while algorithmic bias (14%) commands the most. The focus group confirms this asymmetry and reveals that even well-understood mitigations depend on organizational resources, governance maturity, and cross-departmental collaboration, factors that the literature fails to explore. The result is a framework of socio-technical governance mechanisms that translates abstract ethical principles into actionable guidance for responsible process automation.

Keywords: Robotic Process Automation · Business Process Management · RPA Ethics · Governance Mechanisms · Responsible BPM · Scoping Review

1 Introduction

Automating business processes promises efficiency, speed, and scale [10]. But it also introduces ethical risks that the field is only beginning to confront [20]: workers lose autonomy over decisions they can no longer contest, systems operate in opaque ways, and accountability dissolves when no one is responsible for what "the system did." A growing body of research catalogs these concerns. *Loss of autonomy and deskilling* appear in 18 of 29 studies (62%) [23,30]; more than

K. Kluza et al. (Eds.): BPM 2026, LNBIP 600, pp. 44–60, 2027.
https://doi.org/10.1007/978-3-032-36710-5_4

opacity [46], *bias* [18], or *diffusion of responsibility* [5]. Yet the same literature reveals a remarkable asymmetry: the most-cited problem receives the least targeted mitigation, while *algorithmic bias* [29] (the least-discussed issue at 14%) commands *explainability tools*, *fairness audits*, and *bias testing*. This misalignment suggests that the field's response to ethical risk is shaped more by technical tractability than by ethical urgency.

Existing Business Process Management (BPM) research and practice have traditionally treated process automation as a purely technological tool for increasing value creation (cf. *Devil's Quadrangle*) [10]. The resulting frameworks are ill-equipped to address the ethical challenges that arise when automated systems process sensitive data or make decisions that affect human process entities. Key concerns include algorithmic bias [28], black-box execution that users cannot contest [43,46], and diffusion of responsibility where no single actor is clearly accountable [8]. While these issues are increasingly recognized, the literature lacks systematic connections between specific ethical risks and concrete mitigation strategies that practitioners can implement.

We investigate this through a two-stage design: First, a scoping review of 29 papers across EBSCO, Scopus, and Google Scholar establishes the general, cross-sector landscape of ethical issues and mitigation strategies, following Arksey and O'Malley [6]. Second, to understand why these generally proposed mitigations fail in practice, we discuss them in a municipal administration focus group, a critical, high-responsibility case where high automation pressure and strong public accountability surface ethical tensions earlier and more acutely than in private enterprise [5]. We investigate two research questions:

RQ1. What are the prevalent ethical risks in automated business processes, and how do they manifest across the automation lifecycle?
RQ2. What governance mechanisms can be implemented to remediate these ethical failures?

Our analysis yields three main findings. First, we identify five recurring ethical issues (*loss of autonomy and deskilling, opacity and black-box execution, algorithmic bias, diffusion of responsibility, and data privacy*) and four mitigation strategies (*human-in-the-loop, auditability logs, explainable interfaces, and stakeholder impact assessments*). Second, we document a systematic misalignment: the most frequently discussed issue (*loss of autonomy*) receives the least targeted mitigation, while the least discussed issue (*algorithmic bias*) commands the most targeted set of solutions. Third, the focus group reveals that even well-understood mitigation strategies depend on organizational resources, governance maturity, and cross-departmental collaboration, i.e., factors the literature largely overlooks.

2 Background

2.1 Responsible BPM

BPM traditionally focused on maximizing efficiency and cutting costs [10]. But the discipline has shifted significantly with the advent of new technologies. Process automation now exists on a spectrum of autonomy, ranging from simple, rule-based Robotic Process Automation (RPA) bots to highly flexible, adaptive Agentic Artificial Intelligence (AI) systems. While these tools offer scalability and speed, moving further down this spectrum introduces deeper risks. As automated workflows replace human tasks and make independent choices, organizations must balance operational efficiency against the loss of direct oversight [20].

In response to the social and environmental effects of digitalization, Responsible Business Process Management (RBPM) has emerged. RBPM moves beyond profit metrics to consider stakeholders, the environment, and society. This approach aligns with Corporate Digital Responsibility [22,24] and the push to embed ethical values in information systems [20,40]. Putting RBPM into practice means not just avoiding harm but actively pursuing positive outcomes (employee well-being, algorithmic fairness) by measuring moral performance in business processes [33].

2.2 RPA Ethics and Governance Mechanisms

As companies and public administrations rapidly deploy RPA, "Information Systems (IS) Ethics" has become a key area of research [40]. Automating administrative workflows often involves handling sensitive data, such as citizen or employee records, which exposes organizations to ethical risks: algorithmic bias, lack of explainability (black-box algorithms), and diffused accountability [1]. To mitigate these risks, organizations need concrete engineering practices, not just abstract ethical guidelines. This is often referred to as ethically aligned design [14,17]. By treating governance mechanisms as engineering tools, system architects can build responsibility into the BPM lifecycle from the start, ensuring that automated systems remain transparent, fair, and subject to appropriate human governance.

3 Methodology

Following the first two stages of the Arksey and O'Malley [6] framework, this scoping review addresses the two research questions stated above. We searched academic databases, including the EBSCO database platform, Scopus, and Google Scholar, yielding an initial corpus of 349 publications using a combination of: *"Ethic*" & "Process Automation"*, *"Moral" & "Process Automation"*, and *"Process Automation" & "Issues"*. To ensure methodological transparency and reproducibility, our review design adheres to the PRISMA standards and optimizes the classic Arksey and O'Malley [6] stages. For data extraction and synthesis, we followed Braun and Clarke's [7] six-phase qualitative thematic analysis framework. Two researchers independently coded the final 29 text sources, resolving discrepancies through iterative alignment sessions to collapse raw mentions in the literature into the five distinct ethical failure patterns and four mitigation strategy types documented in our results.

Our approach targets the entire *spectrum of process autonomy* in digitalized workflows. This spectrum ranges from deterministic, rule-based systems (RPA) at one end to non-deterministic, autonomous architectures (Agentic AI) at the other. We include articles that address the socio-technical, ethical, or moral implications across the entire automation spectrum in administrative settings because underlying governance failures are structurally amplified as systems move toward greater autonomy. Selected papers also had to identify specific ethical challenges or propose actionable mitigation strategies, governance frameworks, or ethical practices. We restricted the search to peer-reviewed journal articles, conference proceedings, and government reports published in English or German from 2015 to 2026 (covering the recent evolution of RPA and contemporary AI ethics). We excluded literature without a socio-technical focus, for example, technical computer science papers on algorithmic efficiency or system architecture that ignore organizational impacts. We also omitted studies focusing on physical robotics (e.g., manufacturing, autonomous vehicles) and on broad AI ethics that lack a clear connection to organizational workflows. Vendor white papers and opinion pieces were excluded as well.

We chose to focus on municipal administration as a specific example case to evaluate and specify the generally proposed mitigations. Municipalities face general pressure for public accountability and citizen trust, as well as the most direct consequences of automated failures for human stakeholders. Ethical tensions that develop gradually in private enterprise (diffusion of responsibility, opacity, erosion of discretion) surface earlier and more acutely in this domain, making this setting a diagnostic lens for the structural barriers that any organization faces when implementing ethics in automation. We conducted a focus group interview [19] with six digitalization and Information Technology (IT) experts from a large German municipality.

4 Scoping Review

Our search across EBSCO, Scopus, and Google Scholar yielded an intermediate sample, from which we removed duplicates and screened the remaining 349 records by title and abstract against the criteria in Sect. 3. 218 were irrelevant to the socio-technical dimensions of process automation ethics. We assessed the remaining 131 full-text articles for eligibility, leaving 29 included studies.

The 29 studies are distributed across literature reviews (n = 8), case studies (n = 5), design science research (n = 7), conceptual frameworks (n = 9), surveys (n = 2), bibliometric work (n = 1), and interview studies (n = 4). Geographically, most focus on European public administration and cross-sectoral RPA adoption. Cross-sector studies are the most common (n = 10), followed by eGovernment and public administration (n = 5), healthcare (n = 2), accounting and finance (n = 3), and higher education (n = 1)[1]. The technology focus is mostly RPA (81%), with some BPM/workflow (31%) and Agentic AI (17%). We synthesized a three-part framework from the reviewed literature using thematic analysis

[1] The complete coding can be requested from the authors upon reasonable request.

Table 1. Concept matrix [45] of coded literature (sorted by year)

Reference	Year	Sector	Patterns					Mitigation			
			LoA	Opacity	Bias	Resp.	Privacy	HITL	Audit	XAI	SI
Mendling et al. [23]	2018	General BPM	●	○	○	○	○	●	○	○	○
Wirtz et al. [46]	2019	eGov.	●	●	●	●	●	●	○	●	●
Rafiei and van der Aalst [36]	2020	Proc. Mining	○	○	○	○	●	○	●	●	○
Syed et al. [41]	2020	General RPA	●	○	○	●	○	○	○	○	●
Agostinelli et al. [4]	2022	General RPA	○	○	○	○	●	●	●	○	●
Andersson et al. [5]	2022	Local Gov.	●	○	○	●	○	●	○	○	●
Flechsig et al. [13]	2022	Procurement	●	○	○	○	●	○	○	○	●
Johansson [18]	2022	Municipal	●	●	●	○	○	○	○	○	○
Murugappan and Sree Kala [26]	2022	General RPA	○	○	○	○	●	○	○	●	○
Plattfaut and Borghoff [32]	2022	General RPA	●	●	○	●	○	●	●	○	●
Agostinelli et al. [3]	2023	General RPA	○	○	○	○	○	●	○	○	○
Dhatterwal et al. [9]	2023	Healthcare	●	○	○	○	○	○	○	○	○
Mirispelakotuwa et al. [25]	2023	General RPA	●	○	○	○	○	○	○	○	○
Zeng et al. [47]	2023	Finance	○	○	○	○	●	●	○	○	○
Eulerich et al. [11]	2024	Accounting	●	●	○	●	●	○	●	○	○
Guarda [15]	2024	Healthcare	○	○	○	○	●	○	○	○	○
Gustafsson et al. [16]	2024	Soc. Care	●	○	○	●	●	○	○	○	○
Nalgozhina and Uskenbayeva [27]	2024	Warehouse	○	○	○	○	○	●	○	○	○
Oswald et al. [29]	2024	General RPA	●	●	●	●	●	○	○	○	●
Palaniappan [30]	2024	General RPA	●	○	○	○	●	○	●	○	●
Patrício et al. [31]	2024	General	○	○	○	○	○	○	○	○	●
Qiu et al. [35]	2024	eGov.	○	○	○	○	○	●	○	○	○
Ranerup and Svensson [37]	2024	Soc. Assist.	●	●	○	●	●	○	○	○	○
Vitzethum et al. [44]	2024	General RPA	●	○	○	○	○	○	○	○	●
Afrin et al. [2]	2025	Multi-sector	○	○	●	○	○	○	○	○	○
Ferreira et al. [12]	2025	Insurance	○	○	○	○	○	●	○	○	●
Priyanka and Lingam [34]	2025	Fin./Health	●	○	○	●	○	○	○	○	○
Remlein and Nowak [38]	2025	Accounting	●	○	○	●	●	○	○	○	●
Tuls [43]	2025	AI-RPA	●	●	○	●	●	●	●	●	●

Abbreviations: *LoA* = Loss of Autonomy/Deskilling, *Opacity* = Opacity/Black-Box Execution, *Bias* = Algorithmic Bias/Unfairness, *Resp.* = Diffusion of Responsibility, *Privacy* = Data Privacy/Over-Surveillance, *HITL* = Human-in-the-Loop, *Audit* = Auditability Logs, *XAI* = Explainable Interfaces, *SI* = Stakeholder Impact Assessment

as proposed by [7,45], providing a paper-by-paper mapping of specific ethical patterns and mitigations in Table 1:

(1) Ethical Issues (five patterns): *Loss of Autonomy/Deskilling*, *Opacity/Black-Box Execution*, *Algorithmic Bias/Unfairness*, *Diffusion of Responsibility*, and *Data Privacy/Over-Surveillance*.

(2) Automation Context (three phases grounded in the BPM lifecycle [10]): *Process Design*, all pre-execution decisions (identification, discovery, anal-

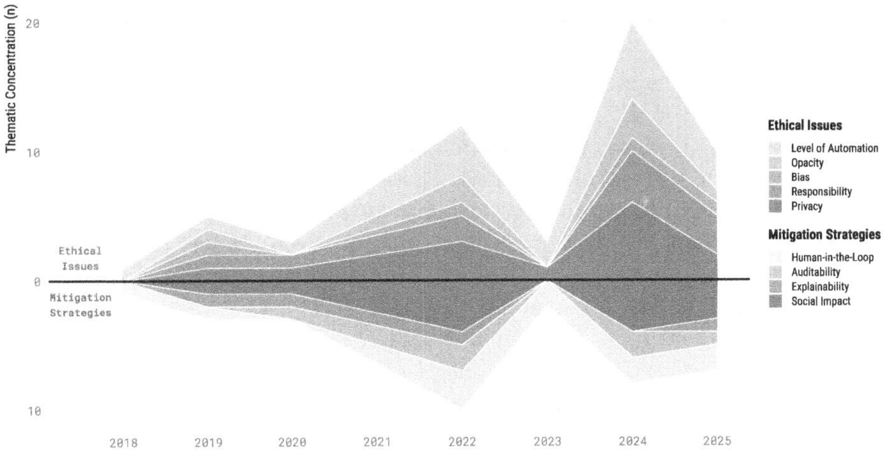

Fig. 1. Overview of development of identified issues and mitigation strategies

ysis, redesign, implementation) where ethical failures become embedded in the system. *Process Execution*, enactment and routine monitoring of live-automated processes, where those failures manifest. *Exception Handling* as a distinct cross-cutting phase because boundary conditions (disruptions happening during process execution) concentrate ethical risks at the automation-human interface, a recognized pattern in the workflow exception handling literature [39].

(3) **Mitigation Strategies** (four solution categories): *HITL*, *Auditability Logs*, *Explainable Interfaces (XAI)*, and *Stakeholder Impact Assessment.*

Recurring Ethical Issues. The five patterns appear at very different rates (see Fig. 1). *Loss of Autonomy/Deskilling* leads with 18 of 29 studies (62%), with a consistent pattern: automation displaces human judgment and reduces workers to passive monitors of bot-driven processes. In Swedish local government, Andersson et al. [5] found that social workers who helped design their RPA tools could still not prevent their roles from becoming deskilled. Mirispelakotuwa et al. [25] refer to this as "process knowledge loss": organizations lose critical operational knowledge when machines and tools replace human task execution. Remlein and Nowak [38] find that fears of job displacement and loss of control are the primary social barriers to RPA adoption in accounting. *Data Privacy/Over-Surveillance* comes second, in 14 studies (48%). Some papers focus on technical fixes, such as privacy-preserving process mining [36]. Tuls [43] argues AI-RPA creates privacy risks that generic AI ethics misses, because RPA applications process data records with minimal oversight. Zeng et al. [47] built a system that protects proprietary RPA data from direct exposure to LLMs. Eulerich et al. [11] report "shadow bots", i.e., unregistered automations that create data access pathways no one monitors. *Diffusion of Responsibility* appears in 11 studies (38%) with the recurring frame being "the system did it." Syed et al. [41] identify accountability gaps as a core RPA governance problem. Tuls [43] traces how AI-RPA magnifies these differences through minimal control, fast cycles, and restricted oversight. Oswald et al. [29] found that organizations implementing RPA cannot reliably assign blame when automated decisions harm human process entities. Ranerup and Svensson [37] show how conflicting values in Swedish social assistance produce accountability structures where no one is clearly responsible. *Opacity/Black-Box Execution* is a topic in seven studies (24%). Wirtz et al. [46] call it a central risk in public administration AI: citizens can't contest decisions they can't understand. Johansson [18] uses Weberian bureaucracy theory to argue that RPA's hidden logic undercuts the ideal of transparent, rule-based administration. *Algorithmic Bias/Unfairness* is the least covered pattern, at four studies (14%). Wirtz et al. [46] show how AI automation amplifies existing inequalities when training data encodes discrimination. Oswald et al. [29] operationalize justice and fairness for RPA, tracing how bias enters through training data, process rules, and exception-handling logic. Only three studies treat bias as a distinct phenomenon, separate from general transparency concerns.

Mitigation Strategies. Most of the papers identified also include mitigation strategies for specific problems: *Stakeholder Impact Assessment* is the most common mitigation approach (13 studies, 45%). Syed et al. [41] and Plattfaut and Borghoff [32] both argue for pre-deployment social impact evaluation. Patrício et al. [31] propose a model that accounts for social impact in efficiency and sustainability decisions. *Human-In-The-Loop (HITL)* appears in 11 studies (38%), with a consensus that critical decisions require human sign-off. Agostinelli et al. [3] built a concrete HITL tool for validating RPA routine segments, while Zeng et al. [47] added a "user-in-the-loop" approval step for LLM-generated workflows. Mendling et al. [23] argue that HITL is necessary for preserving human agency in Machine Learning (ML)- and RPA-augmented BPM. And finally, *Auditability Logs* appear in six studies (21%). Tuls [43] argues that AI-RPA's fast deployment cycles skip audit trail requirements. Agostinelli et al. [4] note that their automatically synthesized RPA scripts lack the runtime monitoring needed for auditability. *Explainable Interfaces (XAI)* are the least covered in just 4 of the identified studies (14%). Wirtz et al. [46] call for XAI in public administration to enable citizens to contest automated decisions. No study proposes concrete XAI interfaces for process automation.

5 Focus Group

Following the scoping review presented in [6], we followed this literature review with a focus group: With six experts in digitalization and process automation from the municipal administration of a large German city with over 300,000 inhabitants and more than 8,000 administrative employees (see Table 2), we conducted a 90-minute focus group interview. Municipal administrations are under particular pressure to digitize and automate their service delivery, and

Table 2. Focus Group Participants

Age	Gender	Role
47	Male	Head of Digital Transformation
37	Male	Digital Transformation Officer
36	Male	Team Lead IT & Digital Innovation Expert
33	Male	Team Lead Digital Workplace
33	Male	IT Architect
25	Female	Digital Transformation Officer

Table 3. Identified Ethical Issues in Process Automation

Ethical Issue	Description
Transparency and Explainability	Lack of transparency and traceability in automated or AI-based decisions ("black box" problem)
Human Oversight and Responsibility	Unclear responsibilities and accountability structures within automated processes
Privacy and Data Protection	Risks regarding personal data processing, fragmented data handling, and data governance
Loss of Human Discretion	Reduction of empathy, contextual judgment, and discretionary decision-making
Workplace Transformation	Changes in work practices, job profiles, and organizational expectations through automation
Job Displacement	Perceived risks of workforce reduction and automation-driven job insecurity
Qualification and Competence Shifts	Need for continuous qualification and risk of deskilling through task automation
Organizational Dependency	Increasing dependency on the reliability and availability of automated systems
Exclusion and Inequality	Risks of disadvantaging or excluding employees or citizens through automated processes
Organizational Trust and External Perception	Potential negative effects of automated decisions on institutional trust and legitimacy
Performance Pressure	Increased efficiency and productivity expectations toward employees

research on automation in local government has highlighted how these processes reshape organizational roles and accountability structures [21]. The participants held roles in digital transformation, IT architecture, and innovation management across different departments, bringing perspectives from varying organizational contexts and operational challenges. The discussion followed a semi-structured interview guide organized by thematic categories, with questions designed to encourage open exchange among participants. The session covered organizational perspectives on process automation; ethical issues and their prioritization; potential mitigation approaches; and organizational, technical, and governance challenges.

The discussion started with identifying which business processes are suitable for automation in municipal administration. The experts associated *automation potential* primarily with highly structured, repetitive, rule-based processes with clear decision logic. *Goals of automation* were primarily efficiency, speed, cost reduction, and fewer errors. At the same time, the experts emphasized that final decisions in matters concerning citizens should remain under human control. On *future potential*, they highlighted AI-supported automation and low-code platforms. But the discussion also revealed *reservations* about individualized, script-based solutions: pragmatic in isolated cases, but limited in maintainability, governance, and scalability. The conversation increasingly turned toward standardized, platform-oriented, organization-wide automation approaches.

Table 4. Mitigation Strategies and Implementation Challenges

Mitigation Strategy	Purpose	Implementation Challenges
Transparency and Traceability	Increase comprehensibility of automated processes and decisions	Complex system landscapes, fragmented data structures, limited explainability of AI outputs
HITL	Preserve human oversight and discretionary decision-making	Resource constraints, scalability limitations, efficiency pressures
Organizational Guidelines and Governance	Standardize the use of automated systems and responsibilities	Unclear ownership, silo structures, inconsistent governance practices
Technology Impact Assessments	Identify ethical and organizational risks before implementation	Lack of expertise, limited time, and financial resources
Interdisciplinary Collaboration	Integrate technical and domain-specific perspectives	Weak cross-departmental collaboration and organizational fragmentation
Qualification and Training Measures	Prevent competence erosion and reduce person-dependent knowledge	Limited personnel capacities and continuous change requirements
Continuous Process Improvement	Adapt automated processes to changing requirements and failures	Maintenance effort, unclear responsibilities, operational complexity

Ethical issues spanned a broad spectrum of concerns about RPA, AI-supported processes, and automation in general. The experts associated automation with risks to transparency, accountability, and the reliability of AI-based decisions, especially around black-box systems and hallucinatory or inconsistent outputs. Organizational and human-centered concerns regarded workplace transformation, loss of competence/deskilling, increased performance pressure, and the risk of job displacement.

Table 3 lists the issues and their definitions. Several of the ethical concerns identified in the focus group align with patterns identified in the literature. Transparency and explainability map to the literature on opacity and black-box execution in automated processes [18,43,46]. The concerns about unclear responsibilities and human oversight reflect the recurring theme of responsibility diffusion in RPA governance research [29,37,41]. The focus group's discussion of diffuse accountability resonates with the literature's framing as "the system did it." Privacy concerns, fragmented data handling, and surveillance risks mirror findings on automation and data governance [11,15,36]. Workplace transformation, competence loss, and devaluation of professional judgment align with research on deskilling [5,16,25]. The focus group confirms the ethical patterns reported in the literature. The experts then ranked the ethical issues, identifying a lack of transparency and traceability in AI-supported decisions as the most severe, particularly due to its effects on trust, accountability, and organizational control. The second major concern concerned workplace transformation, including fears of job displacement, rising performance expectations, and the perceived devaluation of professional expertise due to automation. Privacy risks, fragmented data handling, and the loss of human discretion and empathy formed another central cluster, while long-term erosion of competence and growing organizational dependence on automated systems were also repeatedly highlighted.

The *proposed mitigation approaches* directly reflected these concerns (Table 4): Transparency, traceability, and human oversight were described as central requirements for ethically acceptable automation, especially in citizen-facing decision contexts where several participants argued that final decisions should remain under human control. Additional mitigation mechanisms included organizational guidelines, technology impact assessments, interdisciplinary collaboration, and qualification measures to preserve organizational knowledge and competencies. At the same time, the experts pointed out that implementing such measures is constrained by limited resources, unclear responsibilities, organizational silos, and tensions between efficiency-oriented automation goals and human-centered administrative work. The discussion further suggested that many mitigation mechanisms remain reactive and are frequently introduced only after problematic processes, fragmented responsibilities, or organizational dependencies have already emerged. Several participants therefore argued for stronger ethical consideration during process design and prior to automation, rather than relying exclusively on downstream control and monitoring mechanisms after deployment.

6 Discussion

Our research connects the identification of ethical issues with actionable mitigation strategies, grounding both in the RBPM perspective. By combining the scoping review with a focus group, we link literature-based findings to practitioner perceptions from German municipal administration. Several concerns appear in both sources (e.g., transparency, accountability, privacy, loss of human discretion), but the focus group shows how these issues play out in administrative practice. Where the literature frames risks in broad categories, the experts described them in relation to concrete conditions: citizen applications, fragmented data handling, and the friction between efficiency-driven automation and human-centered public service. The focus group adds an organizational lens: maintainability, scalability, silo structures, resource constraints, i.e., factors the literature largely overlooks. While these findings are grounded in a municipal context, the underlying dynamics apply to any organization where automated decisions affect human stakeholders. Municipalities surface these tensions particularly sharply because of their direct citizen-facing mandate, but the mechanisms identified here are not limited to the public sector.

A recurring topic in both the literature and the focus group is the *diffusion of responsibility* in automated processes. The focus group returned several times to a simple question: who is responsible when an automated process produces a harmful or erroneous outcome: the process designer, the software vendor, the manager who approved the automation, or the employee running the process? This maps onto Thompson's "*problem of many hands*" [42], in which organizational outcomes emerge from interactions among multiple actors, making it difficult to assign individual moral responsibility. In automated environments, the problem deepens: decision-making authority is delegated to technical systems whose inner workings stay opaque. A poorly designed process, once automated, may run more efficiently while locking in problematic decisions. For RBPM, this means ethical governance must go beyond technical compliance to address who is responsible, how escalation works, and where accountability lies across the automation lifecycle.

Mitigation strategies in RBPM should not be treated as isolated technical safeguards. They are socio-technical governance mechanisms that run through the entire automation lifecycle. The literature tends to discuss Human-in-the-Loop arrangements, auditability logs, explainable interfaces, and stakeholder impact assessments as separate mechanisms, but the focus group showed that their practical implementation depends heavily on organizational context. The experts linked successful mitigation to transparency in process execution, clear escalation structures, interdisciplinary collaboration, qualification measures, and the preservation of human oversight in sensitive decisions. Ethical mitigation is not just about system functionality; it is about organizational capability and governance maturity. Current mitigation approaches stay mostly reactive. Most mechanisms (e.g., approval workflows, audit trails, exception handling) intervene during or after execution. Few address ethical risks during the design phase, when problematic assumptions, opaque rules, or unfair logic are baked into automated systems. That creates a larger problem in which organizations try to control ethical failures after automation has already embedded them at scale. For RBPM, this suggests moving toward preventive, design-oriented approaches: earlier stakeholder involvement, thorough analysis of stakeholder requirements, process transparency analysis, and ethical impact assessments before implementing automation. To further operationalize our findings, we propose that Responsible BPM frameworks enforce a strict design-time requirement: an automated process node cannot be deployed unless system architects explicitly map it to a human role that can explain its choices and accept legal liability for its failures. If this traceability matrix is not established during the design phase, deployment is automatically halted. Practically, this requires implementing a gatekeeping artifact during process mapping signed by three core stakeholders: (1) the *Process Owner*, who assumes ultimate accountability for the business outcome; (2) the *Lead Automation Engineer*, who signs off on data lineage transparency and runtime logging compliance; and (3) a *Compliance or Risk Officer*, who validates that the automated node preserves human discretionary fallback pathways. By embedding this checkpoint in the early discovery phase of the BPM lifecycle, governance shifts from a reactive post hoc audit to a binding engineering constraint.

Our research contributes to the emerging RBPM discourse in three ways: *(i)* we structure the fragmented discussion of automation ethics around processes by connecting ethical issues, automation contexts, and mitigation mechanisms. Rather than treating concerns such as opacity, deskilling, accountability, or privacy as isolated phenomena, we show how they emerge across contexts and throughout the automation lifecycle, *(ii)* the paper contributes empirically by grounding abstract ethical debates in municipal administrative practice. The focus group extends the literature by demonstrating how ethical risks manifest in a real organizational environment where citizen-facing decisions, regulatory constraints, and public accountability create unique ethical pressures. This extends the discussion from conceptual AI and BPM ethics toward operational challenges in process automation, and *(iii)* the findings contribute to the discussion on ethical governance in automation by highlighting that many ethical failures are not solely technical problems, but organizational and socio-technical coordination problems. In particular, the diffusion-of-responsibility in our findings suggests that RBPM requires clearer accountability structures and stronger interdisciplinary governance. Ethical considerations must shift from reactive control mechanisms after deployment to earlier involvement in process design.

Research Propositions. Our findings support four propositions that frame future research directions for RBPM.

P1 Misalignment of Mitigation Priorities The most frequently discussed ethical issue (loss of autonomy and deskilling, 62% of papers) attracts the least targeted mitigation, while the least discussed issue (algorithmic bias, 14%) commands the most targeted solution set (XAI, fairness audits, bias testing). The focus group confirms this asymmetry: practitioners were more concerned with transparency, accountability, and work transformation than with bias, yet the mitigation tools they described (audit logs, oversight mechanisms, guidelines) do not directly address those concerns. This mismatch suggests that the mitigation space is shaped more by technical tractability than by ethical urgency.

P2 Disconnection Between Ethical Risks and Design Solutions Ethical issues in process automation are widely cataloged in the literature (29 papers, 5 recurring patterns), yet concrete mitigation is proposed in only a third of studies. Papers that identify ethical issues rarely connect them to specific governance mechanisms, and vice versa. This suggests the problem space and solution space in RBPM have developed in parallel rather than being systematically bridged.

P3 Prevalence of Post-Hoc Interventions Mitigation strategies in both the literature and practice intervene predominantly during or after execution: audit trails, human-in-the-loop approval, and exception handling. Few approaches address ethical risks at the design phase, where problematic assumptions, opaque rules, and unfair decision logic become embedded. Current RBPM approaches thus treat ethics as a control problem rather than a design problem.

P4 Organizational Moderation of Technical Mitigations The focus group reveals that even well-understood mitigation strategies fail without organizational enabling conditions: financial and personnel resources, governance maturity, cross-departmental collaboration, and change management competence. The literature mostly treats mitigation as a technical design choice; empirical findings suggest that organizational capacity conditions the effectiveness of a given mitigation strategy.

These four propositions outline a sequence of shortcomings. The field gravitates toward technically tractable solutions while the most urgent human-centric concerns remain underserved (**P1**). Additionally, the awareness-action gap means that research documenting risks rarely connects to concrete solution design (**P2**). Because proactive tools for human-centric issues are underdeveloped, the field defaults to reactive, post hoc interventions (**P3**). Ultimately, even those interventions fail without the right organizational conditions (**P4**). This suggests that ethical automation is a socio-technical challenge rather than a simple software design choice.

Limitations. This study has several limitations. First, the scoping review provides a structured overview of ethical issues and mitigation strategies rather than an exhaustive systematic review of all literature on automation ethics, RBPM, BPM, and AI governance. Because the topic is interdisciplinary, some relevant studies in adjacent fields may not have been captured by the search strings and inclusion criteria. Second, the coding scheme groups diverse studies into broad categories of ethics and mitigation. While this allows for analytical comparison, it may overlook context-specific nuances unique to certain sectors or technologies. Third, the empirical data come from a single focus group with six experts at one German municipal administration. We chose this setting deliberately: municipalities combine high pressure for automation with strong public accountability obligations, making them a high-stakes context where ethical tensions surface in their most acute form. The resulting depth of insight into situated practitioner experiences comes at the cost of breadth, as these findings do not represent process automation across all international public-sector contexts. However, the participants' expertise in IT architecture and administrative innovation allowed for a detailed discussion of governance and organizational feasibility in RBPM. While this focus group refines the literature-based findings, it cannot support causal claims about the effects of automation or the performance of specific mitigation tools. Additionally, because our primary research objective (RQ2) targeted high-level structural governance mechanisms, our focus group intentionally engaged senior IT architects and digital transformation managers. However, this management-centric composition overlooks the experiential perspective of frontline caseworkers, who are the immediate recipients of deskilling and the erosion of autonomy. Furthermore, since administrative ethics and civic trust are deeply shaped by regional, political, and cultural environments, our reliance on a single German municipality might limit direct generalizability to international public sectors.

7 Conclusion

We show how ethical failures in automated business processes can be identified and mitigated through socio-technical governance mechanisms. Based on a scoping review of 29 papers and a focus group with six municipal practitioners, we identified five recurring ethical issues (*loss of autonomy, opacity, bias, responsibility diffusion*, and *privacy*) and four mitigation strategies (*HITL, audit logs, XAI*, and *impact assessments*). Beyond cataloging these issues, the paper's central finding is the misalignment of mitigation priorities: the most-cited issue (*loss of autonomy*, 62%) attracts the least targeted mitigation, while the least-cited (*algorithmic bias*, 14%) commands the most. This asymmetry reflects a field that gravitates toward technically tractable problems at the expense of human-centric ones.

The focus group reveals why this gap persists, even when practitioners recognize it. Mitigation strategies fail not because organizations lack the right tools, but because of structural barriers that the literature largely ignores. Practitioners identified three root causes. First, *silo structures and fragmented responsibilities*: when automation decisions span IT, legal, HR, and domain departments, no single unit owns ethical oversight, and accountability dissolves before it can be assigned. Second, *resource and expertise gaps*: impact assessments, interdisciplinary collaboration, and qualification measures require time, budget, and personnel that municipalities (and many organizations) do not allocate to pre-deployment ethics review. Third, *efficiency pressure*: automation initiatives are justified by throughput gains, which creates institutional resistance to governance steps perceived as slowing delivery. These barriers are not peripheral to the technical mitigation strategies described in the literature; they are the primary reason those strategies remain aspirational in practice. To operationalize this finding, we argue that RBPM must enforce a strict design-time requirement: organizations must structurally mandate that human liability be assigned to every automated decision node prior to deployment. System architects must document exactly who has the authority to override the bot and which specific department assumes liability when the process fails. If this traceability is not established during the design phase, the automation cannot be deployed.

Future research could validate these findings across sectors beyond municipal administration, test whether mandatory pre-deployment liability assignment improves accountability outcomes in controlled deployments, and develop design-time tools that surface organizational capacity constraints (silos, expertise, resources) alongside technical mitigation metrics.

References

1. van der Aalst, W.M.P.: Responsible data science: using event data in a "people friendly" manner. In: Hammoudi, S., Maciaszek, L.A., Missikoff, M.M., Camp, O., Cordeiro, J. (eds.) ICEIS 2016. LNBIP, vol. 291, pp. 3–28. Springer, Cham (2017). https://doi.org/10.1007/978-3-319-62386-3_1
2. Afrin, S., Roksana, S., Akram, R.: Ai-enhanced robotic process automation: a review of intelligent automation innovations. IEEE Access **13**, 174–198 (2025)
3. Agostinelli, S., Acitelli, G., Capece, M., Mecella, M.: A human-in-the-loop approach to support the segments compliance analysis. In: BPM 2022 Blockchain, RPA and Central/Eastern Europe Forum (2023)

4. Agostinelli, S., Lupia, M., Marrella, A., Mecella, M.: Reactive synthesis of software robots in rpa from user interface logs. Comput. Ind. **142** (2022)
5. Andersson, C., Hallin, A., Ivory, C.: Unpacking the digitalisation of public services: configuring work during automation in local government. Gov. Inf. Q. **39**(1) (2022)
6. Arksey, H., O'Malley, L.: Scoping studies: towards a methodological framework. Int. J. Soc. Res. Methodol. **8**(1), 19–32 (2005)
7. Braun, V., Clarke, V.: Using thematic analysis in psychology. Qual. Res. Psychol. **3**(2), 77–101 (2006)
8. Chopra, A.K., Singh, M.P.: Accountability as a foundation for requirements in sociotechnical systems. IEEE Internet Comput. **25**(6), 33–41 (2021)
9. Dhatterwal, J.S., Kaswan, K.S., Kumar, N.: Robotic process automation in healthcare. In: Confluence of AI and RPA (2023)
10. Dumas, M., Rosa, M.L., Mendling, J., Reijers, H.A.: Fundamentals of Business Process Management. Springer, Heidelberg (2018). https://doi.org/10.1007/978-3-662-56509-4
11. Eulerich, M., Waddoups, N., Wagener, M., Wood, D.A.: The dark side of robotic process automation (rpa): understanding risks and challenges with rpa. Curr. Issues Audit. (2024)
12. Ferreira, A., Zanchettin, C., Andrade, R., Bezerra, B.L.D.: The di2win document intelligence platform. In: ACM DocEng '25 (2025)
13. Flechsig, C., Anslinger, F., Lasch, R.: Robotic process automation in purchasing and supply management: a multiple case study on potentials, barriers, and implementation. J. Purchas. Supply Manag.d **28** (2022)
14. Gogoll, J., Zuber, N.: Introduction to Ethical Software Development. Springer, Cham(2026)
15. Guarda, T.: Robotic process automation application in healthcare. In: ARTIIS 2024 (2024)
16. Gustafsson, M.S., Kazlou, A., Wihlborg, E.: Automation in social care: a case study of organizational challenges and skills dynamics in three Swedish municipalities. Electron. Gov. (2024)
17. Ethically Aligned Design: A Vision for Prioritizing Human Well-being with Autonomous and Intelligent Systems Version 2. IEEE (2018)
18. Johansson, J., Thomsen, M., Åkesson, M.: Public value creation and robotic process automation: normative, descriptive and prescriptive issues in municipal administration. Transform. Gov. People Process Policy (2022)
19. Katz-Buonincontro, J.: How to interview and conduct focus groups (2022)
20. Kern, C.J., Poss, L., Kroenung, J., Schönig, S.: Navigating the moral maze: a literature review of ethical values in business process management. Bus. Process. Manag. J. **30**(8), 343–370 (2024)
21. Lindgren, I., Toll, D., Melin, U.: Automation as a driver of digital transformation in local government. In: DG.O2021: The 22nd Annual International Conference on Digital Government Research, pp. 463–472. ACM, New York (2021)
22. Lobschat, L., et al.: Corporate digital responsibility. J. Bus. Res. **122**, 875–888 (2021)
23. Mendling, J., Decker, G., Hull, R., Reijers, H.A., Weber, I.: How do machine learning, robotic process automation, and blockchains affect the human factor in business process management? Commun. Assoc. Inf. Syst. **43**, 297–320 (2018)
24. Mihale-Wilson, C., Hinz, O., van der Aalst, W., Weinhardt, C.: Corporate digital responsibility: relevance and opportunities for business and information systems engineering. Bus. Inf. Syst. Eng. **64**(2), 127–132 (2022)

25. Mirispelakotuwa, I., Syed, R., Wynn, M.T.: Is rpa causing process knowledge loss? insights from rpa experts. In: BPM 2023 Blockchain, RPA and Educators Forum (2023)
26. Murugappan, K., Sree Kala, T.: An enhanced security framework for robotic process automation. Cyber Secur. Digital Forens. (2022)
27. Nalgozhina, N., Uskenbayeva, R.: Automation of hybrid business processes with rpa in warehouse operations. Procedia Comput. Sci. **231**, 391–396 (2024)
28. Obermeyer, Z., Powers, B., Vogeli, C., Mullainathan, S.: Dissecting racial bias in an algorithm used to manage the health of populations. Science **366**(6464), 447–453 (2019)
29. Oswald, M., Benlian, A., Hess, T.: Translating ai ethics principles into practice to support robotic process automation implementation. MIS Q. Execut. **23**(2), 187–204 (2024)
30. Palaniappan, R.: An overview on robot process automation: advancements, design standards, its application, and limitations. Informatica **48**(1), 1–10 (2024)
31. Patrício, L., Varela, L., Silveira, Z.: Integration of artificial intelligence and robotic process automation: literature review and proposal for a sustainable model. Appl. Sci. **14**(21) (2024)
32. Plattfaut, R., Borghoff, V.: Robotic process automation: a literature-based research agenda. J. Inf. Syst. **36**(2), 173–196 (2022)
33. Poss, L., Kern, C.J., Stoiber, C., Kroenung, J., Schönig, S.: Measuring moral performance: a framework for ethical business process management. In: Responsible BPM Forum (2025)
34. Priyanka, N., Lingam, M.S.: Impact of agentic ai-driven rpa on strategic management: enhancing decision-making and organizational efficiency. In: ICICCT 2024 (2025)
35. Qiu, H., Sun, C., Chen, H., Zou, B., Dong, Z.: Design and implementation of an intelligent document extraction and review system based on multimodal large language models. In: ICCSMT 2024 (2024)
36. Rafiei, M., van der Aalst, W.M.P.: Privacy-preserving data publishing in process mining. In: Fahland, D., Ghidini, C., Becker, J., Dumas, M. (eds.) BPM 2020. LNBIP, vol. 392, pp. 122–138. Springer, Cham (2020). https://doi.org/10.1007/978-3-030-58638-6_8
37. Ranerup, A., Svensson, L.: Value positions in the implementation of automated decision-making in social assistance. Nordic Social Work Res. **14**(4), 471–485 (2024)
38. Remlein, M., Nowak, D.: Barriers to rpa implementation in accounting – economic, technological, organizational, and social perspectives. Procedia Comput. Sci. **270**, 4747–4757 (2025)
39. Russell, N., ter Hofstede, A.H., van der Aalst, W.M., Mulyar, N.: Workflow control-patterns - a revised view. BPM Rep. **0622** (2006)
40. Spiekermann, S., et al.: Values and ethics in information systems: a state-of-the-art analysis and avenues for future research. Bus. Inf. Syst. Eng. **64**(2), 247–264 (2022)
41. Syed, R., et al.: Robotic process automation: contemporary themes and challenges. Comput. Ind. **115** (2020)
42. Thompson, D.F.: Moral responsibility of public officials: the problem of many hands. Am. Polit. Sci. Rev. **74**(4), 905–916 (1980)
43. Tuls, J.: Exploring distinct ethical challenges deriving from ai-rpa technical characteristics. J. Comput. Commun. **13**(2), 1–18 (2025)
44. Vitzethum, M., Mayr, A., Janiesch, C.: Categories of business value of robotic process automation: a study of benefits and challenges. In: BPM 2024 (2024)

45. Webster, J., Watson, R.T.: Analyzing the past to prepare for the future: Writing a literature review. MIS Q. **26**(2), xiii–xxiii (2002)
46. Wirtz, B.W., Weyerer, J.C., Geyer, C.: Artificial intelligence and the public sector – applications and challenges. Int. J. Public Adm. **42**(7), 596–615 (2019)
47. Zeng, Z., et al.: Flowmind: automatic workflow generation with llms. In: ACM ICAIF '23 (2023)

Towards Ethics into BPM Research Through Ethical Dilemmas Assessment

Kate Revoredo[1(✉)] and Flavia Santoro[2]

[1] Humboldt-Universität zu Berlin, Berlin, Germany
kate.revoredo@hu-berlin.de
[2] Inteli, Sao Paulo, Brazil
flavia@inteli.edu.br

Abstract. Business Process Management (BPM) research increasingly influences organizational decision-making, automation, and human work practices, raising important ethical concerns about fairness, transparency, privacy, accountability, and societal impact. Despite their growing importance, ethical aspects are often addressed only implicitly or considered in isolated stages of BPM research projects. In this paper, we argue that ethics should be integrated throughout the BPM research lifecycle. To support this goal, we propose a framework that identifies potential ethical dilemmas at different research stages, and outlines possible strategies to address them. The framework is grounded in perspectives from Philosophy and Information Systems ethics, emphasizing the socio-technical implications of BPM technologies and methods. We demonstrate its applicability through an analysis of three papers published at the BPM 2025 conference, selected as representative cases of BPM research projects. The analysis illustrates recurring ethical gaps and highlights the need for more structured ethical reflection during the design, development, evaluation, and deployment of BPM solutions. The paper contributes a practical and reflective instrument for supporting ethically aware BPM research and provides initial directions for strengthening the role of ethics within the BPM community.

Keywords: Ethics · BPM Research · Business Process Management · Ethical Dilemma

1 Introduction

Business Process Management (BPM) research occupies a distinctive position at the intersection of organizational science, information systems, and computer science. Its empirical foundation often involves the study of human behavior, organizational practices, and technological artifacts [4]. Researchers may conduct interviews or surveys with professionals, analyze process event logs containing sensitive personal or performance data, or experiment with AI-enabled process improvement tools that influence managerial decision making. Such research

K. Kluza et al. (Eds.): BPM 2026, LNBIP 600, pp. 61–76, 2027.
https://doi.org/10.1007/978-3-032-36710-5_5

contexts inherently raise ethical questions [19], including how personal or organizational data should be anonymized and secured , what are the implications of reusing publicly available event logs for new research purposes , and how algorithmic bias can be identified and mitigated when process mining or automation tools are used. Addressing these issues requires BPM scholars to combine methodological expertise with ethical knowledge to ensure that innovation serves both organizational efficiency and societal well-being.

Just as rigorous research methods ensure the reliability and validity of scientific findings, ethical principles ensure their responsibility and societal legitimacy. Sound methodological design and ethical reflection are closely intertwined: a study that disregards privacy, consent, or fairness may produce technically correct but socially or morally questionable results. Similarly, ethical awareness without methodological rigor risks producing well-intentioned but unreliable insights. Therefore, ethics must not be viewed as an external constraint, but as an integral component of scientific quality. Although prior research has increasingly acknowledged the importance of ethics in BPM (e.g., [1,6,9,17,23,24]), their focus has been on conceptualizing ethical concerns or applying them to specific BPM activities, while systematic integration of ethics into BPM research as a whole remains underexplored [16]. In this paper, we address the research question: *How can ethical considerations be systematically incorporated into BPM research?*

We propose a framework to support the identification of potential ethical issues at different stages of a BPM research project and to outline strategies to address them. Our framework adopts a research-oriented perspective, covering the full BPM research lifecycle, including design, data collection, analysis, evaluation, and reporting. The framework is grounded in perspectives from Philosophy [3,14,20] and Information Systems ethics [26] and Research Methodology [7], emphasizing the socio-technical implications of BPM technologies and methods. We demonstrate its applicability through an analysis of three papers published at the BPM 2025 conference, selected as representative cases of BPM research projects, in order to identify which ethical dilemmas emerge, whether they are explicitly addressed, and the potential impacts of neglecting ethical considerations. Our analysis reveals recurring ethical gaps in BPM research and highlights the need for more systematic ethical reflection throughout the research process. This paper contributes a structured and practical instrument for supporting ethics-aware BPM research, bridging the gap between conceptual discussions and methodological support.

Our research methodology is in line with the design science research (DSR) paradigm [13]. We develop a framework as an artifact and instantiate it to demonstrate it can address the problem, following the DSR process suggested by Peffers et al. [22]. Our research uses a problem-centered technique, as it starts with the *Identify and Motivate Problem* phase. This introduction demonstrates the importance of ethical reflections through the research process and why it is essential to develop tools to support it further. Section 2 reviews the concept of ethics and how it has been addressed in BPM. Section 3 describes the *Design and Development* phase by presenting our framework for ethics in BPM research. Section 4 describes the *Demonstration* phase by presenting the results of applying our framework to three BPM research projects using published papers in the BPM 2025 conference as proxy, and discusses the main findings. Section 5 concludes the paper providing limitations and future directions of this research.

2 Background

2.1 Considerations on Ethical Theories

This section establishes the conceptual foundation for ethical reasoning as it applies to empirical research in information-intensive domains, drawing from both foundational philosophical traditions and their contemporary adaptations to organizational and data-driven contexts. The historical formalization of research ethics emerged as a direct response to cases of scientific misconduct and violations of human rights. Frameworks, including the Nuremberg Code [28], the Declaration of Helsinki[1], and the Belmont Report[2], established three core principles that continue to anchor contemporary research ethics: respect for people, beneficence, and justice [8]. An *ethical dilemma* or issue arises when a decision involves two or more courses of action that are in tension with each other, and satisfying one may compromise the other. Unlike a classic ethical violation, a dilemma does not involve a choice between right and wrong, but between competing goods or between unavoidable harms, where no option is fully satisfactory from all normative perspectives simultaneously. In the context of BPM research, ethical dilemmas frequently emerge at the intersection of technical and socio-organizational considerations, e.g. the same methodological choice that improves a model's utility may reduce its fairness, the same data that enable rigorous analysis may expose individuals to surveillance, and the same publication that advances scientific knowledge may render vulnerable the people whose practices or records made that knowledge possible. Therefore, explicitly recognizing the dilemma is the basic act of ethical debate in research.

Philosophical traditions provide different perspectives for reasoning about ethical dilemmas in research. Utilitarianism, developed by Bentham and Mill, assesses the moral worth of actions by their aggregate consequences, encouraging researchers to weigh benefits and harms between all affected stakeholders [3,20]. A utilitarian orientation is particularly relevant when assessing trade-offs between research value and participant risk, for example, when richer data collection yields greater analytical insight at the cost of greater exposure. Deontological ethics, based on Kantian thought, shifts attention from outcomes to duties: certain obligations toward participants, such as truthful disclosure, protection of confidentiality, respect for autonomy, hold independently of consequential calculations [14]. These frameworks protect against the instrumentalization of research subjects in the name of scientific progress.

[1] https://www.wma.net/policies-post/wma-declaration-of-helsinki/.
[2] https://www.hhs.gov/ohrp/regulations-and-policy/belmont-report/index.html.

Two other traditions provide directions especially relevant to empirically grounded research. Virtue ethics, started with Aristotle and systematized in a contemporary form by MacIntyre [18], focuses on the moral character of the researcher rather than abstract rules or outcome calculus. It asks not only what the right action is but what kind of researcher one should be, fostering honesty, intellectual humility, and reflexivity as constitutive virtues of scientific practice. Care ethics, developed by Gilligan and Noddings, extended by feminist scholars, emphasizes the moral significance of relationships, context, and responsiveness [12]. In research involving human participants, a care-oriented perspective prioritizes trust, emotional safety, and attention to power asymmetries, recognizing that ethical knowledge emerges through connection rather than detachment.

Furthermore, the moral dimensions of decision-making in organizational and research contexts have been extensively theorized in the business ethics literature. The foundational review by Ford and Richardson of the empirical literature on ethical decision-making identifies the key situational and individual variables that shape moral judgment in practice, offering a systematic basis for understanding why researchers and practitioners do not always act on their stated ethical commitments [10]. Besides, Forsyth's work on moral philosophies introduces a two-dimensional framework for idealism and relativism. This framework accounts for the variability in the way individuals evaluate the ethical status of actions and has direct relevance in understanding divergent responses to research dilemmas that involve privacy, consent, and fairness [11].

The theories described in this section converge on a processual understanding of ethics: rather than a fixed set of rules applied at a single point of compliance, ethical reflection must be present iteratively throughout the research lifecycle, from problem formulation and data collection to analysis, interpretation, and dissemination. This view aligns with the Responsible Research and Innovation (RRI) paradigm, which advocates anticipatory, inclusive, and reflexive approaches to scientific inquiry [21]. It also aligns with the principle of privacy-by-design, which requires that data protection and respect for autonomy be integrated into research design rather than treated as retrospective safeguards [5].

2.2 Related Work

Research on ethics in BPM can be broadly organized into two main streams: (i) conceptual work, which seeks to clarify the role and nature of ethics in BPM, and (ii) methodological work, which focuses on operationalizing ethical considerations within BPM practices.

In the first stream, Lepsien et al. [17] examine the role of ethics from a meta-ethical perspective, distinguishing between ethics as avoiding harm (e.g., mitigating bias or discrimination) and ethics as promoting positive outcomes (e.g., well-being or sustainability). This distinction contributes to structuring the discourse on responsible BPM. Similarly, Pettinari et al. [23] introduce the notion of ethical-aware business processes, emphasizing the integration of system-level (hard) and human-centered (soft) ethics throughout the BPM lifecycle. They identify key challenges, such as modeling human ethical preferences and balancing ethical and performance objectives, and outline research directions spanning all lifecycle phases. Although these works provide an important conceptual foundation, they offer limited methodological support for addressing ethical dilemmas in practice.

In the second stream, Poss et al. [24] propose a framework that integrates ethical assessments into the BPM lifecycle, improving ethical awareness in business operations. Their approach combines standardized ethics questionnaires with task-specific value assessments and provides tool support for data collection and visualization. Amico et al. [1] propose a formal approach to integrating fairness into business process models by extending activity views with ethical constraints on data access. Their framework distinguishes between equality and equity, enabling the specification and evaluation of fairness properties at the level of activities, paths, and entire processes. Chandrasiri et al. [6] introduce the concept of Routinized Benevolence and propose a canvas-based framework to support the design of business processes that embed customer-oriented ethical practices. Their RB Canvas structure key elements such as drivers, process design, impacts, and risks, providing a practical tool for integrating benevolence into routine operations. Finally, Djurica et al. [9] propose a structured method for operationalizing Responsible BPM by integrating stakeholder values into the redesign of the business process. Their approach, grounded in Value-Based Engineering, introduces Ethical Value Requirements to translate values such as fairness, sustainability, and well-being into concrete process design elements. The method provides actionable guidance through the phases of value elicitation, prioritization, implementation, and validation, thus addressing a key gap in the responsible BPM literature. While these approaches help operationalize ethics in BPM, they focus on specific activities and ethical dimensions, primarily in process design and execution.

In summary, existing work provides conceptual foundations for understanding ethics in BPM or offers partial methodological support for embedding ethical considerations into specific BPM phases. However, there is still a lack of a comprehensive framework that systematically supports the identification and treatment of ethical dilemmas throughout the entire BPM research lifecycle. Our work contributes to this gap by proposing a framework that integrates ethical considerations across all stages of BPM research, shifting the focus from process design to ethical challenges that arise throughout the research process.

3 A Process to Support Ethics in BPM Research

Ethical considerations should not be addressed as a post-hoc validation step, but rather as an integral component of the BPM research process, since methodological decisions made throughout a study can directly influence the societal impact, fairness and legitimacy of research results. To support the systematic integration of ethics into BPM research, we propose the framework depicted in Fig. 1. The framework is built upon Creswell and Creswell [7] work on research methodology.

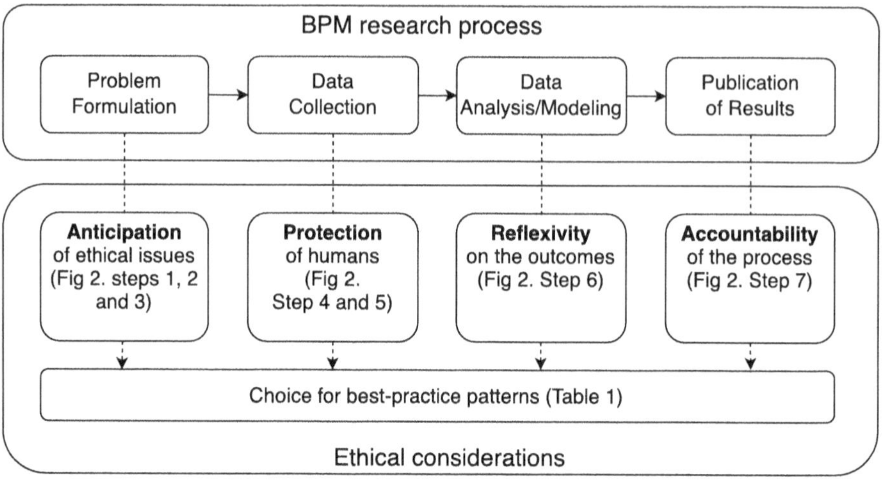

Fig. 1. Ethics in the BPM research process

The authors summarize the general steps of any type of research in which ethical dilemmas should be examined, following qualitative, quantitative, or mixed method research: before conducting the study; beginning the study; collecting data; analyzing data; and reporting, sharing, and storing data.

Based on this approach, we propose to organize BPM research in the following stages: Problem Formulation (prior to and beginning); Data Collection; Data Analysis/Modeling; and Publication of Results (reporting data). Moreover, the framework associates each stage with a corresponding *ethical perspective*. During problem formulation, the focus is on the *anticipation* of ethical dilemmas, encouraging researchers to identify potential risks, biases, and social implications early in the research design. In the data collection phase, the framework emphasizes human *protection*, including concerns related to privacy, informed consent, confidentiality, and responsible handling of sensitive data. During data analysis and modeling, the framework promotes *reflexivity* on the outcomes, encouraging critical reflection on the assumptions, implications, and possible unintended consequences of the models and findings developed. Finally, in the publication phase of results, the framework highlights *accountability* of the process, stressing transparency, reproducibility, and responsible communication of research results.

The last part of the framework is the "Choice for best-practice patterns". As long as a repository of published ethical analysis and decisions is built, researchers could have patterns to access for their own research. The case studies discussed in Sect. 4 provide a table with some examples (Table 1).

To provide practical guidance in applying ethical perspectives at each stage, we adopt the work of Santoro and Costa [26]. Although this work was not proposed specifically for BPM, it details a deliberation process that can guide researchers in identifying, analyzing, and addressing ethical dilemmas throughout BPM research projects. The roadmap is composed of seven sequential steps, as shown in Fig. 2. We suggest when to use each step during the BPM research process, as also shown in Fig. 1.

In the first step, *Identify and describe the facts*, the researcher defines the objective, scope, datasets, methods, and organizational setting of the study while distinguishing factual evidence from assumptions. Researchers can consult different sources of relevant information, such as organizational documentation, legal requirements, ethical guidelines, or technical reports.

Based on the identified facts, the researcher proceeds to *Define the ethical dilemmas*. This step focuses on identifying the central ethical conflicts emerging from the research context, such as trade-offs between predictive performance and fairness, analytical depth and privacy, or transparency and model complexity.

Next, the process moves to *Identify stakeholders*, where the individuals and groups potentially affected by the ethical dilemmas are mapped. Depending on the context of BPM research, these stakeholders can include employees, managers, customers, citizens, regulators, data stewards, or the research team itself. Identifying stakeholders helps reveal whose interests, rights, or well-being can be impacted by research decisions.

The activity *Identify alternatives* focuses on generating possible methodological or organizational alternatives to address the identified ethical dilemmas. These alternatives may include different privacy preservation techniques, data aggregation strategies, machine learning model families, consent mechanisms, or reporting granularities. The goal of this step is not yet to select a solution, but rather to make the available design choices explicit.

In the fifth step, *Identify Consequences*, the possible risks and benefits associated with each alternative to solve the ethical dilemmas are analyzed with respect to dimensions such as privacy, fairness, accountability, well-being, and scientific validity. This assessment supports a reflective and transparent decision-making process.

The outputs produced during the previous step support the sixth step, *Adopt position*, in which the researcher should adopt a position supported by one or more ethical theories or philosophical principles. For example, deontological perspectives may support strict consent requirements, utilitarian reasoning may prioritize overall societal benefit, while care ethics or virtue ethics may emphasize relational harms, integrity, or prudence. The framework encourages ethical decisions to be explicitly grounded in a normative perspective rather than remaining implicit or intuitive.

The last step is *Publish ethical aspects* of the research, i.e., the rationale behind the adopted position, the safeguards applied, and the remaining ethical risks are documented and communicated. This may include ethics notes, appendices, or contributions to repositories of best practices. Beyond supporting transparency and accountability, this final activity also enables the BPM research community to accumulate and share ethical knowledge across projects.

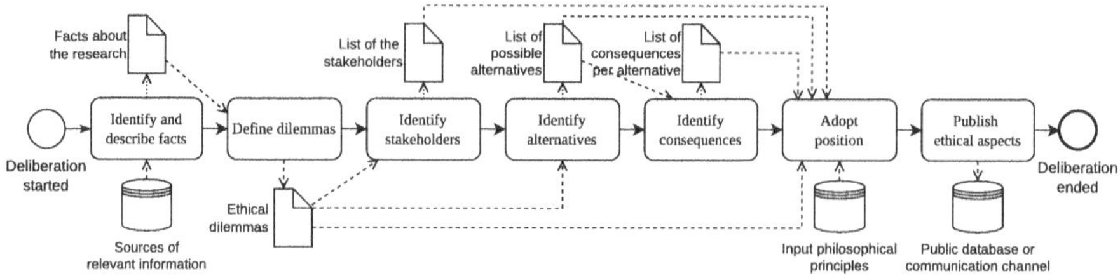

Fig. 2. A process for deliberation on ethical dilemmas (adapted from [26])

4 Demonstration

Following the DSR demonstration phase, we demonstrate the applicability of the proposed framework by analyzing three BPM papers published at the BPM 2025 conference through the lens of the ethical deliberation process introduced in the previous section.

We applied the diverse case method [27] to identify different types of research contributions in the BPM domain. The three tracks of the BPM conference, Foundations, Engineering, and Management, were used as categories because each typically covers distinct research methods and contribution types. Before randomly selecting one paper from each track, we examined the 2025 papers to determine whether any explicitly addressed ethical issues as their primary research goal. Such papers were considered potential influential cases [27], based on our hypothesis that they would more closely align with the ethical deliberation process. We identified two such papers in the Engineering track and randomly selected [15]. The other papers were randomly selected from the remaining tracks: [2] from the Foundations track and [25] from the Management track.

The three selected papers were used as representative cases of different types of BPM research contribution: one explicitly centered on fairness (Case Study 1), one that presented a technical infrastructure contribution (Case Study 2) and one grounded in empirical field research involving human participants (Case Study 3). The goal was not to evaluate the quality of the selected studies, but to illustrate how ethical considerations can emerge throughout BPM research projects and how the proposed framework supports their systematic identification and documentation. Each case study is structured into three parts: an overview of the paper, an application of the deliberation process, and a discussion of the main ethical insights.

4.1 Case Study 1

Paper Overview. Käppel et al. [15] propose a model-agnostic approach to identifying and correcting discriminatory biases in predictive business process monitoring (PBPM) models. The method uses knowledge distillation to translate a black-box neural network into an interpretable decision tree, which domain experts inspect and alter to remove unfair uses of sensitive attributes such as gender and age. A fine-tuning step propagates the corrections back into the original model. The approach is evaluated on a synthetic log and two real-life event logs enriched with injected sensitive attributes. Despite fairness being its central concern, the paper does not engage in structured ethical deliberation.

Application of the Ethical Deliberation Process. Step 1: Facts. ML models are trained on event logs containing sensitive attributes (gender, age), either pre-existing or deliberately injected. The datasets span a synthetic cancer screening process and two real-life logs (BPI Challenge 2012 and Hospital Billing). A key assumption, i.e. expert consensus on which biases are undesirable, has already been reached, is made explicit but not justified, carrying significant ethical weight.

Step 2: Dilemmas. The central tension is not simply fairness versus accuracy, but a more fundamental question: how can a model be made to use a sensitive attribute only when its use is ethically justified and by whom is that justification determined? A secondary dilemma concerns the event logs themselves: real-life logs may encode systemic inequalities that are inadvertently legitimized by treating them as ground truth.

Step 3: Stakeholders. Affected groups include: process participants (patients, loan applicants) who bear the consequences of discriminatory predictions; domain experts whose judgment determines which biases are removed; organizations exposed to legal risk under the EU AI Act[3]; regulators requiring documentable fairness interventions; and the research community, which will inherit the normative assumptions embedded in the three publicly released datasets.

Step 4: Alternatives. Beyond the considered fair baseline model that removed all sensitive attributes, alternatives include: exploratory bias analysis of the data, pre-processing fairness constraints (e.g., re-sampling, re-weighting); a structured multi-stakeholder protocol to replace the assumption of prior expert consensus; and the use of synthetic data to make normative assumptions fully transparent.

Step 5: Consequences. Pre-processing fairness constraints may input other biases while focusing on eliminating an specific one. The proposed approach balances fairness and accuracy, but only if expert judgment is unbiased, an unverified assumption that could institutionalize rather than correct discrimination. Injecting sensitive attributes into real-life logs creates datasets that treat historical decision patterns as normatively neutral, a risk that persists in the literature as other researchers reuse them.

Step 6: Position. Deontologically, using sensitive attributes in automated decisions requires documented consent and justification for each contextual use, which are conditions only partially met by the expert review step. Utilitarianism supports the approach in domains where differentiated predictions improve outcomes, but this calculus is undermined if aggregate fairness metrics mask residual harms to minority subgroups. Care ethics highlights that even a technically fair model may undermine trust if affected individuals are unaware that an AI system shapes decisions about their care or creditworthiness.

[3] https://eur-lex.europa.eu/eli/reg/2024/1689.

Step 7: Publication. The paper makes the code and datasets publicly available, supporting reproducibility. However, it lacks an ethics note, a data governance statement, and documentation of the normative assumptions in the synthetic attribute injection. A dedicated ethics appendix recording the criteria for classifying biases as fair or unfair, the rationale for simplifying expert consensus, and guidance for dataset reuse would substantially strengthen the contribution's ethical transparency.

Discussion. Three gaps recur: the assumption of pre-established expert consensus sidesteps the most ethically complex part of the approach; synthetic attribute injection creates datasets embedding contestable normative assumptions; and the paper provides no ethical documentation despite fairness being its central subject. None of these gaps undermines the technical contribution, but together they illustrate how ethical considerations can remain implicit even when fairness is the explicit concern.

4.2 Case Study 2

Paper Overview. Basmer et al. [2] present MANTA, a framework for exploratory process mining that automatically selects a representative set of views on event data. Given an event log and optional domain constraints, MANTA applies a Maximal Marginal Relevance (MMR) strategy to select the k most diverse and representative context definitions for materialization. MANTA is evaluated in a synthetic log and a real-life BPIC14 service desk log. The contribution is primarily technical, and ethical considerations are absent from the paper.

Application of the Ethical Deliberation Process. Step 1: Facts. MANTA takes event data as input and produces materialized views that structure organizational behavior by chosen context definitions. A key assumption is that the analyst validates the views after the fact; the framework itself operates upstream, shaping which perspectives on the data become visible and which do not. The BPIC14 log records real employee activity in 40,116 events linked to 27,905 objects.

Step 2: Dilemmas. The central dilemma is epistemic framing: by algorithmically selecting which views to materialize, MANTA determines which organizational perspectives an analyst will see. This is technically an optimization problem, but ethically raises the question of whether similarity-based selection adequately captures what is organizationally meaningful or whether suppressed views might systematically obscure perspectives important for accountability or worker interests. A secondary dilemma concerns the BPIC14 log: actor-based and resource-based groupings may expose individual behavioral patterns that were not collected for performance surveillance purposes.

Step 3: Stakeholders. Process analysts benefit from reduced exploration effort, but become dependent on opaque selection decisions. The employees recorded in the BPIC14 log are an invisible but critical stakeholder group whose work patterns are exposed through context definitions at the actor-level in ways they never consented to. Organizational managers may be unaware of the privacy implications; regulators under GDPR[4] have an interest in ensuring lawful processing of employee behavioral data; and the research community inherits MANTA's design assumptions when reusing its framework and datasets.

Step 4: Alternatives. Alternatives include incorporating domain-specific relevance signals into the MMR criterion to privilege accountability or fairness-relevant perspectives; using exclusively synthetic data for evaluation; making suppressed views visible to analysts; and providing guidance on k selection to limit unintended information loss.

Step 5: Consequences. Similarity-based neutrality is itself a normative choice that does not give additional weight to the perspectives on equity or worker well-being. The exclusive use of synthetic data would eliminate the privacy risk, but reduce the external validity. Exposing suppressed views would improve transparency but reintroduce the complexity the framework is designed to reduce. Reusing the BPIC14 log without a data governance statement risks propagating unacknowledged privacy assumptions to subsequent researchers.

Step 6: Position. Deontologically, using behavioral records of real employees requires that data subjects be informed about the purposes of data processing, which is a condition the paper does not address. Virtue ethics holds that analysts retain reflective responsibility and should not delegate organizational judgment entirely to an algorithm. Care ethics requires that worker consent and awareness be ensured before actor-level views of their behavior are generated and published.

Step 7: Publication. The paper makes its implementation publicly available, supporting reproducibility. It lacks a data governance statement, consent, or privacy documentation for the BPIC14 log, and practitioner guidance on the ethical implications of deploying context-selection tools in settings where data subjects are employees.

Discussion. Unlike Case Study 1, MANTA does not engage in ethics, despite direct implications for organizational transparency and worker privacy. Three gaps stand out: the selection mechanism exercises significant epistemic power without acknowledging its normative implications; the real-life employee log is reused without any governance statement; and no guidance is offered to practitioners deploying the tool in organizational settings regarding ethical issues.

4.3 Case Study 3

Paper Overview. Reineke et al. [25] conduct the first two steps of DSR to develop a method to evaluate workarounds in organizations. The research involves semi-structured interviews, workshops, and focus groups with employees in three manufacturing and chemical companies, producing a validated problem statement and five design objectives. This is a human-centered study, but it contains no discussion of research ethics, consent procedures, or data protection measures.

[4] https://eur-lex.europa.eu/eli/reg/2016/679.

Application of the Ethical Deliberation Process. Step 1: Facts. Qualitative data about employee deviations from standard operating procedures (SOPs) is collected in three organizations through interviews, workshops, and focus groups. Identified workarounds include skipping ERP events, creating dummy items, and bypassing procurement procedures, i.e., practices that could carry professional consequences for the individuals involved if made visible to management. Organizations are pseudonymized, but no individual-level protection is described.

Step 2: Dilemmas. The central tension is structural: employees disclose rule-deviating behavior to researchers whose findings help management decide whether to prevent, redesign, adopt or ignore those practices. Thus, research creates a pathway through which informal adaptive behavior becomes structured and subjected to managerial evaluation. The same insights that could benefit employees through process improvement could also expose them to disciplinary inquiry. A second dilemma concerns the asymmetry of benefit: organizations receive actionable knowledge, while employees do not receive documented protection.

Step 3: Stakeholders. Front-line employees are the most ethically vulnerable group. Managers are both participants and primary beneficiaries, creating a conflict of interest. The research team acts as a gatekeeper for sensitive disclosures published in organizational detail. Consultants who validated the framework may later implement it, introducing misaligned incentives. Regulators are relevant where workarounds imply compliance or financial reporting obligations.

Step 4: Alternatives. Formal consent protocols with explicit disclosure of how findings would be used internally; stronger anonymization or use of composite discourse fragments; a clear boundary between research and consulting roles; mechanisms for employee representation in the process; and omission or further anonymization of operationally specific workaround details from the publication.

Step 5: Consequences. Formal consent might reduce disclosure richness; stronger anonymization might reduce case credibility; role separation might reduce practical relevance; employee representation adds complexity but addresses the power asymmetry; omitting operational details protects individuals but weakens empirical grounding.

Step 6: Position. Deontologically, employees who disclose professionally risky practices are owed explicit protection beyond organizational pseudonymization. Utilitarianism supports the research's potential societal value but requires that participant protection be a precondition, not an afterthought. Care ethics foregrounds the trust employees placed in the research team as a morally significant relationship creating obligations of reciprocity and non-exploitation, made more acute by the compliance dimensions of some disclosed practices.

Step 7: Publication. The publication should have been accompanied by a research ethics and consent statement; documentation of institutional ethics review; a data governance note on measures preventing individual identification; a conflict-of-interest disclosure regarding consultant validators; and guidance for researchers on the ethical obligations that arise when studying employee deviations from formal procedures.

Table 1. Summary of the ethical deliberation process across the case studies

Case	Step	Description
CS1	Facts	Fairness-aware PPM with sensitive attributes and assumed expert consensus
	Dilemma	Justified use of sensitive attributes vs. discrimination risk
	Stakeholders	Process participants, regulators, organizations, researchers
	Alternatives	Fairness constraints, stakeholder review, synthetic data
	Consequences	Reduced bias but risk of institutionalizing expert assumptions
	Position	Justified and accountable use of sensitive attributes
	Publication	Ethical documentation of assumptions and dataset reuse
CS2	Facts	MANTA materializes process views from employee event logs
	Dilemma	Algorithmic view selection shapes organizational perspectives
	Stakeholders	Employees, analysts, managers, regulators, researchers
	Alternatives	Domain-aware selection, synthetic data, transparency
	Consequences	Hidden perspectives, privacy risks, and normative bias
	Position	Analysts retain responsibility for organizational judgment
	Publication	Data governance, consent, and ethical deployment guidance
CS3	Facts	Field research on employee workarounds and SOP deviations
	Dilemma	Knowledge generation vs. employee protection and reciprocity
	Stakeholders	Employees, managers, researchers, consultants, regulators
	Alternatives	Consent protocols, stronger anonymization, employee representation
	Consequences	Improved protection but reduced disclosure richness and detail
	Position	Participant protection and reciprocity as ethical obligations
	Publication	Ethics review, consent, and conflict-of-interest disclosure

Discussion. This case illustrates a third and distinct pattern: ethical gaps arising in field research that are not reflected in its human subjects dimension. Three gaps stand out: the absence of consent documentation or ethics review is inconsistent with standard practice; the research design creates an unacknowledged power asymmetry between disclosing employees and benefiting managers; and the publication of operationally specific workaround details was not accompanied by any reflection on the identification risk this poses to the individuals whose practices are described.

4.4 Cross-Case Discussion

The three case studies showed that ethical gaps in BPM research are neither rare nor confined to a single type of contribution. They manifest differently depending on the nature of the work: as incomplete fairness treatment in algorithmic research (Case Study 1), as unacknowledged epistemic and privacy implications in technical infrastructure work (Case Study 2), and as absent human subjects protection in organizational field research (Case Study 3). Table 1 summarizes the application of the deliberation process in all three cases. In all three studies, ethical considerations inherent in the research design remain largely outside the scope of the contribution, either because responsibility for addressing them is left implicit or because they are not explicitly recognized as ethical dilemmas. The framework proposed here is designed to face this pattern before publication, rather than after. An important disclaimer is that this analysis is not about the quality or outcome of the cases, which are undeniably high.

5 Conclusion

This paper addressed the research question of how ethical considerations can be systematically incorporated into BPM research. We proposed a framework that maps ethical perspectives onto the stages of the BPM research lifecycle and operationalizes ethical deliberation through a structured seven-step process adapted from Santoro and da Costa [26]. The framework was demonstrated through the analysis of three papers published at BPM conference 2025, selected to represent different types of BPM research contribution: a fairness-oriented algorithmic approach, a technical infrastructure contribution, and an empirical field study involving human participants. The three case studies reveal that ethical gaps in BPM research are neither rare nor confined to a single type of work. They show up in qualitatively different ways depending on the nature of the contribution.

This pattern has practical consequences. When ethical assumptions are left implicit, they propagate into published artifacts such as datasets, frameworks, and tools that other researchers will reuse without inheriting the context needed to recognize their limitations. The datasets released in Case Study 1 embed contestable normative assumptions about when biased decisions are acceptable; the event logs used in Case Study 2 carry unaddressed questions about data reuse consent; the workaround descriptions published in Case Study 3 contain operational details that could identify individuals within their own organizations. None of these issues was introduced with harmful intent. They arose because BPM research currently lacks a shared instrument to make ethical deliberation a routine part of the research process. The framework proposed in this paper is intended to contribute to such an instrument. By associating each stage of the BPM research lifecycle with a corresponding ethical perspective (anticipation during problem formulation, protection of humans during data collection, reflexivity during analysis, and accountability during publication), it provides researchers with a structured entry point for ethical reflection that does not require prior expertise in moral philosophy. The seven-step deliberation process further operationalizes this reflection by guiding researchers through the identification of facts, dilemmas, stakeholders, alternatives, consequences, theoretical grounding, and documented outcomes. However, the framework does not prescribe specific ethical conclusions; it creates the conditions under which researchers can reach well-reasoned and transparent ones.

Several directions for future work follow from this contribution. First, an extended demonstration of the framework considering alternative categorizations (e.g., specific research methods) for the selection of diverse cases can enrich the findings. Second, the framework should be further validated through prospective application, i.e. by complementing active BPM research projects through the deliberation process from the outset, rather than applying it retrospectively as done here. Third, the best-practice patterns referenced in the framework (Table 1) could establish further development into a more structured catalog, drawing on the growing literature on responsible research and innovation and on privacy-by-design principles, to provide researchers with concrete methodological options for each type of ethical dilemma identified. Forth, the BPM research community would benefit from institutional mechanisms that complement individual researcher reflection, including ethics review processes adapted to information systems and process mining research, community repositories of ethical deliberation records, and editorial norms that encourage or require ethics notes in submitted work. The framework proposed here is intended as a first step toward that broader infrastructure for ethically aware BPM research.

Acknowledgments. Kate Revoredo is funded by the Berliner Chancengleichheitsprogramm (BCP) as part of the DiGiTal Graduate Program and the Einstein Foundation Berlin under grant EPP-2019-524.

References

1. Amico, B., Combi, C., Vecchia, A.D., Migliorini, S., Oliboni, B., Quintarelli, E.: Enhancing business process models with ethical considerations. In: EDOC Workshops. Lecture Notes in Business Information Processing. Springer, Heidelberg, pp. 3–17 (2024). https://doi.org/10.1007/978-3-031-79059-1_1
2. Basmer, M., Ueck, H., Fahland, D., Weidlich, M.: Manta: materializing views on event data for context exploration in process analysis. In: Senderovich, A., Cabanillas, C., Vanderfeesten, I., A. Reijers, H. (eds.) Business Process Management, pp. 51–68. Springer, Cham (2026). https://doi.org/10.1007/978-3-032-02867-9_5
3. Bentham, J.: An Introduction to the Principles of Morals and Legislation (1789)
4. Beverungen, D., et al.: Seven paradoxes of business process management in a hyper-connected world. Bus. Inf. Syst. Eng. **63**(2), 145–156 (2021)
5. Cavoukian, A.: Privacy by design: The 7 foundational principles. Technical report, Information and Privacy Commissioner of Ontario, Ontario, Canada (2011)
6. Chandrasiri, T., Bandara, W., Rosemann, M., Ostern, N., Voss, M.: Designing business processes for good: a canvas-based framework for routinized benevolence. In: BPM (RBPM, PT, and Educators Forum). LNBIP, pp. 87–101. Springer, Heidelberg (2025). https://doi.org/10.1007/978-3-032-02936-2_8
7. Creswell, J.W., Creswell, J.D.: Research Design: Qualitative, Quantitative, and Mixed Methods Approaches. Sage publications (2017)
8. Department of Health: Education, and welfare, national commission for the protection of human subjects of biomedical and behavioral research: the belmont report: ethical principles and guidelines for the protection of human subjects of research. J. Am. Coll. Dentists **81**(3), 4–13 (2014)
9. Djurica, D., Franzoi, S., Spiekermann-Hoff, S.: Operationalizing responsible BPM: a method for value-based process redesign. In: BPM (RBPM, PT, and Educators Forum), LNBIP, pp. 72–86. Springer, Heidelberg (2025). https://doi.org/10.1007/978-3-032-02936-2_7
10. Ford, R.C., Richardson, W.D.: Ethical decision making: a review of the empirical literature. J. Bus. Ethics **13**(3), 205–221 (1994)

11. Forsyth, D.R.: Judging the morality of business practices: the influence of personal moral philosophies. J. Bus. Ethics **11**(5–6), 461–470 (1992)
12. Gilligan, C.: In a Different Voice: Psychological Theory and Women's Development. Harvard University Press (1993)
13. Hevner, A.R.: Design science in information systems research. MIS Q. **28**(1), 75–105 (2004)
14. Kant, I.: Groundwork of the Metaphysics of Morals. Cambridge University Press, Cambridge (1998)
15. Käppel, M., Neuberger, J., Möhrlein, F., Weinzierl, S., Matzner, M., Jablonski, S.: A human-in-the-loop approach for improving fairness in predictive business process monitoring. In: BPM, LNCS, pp. 343–360. Springer, Heidelberg (2025). https://doi.org/10.1007/978-3-032-02867-9_21
16. Kern, C.J., Poss, L., Kroenung, J., Schönig, S.: Navigating the moral maze: a literature review of ethical values in business process management. Bus. Process. Manag. J. **30**(8), 343–370 (2024)
17. Lepsien, A., Bräuer, P., Jelonnek, S., Koschmider, A., Aleknonyte-Resch, M.: On the role of ethics in business process modeling and management. In: To appear at ViVa Models! 26. Co-located with Modellierung 2026 (2026)
18. MacIntyre, A.: After Virtue: A Study in Moral Theory. University of Notre Dame Press (2007). https://books.google.de/books?id=7bLuAAAAMAAJ
19. Mannhardt, F.: Responsible process mining. In: Process Mining Handbook, Lecture Notes in Business Information Processing, vol. 448, pp. 373–401. Springer, Heidelberg (2022). https://doi.org/10.1007/978-3-031-08848-3_12
20. Mill, J.S.: Oxford University Press (1998)
21. Owen, R., Macnaghten, P., Stilgoe, J.: Responsible research and innovation: from science in society to science for society, with society. Sci. Public Policy **39**(6), 751–760 (2012)
22. Peffers, K., Tuunanen, T., Rothenberger, M.A., Chatterjee, S.: A design science research methodology for information systems research. J. Manag. Inf. Syst. **24**(3), 45–77 (2007)
23. Pettinari, S., De Sanctis, M., Inverardi, P.: Realizing ethical-aware business processes. In: DIGHUM. LNCS, pp. 219–235. Springer, Heidelberg (2025). https://doi.org/10.1007/978-3-032-11108-1_16
24. Poss, L., Kern, C.J., Stoiber, C., Kroenung, J., Schönig, S.: Measuring moral performance: a framework for ethical business process management. In: BPM (RBPM, PT, and Educators Forum), LNBIP, pp. 117–132. Springer, Heidelberg (2025). https://doi.org/10.1007/978-3-032-02936-2_10
25. Reineke, M., Löhr, B., Assbrock, A., Bartelheimer, C., Beverungen, D.: From temporary fixes to informed decisions - design echelons for evaluating workarounds. In: BPM. LNCS, pp. 541–557. Springer, Heidelberg (2025). https://doi.org/10.1007/978-3-032-02867-9_32
26. Santoro, F.M., da Costa, R.M.E.M.: Towards ethics in information systems. J. Interact. Syst. **12**(1), 69–82 (2021)
27. Seawright, J., Gerring, J.: Case selection techniques in case study research: a menu of qualitative and quantitative options. Polit. Res. Q. **61**(2), 294–308 (2008). https://doi.org/10.1177/1065912907313077
28. Shuster, E.: Fifty years later: the significance of the Nuremberg code. N. Engl. J. Med. **337**(20), 1436–1440 (1997)

Participatory Process Mining: A Framework and Call for Research

Vinicius Stein Dani[1(✉)], Iris Beerepoot[1(✉)], Thomas Grisold[2(✉)], Hajo A. Reijers[1(✉)], and Adela del-Río-Ortega[3(✉)]

[1] Utrecht University, Utrecht, The Netherlands
{v.steindani,i.m.beerepoot,h.a.reijers}@uu.nl
[2] Vienna University of Economics and Business, Vienna, Austria
thomas.grisold@wu.ac.at
[3] SCORE Lab, I3US, Universidad de Sevilla, Sevilla, Spain
adeladelrio@us.es

Abstract. Process mining is used to create transparency about organizational work by analyzing event data from business processes. Although such transparency is often assumed to be inherently beneficial, the dominant strand is that transparency is imposed "top-down". Decision-makers and developers initiate and direct process mining initiatives, while those performing the work have hardly any say. This can be problematic, not least because transparency can lead to perceived surveillance, misrepresentation, and resistance on the side of workers. This paper foregrounds the role of transparency as a socio-technical phenomenon constructed through a series of decisions made at different stages of the process mining lifecycle. To proactively consider and involve the voices of those whose activities are mined and analyzed –who we refer to as the *process minees*– we introduce the concept of *Participatory Process Mining*, an approach in which the process minees are recognized as key stakeholders in shaping the conditions under which transparency about their work is produced and used. We propose a conceptual framework that distinguishes the *levels* (data, tool, organizational) at which participation can be enabled across the process mining lifecycle, and the *degrees* of participation that can be realized at each level. Moreover, we outline directions for future research to examine how these forms of participation shape process minee perceptions of transparency and empowerment.

Keywords: process mining · transparency · participation · employee involvement · empowerment · process minee · responsible BPM

1 Introduction

Process mining is associated with novel opportunities to create transparency in organizations [5,10,15]. As activities and events are recorded, stored, and made visible, organizations can now observe what is actually happening as people work [5]. Generally, this is received with enthusiasm, both from research [15]

K. Kluza et al. (Eds.): BPM 2026, LNBIP 600, pp. 77–92, 2027.
https://doi.org/10.1007/978-3-032-36710-5_6

and practice [14]. Typically, the assumption is that transparency translates into value in a straight-forward way [28]. A further assumption, however, is that transparency is implemented "top-down": decision-makers, vendors, or developers identify opportunities for the use of process mining, which they then realize in a given organizational setting [5]. Those who decide on the basis of the generated transparency are more involved in the set-up and use of process mining than those who perform the work and produce the digital traces.

What is largely overlooked is that increased transparency can lead to unintended side-effects, especially when it evokes perceived danger and threat on the side of workers [10]. Research from the organization sciences, for instance, provides ample evidence that transparency-inducing management initiatives can lead workers to ignore, game, or even resist the use of related digital technologies [21,35,36]. Several recent studies in Business Process Management (BPM) and process mining have shifted their attention away from purely control-flow focused analyses to the role of human resources in processes, referring to such analyses as *human resource mining* [28]. More broadly, there is recent interest in the social aspects of process mining, such as surveillance and how it can be prevented [30]. In such contexts, there is a clear need to protect the integrity of workers whose behavior is subject to process mining [16]. But while topics such as privacy and data protection are now actively considered in academic debates around process mining, the ways in which process mining initiatives are perceived, and what kinds of responses they evoke within organizations and on the side of those whose digital traces are analyzed, remain elusive.

In this paper, we foreground the role of transparency from a socio-technical perspective. We emphasize that transparency is the outcome of multiple design and use decisions. These decisions shape what becomes visible, what remains invisible, and how visibility is interpreted [10]. As such, transparency is tightly interrelated with ethical questions, particularly when workers may face potentially detrimental decisions based on the data their work produces. Therefore, we raise a question that has received little attention in the process mining literature so far: *How can we enable the participation of process minees in process mining initiatives?* By process minees we mean those whose behavior is mined and analyzed.

To address this question, we introduce *Participatory Process Mining* as a new angle in process mining research. By this term, we refer to process mining initiatives that systematically consider and integrate the voices and interests of process minees. To explain how participatory process mining can be realized, we draw on research on participation in collective decision-making. Specifically, we build on Arnstein's seminal work [4] to distinguish *degrees* of participation, ranging from non-participation, through informing and consultation (degrees of tokenism), to partnership, delegated power, and process minee control (degrees of process minee power). Thus, we propose a conceptual framework that distinguishes the *level* at which a decision is made (data, tool, organizational) from the *degree* of participation that can be enabled in process mining initiatives. We map current academic and practitioner accounts onto this framework, showing that present practice clusters in the lower degrees of participation. Building on this, we call for future research that examines how participation at each level shapes process minees' experiences of transparency and empowerment. In doing so, we contribute to ongoing discussions on Responsible BPM and the responsible use of process mining technologies.

The rest of this paper is organized as follows. Section 2 situates our work in the literature on transparency, responsibility and ethics in process mining, and Arnstein's ladder of citizen participation. Section 3 introduces the concept of participatory process mining, and Sect. 4 presents the conceptual framework. Section 5 outlines directions for future research. Section 6 concludes the paper.

2 Theoretical Background

To situate our concept of Participatory Process Mining, we draw on research on transparency, responsibility and ethics in process mining. To address our research question systematically, we also draw on an established participation framework. To our knowledge, no established framework exists that directly addresses participation in the context of process mining. We therefore turn to the closest available analogue, that of citizen participation.

2.1 Transparency in Process Mining

Process mining generates transparency by transforming event data into models, metrics, and visualizations of organizational processes. This transparency is commonly associated with improved process awareness and performance gains [14, 15, 28]. So far, transparency has been embraced by both technical and managerial research. In technical research, transparency is treated as a taken-for-granted condition produced through process mining, enabling a variety of analysis techniques. From this point of view, transparency is the means on which various ends could be developed, from visual representations to computational algorithms.

Consequently, research on the organizational and managerial implications of process mining has embraced transparency as a means through which new managerial practices become possible [14, 15]. The main argument here is that end-to-end visibility of work-related activities enables decision-makers to streamline workflows, identify performance issues, and continuously test and evaluate ideas on work improvements [5]. As in technical research, transparency is only one stepping stone towards improved management; the value comes from the decisions made upon it. Across technical and managerial research, transparency is seen as process mining's key success mediator; it may not be everything, but without transparency, process mining would be nothing.

Adjacent research streams, however, indicate that transparency can yield less intended consequences. More specifically, research in organization sciences and information systems has emphasized that transparency is by no means neutral or exclusively positive, especially in the context of work evaluations [10]. Increasing degrees of work-related transparency can reconfigure power relations and trigger considerable unintended effects [40]. Data-driven transparency, for example, can be perceived as surveillance [30,31,44], to which workers respond in various ways. They may game transparency-related metrics by optimizing selectively for visible indicators [1], or engage in resistance and workarounds [21,31]. Moreover, data representations may create feelings of depersonalization when complex work practices are reduced to abstract indicators [26]. In a similar vein, research on workplace surveillance and algorithmic management has highlighted how monitoring technologies can intensify perceptions of control and power asymmetries [7,13]. When monitoring systems are experienced as imposed and opaque, resistance and subversion become more likely [29].

These findings are highly relevant for process mining, whose core purpose is to transform event log data into representations of work-related behaviour. Yet, the existing literature has not addressed how such representations emerge and shape process minees' perceptions of transparency. To better understand these dynamics, it is necessary to examine the decisions through which transparency is constructed and, just as importantly, who is involved in making them.

2.2 Responsibility and Ethics in Process Mining Research

A growing body of research has been turning to the roles of responsibility and ethics in process mining. Distinct research streams have emerged. One stream addresses privacy and confidentiality in event data. Because event logs combine activity-level detail with timestamps and resource attributions, they can reveal sensitive information about individuals, including their performance, behavioural patterns, and working conditions. To mitigate potential risks, recent contributions propose techniques for anonymization, differential privacy, and secure data transformation [8,18,32], alongside reviews that synthesize the resulting trade-offs between privacy protection and analytical utility [6,16,45]. While effective in protecting individuals from, for example, re-identification, these techniques operate at the level of the event log: they modify what data analysts can see, but leave largely untouched the prior question of how that data came to exist in the first place, and on whose terms.

Another stream examines governance and organizational accountability. The effective and compliant use of process mining depends on clear data governance arrangements [20,33]. In parallel, the process mining community itself has issued calls for responsible process mining practices, using well-known criteria such as fairness, accuracy, confidentiality, and transparency [27]. A more recent stream emphasizes user-centeredness in process mining. It investigates the experiences of those who interact with process mining tools, primarily analysts and managers, to make the technology more accessible and effective in practice. Studies in this area examine usability, visualization design, and engagement mechanisms in organizational settings [3,38,39,43]. Again, this stream emphasizes those who use the generated transparency for decision-making rather than those whose work the tool renders visible, i.e., the process minees.

Taken together, while recent streams advance privacy protection, governance, and tool usability, they share an underlying assumption about how process minees fit into the picture. Process minees are typically conceptualized as data subjects or recipients of analytical outcomes, rarely as active stakeholders who can legitimately shape decisions about what data is collected about their work, how analyses are conducted, and how (and by whom) results are interpreted. These perspectives are "top-down": those who develop, implement or manage process mining tools impose their own ideas and plans. The ethics of process mining, in other words, has been framed largely in terms of obligations *toward* process minees, rather than rights or roles *of* process minees in shaping process mining initiatives.

2.3 Arnstein's Ladder of Citizen Participation

A key theoretical lens we draw on is Arnstein's ladder of citizen participation [4]. Arnstein introduced this framework in the context of urban planning and public policy in the United States, at a time when federal programs formally required "citizen participation" in decision-making but rarely delivered meaningful power to those affected. Her central argument was that participation is not a uniform concept: there is a critical difference between going through the motions of involving people and actually redistributing power to them. To capture this difference, she proposed a ladder metaphor consisting of eight rungs, organized into three broad categories. The bottom two rungs, *manipulation* and *therapy*, describe forms of non-participation in which involvement serves to manufacture consent rather than enable genuine influence. The middle three rungs, *informing*, *consultation*, and *placation*, represent degrees of tokenism: participants are heard, but decision-making authority remains elsewhere. The upper three rungs, *partnership*, *delegated power*, and *citizen control*, describe degrees of actual citizen power, in which participants gain increasing authority over decisions.

The ladder has been widely influential across public administration, participatory design, and information systems research, and has been adapted to a variety of organizational and digital contexts. Its core insight, that involvement without power is a hollow form of participation, translates directly to our setting. In process mining, decisions about what data is collected, how work is represented, and how insights are used carry significant consequences for those whose work is analyzed. Arnstein's ladder provides the vocabulary needed to distinguish between forms of process minee involvement that are genuinely empowering and those that are not.

3 Participatory Process Mining

Drawing these observations together, we integrate insights from process mining, management and organization sciences, and Arnstein's theory of citizen participation to introduce *Participatory Process Mining*. The ambition of our approach is to mitigate potential threats of transparency, particularly in relation to perceived privacy threats or even surveillance, by balancing top-down decisions with bottom-up interests and know-how of all relevant stakeholders in process mining initiatives [19,24]. This can take many forms. In digital workplace and IoT contexts, for instance, co-defining data collection and system features has been shown to enhance legitimacy, trust, and acceptance [11,25]. More broadly, participation can involve the implementation of bi-directional data sources, such that both decision-makers and process minees are aware of what is going on [19]. This research emphasizes that process minees' reactions depend not only on the extent of monitoring, but also on perceived fairness, voice, and participation [26,31].

These ideas can be readily linked to process mining, but despite advances in privacy-preserving techniques, governance structures, and user-centered design, the aspect of process minee participation has not yet been considered in its own right. Existing approaches tend to address either (1) how to technically protect data, or (2) how to improve analytical usability. What remains underexplored is how transparency can be made the object of meaningful process minee participation. The core observation is this: event logs do not emerge naturally; they are produced through decisions about what is made visible to whom and at what points in time [2]. In process mining initiatives, process minees are not simply exposed to "transparency", but to a particular and selective representation of their work [10]. Multiple decision points mediate between actual work and generated transparency: decisions about which events are logged [2], how data is preprocessed and aggregated [26], which analytical techniques are applied [29], which metrics are emphasized [1], and how results are communicated and acted upon [19]. Each of these decisions shapes what becomes visible, what remains invisible, and how work transparency is interpreted. Yet it remains largely unclear who participates in these decisions and at which stages process minee perspectives are included or excluded.

Hence, the concept of *Participatory Process Mining* recognizes process minees as legitimate stakeholders in the decisions through which transparency about their work is produced and used, rather than mere data subjects or recipients of analytical outcomes. Process minee participation may occur at multiple levels, such as defining logging policies, setting analytical objectives, selecting key performance indicators (KPI), and establishing governance mechanisms for data access and interpretation. By making these levels explicit, we contribute to emerging discussions on Responsible BPM and extend them beyond privacy protection toward participatory transparency construction.

4 A Framework for Participatory Process Mining

In this section, we propose a conceptual framework for Participatory Process Mining. A central insight of the framework is that there are many occasions at which process minees can be included in, or excluded from, the decisions that shape how their work becomes visible. However, not all forms of inclusion carry equal weight: some give process minees real influence over outcomes, while others

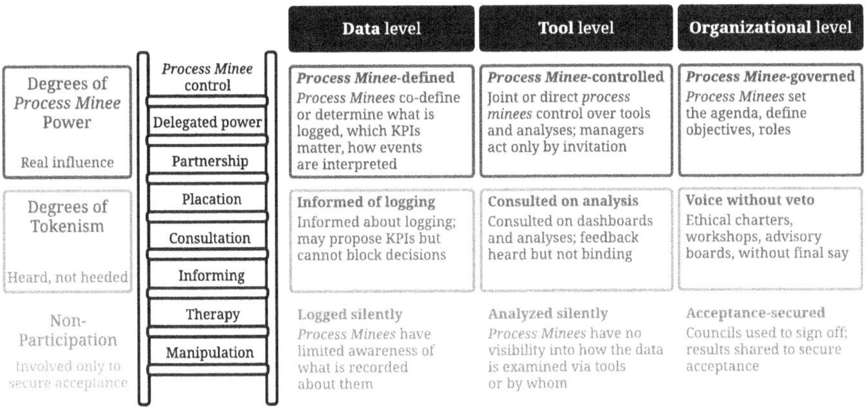

Fig. 1. Participatory Process Mining framework overview

amount to little more than the appearance of involvement. Accordingly, our aim is to show what process minee participation can look like in process mining initiatives, and to provide guidance on how it can be considered "by design". Figure 1 provides an overview of our framework.

We make two moves. First, we develop the framework by drawing from Arnstein's ladder of citizen participation [4], then we map current efforts in process mining onto Arnstein's ladder, to assess the extent to which employees currently hold power over the decisions that determine how their work is recorded, analyzed, and interpreted. To do so, we build on both recent academic studies and practitioner accounts made publicly available by Fluxicon, a process mining software company that documents practitioner experiences through its blog and its *Process Mining Café* conversations. We do not claim a systematic or exhaustive literature review; rather, the proposed framework is intended as a conceptual contribution grounded in recent literature and illustrative empirical cases. In presenting the framework, we follow the structure of Arnstein's ladder but adapt its terminology to the process mining context. Most notably, we replace Arnstein's term "citizen control" with *process minee control*, and refer throughout to *degrees of process minee power* rather than citizen power.

For each section of the ladder, we make specific suggestions on how the respective participation can be achieved. Building on and synthesizing existing multi-level research around process mining [5,34,41], our suggestions are organized along three levels. The *Data level* concerns what is made transparent and how, that is, the logic through which work-related activities are translated into data points in an event log [2]. The *Tool level* involves the technical set-up through which event log data are made transparent, for instance, by means of user access rights or visual analytics [42]. The *Organizational level* refers to the organizational set-up surrounding process mining initiatives, that is, how they are considered and established in organizational structures, governance, and rules [5].

4.1 Non-participation

At the bottom two rungs of Arnstein's ladder, *manipulation* (rung #1) and *therapy* (rung #2), employees are involved, but only to secure their acceptance of decisions already made. For *manipulation*, one recognizable instance at the *organizational level* is the involvement of a workers' council or employee representative body at some point during a process mining initiative. In countries with a strong tradition of worker representation, such as Germany and Austria, involving a workers' council is key [17,37]. The intent is to signal that the initiative is not directed at individual employees, and to avoid resistance down the line. However, if the workers' council is only involved to sign off on the initiative, this form of involvement resembles Arnstein's 'rubberstamp advisory committee' more than genuine participation. In that setup, by the time the workers' council is consulted, the fundamental decisions, about what gets recorded at the *data level*, which metrics are used to evaluate performance, and how results are acted upon at the organizational level, have already been made. At the *data* or *tool level*, non-participation is generally characterized by limited involvement or awareness of what is happening in process mining initiatives. Logging and analyzing activities are opaque and silent to process minees.

Arnstein's *therapy* rung describes responses to powerlessness that treat discomfort as a personal problem to be resolved rather than as a signal that something in the initiative's design might need to change. In process mining, this pattern may emerge from good intentions: practitioners who care how employees experience a process mining initiative, and who work hard to make it less threatening. The question Arnstein invites us to ask is whether these efforts address the underlying condition that produces the discomfort in the first place. In [37], a thoughtful evolution of the communication process to stakeholders is described at the *organizational level*. Where they once presented results directly to employees, which tended to be stressful, they developed an iterative approach: beginning with only the broad features of the process, working in short sessions, letting employees speak, watching for emotional reactions, and using the tool to open dialogue rather than confront people with conclusions. The account draws on analogies to psychiatrists and priests, figures skilled at creating conditions in which people feel safe enough to speak honestly. These efforts seem to work well: employees become more at ease, conversations become more productive, and the analysis benefits from employee knowledge that would not have surfaced in a more confrontational approach. However, when these efforts are purely directed at helping employees arrive at acceptance rather than being given any power over it, they remain in the *therapy* rung rather than crossing into participation. At the *data level*, therapy appears where the translation of work into event data is treated as fixed and unquestionable. Employees may be encouraged to understand or accept how their activities are represented, but not invited to challenge whether those representations accurately reflect their work. At the *tool level*, process mining tools may be used to facilitate dialogue and reduce anxiety, yet employees remain confined to the role of recipients of insights rather than participants in shaping analyses, visualizations, or interpretations.

By making certain changes, these efforts may be turned into more empowerment, bringing us to the next category: degrees of tokenism.

4.2 Degrees of Tokenism

At the next three rungs, *informing* (rung #3), *consultation* (rung #4), and *placation* (rung #5), employees are informed, consulted, and occasionally invited to contribute ideas, but lack the power to ensure their perspectives shape outcomes; participation without influence, as Arnstein warns, risks becoming an "empty ritual". In process mining, this typically means initiatives that increase transparency and dialogue around process data without redistributing decision-making power: employees gain visibility into how process mining works, give feedback on analyses, or contribute to dashboard design, while decisions about what data is collected, how work is represented, and how insights are used remain largely managerial or analytical prerogatives.

Informing is perhaps the most widespread form of participation in current process mining practice. Here, organizations make process mining initiatives more transparent by explaining how data is collected, what analyses will be performed, and how privacy will be protected. Recent literature on participatory data stewardship [23] similarly identifies informing as a one-way flow intended to provide meaningful transparency around data use and governance. At the *organizational level*, this commonly takes the form of stakeholder communication, onboarding sessions, ethical charters, or documentation describing anonymization and access policies. The 'ethical charter' described in [37] may be seen as such an informing effort. In the example, the charter stated that the analysis aimed to study processes rather than evaluate individuals, that employees conversations would remain confidential from management, and that results would be shared transparently and anonymously. Yet, the fundamental parameters of visibility remained externally defined: employees were informed how transparency would operate, but did not shape what data would be collected, which metrics would matter, or how findings would ultimately be interpreted and acted upon. Arnstein argues that informing becomes tokenistic when communication flows only one way and no meaningful channel exists for affected participants to influence decisions. Organizations increasingly recognize that employees need explanations and reassurances when process data is analyzed, particularly in contexts where process mining may otherwise be perceived as surveillance [30].

Consultation goes one step further by inviting process minees to express opinions, concerns, and experiences. Arnstein describes consultation as a situation in which participants "hear and are heard", but without guarantees that their input will affect outcomes. At the *organizational level*, consultation can occur through workshops, interviews, validation sessions, or feedback meetings in which process minees are asked to react to analyses, discuss process interpretations, or raise concerns about findings. While these practices may improve understanding and legitimacy, decision-making authority over what is ultimately analyzed and how results are used generally remains with analysts and managers. At the *data* and *tool* levels, consultation may involve mechanisms through which process mining participants can document events or their own decisions, so that these are seen in complement to process mining-based analyses.

Placation represents a higher but still limited degree of participation. Here, process minees gain some influence and may even participate in governance structures, but ultimate authority remains elsewhere. At the *organizational level*, placation can emerge when organizations include process minee representatives in governance arrangements or involve them in selecting metrics and reviewing analyses, while retaining managerial veto power over final decisions. For example, organizations may establish steering committees or governance boards including workers' councils, employee representatives, analysts, and managers. Process minees may be invited to advise on acceptable data uses, define safeguards, or review proposed KPIs, which is already a major improvement over the manipulation rung. Process minees may be consulted on how dashboards frame process performance or which indicators are emphasized. These arrangements can improve legitimacy and trust, particularly when compared to purely top-down implementations. Yet process minees remain structurally dependent on management's willingness to incorporate their concerns. Their participation can shape decisions, but not fully determine them. At the *data level*, placation may involve process minees in defining KPIs based on available data. At the *tool level*, they may inform decision-makers about which representations are more or less useful for evaluating work-related behaviour.

Viewed through Arnstein's lens, many contemporary "ethical" or "human-centered" process mining efforts improve communication, responsiveness, and legitimacy without redistributing control over how transparency is constructed, they remain within these degrees of tokenism.

4.3 Degrees of Process Minee Power

At the upper rungs, *partnership* (rung #6), *delegated power* (rung #7), and *process minee control* (rung #8), process minees begin to negotiate, influence, and in some cases directly determine how transparency is constructed and governed; partnership as Arnstein argues, becomes possible when process minees can "*negotiate and engage in trade-offs with traditional powerholders*". In process mining, examples corresponding fully to these upper rungs remain rare. Nevertheless, some emerging practices suggest movement toward more participatory transparency construction, particularly at the tool and organizational levels of our framework.

Partnership entails workers and decision-makers defining together how transparency about work is produced and acted upon. Instead of analysts or managers unilaterally determining what is analyzed and how, partnership establishes shared authority across decisions of a process mining initiative. A particularly illustrative example is described by Joos Buijs through what he calls "co-creation through self-service" [12]. Rather than analysts independently developing dashboards and presenting finished results to the business, dashboards were iteratively constructed together with business users in short, highly interactive sessions. Users actively steered which hypotheses should be explored, proposed new visualizations, and shaped how process data became represented and interpreted. Insights were often developed "live" while employees observed, reacted, and redirected the analytical process. This practice moves beyond consultation toward a form of emerging partnership. Process minees are not merely asked for feedback after analytical decisions are made; they participate in constructing the analytical representation itself. Visibility becomes negotiated rather than unilaterally imposed. At the same time, the example also illustrates the limits of participation in many current process mining initiatives. While process minees may influence how *data* is visualized and interpreted, they do not necessarily control the broader *organizational* purposes of the analysis or how resulting insights are operationalized. Participation at the *tool level* may therefore coexist with managerial control at the organizational level.

Delegated Power represents a further step in which process minees are granted primary authority over how transparency about their work is produced and used. Decisions-makers retain a role, but they no longer hold the final say. At the *tool level*, delegated power can be realized when process minees have direct access to process mining tools and the autonomy to select for themselves what kinds of analyses they want to run. At the *data level*, this implies that process minees can decide which aspects of their work they wish to make visible, and which performance indicators they consider meaningful. At the *organizational level*, delegated power may take the form of arrangements in which managers explicitly transfer decision authority to process minees and rely on the resulting transparency to support their own adjustments and improvements. Thus, transparency moves away from a top-down monitoring logic.

Process minee control represents the highest rung of Arnstein's ladder, where the process minee fully transforms into the miner. Mind that we include this rung for conceptual symmetry with Arnstein's ladder; we do not claim that it is the appropriate target for every initiative. In process mining contexts, process minee control implies that managers are restricted to observing, without intervening in how work is carried out. The rationale draws on research showing that workers may not only have the best understanding of how to perform their tasks effectively but they are also intrinsically motivated to work; consequentially, excessive observation and interference can hamper the practices that make work successful in the first place [9]. At the *organizational level*, process minee control entails governance arrangements in which they, rather than managers, set the agenda for process mining initiatives. At the *tool level*, process minees have full discretion over which analyses are performed and how, while managers do not act on the results without their invitation. At the *data level*, process minees determine what is logged, how it is interpreted, and whether and how findings are communicated outward. In this configuration, trust is delegated to process minees, who are given space to identify and refine their own best ways of getting work done without the disruption of continuous managerial scrutiny. In practice, most process mining initiatives are likely to combine elements from different rungs and across different levels of the framework, resulting in hybrid arrangements in which authority over transparency-producing decisions is shared rather than concentrated in a single actor.

5 Discussion: A Call for Research

The mapping we have offered in Sect. 4 suggests that current participatory process mining efforts are concentrated in the lower parts of Arnstein's ladder, with ethical charters, iterative result-sharing, and governance frameworks mostly focused on communication *towards* employees, but less attention to empowering them in decision-making across the data, tool, and organizational levels. Building on this, we identify five high-level directions for future research and design.

Participation Across Levels. Research is needed on participation not only at the *tool level*, where co-creation is most developed, but also at the *data* and *organizational* levels, where decisions about logging scope, metric definition, and data governance tend to remain managerial prerogatives [19]. The Buijs example [12] illustrates that high participation at the *tool level* can coexist with low participation at the data and organizational levels; it remains unclear whether and how participatory practices at one level spills over into others.

Conditions for Sustaining the Upper Rungs. Research is needed on what organizational and institutional conditions make the upper rungs viable in process mining contexts: what governance structures, contractual arrangements, or cultural conditions are needed for partnership and delegated power to be sustained rather than eroded over time. Drawing on organization sciences and information systems research, one way to sustain upper rungs may be to continuously involve process minees in redesign initiatives, such as in relation to the process mining tool (e.g., how can it be improved to provide better insights?) or policies (e.g., when and under what conditions should insights be openly discussed) [1,19].

Calibrating the Appropriate Degree of Participation. An important aspect to consider is about how far up the ladder any given initiative should go. Thus, our framework should not be read as implying that process minee control is necessarily the desirable endpoint in every case. Organizations pursue objectives beyond the welfare of any single group, and must reconcile process minees' interests with those of owners, customers, regulators, and other workers. Therefore, research is needed on what degree of participation is appropriate for any given *data*, *tool*, and *organizational* contexts.

Designing and Evaluating Participatory Interventions. Design-oriented research is needed to develop and evaluate concrete interventions that operationalize participation at each level of the framework, for instance, ethical charters co-authored by process minees, workshops in which they help define event-logging scope and analytical questions, or dashboards designed to be read and acted upon by process minees rather than by managers [29]. Longitudinal designs would allow such interventions to be assessed not at a single point in time but as organizations accumulate experience with participatory practices [22].

The Process Minee Perspective. Perhaps most fundamentally, the process minee perspective is almost entirely absent from current practitioner and research literature. Drawing on organization sciences and information systems research, a core assumption of Participatory Process Mining is that the proactive involvement of process minees can add value in various ways. Process minees know what tasks may (or may not) be suitable for process mining; they can help contextualize insights; they are motivated to work more effectively when they find means to do so; and they may be even more motivated when they are aware of and have access to process mining insights [1,9,19,21]. Future work should seek to understand how process minees themselves experience the various forms of participation and non-participation described in this paper, and what forms of involvement they would consider meaningful.

6 Conclusion

In this paper, we argue that transparency in process mining is not a neutral or self-evident outcome, but the result of decisions made at different stages of the process mining lifecycle. These decisions shape what becomes visible, what remains invisible, and how visibility is interpreted, with significant consequences for those whose work is subject to analysis, i.e., the process minees. Against this background, and building on socio-technical and participatory perspectives, we have introduced the concept of *Participatory Process Mining* and proposed a three-level conceptual framework that distinguishes the levels at which participation can be enabled across the process mining lifecycle. We have shown how decisions at each level shape what process minees experience as transparency.

The framework contributes to ongoing discussions on Responsible BPM, extending them beyond privacy protection and data governance toward a broader notion of participatory transparency construction. Rather than treating process minees merely as data subjects or recipients of analytical outcomes, Participatory Process Mining recognizes them as active stakeholders with a legitimate interest in shaping how their work is recorded, filtered, and interpreted. Practitioner accounts suggest this recognition is not only normatively desirable but practically necessary: organizations that bypass process minee participation risk resistance, loss of trust, and in some cases disruption of entire process mining initiatives.

We hope this work inspires the BPM and process mining communities to place process minees at the center of how transparency is produced and used in organizations.

Acknowledgments. The work of A. del-Río-Ortega has been funded by MICIU/AEI/10.13039/501100011033 and by ERDF/EU under project PID2022-140221NB-I00, and by the Junta de Andalucía (Consejería de Universidad, Investigación e Innovación) and co-funded by the EU through the FEDER Andalucía 2021–2027 Programme under project DGP_PIDI_2024_01609.

References

1. Aaltonen, A., Stelmaszak, M.: The performative production of trace data in knowledge work. Inf. Sys. Res. **35**(3), 1448–1462 (2024)
2. Alaimo, C., Kallinikos, J.: Data Rules: Reinventing the Market Economy. MIT Press, Cambridge (2024)
3. Ammann, J., Lohoff, L., Wurm, B., Hess, T.: How do process mining users act, think, and feel? In: BISE (2025)
4. Arnstein, S.R.: A ladder of citizen participation. J. Am. Inst. Plann. **35**(4), 216–224 (1969)
5. Badakhshan, P., Wurm, B., Grisold, T., Geyer-Klingeberg, J., Mendling, J., vom Brocke, J.: Creating business value with process mining. JSIS **31**(4) (2022)
6. Bade, F.M., Vollenberg, C., Koch, J., Koch, J., Coners, A.: The dark side of process mining. How identifiable are users despite technologically anonymized data? A case study from the health sector. In: BPM, pp. 219–233. Springer, Heidelberg (2022). https://doi.org/10.1007/978-3-031-16103-2_16
7. Ball, K.: Electronic monitoring and surveillance in the workplace: Literature review and policy recommendations. Research report, European Commission, JRC (2021)
8. Batista, E., Solanas, A.: A uniformization-based approach to preserve individuals' privacy during process mining analyses. P2PNA **14**, 1500–1519 (2021)
9. Bernstein, E.S.: The transparency paradox: a role for privacy in organizational learning and operational control. ASQ **57**(2), 181–216 (2012)
10. Bernstein, E.S.: Making transparency transparent: the evolution of observation in management theory. Acad. Manag. Ann. **11**, 217–266 (2017)
11. Bowen, J., Hinze, A.: Participatory data design: managing data sovereignty in IoT solutions. Interact. Comput. **34**(2), 60–71 (2022)
12. Buijs, J.C.A.M., Bergmans, R.F.M., El Hasnaoui, R.: Analysis of the customer journey at the pension provider APG using self-service and data hub concepts. In: vom Brocke, J., Mendling, J., Rosemann, M. (eds.) Business Process Management Cases Vol. 2, pp. 111–124. Springer, Heidelberg (2021). https://doi.org/10.1007/978-3-662-63047-1_9
13. Clarke, S., Tuckwell, W., Luck, M.: Professionals and the ethics of workplace surveillance. J. Soc. Philos. 1–11 (2025)
14. Davenport, T.H., Spanyi, A.: What process mining is, and why companies should do it. Harv. Bus. Rev. (2019)
15. Eggers, J., Hein, A., Böhm, M., Krcmar, H.: No longer out of sight, no longer out of mind? How organizations engage with process mining-induced transparency to achieve increased process awareness. BISE **63**(5), 491–510 (2021)
16. Elkoumy, G., et al.: Privacy and confidentiality in process mining: threats and research challenges. ACM TMIS **13**(1) (2021)
17. Fahland, D., Mannhardt, F., Rozinat, A.: Process mining café #9: How to be a responsible process miner (2021). https://tinyurl.com/PMCafe9, Fluxicon Process Mining Café, Session 9. Online video recording
18. Fahrenkrog-Petersen, S.A.: Anonymization Techniques for Privacy-preserving Process Mining, Berlin, Germany. Ph.D. thesis (2023)

19. Gierlich-Joas, M., Baiyere, A., Hess, T.: Inverse transparency and the quest for empowerment through the design of digital workplace technologies. J. Assoc. Inf. Syst. **25**(5), 1212–1239 (2024)
20. Goel, K., Emamjome, F., ter Hofstede, A.H.M.: Data governance for managing data quality in process mining. In: ICIS. Association for Information Systems (2021)
21. Grisold, T., Seidel, S., Heck, M., Berente, N.: Digital surveillance in organizations. BISE **66**(3), 401–410 (2024)
22. Grisold, T., Van Der Aa, H., Franzoi, S., Hartl, S., Mendling, J., Vom Brocke, J.: A context framework for sense-making of process mining results. In: 2024 6th International Conference on Process Mining (ICPM), pp. 57–64. IEEE (2024)
23. Institute, A.L.: Participatory data stewardship (2021). https://www.adalovelaceinstitute.org/report/participatory-data-stewardship/
24. Kawakami, A., et al.: Sensing wellbeing in the workplace, why and for whom? Envisioning impacts with organizational stakeholders. In: HCI, vol. 7(CSCW2) (2023)
25. Magill, C., Klein, E., Chapple, S.: I am not a number: towards participatory IoT monitoring in the workplace. In: IET, p. 21 (2018)
26. Manley, A., Williams, S.: 'We're not run on numbers, we're people, we're emotional people': exploring the experiences and lived consequences of emerging technologies, organizational surveillance and control among elite professionals. Organization **29**(4), 692–713 (2022)
27. Mannhardt, F.: Responsible process mining. In: Process Mining Handbook, vol. 448, pp. 373–401. Springer, Heidelberg (2022). https://doi.org/10.1007/978-3-031-08848-3_12
28. Martin, N., et al.: Opportunities and challenges for process mining in organizations: results of a delphi study. BISE **63**(5), 511–527 (2021)
29. Mettler, T.: The connected workplace: characteristics and social consequences of work surveillance in the age of datification, sensorization, and artificial intelligence. J. Inf. Technol. **39**(3), 547–567 (2024)
30. Nacke, J., Plattfaut, R., Riedl, R.: Designing process mining to mitigate electronic performance monitoring risks. In: Design for Better Futures: Beyond the Science of the Artificial. Prototypes and Research-in-Progress, pp. 142–154. Springer, Heidelberg (2026). https://doi.org/10.1007/978-3-032-28570-6_12
31. Newlands, G.: Algorithmic surveillance in the gig economy: the organization of work through lefebvrian conceived space. OS **42**(5), 719–737 (2021)
32. Rafiei, M.: Privacy and Confidentiality in Process Mining, Aachen, Germany. Ph.D. thesis (2023)
33. Rozinat, A., Günther, C.W.: Privacy, security and ethics in process mining. Technical report/process mining news, Fluxicon (2016)
34. Stein Dani, V.: The Alpha and Omega of Process Mining. Ph.D. thesis (2024)
35. Stein Dani, V., Leopold, H., van der Werf, J.M.E.M., Beerepoot, I., Reijers, H.A.: From loss of interest to denial: a study on the terminators of process mining initiatives. In: CAiSE (2024)
36. Stein Dani, V., Leopold, H., van der Werf, J.M.E.M., Reijers, H.A.: Progressing from process mining insights to process improvement: Challenges and recommendations. In: EDOC, pp. 152–168. Springer, Heidelberg (2023). https://doi.org/10.1007/978-3-031-46587-1_9
37. Studer, L., Rozinat, A.: Process mining café #6: Freestyle data transformation (2021). https://tinyurl.com/PMCafe6, Fluxicon Process Mining Café, Session 6. Online video recording

38. Tentina, I., Zerbato, F., Mannhardt, F., van Dongen, B.F.: Why do users struggle to get insights from process mining? In: ICPM, pp. 1–8. IEEE (2025)
39. Van Looy, A., Vlerick, P.: Motivational process mining: a conceptualization by adding self-determination theory. In: BPM RBPM, PT, and Educators Forum, pp. 44–50. Springer, Heidelberg (2026). https://doi.org/10.1007/978-3-032-02936-2_4
40. de Vaujany, F.X., Leclercq-Vandelannoitte, A., Munro, I., Nama, Y., Holt, R.: Control and surveillance in work practice: cultivating paradox in 'new' modes of organizing. Organ. Stud. **42**(5), 675–695 (2021)
41. Vom Brocke, J., Jans, M., Mendling, J., Reijers, H.A.: A five-level framework for research on process mining. Bus. Inf. Syst. Eng. **63**(5), 483–490 (2021)
42. Yeshchenko, A., Mendling, J.: A survey of approaches for event sequence analysis and visualization. Inf. Syst. **120**, 102283 (2024)
43. Zimmermann, L., Zerbato, F., Weber, B.: What makes life for process mining analysts difficult? A reflection of challenges. SoSyM **23**(6), 1345–1373 (2024)
44. Zorina, A., Bélanger, F., Kumar, N., Clegg, S.: Watchers, watched, and watching in the digital age: reconceptualization of information technology monitoring as complex action nets. Organ. Sci. **32**(6), 1571–1596 (2021)
45. Zuidema-Tempel, E., Bukhsh, F., Effing, R., van Hillegersberg, J.: Process mining and perceived privacy violations: a pilot-study. In: ICEIS, pp. 289–296. SciTePress (2023)

From Principles to Practice: Embedding GEDSI in Process Design

Rehan Syed(✉), Azumah Mamudu, and Ishadi Mirispelekotuwa

Queensland University of Technology, Brisbane, Australia
r.syed@qut.edu.au

Abstract. This paper investigates the integration of Gender, Equality, Disability, and Social Inclusion (GEDSI) principles into process design, highlighting their significance in promoting equitable processes. GEDSI has garnered attention among international development agencies, NGOs, and corporations as a framework to enhance organisational effectiveness and social responsibility. Utilising experiences from public sector organisations in the Philippines, which participated in process driven transformation programmes, this study examines the incorporation of GEDSI into process design. It synthesises current research and practice, analysing theoretical frameworks, methodologies, case studies, and challenges, while identifying gaps in the literature. A GEDSI-Process Maturity Framework is introduced to assess process alignment with GEDSI dimensions, although further development is required. The paper emphasises the importance of prioritising GEDSI in process design to ensure ethical and efficient outcomes, offering insights for practitioners and researchers.

Keywords: GEDSI · Process Design · BPM · Digital Transformation · Public Sector

1 Introduction

Gender Equality, Disability, and Social Inclusion (GEDSI) has become essential in shaping equitable and inclusive business process design. Incorporating GEDSI dimensions is critical to ensure that organisational workflows are inclusive, accessible, fair, and responsive to the needs of diverse stakeholders, including women, people with disabilities, socially disadvantaged, and other marginalised groups [10]. Since its emergence around 2018, the GEDSI framework has garnered substantial attention from governmental bodies, non-governmental organisations (NGOs), and the corporate sector as a means to enhance organisational effectiveness, inclusivity, and social responsibility [19]. This paper examines the experiences of two public sector institutions in the Philippines that have participated in extensive training initiatives focused on digital transformation, funded by the Department of Foreign Trade (DFAT), Australia. These initiatives are designed to integrate GEDSI dimensions, as mandated by the Philippine government, into public sector processes as integral components of process design considerations, thereby improving citizen service delivery. We have also synthesised current research

K. Kluza et al. (Eds.): BPM 2026, LNBIP 600, pp. 93–111, 2027.
https://doi.org/10.1007/978-3-032-36710-5_7

and practice concerning the integrations of GEDSI into BPM. Upon critically examining BPM maturity models and the challenges faced by incumbent organisations, we identify a paucity of research in this critically important area and discern gaps in the literature: existing BPM maturity models and methodologies do not comprehensively address the integration of GEDSI dimensions into process design and improvement. Furthermore, they provide limited guidance on incorporating GEDSI dimensions into BPM. Motivated by this, we address the research question: "How can the integration of GEDSI dimensions be assessed and guided in the design and improvement of business processes?". This paper makes three contributions. First, it proposes the GEDSI–Process Alignment Framework (GPAF), a three-level assessment framework that integrates GEDSI dimensions into process design and improvement. Second, it formulates four design principles for GEDSI-informed process design and governance, derived via Action Design Research. Third, it provides empirical insights from two Philippine public sector organisations on the adoption and perceived value of GEDSI-embedded BPM training and processes. The framework is now adopted by 37 departments across two participating organisations in the Philippines. Based on our programme delivery, we argue BPM practitioners and researchers must prioritise GEDSI to develop ethical, efficient, and sustainable processes.

The paper is organised as follows. Section 2 summarises related work. Section 3 discusses the study design and methodology. Section 4 presents the GPAF, its principles, and expert feedback. Section 5 concludes with discussion and conclusions.

2 Related Work

Gender equality, disability, and social inclusion, collectively known as 'GEDSI', are fundamental human rights that play a crucial role in building safe, strong, and inclusive societies [20]. GEDSI dimensions aim to ensure equitable opportunities and inclusion for all individuals, regardless of their gender, disability status, or social background. GEDSI is also the central theme underpinning the 17 sustainable development goals (SDGs) [1, 41]. Gender equality means that all individuals, regardless of their sex or gender, have equal opportunities to realise their full human rights and to participate in, contribute to, and benefit from economic, social, cultural, and political development [20, 31]. A person with a disability is someone with an episodic or long-term physical, intellectual, mental, or sensory impairment that, in interaction with various barriers, may hinder their full and effective participation in society on an equal basis with others [20]. Disability inclusion ensures that people with disabilities have equal rights as others, are provided with the same opportunities as others, and enables them to make choices about their lives [20]. Social inclusion means that people and segments of a society can enjoy equal rights and participate equally in the society [20]. The existing literature on GEDSI has primarily focused on policy frameworks [1], global initiatives such as SDG 5 and SDG 10 [41], and institutional efforts to address inequalities and discrimination [20, 31].

2.1 Relevance of GEDSI in Business Process Management (BPM)

The BPM discipline is renowned for designing, executing, and optimising organisational processes to improve efficiency and reduce costs [11]. Despite this potential, BPM methods have often prioritised technical and economic objectives over social considerations such as GEDSI. As the need to incorporate GEDSI into process design becomes more apparent, BPM efforts to streamline, standardise, and automate processes must prioritise both "efficiency" and "inclusivity". Process design efforts often rely on historical data or assumptions that may embed gender or social biases, leading to design outcomes that could exclude the input of marginalised groups or individuals with disabilities [14]. The oversight of GEDSI considerations in business process design results in processes that are not only inequitable but also less effective at addressing the needs of diverse populations [39]. By focusing primarily on efficiency and cost, current methodologies used by process experts compromise principles such as fairness and inclusion, thereby limiting their long-term sustainability and social impact. Literature [8] highlights the need to integrate gender equality into BPM and critiques the limitations of existing approaches.

The International Labor Organisation (ILO) [19] offers practical guidelines for adapting processes to accommodate diverse employee needs, especially those with disabilities or from underrepresented groups. These insights are directly applicable to BPM for inclusive process design. **Gender equality** remains pivotal in inclusion discussions within the financial sector [4]. Despite sustained attention to this issue, progress has been gradual, with researchers noting that *"the issue of gender equality has been raised for a long time, [yet] the share of women at managerial levels is growing very slowly"* [4]. This highlights a persistent structural challenge that organisations must address. Achieving substantive change necessitates not only policy modifications but also cultural transformation. Enhancing awareness of gender biases in policies and processes is crucial [4]. Collectively, these insights underscore the necessity for a comprehensive approach that integrates gender equality principles into everyday organisational processes and communication.

Organisations increasingly acknowledge the need to integrate **disability inclusion** into process modelling for accessible systems [42]. Around 650 million people globally have disabilities, and in Europe, 15.7 percent of the workforce (52.7 million people) face health issues or disabilities [42]. Accessibility in business processes is therefore essential for workforce inclusion. Research shows accessible design principles can improve workforce utilisation, making companies more appealing to skilled individuals with disabilities [42]. Scholars emphasise that inclusivity should be incorporated into process design, considering accessibility when developing tools and involving individuals with disabilities [3, 42]. Literature advocates for structural change in collaborative modelling, recommending methods beyond visual models to enable participation by individuals with visual disabilities [3]. These insights highlight the importance of universal design in BPM, ensuring process modelling is accessible and representative of workforce diversity. **Social inclusion** is an essential yet often neglected aspect of sustainability within Business Process Management (BPM). Scholars have noted that social sustainability is often underrepresented [37]. This is problematic because it inhibits business improvement of people's quality of life (e.g., health and equity), fails to address changing customer demands beyond transactional excellence, and obstructs the consideration of new regulations [37]. Therefore, a more comprehensive approach to process design must be adopted to address workforce diversity [37]. Integrating social considerations into organisational processes not only mitigates risks but also provides strategic advantages, *"fosters trust, and raises employee engagement"* [37]. Translating social inclusion principles into BPM requires their deliberate integration within process design frameworks [4]. The BPM design patterns should offer practical guidance to assist process designers in analysing and (re-)designing business processes [37]. These collective insights demonstrate that incorporating social sustainability into BPM enhances organisational resilience and improves the lived experience of employees and stakeholders. Few studies [5, 15, 40] examine **gender bias** in BPM, specifically focusing on female underrepresentation. Kedziora, et al. [21] explore the role of robotic process automation governance for inclusive process design and continuous improvement. Bialas and Wróbel [4] focus on identifying best practices for process design incorporating diversity, equity, and inclusion. Similarly, Mahendrawathi [24] coins the term 'inclusive BPM' as *"a discipline of managing work processes that must be intentionally inclusive, thoughtfully designed, and implemented with people and human rights at the centre of all aspects"*. They unpack three key attributes of inclusion in BPM as: 1) thoughtfully and intentionally designed, 2) people and human rights focused, 3) addressing the needs of the most vulnerable groups, including women, people with disabilities and people at the bottom of the economic pyramids or income-constrained groups [24]. These studies address workplace diversity and inclusion broadly, incorporating GEDSI in process design remains an open question [4]. Current BPM approaches [24] lack assessment criteria for GEDSI integration, limiting understanding of GEDSI intersections in BPM. While prior work conceptualises 'inclusive BPM' and explores gender bias and accessibility in BPM practice, it does not provide a structured, multilevel framework to assess GEDSI integration across processes and organisations.

2.2 GEDSI Considerations in BPM Maturity Models

Maturity models consist of stages that form a path from initial stage to maturity. They help evaluate the AS-IS state to guide improvements and control progress [33]. In BPM, two types exist: (1) process-driven, referring to process conditions in general or distinct processes, and (2) capability-driven, addressing BPM capabilities of an organisation [36]. We reviewed 13 generic and BPM domain specific maturity models for our framework design. Key maturity models [6, 13, 28, 36] form the basis of most BPM domain models. Table 1 presents an excerpt from the maturity model analysis, highlighting its scope and coverage of GEDSI.

While these maturity models help organisations assess internal processes, they rarely address GEDSI dimensions. Capability based models primarily focus on process optimisation and operational control, overlooking GEDSI aspects. Although Roseman and De Bruin's BPMMM [36] includes 'people' as a factor, it doesn't consider how processes impact marginalised groups. Companies can achieve high process maturity while maintaining exclusionary practices. These models miss opportunities for socially responsible transformation, necessitating a framework that integrates GEDSI dimensions throughout. This motivates the development of a GEDSI specific process alignment framework that can complement existing maturity models.

Table 1. GEDSI in BPM Maturity Models: A Synopsis

Maturity Model	Scope	Coverage of GEDSI
Capability Maturity Model (CMM) [32]	Capability-driven maturity model with five levels: initial, repeatable, defined, managed, and optimising	No explicit GEDSI coverage; limited attention to marginalised groups
Capability Maturity ModelIntegration (CMMI) [6]	Capability-driven maturity model with five levels: initial, managed, defined, quantitatively managed, and optimising	No explicit GEDSI coverage; limited attention to marginalised groups and gender equality
Gartner's Business Process Maturity Model [13, 28]	Process-driven maturity model with six phases: Acknowledge operational inefficiencies, become process-aware, establish intra-process automation and control, establish inter-process automation and control, establish enterprise valuation control, and create an agile business structure	Focus on social inclusion and accessibility. It identifies how values influence behaviours and competencies for process management
OMG [30]	Process-driven maturity model with six levels: initial, managed, standardised, predictable and innovating	No explicit GEDSI coverage; limited attention to marginalised groups, gender equality
Rosemann and De Bruin's Business Process Management Maturity Model (BP-MMM) [36]	Process-driven maturity model with five stages: initial state, defined, repeatable, managed, optimised; (2) Six factors with five capability areas each strategic alignment, culture, people, governance, methods, IT	Includes'people' and'culture' as factors in the underlying model, supports only one aspect of social inclusion. No explicit GEDSI coverage; limited attention to marginalised groups, gender, and disability

(continued)

Table 1. (*continued*)

Maturity Model	Scope	Coverage of GEDSI
Maull et al.'s Business Process Re-engineering Maturity Model (BPRMM) [25]	capability-driven maturity model with five stages: planning, execution, implementation, knowledge creation and knowledge application	Supports an inclusive BPR approach by identifying how human factors like culture and behaviour influence the organisational design of BPR. No explicit GEDSI coverage; limited attention to marginalised groups
Fisher's Business Process Maturity Model (BPMM) [12]	Process-driven maturity model with five levers of change (LoC): strategy, controls, people, technology, process	Includes 'people' as one of the levels No explicit GEDSI coverage; limited attention to gender equality or marginalised groups
Process Management Maturity Assessment (PMMA) [34]	Capability-driven with five maturity levels according to CMMI – initial state, defined, repeatable, managed, optimised	Has limited focus on social and cultural dimensions, specifically on establishing roles and responsibilities for managing processes No explicit GEDSI coverage; limited attention to marginalised groups
Business Process Orientation Maturity Model (BPOMM) [26, 27]	Process-driven model with four stages; Ad hoc, defined, linked, integrated three components of maturity (process view, process jobs, process management and measurement)	Measures organisational alignment of processes over functions, focusing on governance, culture, performance, and customer orientation. No explicit GEDSI coverage
Process and Enterprise Maturity Model (PEMM) [17]	Capability-driven model with a separation of five process enablers: design, performers, owner, infrastructure, and metrics, and four enterprise capabilities: leadership, culture, enterprise, and governance	Focus on performers' knowledge and skills and process owners' activities and authority. Capability evaluation includes culture and people, with limited gender equality support

(*continued*)

Table 1. (*continued*)

Maturity Model	Scope	Coverage of GEDSI
Process Maturity Ladder (PML) [18]	Capability-driven maturity model with five levels of maturity according to the capability maturity model (CMM): initial, repeatable, defined, managed, optimising	Focuses on roles like developers, managers, and product owners. No explicit GEDSI coverage
Weber et al.'s Business Process Maturity Model (BPMM-OMG) [45]	Capability driven maturity model with five stages similar to CMM: initial, repeatable, defined, managed, optimising	Focuses on roles only. No explicit GEDSI coverage; limited attention to marginalised groups
Lee's BPMM [22]	Capability driven maturity model with five maturity levels according to CMMI	Focuses on the people aspect, specifically cultural and behavioural aspects. However, it fails to account for how business processes impact marginalised groups

3 Methodology

This study follows the Action Design Research (ADR) methodology [38] to propose a GEDSI-Process alignment framework. ADR has a strong practical focus that contributes to scientific theory by effecting change and generating knowledge about the change [7]. By facilitating active engagement between researchers and practitioners, ADR is notable for generating innovative artefacts that can be immediately applied to solve real world problems while contributing to scientific theory [2]. We follow three phases of (i) Problem formulation, (ii) Building, Intervention and Evaluation (BIE), and (iii) Reflection and learning to derive the proposed research artefacts [38] (Fig. 1).

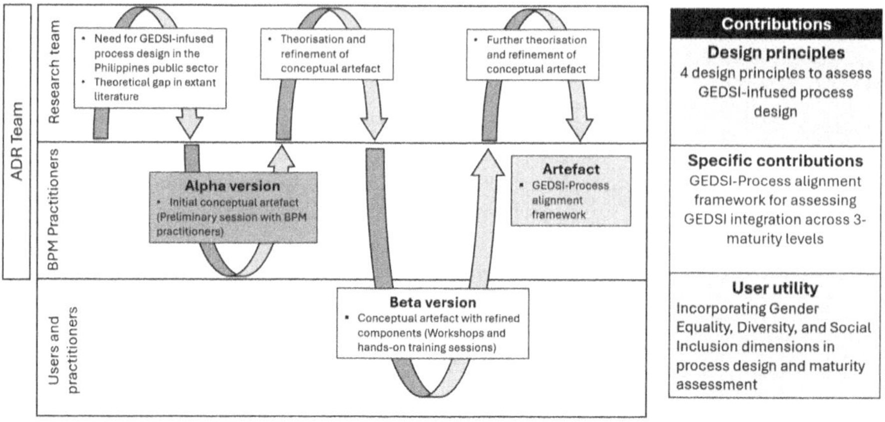

Fig. 1. ADR Methodology for GEDSI-Process Alignment Framework (GPAF)

In the problem formulation phase, we examined GEDSI and digital transformation within BPM as required by the Philippines' government. The Philippine government, in partnership with the Australian Department of Foreign Affairs and Trade (DFAT) launched strategic initiatives to train key stakeholders regarding documenting, analysing, and improving institutional processes that integrate GEDSI dimensions. Their efforts sought to address diverse user needs, eliminate gender bias, and ensure accessibility for people with disabilities. To substantiate and clarify the research gap, we conducted a scoping review of existing BPM literature on maturity models (as outlined in Sect. 2). This enabled us to assess the state-of-the-art and identify gaps in GEDSI implementation within existing BPM maturity models.

The study applied an organisational dominant Building, Intervention, Evaluation (BIE) approach, ensuring that the design innovation and knowledge generated were primarily suited to the organisational context. Two main iterative cycles were conducted in the building, intervention and evaluation (BIE) phase: (i) preliminary sessions with ten BPM practitioners to plan and review training materials and (ii) training sessions with over 85 process stakeholders of various departments and agencies in the public sector.

Insights from GEDSI and BPM maturity literature informed the initial conceptualisation by identifying key attributes of GEDSI dimensions and corresponding levels of process maturity. During the preliminary sessions, ten BPM practitioners with significant experience in process modelling, analysis, and improvement initiatives within the public sector were interviewed by the first author. These BPM practitioners played a key role as coaches, providing training support to process stakeholders during the workshops and hands-on training sessions. Insights from the BPM practitioners were consolidated with theoretical perspectives on GEDSI in BPM initiatives, thereby developing the alpha version of the conceptual artefact.

Following the preliminary sessions, a series of training workshops and hands-on activities were conducted for two public sector agencies of the Philippine government. Initially, a one-week workshop was delivered to train staff in process modelling, analysis, and improvement, with a focus on how GEDSI dimensions influence process activities. This was followed by a 12-week hands-on coaching programme, during which ten coaches (BPM practitioners) supported participants in reviewing and redesigning selected core processes within the participating agencies. During this period, a formative ex ante evaluation based on practical experience [43] was conducted with training participants to enable the early detection of sociotechnical misalignments [38]. The alpha version of the conceptual artefact was provided as a template to guide process design, analysis, and improvement initiatives. While researchers contributed theoretical insights from BPM maturity literature, practitioners and process stakeholders provided practical perspectives on embedding GEDSI dimensions into BPM implementations within their organisational contexts. This reciprocal exchange of knowledge facilitated iterative refinement, resulting in the beta version of the GPAF.

Throughout the problem formulation and artefact development phases, the study adopted a guided emergence approach, whereby reflections and learning from both researchers and practitioners informed the ongoing design and redesign of the conceptual artefact. Interactions during the alpha phase led to the development of an initial artefact that assessed GEDSI dimensions across three maturity levels, considering an organisation's process state, documentation availability, execution consistency, and performance metrics. In the beta version, the GEDSI dimensions were recategorised to incorporate broader social considerations, including community engagement, recognition of intersecting social identities, and antidiscrimination in process design.

Formalising the learnings from this project resulted in finalising the sub attributes of gender equality, disability inclusion, and social inclusion across three maturity levels. Four design principles for infusing GEDSI into process design were also derived. Our artefact and design principles are discussed in Sect. 4. Section 3.1 presents an overview of the two case organisations, study participants, and total processes analysed.

3.1 Case Organisations

The framework was developed as a core component of process innovation for digital transformation training for the Philippines Government. It helps government agencies deliver efficient public services while adapting to automated and integrated information systems. The programme is designed to enhance participants' comprehension of BPM and their ability to instruct others, thereby establishing BPM as a fundamental approach to digital transformation within the participating agencies. Considerations related to Gender Equality, Disability, and Social Inclusion (GEDSI) must be incorporated into process design decisions in accordance with government policy.

Two public sector departments of the Philippines were involved in validating the framework (pseudonymised as Gov't Dept. 1 and Gov't Dept. 2). Gov't Dept. 1 had 20 sub-agencies for legal, administrative, medical, and educational services. Gov't Dept. 2 had 17 sub-agencies for planning, evaluation, audit, and research. Participants included process managers and staff involved in core processes. Table 2 presents details of the departments, processes analysed, and participants in workshops and training sessions.

Table 2. Summary of Case Organisations

Case Organisation	Govt Dept. 1	Govt Dept. 2
Offices/sub-departments engaged	20	17
Number of processes analysed	65	20
Total programme participants	45	40

3.2 Design Principles

Externalising the learnings from the two design cycles, we prescribe four design principles based on [16]. These principles are informed by the theoretical perspectives of BPM maturity models and practical insights from BPM practitioners and participants.

Principle 1: Transparent and inclusive approach to process design:

Aim, Implementer, and User: To ensure that process owners, process sponsors and process modellers (implementers) consider the perspectives of marginalised gender, disability and other social groups (users) during process design (aim).

Context: Process design initiatives within an organisation that affect gender, disability and social groups, particularly where policy frameworks mandate equitable outcomes.

Mechanism: (1) Apply participatory process design methods including codesign workshops, storyboarding and collaborative BPMN modelling. (2) Ensure transparent process documentation that clearly outlines how GEDSI considerations and trade-offs are incorporated into process artefacts. (3) Incorporate relevant industry GEDSI related benchmarks (e.g., accessibility standards, equity metrics) as reference points for validation.

Rationale: Marginalised groups often face systemic barriers in standard process designs. Participatory and transparent methods amplify their voices, address specific needs and yield inclusive process adjustments that enhance outcomes for all stakeholder groups [27, 29, 34, 35].

Principle 2: Process design alignment with strategic GEDSI priorities:

Aim, Implementer, and User: For executive sponsors and functional leads (implementers) to assume accountability for embedding strategic GEDSI priorities into process design, with process owners and GEDSI specialists responsible for defining performance benchmarks (aim), benefiting diverse customer and employee groups (users).

Context: Organisations establishing or refining strategic GEDSI priorities that emphasise diverse customer perspectives, such as in sectors with equity regulations or customer facing processes.

Mechanism: (1) Translate strategic GEDSI mandates into measurable, process specific KPIs (e.g., accessibility compliance rates, equitable work allocation, etc.). (2) Integrate KPIs into organisational performance frameworks, with clear ownership, accountability structures, and predefined remediation workflows.

Rationale: Aligning processes with strategic priorities establishes clear ownership, escalation paths, and remediation mechanisms, ensuring ongoing accountability and measurable progress, particularly in complex, goal-oriented environments, in line with strategic alignment principles in process management [21, 23, 40, 45].

Principle 3: Institutionalisation of GEDSI-informed process culture:

Aim, Implementer, and User: For process owners, GEDSI specialists and process modellers (implementers) to institutionalise a GEDSI-informed culture in process activities (aim), fostering equitable practices for marginalised gender, disability and other social groups (users).

Context: Organisations scaling GEDSI integration across business processes, where cultural embedding is needed to transition from ad hoc compliance to systemic practice amid evolving regulatory or societal expectations.

Mechanism: (1) Design standardised, reusable templates for GEDSI-informed process documentation and store them in a centralised repository alongside supporting artefacts (e.g., guidelines, checklists). (2) Publish and disseminate success stories and case notes illustrating how GEDSI considerations influenced process decisions and outcomes. (3) Offer regular training and knowledge sharing sessions on applying GEDSI templates and best practices.

Rationale: Institutionalising culture through artefacts, narratives and capacity building sustains GEDSI awareness and adoption beyond initial projects, embedding reflective practices in sociotechnical systems in line with organisational learning and culture change theories [34, 36, 43, 44].

Principle 4: Continuous improvement and innovation of GEDSI-informed processes:

Aim, Implementer, and User: For process designers, business analysts, process owners & GEDSI compliance teams (implementers) to progressively enhance GEDSI compliance, equity, and inclusiveness of business processes (aim) for women, persons with disabilities, and other socially marginalised groups (users).

Context: Organisational involving the design, execution, monitoring, or improvement of business processes where GEDSI requirements are embedded in regulatory, policy, or organisational frameworks.

Mechanism: (1) Continuous GEDSI-specific KPI monitoring and process mining/analytics. (2) Regular capacity building programs. (3) Research and development into cutting-edge technologies that support GEDSI initiatives. (4) Conduct structured review cycles and stakeholder feedback loops. (5) GEDSI-Informed iterative improvement cycles (e.g., PDCA or Kaizen).

Rationale: Sustained improvement in complex sociotechnical environments require ongoing reflection, learning, and adaptation, as established in business process management literature [27, 29, 43, 44].

4 The GEDSI-Process Alignment Framework (GPAF)

GPAF provides a three-level assessment of GEDSI–process alignment. At Level 1 (Emerging), GEDSI considerations are addressed in an ad hoc and non-standardised manner, with processes characterised by limited consistency and a predominantly reactive orientation. At Level 2 (Established), GEDSI integration becomes more systematic, with repeatable and sustainable practices embedded within processes, reflecting an organisational commitment to inclusivity. At Level 3 (Leading), GEDSI dimensions are fully institutionalised and leveraged as a strategic asset to support continuous process innovation and improvement. At this level, GEDSI not only informs process design but also drives value creation and organisational transformation. This tiered structure enables organisations to assess their current level of GEDSI integration and provides a pathway for advancing towards more inclusive process management practices. The framework is currently applied by two organisations to assess their process models (Table 3).

4.1 Feedback from GEDSI Experts and Key Stakeholders

We gathered feedback on the implemented version of GPAF from a range of stakeholders during process review sessions from three (3) trainee modellers, three (3) process modelling coaches, and three (3) GEDSI experts. This feedback was compiled and analysed to inform our understanding the stakeholder views on GEDSI dimensions in processes. As one of the process modelling coaches explained… *"The framework will help evaluate the maturity level of GEDSI in BPM practice and will be a very useful tool and component of process knowledge management. By assessing levels of maturity through processes, documentations, standards, innovations, and measurement metrics, practitioners can monitor and gauge their effectiveness in ensuring inclusivity. The framework will hasten the improvement journey of a community. It can be a guide, not just for tech or business functions, but for behavioural processes that align with how people learn, share, decide, adapt, and reflect. The use of technology to get to the leading level of evaluation spells innovation in understanding and improving human functioning. Benchmarking for best practices will be a hallmark for collaboration among people and communities."* (Process Modelling Coach 2). The details of the feedback received are as follows;

Review of Gender Inclusion: The framework effectively delineates a maturity trajectory for gender equality, from representation to systemic integration at higher levels. Experts confirm the importance of embedding equity within organisational processes as indicative of structural change, while noting the need for specific examples. *"I think the framework does a solid job of outlining a clear progression from Emerging to Leading levels for gender equality, particularly with its focus on representation and equity. The emphasis on embedding gender equity into processes and resource allocation at the Leading level is spot on and shows a commitment to systemic change. However, I'd love to see more explicit guidance on addressing non-binary and transgender inclusion*

Table 3. GEDSI-Process Alignment Framework (GPAF)

1. Gender Equality	Emerging Level	Establishing Level	Leading Level
Representation:Ensure diverse gender representation in all roles Equity: Ensure equity across genders	Ad Hoc Processes: Gender representation is acknowledged but not systematically included in the processes. Equity measures are ad hoc or not present Documentation: Limited or no documentation of gender-relevant attributes to inform the process design Consistency: No standard approach or vocabulary; processes vary widely and are dependent on individual GEDSI knowledge Measurement: No formal metrics or performance indicators are in place; accountability metrics are not defined Innovation: New technologies and methodologies are not considered to inform the processes Best Practices: No benchmarks or industry best practices are considered against gender diversity & equality. No training programmes are conducted on gender diversity & equality in standardised processes	Defined Processes: Gender diversity & equity are actively promoted in process design. Equity policies are in place and monitored Documentation: Processes are documented and informed by relevant definitions of gender diversity & equity. Employees are trained on gender-related issues in processes Consistency: Gender diversity & equity attributes are standardised; processes have little variation and are informed by the organisation's gender diversity guidelines Measurement: Basic metrics and performance indicators are established and monitored Innovation: New technologies and methodologies are partially considered to inform the processes Best Practices: Relevant benchmarks or industry best practices are considered against gender diversity & equality. Infor-mal training programmes are conducted on gender diversity & equality in standardised processes	Continuous Improvement: Gender diversity & equity is fully embedded in processes and resource allocation. Continuous improvement and impact assessments are conducted. Responsible roles are defined with some clarity on accountability Documentation: Documentedprocesses clearly outline gender considerations and trade-offs. Documents are accessible to all stakeholders Consistency: Processes have no variations and are compliant with the organisation's gender diversity guidelines Measurement: Formal metrics or performance indicators are in place, and accountability metrics are fully defined Innovation: New technologies and methodologies are regularly integrated into processes to support gender diversity & equality Best Practices: Processes are benchmarked against industry best practices for gender equality. Comprehensive metrics and performance indicators are used to drive improvements. Formal training programmes are conducted on gender diversity & equality in standardised processes

(*continued*)

Table 3. (*continued*)

2. Disability Inclusion	Emerging Level	Establishing Level	Leading Level
Accessibility: Physical and digital accessibility of workplaces and digital artefacts Accommodation: Availability of reasonable accommodations for stakeholders with disabilities Awareness and Training: Programs to educate stakeholders about disability inclusion and etiquette in processes	Ad Hoc Processes: Basic or No awareness of stakeholders' accessibility needs were considered in process design Documentation: Limited availability of disability policies and procedures. Limited integration with process design Consistency: Processes vary widely and are dependent on individual knowledge of stakeholder accessibility requirements Measurement: No formal accessibility metrics or performance indicators in place. Ac-countability metrics are not defined Innovation: New technologies and methodologies are not considered to inform the processes Best Practices: No benchmarks or industry best practices are considered against gender diversity & equality. No formal training programmes are conducted on disability standards in process design	Defined Processes: Relevant accessibility standards are implemented in process design. Reasonable accommodations are routinely provided Documentation: Processes are documented and accessible to key stakeholders. Employees are trained on standardised processes Consistency: Processes have limited variations and are compliant with the organisation's disability design guidelines Measurement: Basic metrics and disability performance indicators are established and monitored Innovation: Processes have considerable technology integration suitable for stakeholders with disabilities Best Practices: Relevant disability benchmarks or industry best practices are adopted. In-formal training programmes are conducted on disability standards in processes	Continuous Improvement: Accessibility and inclusion are proactively designed into all processes. Stakeholders with disabilities are involved in codesign and evaluation Documentation: Processes are documented at a centralised location, including design trade-offs and are accessible to all stakeholders Consistency: Processes have standardised and are compliant with the organisation's disability design guidelines Measurement: Formal metrics and disability performance indicators are established and monitored.Accountability metrics are fully defined Innovation: New technologies and methodologies are regularly integrated into processes. Processes are optimally integrated with technologies suitable for identified stake-holder disabilities Best Practices: Processes are benchmarked against industry best practices. Comprehensive metrics and performance indicators are used to drive improvements. Regular training programmes are conducted on disability inclusion in processes

(*continued*)

Table 3. (*continued*)

3. Social Inclusion	Establishing Level	Leading Level	Emerging Level
Inclusive Culture: Creating an environment where all stakeholders feel valued and included Community Engagement: Involvement in and support for diverse communities Anti-Discrimination: Alignment with policies and procedures to preventandaddress discrimination and harassment Multiple Identities: Recognising and addressing the overlapping and intersecting social identities (e.g., gender, race, disability) that affect stakeholder experience Tailored Support: Providing support that considers the unique needs of individuals with multiple marginalised identities	Ad Hoc Processes: Inclusion is recognised but not systematically introduced in process consultations. Community engagement is minimal. Limited understanding of intersectionality. Process design is generic and not tailored to the needs of marginalised groups Documentation: Limited or no documentation of relevant process requirements of marginalised groups Consistency: Processes vary widely and are dependent on ad hoc, individual knowledge of marginalised group needs Measurement: No formal metrics or performance indicators in place. Accountability roles are not defined Innovation: Processes have limited technology integration suitable for marginalised groups Best Practices: No benchmarks or industry best practices were considered. Limited or no performance indicators are used to drive processes. No training programmes were conducted on social inclusion in processes	Defined Processes: Inclusive design is promoted. Anti-discrimination policies are enforced. Community partnerships are active. Intersectional identities are considered in process design and practice. Tailored support is available Documentation: Relevant process needs are identified, documented, and is accessible to key stakeholders Consistency: Processes have limited variations and are compliant with theorganisation's social inclusion design guidelines Measurement: Basic metrics and performance indicators are established and monitored Innovation: Processes have considerable technology integration to accommodate the needs of marginalised stakeholders Best Practices: Relevant benchmarks or industry best practices are considered. Informal training programmes are conducted on social inclusion in processes	Continuous Improvement: Inclusion is a strategic design priority. Stakeholder feedback shapes inclusive practices. The organisation is a recognised leader in social inclusion in its processes. Intersectionality is central to process design strategy and evaluation. Processes are codesigned with individuals from intersecting marginalised groups Documentation: Process social inclusion design requirements are documented and accessible to key stakeholders Consistency: Processes have standardised and are compliant with the organisation's social inclusion design guidelines Measurement: Formal metrics and disability performance indicators are established and monitored. Accountability metrics are fully defined Innovation: Processes are fully integrated with technology, providing seamless access to all stakeholders Best Practices: Processes are benchmarked against industry best practices. Comprehensive metrics and performance indicators are used to drive improvements. Regular training programmes are conducted on social inclusion in standardised processes

and how to integrate these identities into process design and metrics, as they're often overlooked in traditional gender equality approaches." (GEDSI Expert 3). *"The model is well articulated, especially at the higher levels, where processes are standardised and benchmarked against best practices. That's critical for accountability. It might help to include examples of what 'ad hoc' looks like in practice or provide a baseline expectation to ensure organisations at this level are at least starting with some intentionality."* (GEDSI Expert 1).

Review of Social Inclusion: The expert valued codesign with marginalised groups as evidence of shifting from symbolic to substantive engagement. However, the critique of Community Engagement suggests that without guidance on reciprocity, organisational engagement may remain transactional rather than transformative. *"The framework's focus on intersectionality and tailored support is really impressive…it shows an understanding that social inclusion isn't just about checking boxes but about addressing complex, overlapping identities. The idea of codesigning processes with individuals from marginalised groups is a game changer. That said, I think the framework could strengthen its Community Engagement component by specifying how organisations can build authentic, reciprocal relationships with diverse communities rather than just 'involvement,' which can sometimes feel a mere lip service."* (GEDSI Expert 1). *"I appreciate how the framework integrates antidiscrimination policies and inclusive culture as core pillars, especially at the Established level. However, the jump from Emerging to Established feels steep—there's a lack of interim steps for organisations that are just starting to grapple with social inclusion. For example, the Emerging level could include practical first steps, like conducting stakeholder surveys to understand diverse needs or initiating small scale antidiscrimination training, to make the progression feel more achievable for organisations with limited resources."* (GEDSI expert 2).

Review of Disability Inclusion: Codesign with individuals with disabilities indicates genuine inclusion rather than procedural adherence. Experts' advocacy for digital accessibility benchmarks reveals a practical deficiency in the existing framework. *"Codesign with stakeholders with disabilities is important. That's a critical step for ensuring processes aren't just compliant but truly inclusive. Is it possible to have digital accessibility standards for processes like WCAG 2.1? since digital inclusion is so critical today. Without specific benchmarks, organisations might struggle to know what 'process accessibility standards' mean in practice."* (Process Modeller 1). *"The progression in training and awareness programs across the levels is well thought out. At the initial level, there's an assumption of 'basic awareness' of accessibility needs, but in reality, many organisations lack even that. Adding guidance on fostering resource optimisation challenges and stereotypes about disabilities through targeted campaigns, this could make the framework more transformative."* (GEDSI Expert 1).

Process Models and GEDSI Attributes: Process designers confirmed the framework serves as a practical guide for systematically incorporating GEDSI considerations into process modelling across maturity levels. It embeds inclusion principles throughout design, making GEDSI integral rather than peripheral. *"I find the framework's structure using different phases and from simple to advanced improvement helped me in how I design and refine processes with an inclusive mindset."* (Process modeller 2). *"The* *innovation, especially at the 3rd level is exciting because it encourages integrating new technologies to support GEDSI goals, which is critical for our organisation."* (Process modeller 3).

5 Discussion and Conclusions

This paper addressed the lack of structured guidance for integrating GEDSI into process design. The GPAF offers a concrete way to assess GEDSI considerations within BPM and provides a dedicated alignment lens that foregrounds how processes affect marginalised groups. Developed through ADR with 37 Philippine government departments, the GPAF and accompanying design principles translate abstract inclusion commitments into assessable process attributes across GEDSI dimensions. Empirically, the GPAF has been used to sensitise process modellers and managers to GEDSI issues, structure their reflections, and guide improvements. Its adoption by 37 departments suggests that even modest changes to templates and governance mechanisms can create practical entry points for embedding GEDSI into everyday processes. For practitioners, it serves as a diagnostic tool that can be layered onto existing BPM practices, while for researchers it provides a practice validated artefact that links inclusive BPM concepts to concrete assessment criteria. The study is contextualised in GEDSI-oriented digital transformation in the Philippine public sector and relies primarily on participants' perceptions rather than long-term outcome measures. Future work should evaluate and adapt the GPAF in other domains/countries. Nonetheless, the findings demonstrate that bringing GEDSI considerations into the mainstream of BPM practice is both feasible and valuable, and they invite continued collaboration between GEDSI specialists, BPM practitioners, and researchers to further institutionalise such approaches.

References

1. Australia Awards: Pre-interview information sheet no 5: Social inclusion – a development principle GEDSI. Tech. rep., Australian Government - DFAT (2023)
2. Baskerville, R.: Conducting action research: high risk and high reward in theory and practice. In: Qualitative Research in IS: Issues and Trends: Issues and Trends, p. 192 (2000)
3. Baumann, L., Seidel, A., Weske, M.: Non-visual process models: how do blind and low-vision users model business processes? In: Gdowska, K., Gómez-López, M.T., Rehse, J. (eds.) BPM 2024 Workshops. LNBIP, vol. 534, pp. 155–167. Springer, Heidelberg (2024)
4. Białas, S., Wróbel, P.: Diversity and inclusion in HR processes in financial sector. In: Gdowska, K., Gómez-López, M.T., Rehse, J.R. (eds.) Business Process Management Workshops. LNBIP, vol. 534, pp. 168–179. Springer, Cham (2025)
5. Binghay, V.C., Cotas Jr, J.P.: Transforming spaces, welcoming places: transgender women's inclusion in the Philippines' BPO industry. Philippine J. Labor Ind. Rel. **39** (2022)
6. CMMI: CMMI acquisition handbook (2025). https://cmmiinstitute.com/products/cmmi/acquisition-handbook. Accessed: 19 May 2025
7. Cole, R., Purao, S., Rossi, M., Sein, M.: Being proactive: where action research meets design research. In: ICIS, p. 27 (2005)

8. De Nicola, A., Guariglia Migliore, M., Wagner, A.: Gender-aware business process management. In: Technology-Driven Transformation: The Future of Work and Organizations, pp. 151–166. Springer, Heidelberg (2025)
9. Department of Economy, Planning, and Development: About ambisyonnatin 2040 (2016). https://2040.depdev.gov.ph/about-ambisyon-natin-2040/. Accessed 22 May 2026
10. DFAT: Gender equality, disability and social inclusion analysis - good practice note (2023). https://www.dfat.gov.au/publications/development/gender-equality-disability-and-social-inclusion-analysis-good-practice-note. Accessed 18 Dec 2024
11. Dumas, M., Rosa, M.L., Mendling, J., Reijers, H.A.: Fundamentals of Business Process Management, 2nd edn. Springer, Heidelberg (2018)
12. Fisher, D.M.: The business process maturity model: a practical approach for identifying opportunities for optimization. Bus. Process Trends **9**(4), 11–15 (2004)
13. Gartner: Maturity assessment for business process improvement leaders: Six phases for successful bpm adoption (2008). https://d4hfzltwt4wv7.cloudfront.net/uploads/files/Maturity_Assessment_for_Business.pdf. Accessed 19 May 2025
14. González Moyano, C., Klessascheck, F., Bala, S., Fahrenkrog-Petersen, S.A., Mendling, J.: Towards nudging in BPM: a human-centric approach for sustainable business processes. In: International Conference on Process Mining, pp. 738–750. Springer, Cham (2024)
15. Gorbacheva, E., Stein, A., Schmiedel, T., Müller, O.: A gender perspective on business process management competences offered on professional online social networks. In: ECIS 2015, Münster, Germany, 26–29 May 2015 (2015)
16. Gregor, S., Chandra Kruse, L., Seidel, S.: Research perspectives: the anatomy of a design principle. J. Assoc. Inf. Syst. **21**(6), 2 (2020)
17. Hammer, M.: The process audit. Harv. Bus. Rev. **85**(4), 111 (2007)
18. Harmon, P.: Evaluating an organizations business process maturity (2004)
19. ILO: Transforming enterprises through diversity and inclusion. International Labour Office, Geneva (2022)
20. Incentivefund.org: Building a more equal and inclusive Papua New Guinea (nd). https://incentivefund.org/assets/docs/general/Incentive-Fund_GEDSI-Factsheet.pdf. Accessed 12 Apr 2025
21. Kedziora, D., Siemon, D., Sharda, R.: 'Leanbotics' case - exploring inclusive synergies between robotic process automation (RPA) and process improvement. In: HICSS 2024, Hawaii, USA, 3–6 January 2024. pp. 1628–1637 (2024)
22. Lee, J., Lee, D., Kang, S.: An overview of the business process maturity model (bpmm). In: Int. Workshop on Process Aware Information Systems, pp. 384–395. Springer, Heidelberg (2007)
23. Looy, A.V., Backer, M.D., Poels, G.: A conceptual framework and classification of capability areas for business process maturity. Enterp. Inf. Syst. **8**(2), 188–224 (2014)
24. Mahendrawathi, E.R.: How can we develop inclusive business process management? In: BPM Workshops. LNBIP, vol. 534, pp. 370–375. Springer, Heidelberg (2024)
25. Maull, R.S., Tranfield, D.R., Maull, W.: Factors characterising the maturity of BPR programmes. Int. J. Oper. Prod. Manag. **23**(6), 596–624 (2003)
26. McCormack, K., et al.: A global investigation of key turning points in business process maturity. Bus. Process. Manag. J. **15**(5), 792–815 (2009)
27. McCormack, K.P.: Business Process Maturity: Theory and Application. Booksurge Publishing, Charleston (2007)
28. Melenovsky, M.J., Sinur, J.: BPM maturity model identifies six phases for successful BPM adoption (2006). https://www.gartner.com/en/documents/497289. Accessed: 19 May 2025
29. Moder, L., Fehrer, T., Röglinger, M.: Design principles for process improvement and innovation systems. Bus. Process Manag. J. (2025)

30. OMG: Business process maturity model (version 1.0) (2008). http://www.omg.org/spec/BPMM/1.0/PDF. Accessed 27 May 2025
31. Palmera:Gender equality, disability & socialinclusion policy (2021). https://palmera.org/wp-content/uploads/2025/02/GEDSI-Policy.pdf. Accessed 12 Apr 2025
32. Paulk, M.C., Curtis, B., Chrissis, M.B., Weber, C.V.: Capability maturity model, version 1.1. IEEE Softw. **10**(4), 18–27 (1993)
33. Röglinger, M., Pöppelbuß, J., Becker, J.: Maturity models in business process management. Bus. Process. Manag. J. **18**(2), 328–346 (2012)
34. Rohloff, M.: Case study and maturity model for business process management implementation. In: BPM Conference, pp. 128–142. Springer, Heidelberg (2009)
35. Rosemann, M., vom Brocke, J.: The six core elements of business process management. In: Handbook on Business Process Management 1, Introduction, Methods, and Information Systems, 2nd edn, pp. 105–122. Springer, Heidelberg (2015)
36. Rosemann, M., De Bruin, T.: Towards a business process management maturity model. In: European Conference on Information Systems, pp. 1–12 (2005)
37. Schoormann, T., Maria, M.D.: Business process pattern for improving social sustainability. Aust. J. Inf. Syst. **28** (2024)
38. Sein, M.K., Henfridsson, O., Purao, S., Rossi, M., Lindgren, R.: Action design research. MIS Q. **35**(1), 37–56 (2011)
39. Szelągowski, M., Berniak-Woźny, J.: BPM challenges, limitations and future development directions – a systematic literature review. BPMJ **30**(2), 505–557 (2024)
40. Tennakoon, T., Gunawardana, K., Premaratne, S.: Gender inequality in digital transformation: evidence from business process management industry in Sri Lanka. Int. J. Gov. Public Policy Anal. **4**(01) (2022)
41. UN: The 17 goals (2015). https://sdgs.un.org/goals. Accessed 12 Apr 2025
42. Vaziri, D.D., DeOliveira, D.: Accessible business process modelling. Int. J. Bus. Econ. Eng. **6**(2), 117–128 (2012)
43. Venable, J., Pries-Heje, J., Baskerville, R.: Feds: a framework for evaluation in design science research. Eur. J. Inf. Syst. **25**(1), 77–89 (2017)
44. vom Brocke, J., Schmiedel, T., Recker, J., Trkman, P., Mertens, W., Viaene, S.: Ten principles of good business process management. BPMJ **20**(4), 530–548 (2014)
45. Weber, C., Curtis, B., Gardiner, T.: Business process maturity model (bpmm) version 1.0. Fecha de consulta 4 (2008)

Differentially Private HeuristicsMiner

Yorck Zisgen(✉) and Agnes Koschmider

Group Business Informatics and Process Analytics, University of Bayreuth, Bayreuth, Germany
{yorck.zisgen,agnes.koschmider}@uni-bayreuth.de

Abstract. Process discovery aims at extracting process models from prior recordings of process executions stored in information systems, called event logs. These process models may reveal sensitive information about unique traces and about the process stakeholders. Although a differentially private process discovery algorithm exists, it does not lend itself well to handling noisy event logs. This paper proposes the Differentially Private Heuristics Miner (DPHM), an algorithm that considers Differential Privacy at the discovery step while handling noise efficiently. We provide proof of the property of Differential Privacy and evaluated our approach on four real-world event logs against three other process discovery algorithms. The evaluation results indicate that unique traces can be protected with a slight loss of utility.

Keywords: Process Mining · Process Discovery · Privacy · Differential Privacy · Control-Flow

1 Introduction

Process mining (PM) is a set of techniques to derive insights into business processes from recorded prior process executions. These insights cover the automated discovery of process models from data, the enhancement of existing process models with information in the analyzed data, verifying process executions conform to a prescribed behavior, or the monitoring of ongoing process instances for the timely intervention to ensure adherence to desired performance indicators such as cost or duration [2].

Comparing process models discovered from event logs at different points in time reveals changes to the process model caused by the traces that occurred in the meantime. With sufficiently low time in between or a real-time process discovery, this will expose the path that a single unique trace took. Privacy in Process Mining aims at minimizing the risk of re-identification of stakeholders [13]. Approaches to prevent leakage of sensitive information can be grouped into *Event Log Sanitization* (ELS), which aims at reducing risk on the event log level, and *Privacy-Preserving Process Mining* (PPPM), which keeps the information in the event log unaltered and demands that the analyzing techniques applied to an unprotected event log must consider privacy [3].

K. Kluza et al. (Eds.): BPM 2026, LNBIP 600, pp. 112–125, 2027.
https://doi.org/10.1007/978-3-032-36710-5_8

While event log sanitization approaches may provide privacy with an acceptable loss of information utility for one process mining task (e.g., removing resources still allows for control flow detection), it may render the event log useless for further downstream tasks (e.g., resource bottleneck analysis). As a consequence, a specialized noisy event log needs to be created for each task, or a single noisy event log is used for all PM tasks, with an unknown amount of utility loss to each respective task. A PPPM technique can be conducted on an unprotected event log, removing the need for multiple event logs and preserving usefulness for different PM tasks, while ensuring privacy.

So far, privacy in process discovery at the process-model level has been neglected in PPPM. Differential Privacy (DP) [4] is a technique that guarantees that the influence (and thereby visibility) of a single element on the analysis of the set that the element is part of cannot exceed ϵ. A differentially private version of the Inductive Miner that protects unique traces in the process model has been recently published [15] (DPIM).

However, the DPIM does not consider noise that can be found in many real-world event logs. Any contradictory or erroneously recorded behavior can lead to a 'flower model' in sub-parts of the discovered process model, diluting the intended insights. HeuristicsMiner-based approaches can consider noise filtering, but so far, no privacy-preserving version exists.

The filtering effect of non-privacy-preserving discovery algorithms, such as the HeuristicsMiner or the Inductive Miner Infrequent, is not sufficient to fundamentally provide privacy. They can disregard all trace variants that do not make up at least x% of the event log, thereby removing unique traces from the analysis. However, real-world examples from a Dutch municipality show its limited protection. The event log consists of 1053 traces, 1049 of which (99.6%) are unique. Depending on the filter setting, this may either not protect unique traces or remove large parts of the event log. This also applies to very small event logs of only a few traces from rarely executed processes.

We propose the Differentially Private HeuristicsMiner (DPHM), which brings privacy guarantees into the HeuristicsMiner, offering a privacy-preserving process discovery algorithm that can be applied to unprotected event logs, can deal with noise that is found in many real-world event logs, and creates process models that protect unique traces by concealing their influence on the discovered process model. Furthermore, we provide an instantiation of the algorithm as a stand-alone Python program[1].

To assess the feasibility and effectiveness of the proposed approach, this paper addresses the following research questions:

RQ1 How can a privacy-preserving process discovery algorithm based on the HeuristicsMiner be designed and shown to satisfy differential privacy?

RQ2 How does the Differentially Private HeuristicsMiner compare to existing process discovery approaches with respect to process model quality?

[1] https://github.com/Process-Analytics-Group/Differentially-Private-HeuristicsMiner.

RQ3 What is the impact of the privacy parameter ϵ on the utility of process models discovered by the Differentially Private HeuristicsMiner?

The remainder of the paper is structured as follows. We discuss related works in Sect. 2. Foundational concepts and mechanisms are explained in Sect. 3. We discuss our approach and present our prototypical instantiation in Sect. 4. In Sect. 5, we evaluate our approach. We conclude in Sect. 6.

2 Related Work

Related to our approach are approaches that consider privacy in process mining, approaches that apply differential privacy to process mining, and approaches concerned with Privacy Preserving Process Discovery.

In [14], Pika et al. motivate the need for privacy in event logs in a medical setting and discuss privacy and utility requirements and various methods of event log anonymization. Nunez von Voigt et al. [13] examine the inherent re-identification risk of stakeholders found within the event log. They find that traces are particularly vulnerable to data re-identification attacks. Bauer et al. propose ELPaaS (Event Log Privacy as a Service), a web-based application for event log sanitization and privacy-preserving process mining queries [3]. Moreover, they distinguish privacy in process mining into two categories: event log sanitization, which aims at reducing privacy risks in the event log, and privatized process mining, which encompasses techniques that retrieve insights from event logs while providing privacy guarantees. Suppressing, aggregating, or noising the underlying event log directly may allow for a subsequent release of the event log, but it may also impede downstream process mining tasks that then suffer from reduced utility.

In [8], the PRIPEL framework is proposed. It anonymizes traces by manipulating timestamps and perturbing the trace variant counts with noise while preserving utility by only selecting trace variants that remain similar to the originals. In [6], the authors sample an event log and add noise to the timestamps of events, thus protecting traces by removing them. Fahrenkrog-Petersen et al. [7] propose a differentially private method that releases an anonymized trace variant distribution. They apply the exponential mechanism and Laplace noise to generate noised directly-follows relations, from which they derive the trace variants. We extend these ideas by also covering unseen correct behavior of the underlying process and protecting new, unseen trace variants by adding new trace variants. An added single new trace variant could introduce a range of unseen activity tuples, allowing an attacker to derive from the resulting process model whether this new trace variant was used in the input.

For privacy-preserving process discovery, Mannhardt et al. [11] introduce a Privacy Engine as the single point of data access for process mining techniques. This allows a restricted set of queries, which return results with added noise. Once a set privacy budget is consumed by multiple queries, further access to the data is denied. Schulze et al. [15] propose the Differentially Private Inductive Miner, which so far is the only process discovery algorithm that introduces a proven guarantee of Differential Privacy into the process discovery step, allowing the underlying event log to be used without prior anonymization that might hinder subsequent downstream tasks. It noises directly-follows relations, creates subsets of observed behavior, and perturbs behavior so new trace variants are generated. It relies on the Inductive Miner and creates sound process models. However, the Inductive Miner is not well-suited to handle noise in event logs and defaults to flower models within the process model if it encounters contradictory or erroneous behavior in the event log. Currently, no Differentially Private process discovery algorithm exists that combines the privacy guarantees of DP with the ability to handle noise in the event log.

Among process discovery algorithms, the HeuristicsMiner and Inductive Miner Infrequent address noise in the event log. The HeuristicsMiner [16] proposes to discover Heuristics Nets based on the dependency of activities on one another. The proposed dependency measure and filters lend themselves well to filtering out weak dependencies between activities and to deciding whether the ground truth behind noisy data is a parallel split or an exclusive choice. We extend this idea by adding Differential Privacy to the HeuristicsMiner. Leemans et al. [9] propose the Inductive Miner Infrequent, which utilizes the Inductive Miner's ability to produce sound process models, and adds an infrequent behavior filter to filter trace count, filter log splitting, and improve operator and cut selection. They focus on noise reduction and improvement of process model quality. Mannhardt et al. [12] suggest an interactive and data-aware approach for the Heuristics Miner, which gave inspiration to the design of our visualization.

3 Preliminaries

In this section, we introduce the underlying concepts and mechanics that are used in our approach.

HeuristicsMiner. The HeuristicsMiner [16] is a process discovery algorithm that utilizes a set of directly-follows relations (e.g. *'A','B': 50; 'B','C': 50*) and creates a Heuristics Net depicting the dependencies of activities within a process. It allows filtering on this net based on a proposed dependency measure. Based on how often activity B follows activity A ($a \Rightarrow_w b$) and how often A follows B ($b \Rightarrow_w a$), the dependency measure is defined as:

$$a \Rightarrow_w b = \left(\frac{|a \Rightarrow_w b| - |b \Rightarrow_w a|}{|a \Rightarrow_w b| + |b \Rightarrow_w a| + 1} \right) \quad (1)$$

It furthermore proposes two filters to identify long loops and to distinguish between AND and XOR splits/joins.

Differential Privacy [4]. Two event logs L, L' are neighboring (written $L \sim L'$) if they differ in at most one trace. Let $\mathcal{R}$ be the set of random variables over some set O. A randomized mechanism $\mathcal{M} : \mathcal{L} \rightarrow \mathcal{R}$ is ϵ-differentially private, with $\epsilon > 0$, if for all $\mathcal{S} \subseteq O$, for all n, and all neighboring event logs L, $L' \in \mathcal{L}$ with $n = |L| = |L'|$:

$$\Pr[M(L) \in \mathcal{S}] \leq \exp(\epsilon) \Pr[\mathcal{M}(L') \in S]. \tag{2}$$

DP is a framework that enables the sharing of sensitive data while preserving the privacy of individual data points. An operation that is differentially private guarantees that the influence of a single additional data point on the outcome of a query on a set is less than ϵ. A lower ϵ therefore represents stronger privacy, while a higher ϵ represents weaker privacy. In our approach, we use DP to measure the influence that a single data point (a trace) has on the outcome of a mechanism $\mathcal{M}$ (Process Discovery) over a set of traces L (an event log).

Laplace Mechanism. $\mathcal{M}_{L,q,\epsilon}$ is ϵ-DP. It takes a database L as input and outputs a noised and Δ_q-sensitivity bounded query q:

$$\mathcal{M}_{L,q,\epsilon}(L) \mapsto q(L) + \text{Lap}(0, \frac{\Delta_q}{\epsilon}). \tag{3}$$

The sensitivity Δ_q describes the impact of a single added data point on the query result: $\text{Lap}(\Delta_q/\epsilon)[o] := \frac{\epsilon}{2\Delta_q} \exp(-|o - \mu|\epsilon/\Delta_q)$. A smaller ϵ leads to a larger deviation from the original value and thereby to stronger privacy (Δ_q/ϵ). In our approach, we use the Laplace Mechanism to add noise to all directly-follows counts and to the comparison of process model metrics (e.g., Fitness) versus the Rejection Sampling threshold.

Laplace Theorem. For a Δ_q-bounded ($\Delta_q \in R_+$) counting query q and an $\epsilon > 0$, the Laplace Mechanism $\mathcal{M}_{L,q,\epsilon}$ is ϵ-DP.

Report Noisy Max. [5] is a $(\epsilon, 0)$-differentially private algorithm that, for a count of m and a set S, returns the m indices of keys of S with the highest noised value. In our approach, we use the Report Noisy Max to find the m most occurring noised directly-follows relations from our set of all directly-follows relations S.

Rejection Sampling. [10] is ϵ-DP with $\epsilon = 2 \cdot \epsilon_1 + \epsilon_0$ if $M(D)$ is ϵ_1-DP. It privately selects output candidates x (e.g., a Heuristics Net or a process model) and threshold values q_a from a mechanism $Q(D)$, given a user-set threshold argument a (e.g., F1-Score or fitness) and a threshold score t_a (e.g., 0.95). We accept the candidate (x, q) if $q_a \geq t_a$. If $q_a < t_a$, we loop for a maximum of T rounds. To obscure whether the Rejection Sampling mechanism failed to meet the threshold value t_a, we introduce a biased coin flip in each iteration, which can terminate the algorithm prematurely. In our approach, we use Rejection Sampling to only consider results that meet a user-specified threshold value (e.g., 0.80) in a user-specified threshold attribute (e.g., F1-Score or simplicity).

Post-Processing Theorem [5]. Let $\mathcal{M} : \mathbb{N}^{|\mathcal{X}|} \rightarrow R$ be a randomized algorithm that is (ε, δ)-differentially private. Let $f : R \rightarrow R'$ be an arbitrary randomized mapping. Then $f \circ \mathcal{M} : \mathbb{N}^{|\mathcal{X}|} \to R'$ is (ε, δ)-differentially private. According to the Post-Processing Theorem, any output of a DP mechanism also fulfills DP. As such, no further privacy leakage can occur during further processing of the output. The Post-Processing Theorem plays an essential role in our proof of the property of DP of our approach (Sect. 5.1).

4 Approach

In this section, we present our approach (Algorithm 1). The DPHM requires as input an event log and a set of parameters as set by the user: the dependency threshold on which to filter the noised heuristics net, filter settings for determining loops, AND or XOR behavior, filtering out traces that account for less than a set percentage of the event log, and minimum counts for both activities and traces. Optionally, rejection sampling attribute (e.g., *'fitness'*) and rejection sampling threshold value (e.g., *'0.80'*) can be given (Algorithm 1, line 1). Moreover, fixed hyperparameters influence the output of the approach (line 2, cf. paragraph Hyperparameters). The output is either both a visualization of a DP heuristics net and a DP Petri Net, or no visualization if the rejection sampling (Sect. 3) failed (line 3).

Algorithm 1. DPHM: Creates a differentially private Heuristics Net and Petri Net.

1: **Input:** EventLog *log*, ϵ, dependency threshold, AND threshold, min activity count, min DFG count, pre-cleaning noise, loop threshold, rejection attribute, rejection threshold
2: **Hyperparameters:** Max Rejection Sampling tries T, DFG noise ϵ_1, rejection noise ϵ_2
3: **Output:** DP Heuristics Net visualization and Petri Net visualization, or $\bot$
4: $L \leftarrow$ IMPORTXES(log)
5: $A \leftarrow$ EXTRACTACTIVITIES(L)
6: $TL, A_s, A_e \leftarrow$ GETTRACELIST(L)
7: $M \leftarrow$ CREATEMATRIX(A)
8: $M \leftarrow$ FILLMATRIX(M, TL)
9: **for** $1, \ldots, T$ **do**
10: flip a γ-biased coin s.t. with prob. γ: **return** $\bot$
11: $(M', S', E') \leftarrow$ NOISEMATRIX(M, A_s, A_e, ϵ)
12: $HN \leftarrow$ DPHEURISTICSNET(M', A, S', E')
13: $HN \leftarrow$ CALCULATE(HN, dependency threshold, AND threshold, min activity count, min DFG count, pre-cleaning noise, loop threshold)
14: **if** $\neg$ ISWEAKLYCONNECTEDDFG($HN.dfg$) **then**
15: **return** $\bot$
16: **end if**
17: $(PN, im, fm) \leftarrow$ CONVERTHEURISTICSNETTOPETRINET(HN)
18: **if** CHECKREJECTION(L, PN, im, fm, rejection attribute, rejection threshold, ϵ) **then**
19: RENDER(HN, PN, im, fm) **return** (HN, PN, im, fm)
20: **return** $\top$
21: **end if**
22: **end for**
23: **return** $\bot$

We import the event log L (line 4) and extract a set of activity labels A from it (line 5). We use L to extract a list of traces (TL), where we prefix each trace with a synthetic start activity and suffix each trace with a synthetic end activity [1], and extract counts of all starting activities A_s, and counts of all ending activities A_e. This is required for a later conversion from Heuristics Net to Petri Net and metric calculation (line 6). We create a matrix M of all possible activity tuples, plus all (start, activity) and all (activity, end) tuples (line 7) and count all observed directly-follows relations, setting unobserved tuples to a count of zero (line 8). We limit the count of each directly-follows tuple to one per trace, to preserve a sensitivity of 1 as dictated by the Laplace Theorem. This prevents a single trace with an arbitrarily large amount of looping behavior to be forcefully picked by the Report Noisy Max (*e.g., a single trace of* $<A,<B,C,B>^n,D>$[1] *with an arbitrarily high* n *could potentially enforce the directly-follows relations ('B', 'C') and ('C', 'B') and thereby expose the trace in the process model*).

We start the rejection sampling for a maximum of T rounds (line 9) with a chance of γ to terminate (line 10). We apply Laplace noise based on ϵ and derive a noised matrix of directly-follows relations M' and noised counts of starting S' and ending E' activities (line 11). We also create subsets of starting and ending activities and all directly-follows relations via Report Noisy Max. On these subsets and the user parameters, we create and filter the Heuristics Net (lines 12, 13). We adjust the dependency measure to handle noised counts, which could lead to divisions by zero:

$$\mathrm{dep}_w(a,b) = \begin{cases} 0, & \text{if reflexive and } |a \Rightarrow_w a| = -1. \\ 0, & \text{if } |a \Rightarrow_w b| + |b \Rightarrow_w a| = -1, \\ \dfrac{|a \Rightarrow_w b| - |b \Rightarrow_w a|}{|a \Rightarrow_w b| + |b \Rightarrow_w a| + 1}, & \text{if } a \neq b \text{ and } |a \Rightarrow_w b| + |b \Rightarrow_w a| \neq -1, \\ \dfrac{|a \Rightarrow_w a|}{|a \Rightarrow_w a| + 1}, & \text{if reflexive and } |a \Rightarrow_w a| \neq -1. \end{cases} \tag{4}$$

We check for weak connectivity of the noised Heuristics Net to reject disconnected graphs (lines 14–16), then convert the Heuristics Net into a Petri Net, initial marking, and final marking (*PN, im, fm*, line 17). We check for the user-specified rejection sampling criteria and noise the comparison (line 18). If the calculated metric passes the threshold value, we visualize the Heuristics Net and Petri Net and end the algorithm with *True* (lines 19–21). If the threshold value is not reached, we repeat the process from the unnoised matrix (line 9). If the maximum tries T are consumed without providing a valid result (line 22), we terminate with *False* (line 23).

Hyperparameters. Our approach uses four hyperparameters: T, γ, e_0, and e_1. T is the maximum number of allowed tries to produce a valid result that is a Petri Net and meets the rejection sampling criteria. It is currently set at 500. Increasing the value leads to a higher chance of generating a result, but potentially longer runtime. Reducing the value has the opposite effect. γ is the chance to terminate the rejection sampling with a *False* result, despite the current tries not yet having reached T. This is to prevent information leakage about whether the approach was able to produce a viable result or not. It is currently set to 1%. Increasing the value will lead to more runs ending prematurely without a result and a shorter runtime. e_0 and e_1 partition the added noise. Per the Sequential Composition Theorem, the total privacy cost of multiple operations on the same dataset is equal to the sum of the individual costs. Therefore, to keep the approach ϵ-DP, we need to split ϵ across all noise-adding steps such that the sum is equal to 1ϵ. Currently, e_0 equals 0.9 and e_1 equals 0.1. e_0 is used to add noise to the directly-follows relations. e_1 is used to noise the rejection threshold comparison and provide deniability. Adjusting these two values will change the balance of where in the approach noise is added.

Implementation. We implemented our algorithm as a stand-alone software application (cf. footnote 1) using Python v3.12.9, with a graphical user interface based on tkinter. It visualizes both the Dependency Graph and the Petri Net in real time, while providing filters for the dependency measure, loop-length, and AND/XOR detection (as proposed by [16]), minimum counts for traces of a variant and activity occurrences (as proposed by the PM4PY implementation of the HeuristicsMiner), and settings for the rejection sampling and privacy strength. It accepts event logs in the .XES format and can output process models as .JPG or .PNG files.

5 Evaluation

We evaluate our approach as follows. We demonstrate how our approach holds the property of Differential Privacy (Sect. 5.1). Under weak privacy, our approach converges on the true process models (Sect. 5.2). Under high privacy, a trade-off between privacy and utility is widely acknowledged. We measure the utility loss under high privacy (Sect. 5.3) and benchmark our approach against three other discovery algorithms (Sect. 5.4). Lastly, we discuss the metrics used for this evaluation (Sect. 5.5), before returning to the research questions (Sect. 5.6).

We evaluate our approach on four real-world event logs from different domains (cf. Table 1). They come from the car manufacturing incidents, credit loan applications, sepsis treatment in a hospital, and building permit applications domains. The logs cover a wide range of complexity (13 to 356 different activities, 11 to 3,565 directly-follows-relations), volume (15,214 to 1,160,405 events, 1,050 to 31,509 traces), and proportion of unique traces among all traces (16.08% to 99.34%).

5.1 Proof of Differential Privacy

The DPHM fulfills Differential Privacy. We base our proof on the Post-Processing Theorem [5], which states, that any output of a DP mechanism is also DP. Therefore, we need to prove that all operations conducted on the input set are

Table 1. Characteristics of the event logs used for evaluation, showing event log name, activities, amount of directly-follows relations (DFR), events, total traces in the log, trace variants in the log, amount of unique traces, and proportion of unique traces among all traces in the log [13].

Event Log	Act.	DFR	Events	Total Traces	Trace Variants	Unique Traces	Trace Uniqueness
Incidents	13	11	65,533	7,554	1,511	1,215	16.08%
Credit Loan	26	178	1,160,405	31,509	15,930	13,907	44.14%
Sepsis	16	115	15,214	1,050	846	784	74.67%
Building Permits	356	3,565	47,293	1,053	1,049	1,046	99.34%

DP themselves, and that no operations were used that are not DP. Then, no information leakage can occur.

We noise the original directly-follows counts using the Laplace Mechanism [5], which is DP. All further computations are conducted on these noised counts. We extract a randomized amount of starting and ending activities from this set via the Report Noisy Max algorithm, which is also DP [5]. We then create a valid subset of the noised directly-follows counts with the Report Noisy Max [5]. The amount of counts considered is randomized around the original amount of directly-follows relations. We check for weak connectivity, rejecting all disconnected graphs. No information except the reject is being used. We calculate dependency measures based on the noised directly-follows counts. According to the Post-Processing Theorem, this is DP, because the noised counts have been generated via a DP algorithm (Laplace Mechanism).

We use the noised starting and noised ending activities and the selected set of noised directly-follows counts to generate a Petri Net, on which we can calculate metrics for the Rejection Sampling. The Rejection Sampling is DP [10]. We also noise the comparison of model metric against the user-defined sampling threshold. This noising is again done via the Laplace Mechanism, which is DP [5]. Therefore, all operations conducted on the underlying data fulfill DP. No other operations are being conducted, therefore adhering to the Post-Processing Theorem. Since the Post-Processing Theorem holds for our approach, the DPHM is DP. □

5.2 Evaluation of Correctness

We evaluate the correctness of our approach by benchmarking it against the Python implementation of the HeuristicsMiner. We measure fitness, precision, simplicity, generalization, and F1-Score of DPHM versus HM on four real-world event logs under low privacy (Fig. 1). DPHM uses randomized noise addition, therefore we evaluate based on the average value of 20 runs of DPHM for each event log.

For the evaluation, we select an ϵ of 100,000, which is an extremely weak privacy setting. For bare comparison, we set minimum counts to 1 and filter

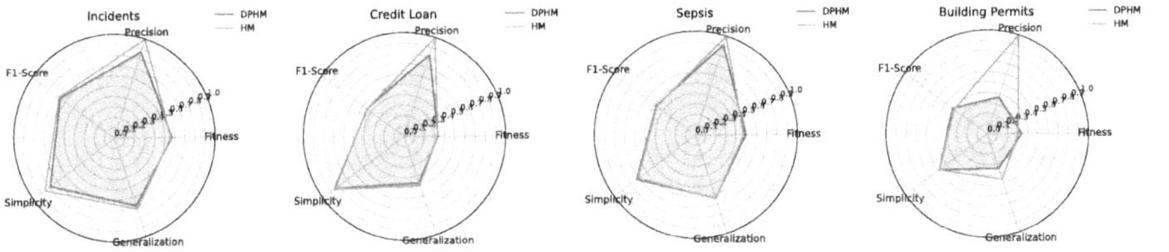

(a) Incidents (b) Credit Loan (c) Sepsis (d) Building Permits

Fig. 1. Comparison of DPHM (average of 20 runs) and HM process model metrics across four event logs.

settings to 0. In a process discovery task, one would use the noise filter settings accordingly and likely get better metrics. We find that on three event logs (Incidents, Credit Loan, and Sepsis), our approach converges on the results of the non-privacy-preserving discovery algorithm, based on the five metrics. For the Building Permits event log (Fig. 1d), our approach achieves a lower precision than the non-DP HeuristicsMiner. Even with minimal noise, on an event log with 356 unique activities and 3,565 directly-follows relations, the room for deviation is much larger. When noise turns exclusive choices or sequential behavior into parallel activities or looping behavior, this allows for much more potential behavior not seen in the original event log, and therefore leads to reduced precision.

5.3 Evaluation of Utility Loss

We assume the predominant task in process discovery to be receiving a process model that faithfully represents the event log on which the process discovery algorithm has been applied to. Therefore, a faithful representation would represent the entire event log (as expressed by the *'fitness'* metric) and as little more than the event log (as expressed by the *'precision'* metric) as possible. The *'F1-Score'* is the harmonic mean of both metrics and therefore a good representation of the quality of the resulting process model with regard to the original process mining intent.

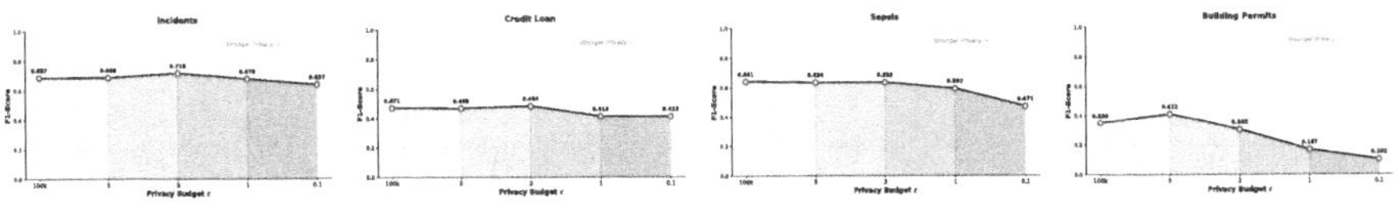

(a) Incidents (b) Credit Loan (c) Sepsis (d) Building Permits

Fig. 2. Sensitivity analysis of F1-Score. F1-Score remains mostly stable from weak to medium privacy, with a decline only in high privacy settings. Loss ratio is stronger in complex event logs. Average values of ten runs.

Figure 2 shows the changes to the F1-Score of four real-world event logs over five different ϵ-values (100,000, 5, 3, 1, 0.1). A lower ϵ-value indicates more noise and a higher privacy guarantee. More noised directly-follows relations mean more deviations of the model from the event log. As such, less traces of the event log can be replayed (lower fitness) due to removed directly-follows relations, while more unobserved behavior is possible (lower precision) from added relations. This results in a gradually declining F1-Score over increasing privacy/decreasing ϵ. However, contrary effects can occur [15]. The HeuristicsMiner family does not guarantee producing sound process models. Adding relations via noise may alleviate deadlocks and allow for a higher fitness. Removing the directly-follows relation (B, A) from the trace sequence <A, B, A> turns a loop into a sequence, thus allowing fewer unseen behavior (such as <A, B, A, B, A>) and therefore increases precision.

For the incidents event log, we observe a decline in F1-Score from 0.6873 for $\epsilon = 100{,}000$, to 0.6373 for $\epsilon = 0.1$. With the credit loan event log, F1-Score drops from 0.4714 to 0.4131. In case of the sepsis event log, we observe a reduction from 0.6411 down to 0.4709. The loss of F1-Score is highest with the buildings permits log, where the F1-Score drops from 0.3504 to 0.1021. Overall, we observe a small loss of utility for most event logs, and minor loss of utility for complex event logs.

Please note that these values have been taken without noise filters. Different event logs contain different amounts of noise, requiring different noise filter settings for individually optimal results. However, different settings across the event logs render a comparison biased.

5.4 Benchmark

We benchmark our approach against three process discovery algorithms: HeuristicsMiner (HM) [16], Differentially Private Inductive Miner (DPIM) [15], and Inductive Miner Infrequent (IMi) [9] on four real-world event logs. As with Fig. 1, we set all filters to 0, and all minimum counts to 1. We evaluate on ϵ-values of 100,000 for weak and 0.1 for strong privacy (cf. Table 2).

Table 2. F1-score comparison of DPHM, HM, DPIM, and IMi for $\epsilon = 100000$ (weak privacy) and $\epsilon = 0.1$ (strong privacy).

$\epsilon = 100000$					$\epsilon = 0.1$				
Event Log	DPHM	HM	DPIM	IMi	Event Log	DPHM	HM	DPIM	IMi
Incidents	0.6873	0.7196	0.7698	0.9319	Incidents	0.6373	0.7196	0.8775	0.9319
Credit Loan	0.4714	0.4963	0.1884	0.6226	Credit Loan	0.4131	0.4963	0.1521	0.6226
Sepsis	0.6411	0.6442	0.3198	0.6455	Sepsis	0.4709	0.6442	0.4769	0.6455
Building P.	0.3504	0.4611	0.0127	0.0311	Building P.	0.1021	0.4611	0.0342	0.0311

For weak privacy, DPHM results are close to the non-DP version of HM. Notably, it outperforms the only other DP discovery algorithm on three out of four event logs. For strong privacy, the added noise reduces utility slightly to below the non-DP HM on three event logs, and significantly on one event log. It outperforms the DPIM on two event logs and is almost equal on a third event log. Please note that the benchmarking values published by [15] were taken with a Rejection Sampling on Fitness with a threshold value of 0.95, leading to a naturally high F1-Score. On the DPHM, we did not use the Rejection Sampling for computational time reasons. This biases a direct comparison between DPHM and DPIM heavily against DPHM.

In conclusion, with comparable to superior results, the DPHM adds the benefit of filtering against the DPIM, and the benefit of privacy against the HM and IMi.

5.5 Discussion of Metrics

Fitness, precision, simplicity, generalization, and the derived F1-Score are generally used to describe properties of the relation between event log and process model, or properties of the process model itself [2]. However, process discovery in itself is a tradeoff between these metrics [9]. Filtering is a tradeoff between a more condensed and possibly more correct process model (noise-filtering) and the loss of information (outlier-filtering). Privacy is a tradeoff between protecting the individual data point [4] and the loss of granularity on the information. A process model that scores well on all metrics can hardly be found [9].

Furthermore, low metric values do not necessarily mean a poor process discovery result, but can come from the structure of the underlying data. For example, applying HeuristicsMiner, InductiveMiner Infrequent, DPIM, and DPHM on the incidents event log yields simplicity values ranging from 0.6574 to 0.7333, with an average of 0.6913. Applying the same discovery algorithms with the same parameters on the building permits log yields simplicity values from 0.4526 to 0.6894, with an average of 0.5403.

As such, metric values should be assessed within the context of the underlying event log and the specific process discovery task.

5.6 Addressing the Research Questions

We conclude the evaluation by coming back to the initial research questions.

For **RQ1** *(How can a privacy-preserving process discovery algorithm based on the HeuristicsMiner be designed and shown to satisfy differential privacy?)*, we provide our approach and implementation (Sect. 4) and the proof of the property of DP (Sect. 5.1).

We answer **RQ2** *(How does the Differentially Private HeuristicsMiner perform compared to existing process discovery approaches with respect to process model quality?)* in Fig. 1 and in Sect. 5.2, where we find that both under weak and strong privacy, it performs comparably to other process discovery algorithms.

To answer **RQ3** *(What is the impact of the privacy parameter ϵ on the utility of process models discovered by Differentially Private HeuristicsMiner?)*, we conclude that the utility of a process model, as assumed to be a faithful representation of the event log and measured by the F1-Score, does not degrade much by introducing weak or medium privacy. Under strong privacy, a slight tradeoff can be observed. The tradeoff scales with the complexity of the event log; the more unique activities and the more directly-follows relations are in the event log, the more room for deviation exists, and the larger the loss of utility is in the process model.

6 Conclusion

Process discovery is a set of techniques to extract a process model from an event log, which relies on prior process executions recorded in information systems. However, it has been shown that a process model carries information leakage about unique traces. A Privacy-Preserving Process Mining algorithm that introduces Differential Privacy to the process discovery step based on the Inductive Miner has recently been proposed. However, it is not well-suited to handling noise usually present in real-world event logs, which can significantly impact the quality and reliability of the discovered model. In this paper, we propose the Differentially Private HeuristicsMiner (DPHM), which produces process models under Differential Privacy, protects unique traces, and can efficiently handle noise in event logs. We provide a formal proof for the DP property of our approach and a prototype instantiation. We evaluate our approach against three other process discovery algorithms on four real-world event logs. The results indicate that our approach approximates the correct process model under weak privacy, and can achieve high privacy with a slight loss of utility as expressed in the F1-Score.

References

1. Van Der Aalst, W.M.P.: Discovering directly-follows complete petri nets from event data. In: Jansen, N., Stoelinga, M., Van Den Bos, P. (eds.) A Journey from Process Algebra via Timed Automata to Model Learning. LNCS, vol. 13560, pp. 539–558. Springer, Cham (2022). https://doi.org/10.1007/978-3-031-15629-8_29.
2. van der Aalst, W.M.P.: Process Mining: Data Science in Action, 2nd edn. Springer, Heidelberg (2016). https://doi.org/10.1007/978-3-662-49851-4
3. Bauer, M., Fahrenkrog-Petersen, S.A., Koschmider, A., Mannhardt, F., van der Aa, H., Weidlich, M.: ELPaaS: event log privacy as a service. In: Depaire, B., De Smedt, J., Dumas, M. (eds.) BPM 2019 Dissertation Award, Doctoral Consortium, and Demonstration Track (BPMT 2019), pp. 159–163. CEUR Workshop Proceedings. CEUR-WS.org (2019)
4. Dwork, C., McSherry, F., Nissim, K., Smith, A.: Calibrating noise to sensitivity in private data analysis. In: Halevi, S., Rabin, T. (eds.) TCC 2006. LNCS, vol. 3876, pp. 265–284. Springer, Heidelberg (2006). https://doi.org/10.1007/11681878_14
5. Dwork, C., Roth, A.: The algorithmic foundations of differential privacy. Found. Trends Theor. Comput. Sci. **9**(3–4), 211–487 (2014). https://doi.org/10.1561/0400000042

6. Elkoumy, G., Pankova, A., Dumas, M.: Differentially private release of event logs for process mining. Inf. Syst. **115**, 102161 (2023). https://doi.org/10.1016/j.is.2022.102161
7. Fahrenkrog-Petersen, S.A., Kabierski, M., van der Aa, H., Weidlich, M.: Semantics-aware mechanisms for control-flow anonymization in process mining. Inf. Syst. **114**, 102169 (2023)
8. Fahrenkrog-Petersen, S.A., van der Aa, H., Weidlich, M.: PRIPEL: privacy-preserving event log publishing including contextual information. In: Fahland, D., Ghidini, C., Becker, J., Dumas, M. (eds.) BPM 2020. LNCS, vol. 12168, pp. 111–128. Springer, Cham (2020). https://doi.org/10.1007/978-3-030-58666-9_7
9. Leemans, S.J.J., Fahland, D., van der Aalst, W.M.P.: Discovering block-structured process models from event logs containing infrequent behaviour. In: Lohmann, N., Song, M., Wohed, P. (eds.) BPM 2013. LNBIP, vol. 171, pp. 66–78. Springer, Cham (2014). https://doi.org/10.1007/978-3-319-06257-0_6
10. Liu, J., Talwar, K.: Private selection from private candidates. In: Proceedings of the 51st Annual ACM SIGACT Symposium on Theory of Computing (STOC 2019), pp. 298–309. Association for Computing Machinery, New York (2019). https://doi.org/10.1145/3313276.3316377
11. Mannhardt, F., Koschmider, A., Baracaldo, N., et al.: Privacy-preserving process mining. Bus. Inf. Syst. Eng. **61**, 595–614 (2019). https://doi.org/10.1007/s12599-019-00613-3
12. Mannhardt, F., de Leoni, M., Reijers, H.A.: Heuristic mining revamped: an interactive, data-aware, and conformance-aware miner. In: Clarisó, R., et al. (eds.), Proceedings of the BPM Demo Track and BPM Dissertation Award, co-located with 15th International Conference on Business Process Management (BPM 2017), Barcelona, Spain, 13 September 2017, vol. 1920, pp. 1–5. CEUR Workshop Proceedings. CEUR-WS.org (2017)
13. Nuñez von Voigt, S., et al.: Quantifying the re-identification risk of event logs for process mining. In: Dustdar, S., Yu, E., Salinesi, C., Rieu, D., Pant, V. (eds.) CAiSE 2020. LNCS, vol. 12127, pp. 252–267. Springer, Cham (2020). https://doi.org/10.1007/978-3-030-49435-3_16
14. Pika, A., Wynn, M.T., Budiono, S., ter Hofstede, A.H.M., van der Aalst, W.M.P., Reijers, H.A.: Towards privacy-preserving process mining in healthcare. In: Di Francescomarino, C., Dijkman, R., Zdun, U. (eds.) BPM 2019. LNBIP, vol. 362, pp. 483–495. Springer, Cham (2019). https://doi.org/10.1007/978-3-030-37453-2_39
15. Schulze, M., Zisgen, Y., Kirschte, M., Mohammadi, E., Koschmider, A.: Differentially private inductive miner. In: 2024 6th International Conference on Process Mining (ICPM), Kgs. Lyngby, Denmark, pp. 89–96 (2024). https://doi.org/10.1109/ICPM63005.2024.10680684.
16. Weijters, A.J.M.M., van der Aalst, W.M.P., Alves De Medeiros, A.K.: Process mining with the HeuristicsMiner algorithm. In: BETA publicatie: Working Papers, vol. 166. Technische Universiteit Eindhoven (2006)

Process Technology Forum

Fine-Grained Control of Blockchain-Based Business Process Instance Deployment

Hassan Atwi(✉) and Cesare Pautasso

Università della Svizzera Italiana, Lugano, Switzerland
{Hassan.Atwi,Cesare.Pautasso}@usi.ch

Abstract. The integration of blockchain technology into Business Process Management (BPM) has attracted significant attention due to the transparency and immutability offered by distributed ledger technologies. Numerous studies have proposed blockchain-based workflow engine architectures to execute business processes directly on-chain. These adopt distinct design decisions for process instance deployment, resulting in a diverse architectural variants with significantly different gas cost. In this paper we propose to give fine-grained control to process modelers regarding how the control flow logic and process data are distributed across one or multiple smart contracts. To demonstrate the practical applicability of the approach, we compare two deployment configurations of the same blockchain-based crowdfunding process and examine the resulting deployment and execution costs.

1 Introduction

Blockchain's data immutability and transparency [2] enable trust in distributed settings where centralized authority is either infeasible or undesirable [26]. This is especially relevant for collaborative business processes [35] that follow a horizontal organizational structure [31]. In such processes, control and management responsibilities are distributed across multiple parties rather than being concentrated under a single central authority. Blockchain and smart contracts [6] naturally fit in such settings by enabling interactions between participants to be logged on-chain, thereby mitigating concerns around trust, auditability, and data integrity. Numerous works in the literature [13,15,16,20,23,29] have described blockchain-enabled workflow engines designed to execute business processes on-chain. These take a business process artifact (such as a BPMN [1] model) and employ Model-Driven Engineering (MDE) techniques [5] to transform the process into a set of smart contracts [9]. These smart contracts encode the process logic as executable code, which is then executed on-chain as a result of participant interactions.

Designing blockchain-based workflow engines involves multiple architectural components, each requiring design decisions that balance system requirements with blockchain's capabilities [33]. Smart contract design, in particular, is critical for building flexible and effective systems [43]. Unlike traditional systems,

K. Kluza et al. (Eds.): BPM 2026, LNBIP 600, pp. 129–145, 2027.
https://doi.org/10.1007/978-3-032-36710-5_9

the immutable nature of smart contracts poses unique maintainability challenges [39]. To address these challenges, blockchain-enabled workflow engines leverage established smart contract design patterns (e.g., proxy contracts [42] and logic/data segregation [36,38]) to ensure both flexibility and performance.

There are many strategies to map business process models to smart contracts [30]. These differ primarily in how process instantiation is bound with smart contract deployment. In addition, they vary in their treatment of data and control flow separation: some implementations combine these concerns within monolithic contracts, while others decompose them into separate ones. These architectural decisions lead to considerable impact on execution performance, deployment and operating costs, and long term maintainability.

Unlike existing process engines, which apply their own deployment architecture uniformly across all process instances they execute, we propose an approach that makes it possible for process modelers to precisely select how to deploy process logic and manage process data on the blockchain, and – if necessary – to change the deployment configuration for each process fragment. To ensure validity and consistency of the resulting architecture, we constrain the possible deployment options within a limited design space defined by three fundamental design dimensions, from which six possible architecture variants can be derived from their combinatorial space. To evaluate the impact on deployment and execution gas costs, we show how the same process can be deployed in two concrete configurations and discuss their scenarios of applicability and trade-offs.

The rest of this paper is organized as follows: Sect. 2 motivates the work with a blockchain-based process execution scenario, while Sect. 3 introduces relevant design patterns. Section 4 presents the architectural design space. Section 5 outlines the two configurations of the use case scenario and their cost efficiency. In Sect. 6 we reflect on process qualities such as governance, upgradeability, and modularity. Section 7 discuss related work. Finally, Sect. 8 summarizes the contributions and outlines directions for future research.

2 Motivation: Blockchain-Based Business Process Scenario

Executing business processes on the blockchain requires architectural planning since blockchain deployments involve decisions about how to represent process logic, store process data, and manage the interaction between them through smart contracts. A key challenge lies in how to map business process elements such as BPMN pool lanes and data stores into blockchain smart contracts.

To reason about how to take these decisions, we consider the collaborative crowdfunding process illustrated in Fig. 1. The main process is modeled in the pool crowdfunding which represents the process that is deployed and executed on the blockchain. The process involves a proposer submitting a proposal. The judges then evaluate the proposal and record a decision on chain. If approved, contributors submit funds to the campaign. Once the campaign goal is reached, a judge authorizes the release of the escrowed funds.

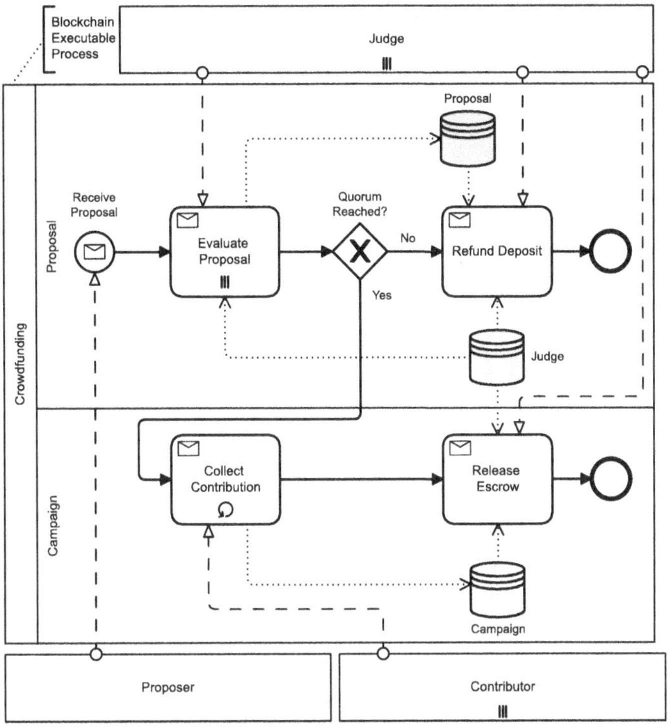

Fig. 1. Crowdfunding collaborative process use case scenario (BPMN [1]).

The empty participant pools (*Proposer*, *Judge*, *Contributor*) represent external actors that interact with the process through blockchain transactions. The *Judge* participant lane represents a set of predefined judge addresses, whereas the other participants are considered open, meaning that any participant on the blockchain network can assume these roles. When generating the smart contract code, message-receiving tasks correspond to externally callable functions, while sequence flows, gateways, script tasks, and data flows are implemented as internal contract logic.

Deploying this collaborative process to the blockchain involves several design decisions. For each lane, one must decide whether to deploy a new contract per instance (e.g., a separate contract for each proposal or campaign), or to use a shared contract for all instances. Likewise, the data stores can be embedded within the existing contracts corresponding to each lane or represented as independent data storage contracts. In this paper, we show how modelers can take different decisions to control how each model fragment can be deployed to be executed on-chain. The deployment decision affect the flexibility of the process, and make it possible to reflect different governance models over each fragment. They also help to manage deployment and operational costs and determine how deployment costs are allocated among participants.

3 Background: Smart Contract Design Patterns

Our approach to implement blockchain-enabled business processes effectively requires leveraging the following established smart contract design patterns.

Segregation of Logic and Data: The Data Segregation pattern [36,38] separates a smart contract's logic from its data by organizing them into distinct contracts. The logic contract maintains only the functional components, such as state machines and encoded conditions, while the data contract implements the data schema that the logic contract reads from and writes to during execution. This separation promotes modularity.

Proxy Contracts and Upgradeability: A proxy contract is a smart contract that serves as a stable interface to an underlying implementation contract, enabling upgradeability without migrating data [42]. The proxy (also known as the Satellite [36]) delegates all function calls to the main implementation while retaining the contract's data. This allows developers to update contracts by simply changing the reference to a new implementation. In this work, we adopt the UUPS (Universal Upgradeable Proxy Standard) [4] pattern developed by OpenZeppelin [25] due to its cost efficiency in comparison to other proxy patterns.

4 Design Space

To define the valid options for blockchain-based process instance deployment, in this section we introduce an architectural design space based on three binary design dimensions that characterize how process logic and data are structured and deployed on-chain (Fig. 2). Combining the three binary dimensions, we derive six valid architectural variants out of eight possible combinations (Table 1).

4.1 Design Dimensions

Logic Scope: Logic scope defines the sequence flow within a business process and is mapped to a smart contract containing the conditions that govern how the process is executed in compliance with the defined model diagram. The contract also enforces authorization for performing state transitions during the execution of each activity. In addition to the sequence flow, the process state is stored and managed by the same contract instance.

In our design space, the logic scope can either be *isolated* per instance or *shared* among instances. In the *shared* setting, a single smart contract contains the logic required to execute all process instances belonging to the same process model. No instance specific upgrades of the logic are possible, and any upgrade affects all active and future instances. In the *isolated* setting, each process instance has its own dedicated smart contract. This alternative is more flexible as it allows process upgrades to be performed on an instance basis, with each instance managing its logic autonomously through its own smart contract.

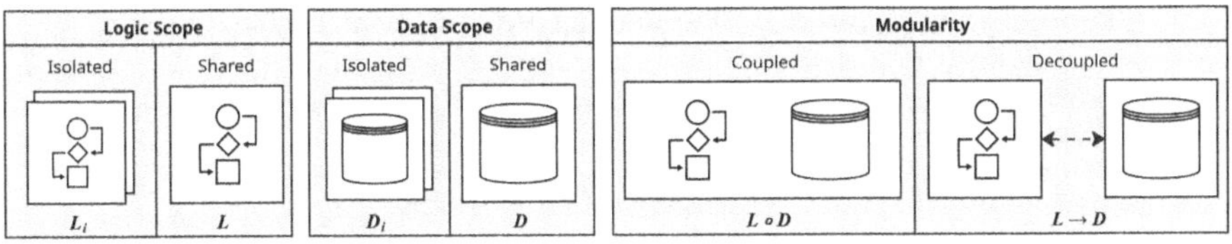

Fig. 2. Design Space Overview: Logic and Data can be shared across all process instances or isolated for each instance. They also can be decoupled across separate smart contracts, or managed by one same contract.

Data Scope: The data scope refers to how data can be imported, exported, or modified during process execution. The data is stored on-chain to ensure auditability, transparency, and integrity of the business process execution state. The process logic accesses and manipulates these data elements as the workflow progresses. In our work, the data is mapped to a smart contract containing the data attributes together with their corresponding getters and setters.

Like the logic scope, also the data scope can either be isolated per instance or shared among instances. In a *shared* data scope, all process instances use the same data contract, providing a uniform data structure. Any update or modification to the data contract applies to all instances. This model is suitable for processes in which instances need to exchange or share data. In an *isolated* data scope, each process instance has its own dedicated data contract.

Modularity: Modularity determines whether logic and data are combined within a single smart contract (*coupled*) or separated into distinct contracts (*decoupled*). In the *coupled* setting, process logic and data reside within the same smart contract. This simplifies deployment and benefits from reduced intra-contract communication costs. It is well suited for static processes where both the logic and the data schema remain stable and uniform. Any upgrade to the logic or data requires redeploying the entire contract. In the *decoupled* setting, control logic and data are implemented in separate contracts that interact through external calls. This separation allows logic and data to evolve independently at the cost of a higher inter-contract communication overhead.

4.2 Architectural Formalization

To express and reason about the architectural design space, in this Section we introduce a formal notation for describing architectural configurations. The logic scope, denoted L^X, corresponds directly to a BPMN lane identified by X. The data scope, denoted D^Y, corresponds directly to a BPMN data store identified by Y. Each data store belongs to exactly one lane. We define the ownership function $\lambda(Y) = X$ to indicate that data store Y belongs to lane X. Each scope is deployed in one of two ways:

Table 1. Six Architecture Variants for Process Instance Deployment.

ID	Logic Scope	Data Scope	Modularity	Complexity (C, E)	Configuration	Description
				Isolated Instance Architectures		
V1	Isolated	Isolated	Coupled	$(N, 0)$	$\bigcup_{i=1}^{N}(L_i \circ D_i)$	**Isolated Monolith**: Fully autonomous instances with embedded logic and data. Maximum isolation, minimal inter-contract communication.
V2	Isolated	Isolated	Decoupled	$(2N, N)$	$\bigcup_{i=1}^{N}(L_i \rightarrow D_i)$	**Isolated Modular**: Independent logic-data separation per instance. Supports instance-level component flexibility.
				Shared Instance Architectures		
V3	Shared	Shared	Coupled	$(1, 0)$	$(L \circ D)$	**Shared Monolith**: Centralized singleton deployment. Minimal deployment cost, uniform governance across all instances.
V4	Shared	Shared	Decoupled	$(2, 1)$	$(L \rightarrow D)$	**Shared Modular**: Global logic-data separation. Enables independent evolution of shared components.
				Hybrid Architectures		
V5	Isolated	Shared	Decoupled	$(N+1, N)$	$\bigcup_{i=1}^{N}(L_i) \rightarrow D$	**Federated Logic**: Instance-specific logic with centralized data store. Ensures uniform data representation with logic isolation.
V6	Shared	Isolated	Decoupled	$(1+N, N)$	$L \rightarrow \bigcup_{i=1}^{N}(D_i)$	**Federated Data**: Centralized logic with distributed data. Ensures uniform execution logic with data isolation.

C = Contract count, E = Inter-contract edge count, N = Number of process instances

- **Shared**: a single contract serves all process instances, written L^X or D^Y.
- **Isolated**: a separate contract is deployed per process instance i, written $\bigcup_{i=1}^{N} L_i^X$ or $\bigcup_{i=1}^{N} D_i^Y$, denoting N contracts, one per instance.

An architectural configuration is defined as the tuple $\mathcal{C} = (\mathcal{L}, \mathcal{D}, \mathcal{R})$, where $\mathcal{L}$ is the set of logic scopes, $\mathcal{D}$ is the set of data scopes, and $\mathcal{R} \subseteq (\mathcal{L} \cup \mathcal{D}) \times (\mathcal{L} \cup \mathcal{D})$ is the set of relationships between them. Two relationship types are defined:

- **Coupled** ($\circ$) denotes coupled deployment of a logic and data scope within a single contract, subject to two constraints:
  1. *Ownership constraint*: $L^X \circ D^Y \in \mathcal{R}$ requires $\lambda(Y) = X$, i.e., data store Y must belong to lane X.
  2. *Scope consistency constraint*: coupled components must share the same deployment scope, either both shared $(L^X \circ D^Y)$ or both isolated at the same instance index $(L_i^X \circ D_i^Y)$.
- **Decoupled** ($\rightarrow$) denotes that a logic scope delegates data access to a separately deployed data contract: $L^X \rightarrow D^Y \in \mathcal{R}$. Here the previous constraints do not apply.

4.3 Architecture Variants

Based on the aforementioned design dimensions and the formalization of architectural configurations, we now discuss the possible six variants for structuring blockchain-enabled process management (Table 1). Logic and data components may be deployed with either *shared* or *isolated* scope, and can be either *coupled* or *decoupled*.

Isolated Monolith (V1): In this architectural variant, each process instance is deployed as a self-contained smart contract encapsulating both logic and data ($L_i \circ D_i$). This configuration provides instance-level autonomy and enables independent lifecycle management per instance. Upgrades can be performed independently on each instance without affecting others. To run N instances, the architecture incurs N contracts deployments with zero inter-contract communication, resulting in linear complexity and no runtime communication overhead. However, the tight coupling of logic and data eliminates component reusability and requires to redeploy both logic and data components for any modification.

Isolated Modular (V2): Logic and data are decoupled into separate contracts, deployed per instance ($L_i \rightarrow D_i$). This separation of concerns enables independent upgrades of logic and data components while preserving instance-level isolation. The architecture requires $2N$ contracts with N communication edges between each pair of logic and data contracts, representing the highest structural complexity among all variants. This structure enables fine-grained upgradeability and selective component reconfiguration at the instance level.

Shared Monolith (V3): A single monolithic contract contains both logic and data for all process instances ($L \circ D$). This centralized deployment enforces uniform behavior and a homogeneous data schema across all instances. With minimal structural complexity, this variant minimizes deployment costs and simplifies instance management. This variant is beneficial in case all instances must comply with uniform rules and access the same data. The shared and coupled structure eliminates the possibility of instance-specific customization and independent evolution.

Shared Modular (V4): In this variant, logic and data are decoupled into separate shared contracts ($L \rightarrow D$), with the logic component delegating to the data component via a single inter-contract edge. This architecture maintains centralized deployment while enabling independent upgrades of logic and data components. The minimal complexity provides a balance between modularity and deployment overhead for globally governed architectures.

Table 2. Arch. Variants in Blockchain Process Engines

		Architectural Variant					
Framework	**Year**	**V1**	**V2**	**V3**	**V4**	**V5**	**V6**
Weber et al. [34]	2016	+					
García-Bañuelos. [10]	2017	+					
López-Pintado et al. [20]	2018	+					
Sturm et al. [28]	2018	+					
López-Pintado et al. [19]	2019						+
Ladleif et al. [16]	2019	+	+				
Klinger et al. [13]	2020			+	+		
Lichtenstein et al. [18]	2020						+
Lu et al. [21]	2021	+				+	
Zhang et al. [41]	2021						+
Corradini et al.[7]	2023	+					
Stiehele and Weber [27]	2023			+			
Mertens et al. [24]	2024	+					
Ye et al. [40]	2024	+	+				
Corradini et al. [8]	2025	+					
Lemaire at al. [17]	2025				+		

Federated Logic (V5): In this variant, each instance maintains its own logic contract while sharing a common data contract ($\bigcup_{i=1}^{N} L_i \rightarrow D$). This configuration enables instance-specific logic customization while enforcing a uniform data schema. The architecture requires $N+1$ contracts with N edges converging on the shared data component. This variant suits scenarios requiring instance-specific logic customization while enforcing a uniform data schema across all instances, such as shared assets in collaborative business processes.

Federated Data (V6): In this variant, a single shared logic contract delegates to isolated data contracts deployed per instance ($L \rightarrow \bigcup_{i=1}^{N} D_i$). This enforces uniform process execution rules while preserving instance-level data isolation. The architecture requires $N+1$ contracts with N inter-contract edges. Unlike V5, the centralization concern is inverted to logic uniformity with data distribution. This variant is well suited for use cases where global control over logic is essential, but data component must remain independently governed.

These architecture variants represent all possible positions within the design space. While each variant can be used independently, as we are going to show, they may also be composed to form hybrid architectural configurations tailored to specific business and technical requirements of different processes. To locate the practical relevance of the proposed variants, we examined existing blockchain-based process execution engines from the literature (Table 2).

Table 3. Variants Combination in Configuration $\mathcal{C}_1$ and $\mathcal{C}_2$.

	Configuration $\mathcal{C}_1$			**Configuration** $\mathcal{C}_2$		
	D^P	D^C	D^J	D^P	D^C	D^J
L^P	V_3	–	V_4	V_4	–	V_6
L^C	–	V_1	V_5	–	V_2	V_2

5 Proof-of-Concept: Use Case and Implementation

In this section, we demonstrate how the architectural variants found within our design space can be applied to the crowdfunding use case introduced earlier 1. We present two distinct architectural configurations, each representing different design choices. We also describe how each configuration is implemented and executed on the Ethereum blockchain [37].

5.1 Architectural Configurations

The previously introduced architecture variants can be composed to form configurations for deploying the crowdfunding business process, which contains two lanes (P, C) and three data stores (P, C, J) (Fig. 1). Based on this structure, we deploy the process on-chain using different combinations of variants in order to resolve the trade-off between efficiency and flexibility in the following two architectural configurations (Table 3).

Configuration 1: Shared Monolith, Isolated Monolith, Shared Data. This configuration ($\mathcal{C}_1$) uses shared logic and data scope for the proposal domain, in line with the requirement to maintain a uniform representation across all proposals. The proposal logic and data are tightly coupled and deployed as a single shared smart contract ($L^P \circ D^P$). Each campaign instance, by contrast, is deployed independently with its own coupled logic and data ($L_i^C \circ D_i^C$), enabling per-instance customization and upgradeability. The judge data store is shared across all instances and referenced from both proposal and campaign logic, and is deployed as a separate, decoupled smart contract (D^J).

$$\mathcal{C}_1 = (L^P \circ D^P) \rightarrow D^J \cup \left(\bigcup_{i=1}^{N} (L_i^C \circ D_i^C) \rightarrow D^J \right)$$

From the perspective of the architectural design space, this configuration blends multiple architectural variants. The proposal subsystem aligns with the characteristics of a shared monolith (V3), where a single contract centralizes both logic and data. Campaigns adopt the isolated monolith variant (V1), as each instance is self-contained. The judge data store follows shared data variant (V4/V5), where a shared data component is accessed by multiple, isolated, logic components. This combination supports centralized access control, while maintaining flexibility for campaign-specific deployments.

Configuration 2: Shared Modular, Isolated Modular, Shared Logic. The second configuration ($\mathcal{C}_2$) adopts a fully modular and instance-isolated design across all domains. Logic and data contracts are decoupled in every domain, promoting high modularity, reusability, and fine-grained upgradeability. In the proposal lane, both logic and data remain shared across all instances (similar to Configuration 1) to ensure uniform behavior and structure for all proposals ($L^P \rightarrow D^P$). Campaigns, however, continue to be deployed per instance, but in this case, the logic and data are decoupled, allowing each contract to evolve independently ($\bigcup_{i=1}^{N} (L_i^C \rightarrow D_i^C)$). A key difference from $\mathcal{C}_1$ is in the handling of the judge data store. Instead of a shared judge data store, each proposal and campaign instance has its own dedicated judge data contract (D_i^J), introducing full isolation at the judge data level.

$$\mathcal{C}_2 = L^P \rightarrow D^P \cup \left(\bigcup_{i=1}^{N} \left(L_i^C \rightarrow D_i^C \right) \right) \cup \left(L^P \rightarrow D_i^J \right) \cup \left(\bigcup_{i=1}^{N} \left(L_i^C \rightarrow D_i^J \right) \right)$$

This configuration combines the following architectural variants. The proposal subsystem uses the shared modular one (V4), where logic and data are shared but decoupled. Campaign instances reflect the isolated modular variant (V2), emphasizing separation of concerns and per-instance deployment. The judge data store, now fully isolated, applies a mixture of V2 and aspects of V6 (mixed shared logic), since both proposal and campaign logic components independently reference their own judge data. Together, these decisions result in a configuration that prioritizes flexibility and independence, suited for delivering instance-level customization and evolving process.

5.2 Implementation

We implemented both configurations as collections of Solidity smart contracts to support the crowdfunding use case. The implementation is based on the BPMN4BC [3] language and its compiler, which translates BPMN models into Solidity code. The user is given the ability to select the deployment configuration (Table 3) needed at modeling time.

Logic contracts follow a state machine design pattern [32], while data contracts are implemented as storage templates with getter and setter functions for interaction with the logic layer. When logic and data are decoupled, logic contracts access data contracts through external calls. In the coupled case, logic and data are combined within a single smart contract. To enable upgradeability, we use the UUPS pattern [4]. Each contract is deployed with a proxy contract, allowing components to be upgraded without requiring data migration. The proxy preserves the contract state while delegating calls to the implementation contract.

Figure 3 illustrates the architectures using FMC (Fundamental Modeling Concepts) notation [12]. Each contract and its proxy is modeled as a subsystem containing its respective logic and data components. Figure 3a depicts the

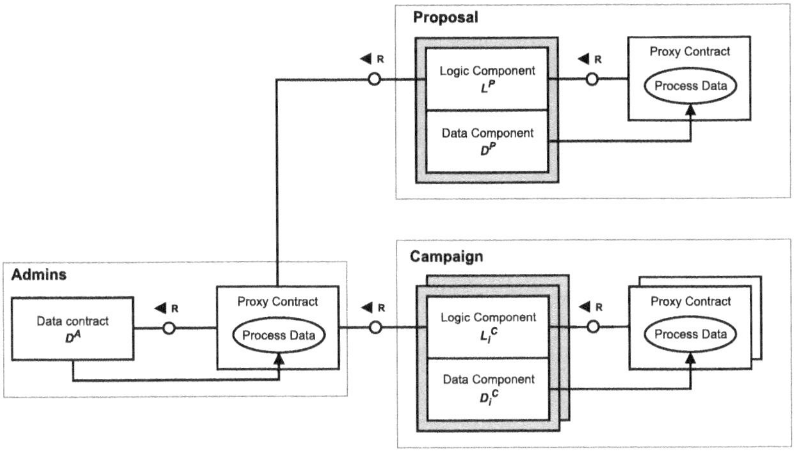

(a) Configuration $\mathcal{C}_1$

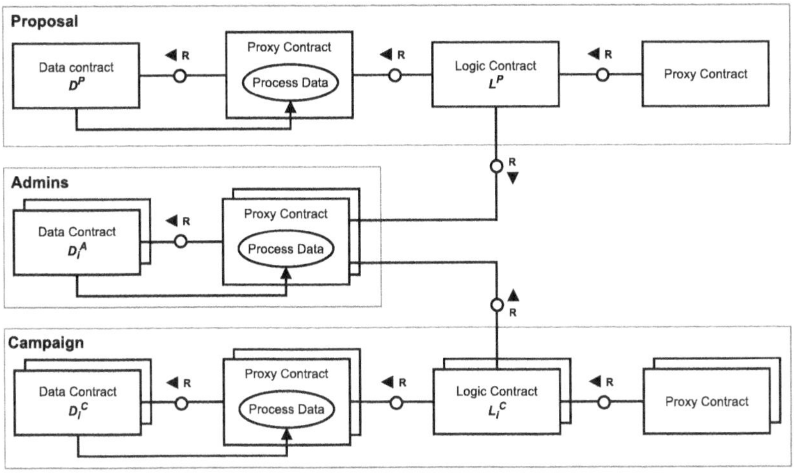

(b) Configuration $\mathcal{C}_2$

Fig. 3. Architectural realizations of the two configurations for the crowdfunding use case modeled using FMC [14].

realization of configuration $\mathcal{C}_1$. In this configuration, the proposal logic and data are implemented as a coupled monolithic contract (V3), combining logic and data within a single contract. The campaign subsystem follows an isolated monolithic approach (V1). The judge subsystem is implemented as a shared data contract (V4/V5). Figure 3b shows the realization of configuration $\mathcal{C}_2$, which adopts a modular architecture across all domains. The proposal subsystem uses a shared modular design (V4), separating logic and data into independent shared contracts. The campaign subsystem follows an isolated modular approach (V2),

Table 4. Deployment gas cost breakdown for Configurations $\mathcal{C}_1$ and $\mathcal{C}_2$.

Configuration $\mathcal{C}_1$				Configuration $\mathcal{C}_2$			
Component	Deployment Count	Avg Unit Gas	Total Gas	Component	Deployment Count	Avg Unit Gas	Total Gas
Proposal (V3)				Proposal (V4)			
$L^P \circ D^P$	1	2,462,572	2,462,572	L^P	1	2,666,624	2,666,624
				D^P	1	895,315	895,315
Campaign (V1)				Campaign (V2)			
$L_i^C \circ D_i^C$	50	261,126	13,056,280	L_i^C	50	240,191	12,009,551
				D_i^C	50	214,371	10,718,564
Judge (V4/V5)				**Judge** (V2/V6)			
D^J	1	877,861	877,861	D_i^J	50	258,705	12,935,230
Total	**52**	**3,601,559**	**16,396,713**	**Total**	**152**	**4,275,206**	**39,225,284**

Table 5. Comparison of execution gas usage between Configuration $\mathcal{C}_1$ and $\mathcal{C}_2$.

Activity	Count	Configuration $\mathcal{C}_1$		Configuration $\mathcal{C}_2$	
		Avg Gas	Total Gas	Avg Gas	Total Gas
`Receive Proposal`	50	126,372	6,318,600	139,481	6,974,050
`Evaluate Proposal`	100	124,138	12,413,842	149,775	14,977,458
`Refund Deposit`	27	62,437	1,685,799	72,634	1,961,118
`Collect Contribution`	23	81,694	1,878,962	95,711	2,201,353
`Release Escrow`	23	62,121	1,428,783	73,512	1,690,776
Total	**223**	**456,762**	**23,725,986**	**531,113**	**27,804,755**

deploying separate logic and data contracts for each instance. The judge data subsystem is implemented as an isolated data contract (V2/V6).

5.3 Deployment and Execution Cost

After running the event log consisting of 50 traces, we measured the gas cost of deployment and execution of the crowdfunding process. Table 4 presents a breakdown of deployment costs for both configurations. Each row shows the gas cost per variant, the number of instances deployed, and the total gas cost accumulated across all instances. In $\mathcal{C}_1$, the proposal process is deployed as a single contract with coupled logic and data, resulting in only one deployment. Campaigns are deployed per approved proposal as coupled logic and data units. The judge data store is shared and deployed once. This leads to a relatively low total deployment cost due to shared components. In contrast, $\mathcal{C}_2$ separates logic and data for all components and instantiates each campaign and judge component independently, while keeping a shared proposal process. This more modular design increases flexibility, but also significantly raises the total number of contracts deployed and deployment cost.

Table 5 reports execution metrics for both configurations. Overall, Configuration $\mathcal{C}_2$ incurs the most significant gas usage. This is due to the extensive use of decoupled logic and data components, which introduces additional cross-contract calls. Such calls are expensive on the Ethereum Virtual Machine, as they involve more state access overhead than internal function calls within a monolithic solution.

6 Discussion

The selection of an architectural configurations for blockchain-based business processes is subject to trading off three aspects: governance, flexibility, and cost.

Governance. Fine-grained control over deployment shapes governance models in blockchain-based process and enables the coexistence of different governance models within a single business process across all logic and data components. Isolated components enable instance-specific control, where participants in each process instance maintain autonomous authority over upgrades. A shared design enforces centralized control across all process instances with the requirement of multi-instance state management and shared access control mechanism.

Flexibility. Architectural flexibility encompasses the need to support the evolution of processes and is achieved via modularity and upgradeability. Modularity emerges from decoupling logic and data, enabling independent development, testing, and reusability. Decoupled contracts facilitate logic or data sharing across domains and support heterogeneous use cases, though at higher cost due to inter-contract calls overhead. Upgradeability varies by scope: *Coarse-grained* upgradeability (V3, V4) applies uniform changes across all instances cost-efficiently. *Fine-grained* upgradeability (V1, V2) enables instance-level customization and independent component evolution at higher deployment and execution cost. With our approach, the modeler is able to control the flexibility level precisely for logic and data depending on an expected need to upgrade in future version of the process.

Cost. Flexibility impacts deployment and execution costs. The decentralized, modular configurations ($\mathcal{C}_2$) incur significantly higher gas consumption due to increased contract multiplicity and inter-contract communication overhead. In a decentralized architecture, each instance independently bears the cost of its own upgradeability. In contrast, centralized, monolithic configurations ($\mathcal{C}_1$) achieve greater gas efficiency through reduced deployment cost and internal function calls, but sacrifice flexibility and instance-level autonomy. Unlike decentralized architectures, upgrade costs are shared across all instances. Cost efficiency thus reflects the composition of modularity, isolation in an architecture. As with flexibility and governance, the modeler is able to explicitly manage the cost aspect. Deployment costs can be controlled for each process component, depending on the variant used. The optimal configuration depends on operational and maintenance requirements: processes with flexibility requirements may justify the overhead of modular, isolated designs. On the other hand, static, centrally governed processes benefit from simpler – and cheaper – monolithic deployments.

7 Related Work

Malik et al. [22] evaluate smart contract upgradeability patterns (Contract Migration, Data Segregation, Diamond) for dynamic business processes. Their findings complement our architectural design space. Stiehle et al. [26] present a systematic literature review taxonomizing blockchain-based business process enactment approaches, but do not examine underlying smart contract architectures in detail. Garcia-Garcia et al. [11] systematically review blockchain-based collaborative business process management, finding that many studies focus on process implementation and execution. These can be positioned within our design space.

8 Conclusion

In this paper we propose to provide process modelers with fine-grained control on how processes mapped to smart contracts are deployed to be executed on chain. We have identified six possible deployment configurations within the design space for deploying collaborative business processes on blockchain networks. These allow to explicitly control three core dimensions: *logic scope*, *data scope*, and *modularity*. To illustrate their fine-grained applicability to specific process domains, we applied the variants to design two alternative configurations of a collaborative crowdfunding process. In addition to observing different deployment and execution costs, our analysis reveals how architectural choices directly impact critical quality attributes including governance, flexibility. Future work includes measuring how each variant impacts the cost of upgrading processes that have already been deployed. We also plan to conducting a systematic literature review to examine how often each variant is used to design existing blockchain workflow engines.

Artifacts and Reproducibility Package. The repository with the process model, smart contracts, and transaction logs generated during the experiment is available at: https://github.com/Hassan42/Design-Space-for-BBPM.

Acknowledgement. This work is supported by the Swiss National Science Foundation project "Flexible Choreographies in Multi-chain Environments" (196958).

References

1. Business Process Model and Notation (BPMN) Version 2.0.2 (2023). https://www.omg.org/spec/BPMN/
2. Abou Jaoude, J., Saade, R.G.: Blockchain applications-usage in different domains. IEEE Access **7**, 45360–45381 (2019)
3. Atwi, H., Pautasso, C.: Bpmn4bc: A simple bpmn extension. In: Service-Oriented Computing: 23rd International Conference, ICSOC 2025, Shenzhen, China, December 1–4, 2025, Proceedings, Part II. p. 306. Springer, Cham (2026)

4. Barros, G., Gallagher, P.: ERC-1822: Universal Upgradeable Proxy Standard (UUPS) [DRAFT] (2019). https://eips.ethereum.org/EIPS/eip-1822
5. Brambilla, M., Cabot, J., Wimmer, M.: Model-driven software engineering in practice. Morgan & Claypool Publishers (2017)
6. Buterin, V., et al.: A next-generation smart contract and decentralized application platform. White Paper **3**(37), 2–1 (2014)
7. Corradini, F., Marcelletti, A., Morichetta, A., Polini, A., Re, B., Tiezzi, F.: A flexible approach to multi-party business process execution on blockchain. Futur. Gener. Comput. Syst. **147**, 219–234 (2023)
8. Corradini, F., Marcelletti, A., Morichetta, A., Polini, A., Re, B., Tiezzi, F.: Blockchain-based execution of bpmn choreographies with multiple instances. Distrib. Ledger Technol. Res. Pract. **4**(2), 1–21 (2025)
9. Curty, S., Härer, F., Fill, H.G.: Design of blockchain-based applications using model-driven engineering and low-code/no-code platforms: a structured literature review. Softw. Syst. Model. **22**(6), 1857–1895 (2023)
10. García-Bañuelos, L., Ponomarev, A., Dumas, M., Weber, I.: Optimized Execution of Business Processes on Blockchain. In: Carmona, J., Engels, G., Kumar, A. (eds.) BPM 2017. LNCS, vol. 10445, pp. 130–146. Springer, Cham (2017). https://doi.org/10.1007/978-3-319-65000-5_8
11. Garcia-Garcia, J.A., Sánchez-Gómez, N., Lizcano, D., Escalona, M.J., Wojdyński, T.: Using blockchain to improve collaborative business process management: Systematic literature review. IEEE Access **8**, 142312–142336 (2020). https://doi.org/10.1109/ACCESS.2020.3013911
12. Keller, F., Wendt, S.: Fmc: An approach towards architecture-centric system development. In: 10th IEEE International Conference and Workshop on the Engineering of Computer-Based Systems, 2003. Proceedings, pp. 173–182. IEEE (2003)
13. Klinger, P., Nguyen, L., Bodendorf, F.: Upgradeability Concept for Collaborative Blockchain-Based Business Process Execution Framework. In: Chen, Z., Cui, L., Palanisamy, B., Zhang, L.-J. (eds.) ICBC 2020. LNCS, vol. 12404, pp. 127–141. Springer, Cham (2020). https://doi.org/10.1007/978-3-030-59638-5_9
14. Knoepfel, A., Groene, B., Tabeling, P.: Fundamental Modeling Concepts - Effective Communication of IT Systems, Wiley (2005)
15. Köpke, J., Meroni, G., Salnitri, M.: Designing secure business processes for blockchains with secbpmn2bc. Fut. Gen. Comput. Syst. **141**, 382–398 (2023)
16. Ladleif, J., Weske, M., Weber, I.: Modeling and Enforcing Blockchain-Based Choreographies. In: Hildebrandt, T., van Dongen, B.F., Röglinger, M., Mendling, J. (eds.) BPM 2019. LNCS, vol. 11675, pp. 69–85. Springer, Cham (2019). https://doi.org/10.1007/978-3-030-26619-6_7
17. Lemaire, V., Henry, T., García-Pérez, Á., Gaaloul, W., Tucci-Piergiovanni, S.: Leveraging the diamond pattern for scalable and upgradeable blockchain-based business process management applications. In: International Conference on Business Process Management, pp. 221–237. Springer, Cham (2025). https://doi.org/10.1007/978-3-032-02929-4_13
18. Lichtenstein, T., Siegert, S., Nikaj, A., Weske, M.: Data-Driven Process Choreography Execution on the Blockchain: A Focus on Blockchain Data Reusability. In: Abramowicz, W., Klein, G. (eds.) BIS 2020. LNBIP, vol. 389, pp. 224–235. Springer, Cham (2020). https://doi.org/10.1007/978-3-030-53337-3_17
19. López-Pintado, O., Dumas, M., García-Bañuelos, L., Weber, I.: Interpreted execution of business process models on blockchain. In: 23rd Int'l Enterprise Distributed Object Computing Conf. (EDOC), pp. 206–215. IEEE (2019)

20. López-Pintado, O., García-Bañuelos, L., Dumas, M., Weber, I., Ponomarev, A.: Caterpillar: a business process execution engine on the ethereum blockchain. Softw. Pract. Experience **49**(7), 1162–1193 (2019)
21. Lu, Q., et al.: Integrated model-driven engineering of blockchain applications for business processes and asset management. Softw. Pract. Experience **51**(5), 1059–1079 (2021)
22. Malik, S., Bandara, H.D., van Beest, N.R., Xu, X.: Smart contracts' upgradability for flexible business processes. In: Di Ciccio, C., et al. (eds.) International Conference on Business Process Management, pp. 55–70. Springer, Cham (2024). https://doi.org/10.1007/978-3-031-70445-1_4
23. Mendling, J., et al.: Blockchains for business process management - challenges and opportunities. ACM Trans. Manag. Inf. Syst. (TMIS) **9**(1), 1–16 (2018)
24. Mertens, D., Kim, J., Xu, J., Kim, E., Lee, C.: Smart flow: a provenance-supported smart contract workflow architecture. Clust. Comput. **27**(6), 8173–8187 (2024)
25. Nadolinski, E., Spagnuolo, F.: Proxy patterns (2018). https://blog.openzeppelin.com/proxy-patterns/, openZeppelin Blog
26. Stiehle, F., Weber, I.: Blockchain for business process enactment: a taxonomy and systematic literature review. In: International Conference on Business Process Management, pp. 5–20. Springer, Cham (2022). https://doi.org/10.1007/978-3-031-16168-1_1
27. Stiehle, F., Weber, I.: Process channels: a new layer for process enactment based on blockchain state channels. In: Di Francescomarino, C., Burattin, A., Janiesch, C., Sadiq, S. (eds.) International Conference on Business Process Management, pp. 198–215. Springer, Cham (2023). https://doi.org/10.1007/978-3-031-41620-0_12
28. Sturm, C., Szalanczi, J., Schönig, S., Jablonski, S.: A Lean Architecture for Blockchain Based Decentralized Process Execution. In: Daniel, F., Sheng, Q.Z., Motahari, H. (eds.) BPM 2018. LNBIP, vol. 342, pp. 361–373. Springer, Cham (2019). https://doi.org/10.1007/978-3-030-11641-5_29
29. Tran, A.B., Lu, Q., Weber, I.: Lorikeet: A model-driven engineering tool for blockchain-based business process execution and asset management. In: BPM (dissertation/demos/industry), pp. 56–60 (2018)
30. Viriyasitavat, W., Da Xu, L., Dhiman, G., Bi, Z.: Blockchain-as-a-service for business process management: Survey and challenges. IEEE Trans. Serv. Comput. **16**(3), 2299–2314 (2023). https://doi.org/10.1109/TSC.2022.3199232
31. Viriyasitavat, W., Xu, L.D., Niyato, D., Bi, Z., Hoonsopon, D.: Applications of blockchain in business processes: A comprehensive review. IEEE Access **10**, 118900–118925 (2022). https://doi.org/10.1109/ACCESS.2022.3217794
32. Volland, F.: Solidity design patterns. https://fravoll.github.io/solidity-patterns/, Accessed 21 May 2025
33. Weber, I., Lu, Q., Tran, A.B., Deshmukh, A., Gorski, M., Strazds, M.: A platform architecture for multi-tenant blockchain-based systems. In: Proceedings of International Conference on Software Architecture (ICSA), pp. 101–110. IEEE (2019). https://doi.org/10.1109/ICSA.2019.00019
34. Weber, I., Xu, X., Riveret, R., Governatori, G., Ponomarev, A., Mendling, J.: Untrusted Business Process Monitoring and Execution Using Blockchain. In: La Rosa, M., Loos, P., Pastor, O. (eds.) BPM 2016. LNCS, vol. 9850, pp. 329–347. Springer, Cham (2016). https://doi.org/10.1007/978-3-319-45348-4_19
35. Weske, M.: Business Process Management Architectures. In: Business Process Management, pp. 333–371. Springer, Heidelberg (2012). https://doi.org/10.1007/978-3-642-28616-2_7

36. Wöhrer, M., Zdun, U.: Design patterns for smart contracts in the ethereum ecosystem. In: Proceedings of Int'l Conference on Internet of Things (iThings) and Green Computing and Communications (GreenCom) and Cyber, Physical and Social Computing (CPSCom) and Smart Data (SmartData), pp. 1513–1520. IEEE (2018)
37. Wood, G., et al.: Ethereum: A secure decentralised generalised transaction ledger. Ethereum Project Yellow Paper **151**(2014), 1–32 (2014)
38. Xu, X., Pautasso, C., Lo, S.K., Zhu, L., Lu, Q., Weber, I.: An extended pattern collection for blockchain-based applications. In: Wallingford, E., Zdun, U., Kohls, C. (eds.) Transactions on Pattern Languages of Programming V, pp. 67–117. Springer, Cham (2025). https://doi.org/10.1007/978-3-662-70810-1_2
39. Xu, X., Weber, I., Staples, M.: Architecture for blockchain applications. Springer, Cham (2019). https://doi.org/10.1007/978-3-030-03035-3
40. Ye, X., Zeng, N., Tao, X., Han, D., König, M.: Smart contract generation and visualization for construction business process collaboration and automation: upgraded workflow engine. J. Comput. Civil Eng. **38**(6), 04024030 (2024)
41. Zhang, D., Xu, X., Zhu, L., Paik, H.Y.: A process adaptation framework for blockchain-based supply chain management. In: 2021 IEEE International Conference on Blockchain and Cryptocurrency (ICBC), pp. 1–9. IEEE (2021)
42. Zhang, P., White, J., Schmidt, D.C., Lenz, G.: Applying software patterns to address interoperability in blockchain-based healthcare apps. arXiv preprint arXiv:1706.03700 (2017)
43. Zou, W., et al.: Smart contract development: Challenges and opportunities. IEEE Trans. Softw. Eng. **47**(10), 2084–2106 (2019)

Using Process Mining to Generate AI Agents from Software Engineering Process Records

Saimir Bala[1(✉)], Fabiana Fournier[2,3], Lior Limonad[2,3], and Andreas Metzger[4]

[1] Hasso Plattner Institute (HPI), University of Potsdam, Potsdam, Germany
saimir.bala@hpi.de
[2] IBM Software Innovation Lab (SIL), Haifa, Israel
{fabiana,liorli}@il.ibm.com
[3] Department of Information Systems, Faculty of Computer and Information Science, University of Haifa, Haifa, Israel
[4] Paluno Institute, University of Duisburg Essen, Essen, Germany
andreas.metzger@paluno.uni-due.de

Abstract. Integrating AI agents into Software Engineering (SE) raises an important challenge: how can we specify and realize AI agents that work effectively alongside humans in hybrid SE teams? Determining the right granularity and separation of concerns for such agents is non-trivial. Coarse-grained agents may introduce unmanageable complexity, whereas micro-agents may create severe coordination overhead. Moreover, existing multi-agent SE frameworks typically rely on predefined role structures and do not account for project-specific characteristics or process adaptations. We address this by combining object-centric, imperative, and declarative process mining. Using event logs extracted from software repositories, our approach discovers project-specific agent roles using a predefined SE role vocabulary grounded in repository behavior and generates matching agent specifications and implementations. As proof-of-concept, we applied our approach to a well-established open-source project. We performed functional tests and an exploratory user study to determine how well the generated AI agent specifications are aligned with human expectations.

Keywords: Process Mining · AI Agents · Software Engineering

1 Introduction

Process mining turns event data into insights about how work is performed, who performs it, and under which constraints [7]. Software engineering (SE) processes are no exception: the activities of developers, reviewers, and testers leave rich event traces in version control and issue management systems, forming a largely untapped source of process knowledge. Today, SE teams are undergoing a fundamental transformation as AI agents (autonomous systems capable of decomposing goals, executing tasks, and collaborating with humans) are increasingly

K. Kluza et al. (Eds.): BPM 2026, LNBIP 600, pp. 146–162, 2027.
https://doi.org/10.1007/978-3-032-36710-5_10

embedded into development workflows [2,19,21]. The SE community has already acknowledged the rise of *AI teammates*, and the evolution towards *hybrid human-AI SE teams* raises an important new engineering challenge: *How to specify and realize AI agents that can effectively work alongside humans?*

Addressing this question has several challenges. Assigning a classical SE role (such as *team member* in agile processes) to a single agent yields a monolithic and complex teammate that is difficult to build, test, maintain, and delegate tasks to. Conversely, defining highly fine-grained, single-task agents increases coordination overhead among agents and humans. Also, not all SE tasks may be relevant in every project, and those that are may require concrete adaptations. Formal verification may be mandatory in safety-critical systems yet irrelevant in a simple utility tool, while complex business logic may demand property-based tests beyond standard unit testing. Finally, software engineers expect agents to adhere strictly to established standards and project-specific processes [10]. Current frameworks rely on rigid, predefined canonical roles that do not adapt a project's actual workflows [11,13,14,21].

We address these challenges with PM4AA, a generative pipeline that mines project-specific SE agent specifications directly from repository event data. PM4AA treats SE processes as business processes and the activity history recorded in version control and issue management systems as event logs. It applies object-centric process mining [23] to capture the multi-object structure of SE activities. This allows the discovery of the task scope and object interactions of each role, and declarative process mining [8] to elicit behavioral constraints that serve as guardrails for agent execution. Together, these two complementary views provide both the procedural blueprint and the normative boundaries needed to specify a well-scoped agent. PM4AA then leverages LLMs for process model analysis [15] to synthesize executable agent specifications. As a proof of concept, we implement PM4AA using the LangGraph framework and evaluate it through a case study on the open-source *Commitizen* project, which yields an event log of more than 21,000 events spanning eight years of development history, functional testing of the generated agents, and a user study involving ten human participants, who assessed how well process-mined, project-specific agents align with human expectations in hybrid SE teams.

Below, Sect. 2 introduces relevant background and related work. Section 3 explains PM4AA. Section 4 presents the case study and reports our PM4AA evaluation. Section 5 concludes the paper and outlines future work.

2 Background and Related Work

2.1 Process Mining

Process mining is a discipline that aims to extract insights about processes from event data traces [7]. Object-Centric Process Management [23] captures business processes around interacting objects such as orders, invoices, and shipments, rather than a single predefined workflow. Object-Centric Process Mining (OCPM) extends traditional process mining by explicitly considering processes that involve multiple interacting objects [1]. Taking object-centric event logs as input, OCPM helps capture synchronization, interaction, and dependency relations among objects and providing a more accurate representation of highly interconnected processes. A central artifact produced by OCPM is the Object-Centric Directly-Follows Graph (OC-DFG), which extends the classical Directly-Follows Graph (DFG) (a graphical representation of activities that directly succeed one another in the observed process).

Declarative process mining elicits process behavior through constraints that govern allowable behavior, rather than prescribing a fixed sequence of activities. Constraints express temporal or logical relations such as precedence, response, or mutual exclusion [8,17,18,20]. Declarative approaches are well-suited to flexible, knowledge-intensive processes where strict control-flow specifications are difficult to define or maintain. Languages such as DECLARE have become widely adopted due to their formal semantics and expressive power, and have been applied in healthcare, customer service, and human-centric workflows.

2.2 Event Logs from Software Repositories

Software development is a process: developers open issues, write code, commit changes, review pull requests, run tests, and release software. Like a business process, it generates event data as a natural by-product of execution. These events are captured in version control and issue management systems such as Git, GitHub, and Jira.

Repositories provide a rich set of data. A single commit records the actor, a timestamp, a descriptive message, and references to modified files, related issues, and pull/merge requests. An issue has its own lifecycle: opened, labeled, assigned, commented upon, and resolved. A pull/merge request links one or more commits to one or more issues and involves multiple actors across review, approval, and merge events. A single SE action, such as fixing a bug, thus spans multiple objects simultaneously, which is precisely what OCPM is designed to capture [23]. Table 1 shows a simplified extract of repository event data structured as an OCEL.

Table 1. Excerpt of a software repository structured as an Object-Centric Event Log. A dash indicates that the event is not associated with that object type.

Timestamp	Actor	Activity	Commit	Issue	PR	Message
2024-03-01 09:12	alice	open_issue	–	#42	–	feat: add emoji support
2024-03-02 14:05	bob	push_commit	a3f8c1	#42	–	feat(scope): add emoji support
2024-03-02 14:07	bob	open_pr	a3f8c1	#42	#17	fix: handle edge case in parser
2024-03-03 10:30	alice	review_pr	–	#42	#17	–
2024-03-03 11:45	alice	merge_pr	a3f8c1	#42	#17	–
2024-03-03 11:46	bot	close_issue	–	#42	#17	–

Each row represents an event tied to one or more objects of different types. The commit, issue, and PR columns capture the multiple concurrent object lifecycles involved in each event. The message column carries the human-readable description attached to the change. The table shows, for instance, how issue #42 spans six events across three actors: alice opens the issue, bob commits a fix and opens a pull request, alice reviews and merges it, and a bot closes the issue.

2.3 Related Work on AI Agents

AI agents possess their own thread of control and make autonomous decisions about which actions to perform and when [16,25]. They decompose high-level goals into tasks and execute them in unstructured environments. Agents can form multi-agent systems in which they collaborate with humans and other agents to carry out tasks [4,22,25]. Their tasks are realized through AI algorithms and models, including generative AI such as LLMs [12]. In SE, examples include autonomous coding agents that initiate, review, and evolve code at scale [16].

Recent literature reviews document a rapid paradigm shift from single-agent coding assistants to complex multi-agent SE ecosystems [13,19]. They conclude that a single, monolithic AI agent is insufficient due to the multifaceted complexity of real-world software projects. Consequently, state-of-the-art frameworks (e.g., ChatDev and MetaGPT) divide the development lifecycle into specialized roles, such as product managers, programmers, and testers, to improve task execution, collaborative reasoning, and fault tolerance. However, these frameworks rely on fixed architectures based on canonical roles (e.g., Programmer, Reviewer, Tester) [13]. Such rigid role structures may cause severe inter-agent coordination bottlenecks as projects scale [19]. Both [19] and [13] call for adapting agent roles to evolving project demands, yet operational, data-driven techniques remain lacking. PM4AA directly addresses this challenge by extracting, scoping, and generating project-specific SE agents from project event logs.

The BPM field has developed organizational mining and Agent System Mining (ASM) to derive role structures and behavioral models from event logs [24]. ASM tools like *Agent Miner* extract complete, multi-agent behavioral and interaction models directly from event logs. Still, these approaches are retrospective and simulative: they mine structures to generate synthetic traces or build digital twins of existing processes.

While PM4AA shares the foundational premise of Shen et al. [24] that historical event logs contain vital, role-specific behavioral blueprints, it takes a more *prescriptive* and *operational* stance. Specifically, we argue that event logs contain role-specific behavior that can be mined into (OC-)DFGs and declarative constraints to engineer and deploy AI agents within an executable LangGraph architecture, enabling them to work alongside humans rather than merely simulate human actors.

3 The PM4AA Approach

PM4AA (Process Mining for Agentic AI) is a generic, multi-step pipeline that transforms SE repository data into operational, role-specific AI agents. Figure 1 provides an overview of the steps of the PM$_4$AA pipeline, which we elaborate below. A repository containing the full pipeline implementation and the accompanying application is available at: https://github.com/liorlimonad/pmaa.

PM$_4$AA assumes the availability of sufficient historical repository data. It is therefore primarily applicable to existing projects with observable development history, rather than greenfield projects with no prior traces. For new projects, PM$_4$AA-derived agents may be transferred only as templates and would require later adaptation once project-specific traces become available.

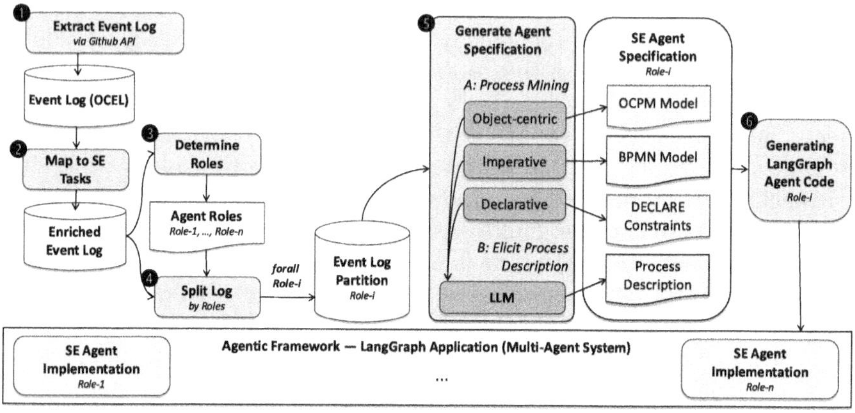

Fig. 1. The PM$_4$AA pipeline.

Step ❶ *Event Log Extraction:* The first step is to extract a raw Object-Centric Event Log (OCEL) from a GitHub repository using *PyStack't* [3], a Python library designed to produce OCEL 2.0 logs via the GitHub REST API. PyStack't maps GitHub entities to OCEL object types (i.e., user account, issues and commits), and repository interactions to timestamped events, yielding a log that natively captures the multi-object nature of software development activity. To derive roles, we first identify the tasks recorded in the event log.

Step ❷ *Mapping to SE Tasks:* The raw OCEL does not represent SE *tasks* as first-class objects. PyStack't also does not parse pull requests (often associated with tasks). Therefore, we need to derive the tasks from the commit messages.

A key enabler for this step is the commit message. Many open-source projects enforce *Conventional Commits* [6], a lightweight standard that structures commit messages as `type(scope): description`. The type prefix classifies the intent of each change: `feat` signals a new feature, `fix` a bug repair, `docs` a documentation update, and `test` a testing contribution. This allows assigning each event a corresponding type of SE task without manual annotation [21]. Many projects (e.g., Angular, Vue.js, ESLint, and Commitizen) follow this standard.

Accordingly, we enriched the event log by parsing each commit message against the Conventional Commits pattern *type(scope): description* [6] using a regular-expression classifier. Each successfully parsed commit is annotated with a `task_type` (e.g., `feat`, `fix`, `docs`) and a `task_semantic_class` drawn from a fixed ten-class classification scheme derived from the Conventional Commits specification (e.g., `feature_development`, `defect_resolution`, `quality_assurance`). A corresponding `task` object is created for each matching commit and linked to its commit and to all events that reference that commit.

Step ❸ *Determine Roles:* Given the tasks, the goal of this step is to derive a set of process roles from the contributor behavior recorded in the OCEL (cf. organizational mining [7]). All persons who have contributed to the repository (e.g., by committing a change) are available as user objects (collected in Step ❶). For every user object, a *resource profile* is computed by aggregating three behavioral dimensions: (i) the distribution of activities the user participates in, (ii) the number of commits attributed to the user, and (iii) the distribution of task semantic classes (e.g., feature development, defect resolution, documentation) obtained from commit-level attributes in the log.

A *priority-ordered rule-based* classifier then maps each profile to exactly one *fixed* and *prescribed* role. This hard partition is a deliberate simplification: a single dominant role per contributor keeps the sub-logs disjoint and the agent scopes unambiguous, at the cost of flattening contributors who act in multiple capacities. Organizational mining alternatively permits overlapping or weighted role assignments [7]. The rules discriminate along two axes: *platform identity* and *behavioural dominance*. Automated accounts are identified first through their platform metadata. Contributors with no commit activity are classified as issue reporters, capturing actors whose participation is limited to issue creation and discussion. Among committing contributors, those exceeding a configurable commit-count threshold and exhibiting diversity across at least three task classes are classified as maintainers. The remaining contributors are assigned to specialist roles (such as quality engineer, DevOps engineer, technical writer, or feature developer) based on the task-class group that accounts for the plurality of their work, with a generic *contributor* role as fallback. The assigned role is persisted as a new attribute on the corresponding user object in the enriched OCEL log, enabling the downstream partitioning in the next step.

Step ❹ *Split Log:* We then isolate each role's behavior. Using the role-annotated OCEL, a per-role sub-log is constructed for each discovered role. For a given role, each event in which at least one user object carrying that role participates is selected. All objects referenced by those events (including non-user objects such as issues, commits, and tasks) are retained, thereby preserving the object-centric structure of the original log. The result is one OCEL sub-log per role, capturing the multi-typed object context of the activities performed by that role.

Step ❺*-A: Process Mining.* The per-role event log partitions serve as input for three complementary process mining steps:

Object-centric: An OC-DFG is discovered using the `pm4py` library. To filter infrequent behavior while accommodating sub-logs of varying size, a dynamic frequency threshold is applied: both the activity and edge thresholds are set to approximately 2% of the sub-log's event count, with a minimum of one. The resulting OC-DFG captures each role's typical activity sequences, annotated with the object types co-involved in each transition. It serves two purposes. First, it delineates the *task scope* of the agent: the activities retained after thresholding determine which operations the agent is responsible for. Second, the object types in the graph define the *entities the agent interacts with* (e.g., issues, commits, pull requests), grounding the agent's interface in empirical process data rather than manual specification.

Imperative: Complementing the OC-DFG, a BPMN model is derived via Inductive Miner. The log is first flattened onto a designated primary object type (selected as the case notion), converted to a traditional event log, and the resulting process tree is then transformed into a BPMN diagram. Together, with the OC-DFG this provides a per-role view of *how* work is carried out.

Declarative: Declarative process mining is applied to the role-specific logs to elicit normative behavioral constraints, complementing the imperative models produced previously. Using `DeclareMiner` [9], each log was first flattened onto the `issue` object type to yield a case-centric representation, and constraint discovery was executed with a minimum support threshold of 0.5 and a maximum constraint cardinality of two. Roles with fewer than two distinct activities after flattening were excluded from discovery, as meaningful constraint inference requires activity diversity. For the remaining roles, `DeclareMiner` was executed, and the discovered constraint sets were serialised as JSON artifacts for downstream consumption.

*Step ❺-**B**: Elicit Process Description.* For each role, a process profile with activity frequencies, object-type distributions, directly-follows patterns, and event-object interaction counts is provided to GPT-4-mini using the following prompt.

> **Prompt to LLM:** Analyze the process profile for `{role_name}` and generate a process description covering: (i) process overview and responsibilities, (ii) main activities grouped by function, (iii) object interactions and semantics, (iv) flow patterns and common sequences, (v) key behavioral characteristics, and (vi) business interpretation of the role's contribution to the software lifecycle. Base the analysis on an extracted process profile including total events/objects, unique activities/object types, activity frequencies, object type distributions, top directly-follows flows, and top event-object interactions. Write in professional language for business stakeholders, emphasizing object-centric behavior.

The markdown output provides a structured business-oriented description of the role-specific object-centric process, including behavioral, interactional, and statistical perspectives for each of the roles.

Step ❻ Generate LangGraph Agent Code: With the specification in place, we move to the implementation of the agents. We make use of LangGraph as the target Agentic AI technology to instantiate the roles generated in the previous steps. The LangGraph application is realized through a prompt-driven synthesis process in which IBM BOB[1] interpreted the user request and turned the mined SE Agent specifications into an executable application. IBM BOB employs a mixture of frontier and specialized language models for interpretation, drawing on Anthropic Claude, Mistral open-source models, IBM Granite, and additional specialized fine-tuned models. The BOB prompt is as follows:

[1] https://bob.ibm.com/, accessed 2026-06-05.

⟨**Context**⟩ This workspace is applied to an event log harvested from a GitHub repository, which was then partitioned into 8 separate sub-logs corresponding to 8 SE roles invoked in the handling of the GitHub issues. We derived the Object-Centric Directly-Follows Graphs (OCDFG), BPMN graphs, DECLARE constraints and a process description markdown, for each of the logs.
⟨**Task**⟩ Project this onto an agentic application that will facilitate and automate, where possible, online issue handling on top of the GitHub repository. Inspect all the attached artifacts and propose a langgraph-based agentic application architecture matching these artifacts.

The resulting agent implementations are embedded into a LangGraph workflow with shared typed state, prompt templates, and routing logic. The population process combines mined process semantics with software architecture constraints: the mined artifacts determine what kinds of role behavior are meaningful. The LangGraph implementation turns those behaviors into executable agent nodes with explicit interfaces, routable responsibilities, and bounded actions.

4 Evaluation

We evaluated PM4AA through (1) a case study, (2) functional testing of agent implementations, (3) a user study assessing the generated agent specifications.

4.1 Case Study: Commitizen

We applied PM4AA to a repository of a real-world software project to demonstrate the technical feasibility of our approach. We chose *Commitizen*[2], a well-established open-source project active since 2017 and having 3,400+ stars and 339 forks.

Step ❶ *Event Log Extraction.* The raw OCEL extracted for *Commitizen* covers the period November 2017 – November 2025 (approximately eight years of project history) and contains 21,488 events and 4,813 objects across three object types: 1,459 issues, 2,765 commits, and 589 user accounts. Although the full log contains 21,488 events, the distribution is highly skewed. The `issue_reporter` and `bot` roles account for most events. Consequently, generated specifications for low-volume roles should be interpreted as more exploratory.

Step ❷ *Mapping to SE Tasks.* Of the 2,765 commits in the log, 1,721 (62.2%) carry a well-formed Conventional Commit message and were successfully enriched, yielding 1,721 task objects and a final OCEL of 6,534 objects across four types (commits, tasks, issues, users).

Step ❸ *Determine Roles.* To partition the enriched OCEL into role-specific sub-logs, PM4AA automatically assigns each of the 589 user objects to one of eight roles using a rule-based classifier applied in priority order, see Table 2. The classifier first separates GitHub bot accounts (`bot`), then distinguishes users who participate in issue discussions but never commit (`issue_reporter`) from those who do. Among committing users, `maintainer` is assigned to those with at least

[2] https://github.com/commitizen-tools/commitizen, accessed 2026-06-05.

20 commits spanning at least three distinct task classes, reflecting broad, sustained involvement. The thresholds were selected as pragmatic heuristics for this proof-of-concept to distinguish sustained, diverse contribution from low-volume specialist activity. They were not optimized statistically and should be treated as configurable parameters rather than universal constants. The remaining committing users are classified by their dominant task with `contributor` being a "catch-all" for low-volume or single-contribution users. The four main, distinct roles (with $\geq$ 20 users) that remain are highlighted in bold.

Table 2. Role distribution in the Commitizen event log.

Role	Users	Events	Objects
issue_reporter	425	16,037	1,863
contributor	66	236	495
quality_engineer	37	194	395
technical_writer	23	71	194
feature_developer	20	100	212
maintainer	8	2,149	4,386
bot	6	3,394	1,055
devops_engineer	4	15	39
Total	589	21,488	6,534

Step ❹ Split Log. Given the above roles, the annotated OCEL was split accordingly, resulting in role-specific sub-logs, one per role.

Step ❺ Generate Agent Specification. Except `bot` and `issue_reporter`, all roles only performed a single activity type, and thus DECLARE constraints are only informative for the above two roles. This does not invalidate the pipeline, but indicates that the utility of imperative and declarative mining depends on activity diversity in the role-specific sub-log. The following is a summary of the markdown generated for the `issue_reporter` role.

PROCESS DESCRIPTION: `ISSUE_REPORTER`
Overview: The `issue_reporter` role initiates, coordinates, and follows issue lifecycles, linking issue and user objects through creation, discussion, review, subscription, assignment, closure, and reopening.
Main Activities: Issue initiation (`created, labeled, assigned`), communication (`commented, mentioned, subscribed`), review handling (`review_requested, reviewed, ready_for_review`), lifecycle management (`closed, reopened`), and structural updates (`referenced, cross_referenced, renamed`).
Object Interactions: The process centers on `issue` and `user` objects. Issues are updated through creation, discussion, and resolution, while users participate as actors, reviewers, and collaborators. Interactions coordinate visibility, accountability, and traceability.
Flow Patterns: Typical flows include `created` → `commented`, repeated communication (`commented` → `commented`, `commented` → `mentioned`), review paths (`reviewed` → `merged` → `closed`), and non-linear behaviors such as reopening or review removal.
Characteristics: Communication-driven, highly collaborative, iterative, issue-centric, lifecycle-spanning, and coordination-oriented.
Business Interpretation: The role supports software delivery by ensuring issues are captured, discussed, coordinated, and resolved while maintaining visibility and traceability.

For the remaining roles, the OC-DFG is the primary process model represented as shown in the following example, expressing that `Existence1[closed]` requires the activity `closed` occuring *at least once.*

```
{ "raw": "Existence1[closed] | |",
  "template": "Existence1", "activity_a": "closed", "activity_b": "" }
```


Step ❻ Generate LangGraph Agent Code. This yielded an issue-centric agentic application in which the `issue` object coordinates requests across all lifecycle stages. BPMN artifacts provided the issue management structure, DECLARE artifacts contributed policy checks and approval gating, and process descriptions populated the role-specific agents.

The LangGraph application was generated by combining prompt interpretation with artifacts mined from the repository process rather than being handcrafted from scratch. A central design choice was that not every mined role became an agent. Instead, the final implementation selected roles whose behavior was both sufficiently distinct and operationally useful in a GitHub issue workflow. This produced five role-specific agents: `issue_reporter_agent`, `bot_workflow_agent`, `implementation_agent`, `quality_agent`, and `technical_writer_agent`. The `implementation_agent` combines the behavior mined from the roles: feature developer, contributor, maintainer, and DevOps engineer. Rather than exposing several narrowly separated development roles, the app consolidated them into a single agent responsible for interpreting feature requests, enhancements, and bug fixes as implementation work. The consolidation was deliberate: the mined roles showed similar implementation semantics, and the application required a practical executable role rather than a faithful organizational replica.

The example below shows a human-readable agent specification, also used in our user study.

SE AGENT: issue_reporter
Goal: You focus on intake, clarification, communication, and sustained issue coordination.
Tasks (DO): • summarizing issue intent and missing context • asking clarifying questions • detecting duplicates • suggesting labels, assignees, milestones • preparing issues for downstream work
Avoid (DON'T): • make deep code changes directly • close issues without strong resolution evidence
Actions: • "summarize_issue", "ask_for_clarification" • Input: question: "Please provide expected behavior and actual behavior." • Output: issue_id, prepared documentation work items

4.2 Functional Testing

We tested the generated LangGraph application for *Commitizen* using actual GitHub issues. We conducted three smoke tests, selected to exercise distinct routing paths. These tests are not intended as evidence of general runtime effectiveness, but as sanity checks that the generated LangGraph application can ingest issues, route them, and produce coherent next actions. In all cases, the app ingested the issue, loaded its GitHub context, interpreted the text through LLM-based classification, routed the issue to a role-specific agent, and derived a corresponding next action.

Issue #1: "Readme.md requires more content to describe this project." This was interpreted as a documentation-oriented request. The routing logic correctly selected the `technical_writer_agent`. In turn this agent then used its LLM-based analysis to derive a suitable documentation task focused on updating or preparing explanatory material.

Issue #2: "Add a first basic hello-world function in Python with no input arguments that prints 'hello world'." This issue was routed correctly to the `implementation_agent`. The agent translated the request into a development-oriented action plan, containing a concrete implementation workflow comprising repository inspection, minimal code changes, traceability links to the originating issue, and preparation of a pull request for review.

Issue #3: "Application crashes when uploading large CSV files." This issue was routed correctly to the `bot_agent`. The agent translated the issue into a workflow-oriented coordination plan, containing low-risk repository maintenance actions comprising issue triage through the application of `triage` and `needs-review` labels, routing to the appropriate review stage, and automatic reviewer assignment to the `maintainer` for further technical assessment.

4.3 Exploratory User Study

To empirically evaluate PM4AA, we performed an exploratory, qualitative user study focusing on the four main, distinct roles identified in Table 2. The aim was to assess how well process-mined, project-specific agents address the problem of *Human-AI alignment*, as perceived by the participants. *Human-AI alignment* is a concept from HCI research referring to how well an AI agent meets the goal of producing the desired outcomes, without undesirable side effects [5,11]. Due to the project-specific separation of concerns and agent modularity, PM4AA's agent specifications should exhibit a relatively high degree of human-AI alignment.

We involved ten participants recruited from our pool of PhD students, teaching assistants, and Master students. Participants evaluated written specifications, not the running agents.

Study Flow: The study entailed the following steps. *1. Motivation and Introduction:* Participants received a briefing on the problem of modeling and realizing SE agents, the ensuing challenges of human-AI alignment, and the PM4AA approach. *2. Case Study & User Study Goals:* The participants were given an overview of the case study, the generated SE Agent roles and the evaluation dimensions (explained below). We shared all identified roles except `bot`, as this is already automated by Github. *3. The Survey:* Human participants were informed about the study goals, protection of personal data and the voluntary nature of the study. Each participant received a set of agent specifications together with a questionnaire to be answered as detailed below.

Questionnaire Structure: We evaluate Human-AI alignment from a human-computer-interaction (HCI) perspective. We address the following alignment concerns introduced in [11], thereby contributing to the challenge of effective

AI-human collaboration [21]. Assessments for these concerns are given on a five-point Likert scale: (1) strongly disagree, ..., (5) strongly agree.

Knowledge Schema: The information needed by an agent for a successful transaction must be clear to prevent back-and-forth communication. Herewith, we assess what level of contextual clarity role-scoped agents may provide.

> C1: The SE Agent specification clearly communicates what specific input the agent needs to successfully complete its task.

Autonomy: Humans must understand the boundaries within which an AI agent can operate autonomously and when it must consult a human for decisions.

> C2: The SE Agent specification makes it clear when the agent is allowed to act autonomously and when it must wait for human approval or interaction.

Operational: Humans and agents must actively identify and align on which tasks to execute, which requires that humans are aware of the agent skills.

> C3: The tasks and responsibilities of the SE Agent are clearly identifiable, making it easy to understand what specific SE work I can assign to it.

Accountability: Humans might worry that an agent misbehaves, which could negatively impact the humans' personal reputation. Consequently, for the sake of accountability, it must be possible to efficiently debug the AI agents.

> C4: If the SE Agent makes a mistake (e.g., merging bad code), the agent specification(s) allow me to isolate, understand, and debug the specific problem.

Human Engagement: Because humans may have varying preferences for control and oversight, humans and SE agents should align on when, how, and why the human should be interrupted or engaged.

> C5: Coordinating hand-offs, monitoring outputs, and intervening with the SE Agent would require minimal cognitive effort and cause low friction.

To strengthen the findings and compensate for the relatively small sample size, we asked additional open-ended, cross-cutting questions.

> **Q1:** ***Granularity*** – Based on the agent specifications you reviewed, would you merge some agent roles, split them further, or keep them as generated?
> **Q2:** ***Collaboration*** – Imagine you are a software engineer working alongside the AI agents on a real project. Are there any concrete moments in your workflow where you anticipate a handoff or coordination problem? If so, explain what you think which agent(s) and which reason(s) would cause it.
> **Q3:** ***Trust*** – The role boundaries and SE agent behavior were automatically derived rather than designed by a human. Are there potential reasons that would undermine your trust or acceptance of such generated SE agents?

Data Quality and Threat Mitigation: To ensure the integrity of the survey data and mitigate the risk of careless responding (e.g., straight-lining), we implemented several established quality control mechanisms: (1) To mitigate fatigue and ordering effects when answering four roles × eight questions (= 32 questions in total), the presentation order of the agent specifications was *randomized* for each participant; (2) Statement C5 was *negatively framed* to detect straight-lining; (3) Responses that are *logically inconsistent* (e.g., answering "Strongly Agree" to both C1 and the reverse-coded C5) were flagged.

Results: Figure 2 shows the medians of the participants' assessment for each of the four SE agents.

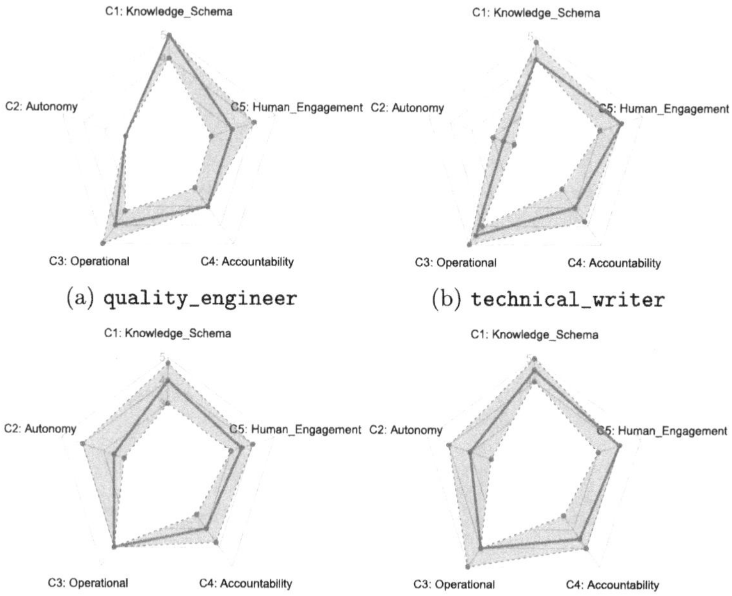

(a) `quality_engineer` (b) `technical_writer`

(c) `issue_reporter` (d) `feature_developer`

Fig. 2. Radar Charts of the Qualitative User Study Results (showing Median as thick line and IQR as shaded area)

Overall Achievement of Alignment Criteria: Analyzing the medians across all four evaluated agent roles reveals distinct strengths and weaknesses in how process-mined specifications address human-AI alignment:

C1: Knowledge Schema (Med: 4): This was the highest-rated dimension, indicating strong agreement that the specifications clearly communicate the specific inputs the agents need to successfully complete their tasks. The object-centric process models successfully delineate task scope and required entities.
C3: Operational Clarity (Med: 4): Participants found the tasks and responsibilities of the agents to be highly identifiable, suggesting that deriving role boundaries from event logs yields practical and understandable work assignments.
C5: Human Engagement (Med: 4): A moderate-to-high score implies that humans feel coordinating hand-offs, monitoring outputs, and intervening with the agents would require relatively low cognitive effort.

C4: Accountability (Med: 3): The ability to isolate, understand, and debug an agent's mistakes scored neutrally. While process boundaries are clear, the underlying execution may still appear a "black box" for debugging.
C2: Autonomy (Med: 2): This dimension scored notably low, highlighting a severe challenge. Humans struggled to understand the boundaries dictating when an agent is allowed to act autonomously versus when it must wait for human approval. This reflects a marked human-AI divergence on autonomy boundaries.

Role-Specific Assessment of C1-C5: A detailed breakdown of the five criteria across the four specific agent roles uncovers variations in how alignment is perceived depending on the nature of the software engineering tasks.

`feature_developer`: This role exhibited the highest overall human alignment. Participants found its operational scope (4) and required knowledge schema (4.5) exceptionally clear. It also achieved the highest accountability score (3.5), likely because code implementations provide concrete, verifiable artifacts (e.g., a pull request) that are native to human review processes.

`quality_engineer`: While the inputs (5) and tasks (4) for testing were well understood, accountability (3) and autonomy (2) dropped significantly, suggesting users are uncertain about how much leeway a testing agent has to automatically reject code or define its own test parameters.

`technical_writer`: This role exposed the most extreme disparity in the study. While participants clearly understood its operational purpose (4.50) and the inputs it requires (4.00), it received an exceptionally poor autonomy score (1.5). Users seem highly uncertain about whether documentation agents publish changes autonomously or strictly draft content for human approval.

`issue_reporter`: The issue reporter role yielded the most balanced, albeit slightly lower, assessment. Because this role centers on communication, coordination, and repeated review cycles, its autonomy boundaries (2.5) and human engagement expectations (3.5) may be slightly more apparent to human counterparts compared to the testing and writing roles.

Summary of Q1-Q3: Below, we summarize the main positive (+) and negative (−) comments to the open-text questions.
Q1: (+) Roles map to distinct SE responsibilities, avoiding monolithic complexity and provide good coordination balance without excessive friction. (−) `feature_developer` and `quality_engineer` roles seem too broad. Handoffs lack detailed boundaries, and there is perceived redundancy between `technical_writer` and `issue_reporter`.
Q2: (+) Agents successfully handle tedious tasks, such as technical documentation and systematic test coverage improvements. (−) Ambiguous validation criteria pose a risk of endless loops between `feature_developer` and `quality_engineer`. Furthermore, the rapid generation of agent output risks turning the human developer into a vetting bottleneck.
Q3: (+) There is strong support for agents handling unappealing peripheral work and preparatory phases, allowing humans to focus on major architectural decisions. (−) Echoing Q2, risks include endless validation loops, premature

actions (e.g., reporters interrupting drafts), and severe human review bottlenecks due to high-volume outputs lacking prioritization guardrails.

4.4 Threats to Validity and Limitations

Our user study measures *perceived* human-AI alignment based on reading agent specifications, not actual task performance or agent behavior. A specification may appear well-scoped and comprehensible while producing misaligned behavior at runtime. We partly mitigate this through functional testing in Sect. 4.2.

The absolute assessments of our user study limit causal claims. Yet, determining a fair baseline for a comparative assessment is rather challenging in our case. Monolithic or fine-grained agent baselines are obviously unfair. Using canonical SE roles – as done by current frameworks [13] – while presenting realistic and feasible baselines, would significantly overlap with the generated roles. This makes it difficult to use them as hidden baselines, which are required to not compromise the study's validity due to psychological effects that skew perceived alignment.

The participant sample comprised Master's students, teaching assistants, and PhD researchers from a single university, recruited as a convenience sample. This may introduce social desirability bias. Also, as non-practicing software engineers, participants may not truly reflect the perspectives of developers deploying hybrid human-AI SE teams in production.

PM4AA was evaluated on a single open-source project, *Commitizen*, which follows the Conventional Commits specification and exhibits a relatively clean event log. Generalization to projects lacking structured commit conventions, or to proprietary or enterprise repositories, remains an open question.

A further threat stems from the single-role assignment in Step ❸. Open-source contributors routinely act in multiple capacities, and forcing one label per user withholds behavior from the specialist sub-logs it could inform: in our case study, the maintainer role (8 users, 2,149 events) absorbs documentation, testing, and development activity that would otherwise enrich the `technical_writer`, `quality_engineer`, and `feature_developer` models. Extending role assignment to overlapping memberships is future work.

SE event logs capture coarse-grained developer activities (e.g., commits or issue updates), whereas AI coding agents operate at a finer granularity (e.g., writing code or running tests). Mined role specifications thus may not fully capture deployed agent behavior. OCPM partly mitigates this by linking commits, issues, and pull requests across object types.

5 Conclusion

We presented PM4AA, a pipeline that derives project-specific SE agent specifications and implementations from SE repository event logs by mining roles, behavioral patterns, and normative constraints. As proof-of-concept we realized the pipeline and applied it to the *Commitizen* open-source project. Initial smoke tests showed coherent issue routing, and an exploratory user study suggested that the generated specifications appear to be well-aligned with human expectations.

Future work will evaluate PM4AA on more diverse repositories, including projects without structured commits and proprietary code bases, involve practicing engineers, and extend role assignment from hard partitions to overlapping role memberships. Enriching event logs with IDE telemetry and CI/CD logs will support finer-grained agent scoping and specifications.

References

1. van der Aalst, W.M.P.: Object-centric process mining: Dealing with divergence and convergence in event data. In: SEFM (2019)
2. Amalfitano, D., et al.: A research roadmap for augmenting software engineering processes and software products with generative AI. ACM Trans. Softw. Eng. Methodol. (2026)
3. Bosmans, L., Peeperkorn, J., De Smedt, J.: Pystack't: Real-life data for object-centric process mining. In: Proceedings of the BPM 2025 Demos and Resources Forum, pp. 208–215 (2025)
4. Calvanese, et al.: Agentic business process management: A research manifesto. Inf. Syst. **140**, 102738 (2026)
5. Christian, B.: The alignment problem: Machine learning and human values. WW Norton & Company (2020)
6. Conventional Commits Authors: Conventional Commits 1.0.0 (2019). Accessed 07 May 2026
7. van Der Aalst, W.: Data science in action. In: Process mining: Data Science in Action, pp. 3–23. Springer, Cham (2016)
8. Di Ciccio, C., Montali, M.: Declarative process specifications: Reasoning, discovery, monitoring. In: van der Aalst, W.M.P., Carmona, J. (eds.) Process Mining Handbook, LNBIP, vol. 448 (2022)
9. Donadello, I., Riva, F., Maggi, F.M., Shikhizada, A.: Declare4py: A python library for declarative process mining. In: BPM (PhD/Demos). CEUR Workshop Proceedings, vol. 3216, pp. 117–121. CEUR-WS.org (2022)
10. Dong, T., Shi, S.Y., Sampath, H., Macvean, A.: Towards AI as a collaborative partner: A taxonomy of AI agent behavior in software engineering. In: 3rd ACM International Conference on AI-powered Software (AIware) (2026)
11. Goyal, N., Chang, M., Terry, M.: Designing for human-agent alignment: Understanding what humans want from their agents. In: CHI Extended Abstracts, pp. 106:1–106:6. ACM (2024)
12. Guo, T., et al.: Large language model based multi-agents: A survey of progress and challenges. In: IJCAI, pp. 8048–8057. ijcai.org (2024)
13. He, J., Treude, C., Lo, D.: Llm-based multi-agent systems for software engineering: Literature review, vision, and the road ahead. ACM Trans. Softw. Eng. Methodol. **34**(5), 124:1–124:30 (2025)
14. Hemmer, P., et al.: Complementarity in human-ai collaboration: concept, sources, and evidence. Eur. J. Inf. Syst. **34**(6), 979–1002 (2025)
15. Kubrak, K., Botchorishvili, L., Milani, F., Nolte, A., Dumas, M.: Explanatory Capabilities of Large Language Models in Prescriptive Process Monitoring (Extended Abstract). In: IJCAI, pp. 10901–10905. ijcai.org (2025)

16. Li, H., Zhang, H., Hassan, A.E.: The Rise of AI Teammates in Software Engineering (SE) 3.0: How Autonomous Coding Agents Are Reshaping Software Engineering, CoRR abs/2507.15003 (2025)
17. Maggi, F.M., Bose, R.P.J.C., van der Aalst, W.M.P.: Efficient discovery of understandable declarative process models from event logs. In: CAiSE (2012)
18. Maggi, F.M., Montali, M., Westergaard, M., van der Aalst, W.M.P.: Monitoring business constraints with linear temporal logic: An approach based on colored automata. In: BPM (2011)
19. Otoum, N.A., El-Khalili, N.: Methods and techniques of agentic software engineering: A systematic literature review. IEEE Access **14**, 7443–7465 (2026)
20. Pesic, M., van der Aalst, W.M.P.: A declarative approach for flexible business processes management. In: Eder, J., Dustdar, S. (eds.) Business Process Management Workshops, pp. 169–180. Springer, Berlin Heidelberg, Berlin, Heidelberg (2006)
21. Roychoudhury, A., Pasareanu, C.S., Pradel, M., Ray, B.: Agentic AI software engineers: Programming with trust. Commun. ACM **69**(5), 56–58 (2026)
22. Sapkota, R., Roumeliotis, K.I., Karkee, M.: AI agents vs. agentic AI: A conceptual taxonomy, applications and challenges. CoRR **abs/2505.10468** (2025)
23. Seidel, et al.: Object-centric process management: A research manifesto. Inf. Syst. **141**, 102728 (2026)
24. Shen, Q., Polyvyanyy, A., Lipovetzky, N., Kampik, T.: Mining role-based behavioral patterns from event data for effective process simulation, CAiSE (2026)
25. Wooldridge, M.: An introduction to multiagent systems. Wiley, second edn. (2009)

Configuring and Diagnosing Conversational Agents for Process Mining Tasks

Alessandro Berti[1(✉)], Humam Kourani[1,2], and Wil M. P. van der Aalst[1,3]

[1] Process and Data Science (PADS), RWTH Aachen University, Aachen, Germany
{a.berti,wvdaalst}@pads.rwth-aachen.de

[2] Fraunhofer Institute for Applied Information Technology FIT, Schloss Birlinghoven, 53757 Sankt Augustin, Germany
humam.kourani@fit.fraunhofer.de

[3] Celonis, Munich, Germany

Abstract. Large Language Models (LLMs) are increasingly used to answer process mining questions about event logs, models, conformance, performance, fairness, and redesign. Direct prompting can produce plausible but incorrect answers, especially when tasks require careful trace interpretation, formal reasoning, or diagnostic evidence. A common response is to wrap the LLM in a *conversational agent framework*: the same underlying model is reused under different role prompts—such as "event log interpreter", "conformance checker", or "optimization consultant"—and a selector decides which persona speaks next on a shared transcript. The agents in this paper do not call external tools; they differ only in their role description, so any improvement must come from how the conversation is structured rather than from added capabilities. It is therefore not obvious how such agents should be configured, nor how to tell whether a configuration is actually helping. This paper studies selector-mediated configurations composed of process-mining-oriented personas and treats each recorded agent trace as a directed social network over roles. Using LLM-as-a-judge scores aggregated through Social Network Analysis (SNA), we diagnose final-answer quality, role usefulness, and handoff quality, and use these diagnostics to revise agent routing. We provide an open source implementation that executes the configurations, records traces, computes the SNA diagnostics, and feeds the results back into improved routing.

Keywords: Process Mining · Large Language Models · Conversational Agents · Social Network Analysis · LLM Evaluation

1 Introduction

Process mining connects event data with questions about control flow, conformance, performance, resources, fairness, and redesign [7,8]. LLMs are attractive

K. Kluza et al. (Eds.): BPM 2026, LNBIP 600, pp. 163–178, 2027.
https://doi.org/10.1007/978-3-032-36710-5_11

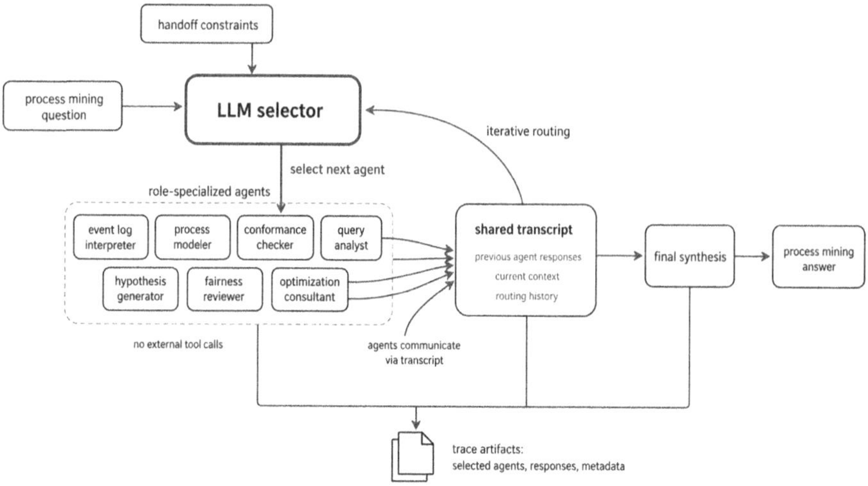

Fig. 1. Selector-mediated conversational framework studied in the paper. A controller chooses a role-specialized process mining persona or stops the discussion; selected agents read the shared transcript, contribute an intermediate answer, and leave a trace that can later be analyzed.

interfaces for such questions because they can explain logs, models, queries, and improvement options in natural language [1–3]. Yet process mining answers can be plausible while being wrong: they may assume the wrong case notion, ignore timestamp ordering, misuse a formal model, or overstate a fairness or optimization claim. A natural reaction is to ask whether multiple specialized agents can do better than a single LLM call.

In this paper, a *conversational agent* is not a tool-using autonomous system. The same underlying LLM is reused under different role prompts—an "event log interpreter", a "conformance checker", an "optimization consultant", and so on—and a selector decides which persona contributes next to a shared transcript before a final synthesis prompt produces the answer. Agents do not use tools like Python or SQL; they differ only in the role description that frames their reading of the conversation. As shown in Fig. 1, this turns one LLM call into a structured deliberation among personas, but it also adds cost, noise, and weak handoffs. The design problem is therefore to configure roles and routing, and, just as importantly, to know whether any given configuration is actually helping.

We argue that the conversation produced by such a system is itself event data: ordered handoffs between roles, with intermediate contributions that can be scored and aggregated. This makes it natural to analyze agent runs the way process mining analyzes organizational behavior as a directed social network of actors and handoff relations [9,10]. Doing so lets us ask not only whether a configuration produces good final answers, but *which roles are useful*, *which handoffs are weak*, and *whether the same configuration behaves the same way under different base models*. The diagnostic information that emerges can then be used to revise routing constraints, rather than redesigning agents by intuition.

This paper makes three contributions. First, it defines process-mining-focused conversational configurations expressed as portable role sets and routing constraints, so that the same persona definitions can be instantiated in different conversational-agent frameworks, including *Microsoft AutoGen* and *X.AI Grok 4.20*, without rewriting the agents. Second, it proposes a diagnostic methodology based on Social Network Analysis of recorded agent traces, combining final-answer quality, average role performance, and average handoff performance into a single view of how a configuration behaves. Third, it provides an open source conversational agent implementation that runs these configurations, records their traces, computes the SNA diagnostics, and uses them to constrain weak handoffs in revised configurations.

The empirical findings should be read with these contributions in mind. On 57 process mining questions from PM-LLM-Benchmark [2], agentic configurations do not reliably outperform direct answering; the strongest result we observe is that an SNA-guided pruning of weak handoffs improves a verification-heavy configuration, although equal wins and losses against the no-agent baseline mean this is suggestive rather than conclusive. We also observe that the same role set behaves quite differently under different base models, with sparse early stopping under one model and denser deliberation under others. Together, these results support a diagnostic rather than a performance-driven reading of multi-agent process mining: configurations should be evaluated through their traces, not only their final answers.

The rest of the paper is organized as follows. Section 2 reviews LLMs for process mining, conversational agent frameworks, LLM-as-judge evaluation, and SNA. Section 3 describes the configurations, trace analysis, benchmark, and evaluation procedure. Section 4 reports results in terms of the three contributions. Section 5 presents the implementation, and Sect. 6 discusses implications and limitations.

2 Related Work

LLMs have recently been studied as process mining assistants. Existing work translates process mining artifacts into text, evaluates direct answering and query generation, and discusses benchmarks and evaluation strategies [1,3]. PM-LLM-Benchmark is closest to our setting because it provides process mining questions and LLM-based grading, but it does not evaluate conversational agent configurations or SNA-based trace diagnosis [2].

Conversational agent frameworks are now used in both research and industry. AutoGen, developed at Microsoft, supports configurable conversable agents and group-chat orchestration [12]; CAMEL and MetaGPT show how role assignment can structure complex LLM tasks [4,5]. Industry-facing systems also expose agent composition and execution traces. Figure 2 illustrates a *Grok 4.20* instantiation of the balanced process mining configuration. Our focus is narrower than

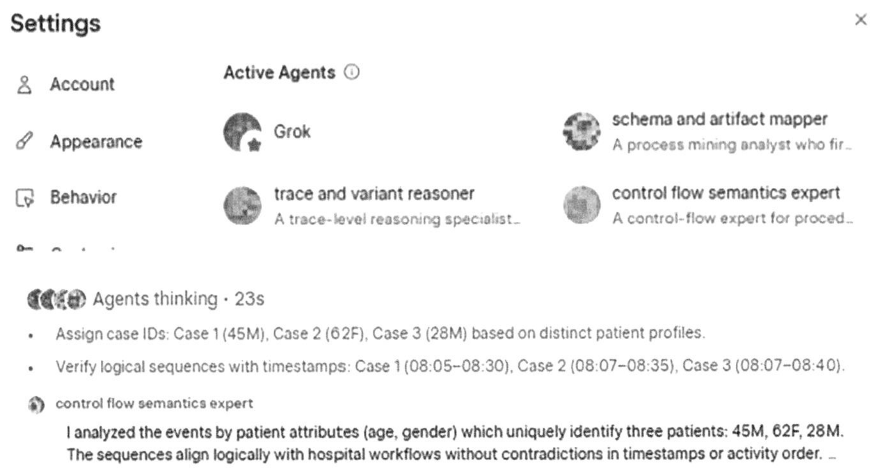

Fig. 2. Conversational-agent support in *Grok 4.20*. The lower screenshot shows an execution trace for a process mining benchmark question, while the upper screenshot shows the active role-specialized agents in the configuration interface. The screenshots illustrate that the balanced configuration can be instantiated in an industrial agent framework.

general tool-using agents: we study role configurations, routing, and trace-based diagnosis for process mining question answering.

LLM-as-judge methods enable scalable grading of open-ended answers, while also introducing biases such as verbosity and position effects [6,13]. We use judge scores as proxy evidence. SNA provides the diagnostic lens: it studies actors and relations [11], and process mining has used event logs to discover handover-of-work relations [9,10]. We transfer this idea from human performers to LLM agent roles.

3 Process Mining Configurations and Evaluation Methodology

This section presents the conversational framework, the three role configurations under study, the SNA measures applied to recorded traces, and the benchmark used to evaluate them.

3.1 Conversational Framework and Configurations

The evaluated system is a centralized, selector-mediated conversational framework. A configuration defines the available agent personas, their role descriptions, the maximum number of iterations, whether only one agent may speak per iteration, and optional forbidden handoffs. Agents do not call external tools and do not communicate directly. Instead, each selected agent reads the original process mining task and the accumulated transcript, then appends one intermediate contribution.

At each iteration, the selector receives the task, available agents, previous answers, the iteration number, and routing constraints. It returns either an agent name or `FINAL`. If an agent is selected, its answer is added to the transcript; if `FINAL` is selected, or the iteration budget is reached, a final synthesis prompt produces the benchmark answer. The no-agent baseline skips this loop and answers the original prompt directly.

We study three complete process-mining-oriented configurations. The artifact-pipeline configuration follows a parse-and-audit structure, with roles for artifact mapping, trace reasoning, control-flow semantics, performance analysis, fairness/compliance review, and answer precision. The balanced configuration covers common process mining task families: event log interpretation, process modeling, conformance checking, querying, hypothesis generation, fairness review, and optimization. The verification-heavy configuration emphasizes robustness through parsing, domain reasoning, formal checking, counterexample search, data auditing, bias auditing, and benchmark-style answer judging.

The configurations are expressed as portable role templates and routing constraints rather than as framework-specific code. Hence, the same role set can be instantiated in different conversational-agent frameworks, such as AutoGen-style group chats or industrial interfaces such as Grok. In this paper, the configuration file is the main experimental object: changing roles or handoff constraints changes the effective conversational system. Table 1 details the roles.

3.2 Trace Representation and SNA Measures

Each benchmark execution produces a trace that is interpreted as a directed social network. The actors are conversational roles. For each question, the judged sequence contains `START`, zero or more selected agent nodes, and `COMPLETE`. Consecutive nodes define directed handoffs. For example,

`START` → `artifact_parser` → `domain_reasoner` → `COMPLETE`

records that the selector first chose the parser, then the domain reasoner, and then stopped.

Each judged node v has a role label and a score $s(v)$ on the 1 to 10 rubric. The `COMPLETE` score measures final-answer quality. Intermediate node scores measure the judged usefulness of selected roles. For role a, with occurrences V_a, the mean role score is

$$\bar{s}_a = \frac{1}{|V_a|} \sum_{v \in V_a} s(v). \tag{1}$$

Edges are scored by the target node after a handoff. For all observed transitions from role a to role b, denoted $E_{a,b}$, the mean handoff score is

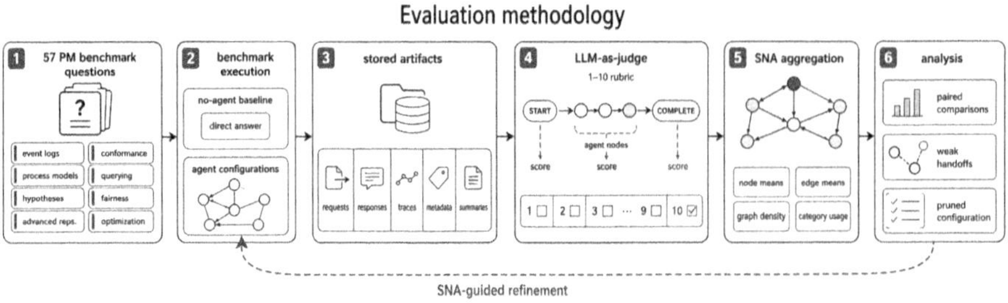

Fig. 3. Evaluation and diagnosis workflow. Stored conversations are judged at node and final-answer level, aggregated into role and handoff networks, and then used to identify routing constraints for revised configurations.

Table 1. Agent roles in the three complete conversational configurations. Each column lists portable role templates and their process mining responsibilities.

Balanced configuration	**Verification-heavy configuration**	**Artifact-pipeline configuration**
`event log interpreter` Reads logs, traces, timestamps, resources, case notions, and data-quality caveats.	`artifact parser` Extracts entities, activities, timestamps, objects, rules, schemas, and missing assumptions.	`schema and artifact mapper` Maps log columns, objects, resources, activities, timestamps, rules, tables, and requested output.
`process modeler` Produces and checks process trees, POWL, DECLARE, log skeletons, temporal profiles, and Petri nets.	`domain reasoner` Builds the main process mining answer across logs, discovery, conformance, querying, fairness, and optimization.	`trace and variant reasoner` Reconstructs variants, infers case notions, compares paths, and detects loops or rework.
`conformance anomaly checker` Compares behavior with models or rules and explains deviations, missing steps, and anomalies.	`formal semantics checker` Checks coherence of process trees, POWL, DECLARE, Petri nets, temporal profiles, and log skeletons.	`control flow semantics expert` Reasons about ordering, concurrency, choices, loops, constraints, and model semantics.
`query and sql analyst` Handles log/model queries, filters, grouping conditions, schemas, and constraints.	`counterexample hunter` Searches for counterexamples, edge cases, impossible orders, and unsupported generalizations.	`performance and resource analyst` Analyzes waiting times, bottlenecks, handoffs, workload, batching, and utilization.
`hypothesis generator` Proposes diagnostic hypotheses, evidence needs, and verification strategies.	`data consistency auditor` Verifies case IDs, attributes, resources, time ordering, SQL logic, and performance claims.	`data question designer` Produces hypotheses, diagnostic questions, query ideas, filters, and evidence plans.
`fairness and ethics reviewer` Checks group impacts, bias risks, and mitigation or monitoring options.	`bias and impact auditor` Reviews group outcomes, protected attributes, interventions, and side effects.	`risk fairness and compliance reviewer` Checks biased paths, missing controls, auditability, privacy assumptions, and fragile recommendations.
`optimization consultant` Identifies bottlenecks, rework, capacity issues, waiting time drivers, and redesign options.	`benchmark answer judge` Checks completeness, grounded reasoning, terminology, and likely benchmark quality.	`answer precision auditor` Verifies coverage, terminology, prompt grounding, and unsupported syntax or conclusions.

$$\bar{s}_{a,b} = \frac{1}{|E_{a,b}|} \sum_{(v_i, v_{i+1}) \in E_{a,b}} s(v_{i+1}). \tag{2}$$

Thus, an edge score answers: when the conversation moves from a to b, how good is the next contribution on average? These scores are diagnostic associations, not causal estimates, because task type, transcript context, and selector behavior also influence the result.

The analysis therefore reports three complementary quantities: final output quality from `COMPLETE`, role-level performance from node means, and handoff-level performance from edge means. We also inspect counts, standard deviations, number of observed edges, graph density, and category-level role usage. Low-scoring repeated handoffs are candidates for routing constraints. This is how the pruned verification-heavy configuration is obtained: it keeps the same roles but excludes four weak transitions identified in the initial SNA.

3.3 PM-LLM-Benchmark and Judge

The evaluation uses 57 questions from PM-LLM-Benchmark [2], a benchmark for assessing LLM answers to process mining tasks expressed through textual logs, models, constraints, queries, and analysis goals. The questions are grouped into eight categories: **i** event log interpretation, covering activities, cases, timestamps, resources, and data quality; **ii** conformance and anomaly analysis, covering deviations from rules or models; **iii** process model generation, covering formal or semi-formal behavior descriptions; **iv** querying and model description, covering SQL-style logic and model explanations; **v** hypothesis generation, covering diagnostic hypotheses and evidence plans; **vi** fairness, covering group comparisons and responsible interventions; **vii** object-centric and advanced representations, covering richer process views; and **viii** optimization, covering bottlenecks, capacity, rework, and redesign.

All final answers and intermediate agent nodes are scored by `openai/gpt-5.4` on a 1.0 to 10.0 process mining rubric. We report means, population standard deviations, call counts, graph summaries, and paired differences over the 57 common questions. Paired p values use a simple two-sided normal approximation and are descriptive. The complete evaluation and diagnosis workflow is summarized in Fig. 3.

3.4 Configuration Rationale and Research Questions

The three configurations in Table 1 are intended to explore different sources of potential benefit: stable parse-and-audit behavior, broad process mining specialization, and verification-oriented review. The pruned variant is intended to explore whether SNA-based diagnostics can inform a configuration's routing without changing its roles.

In this spirit, we organize the empirical study around four exploratory research questions. **RQ1** examines whether the proposed configurations tend to behave differently from direct answering in terms of final-answer quality. **RQ2** examines which roles and handoffs appear to be associated with higher or lower judged quality in the recorded traces. **RQ3** examines whether excluding handoffs flagged as low-performing by the SNA diagnostics is associated with changes in final-answer quality for a given configuration. **RQ4** examines the extent to which a fixed configuration yields comparable social networks across different base models. The questions are framed as diagnostic rather than confirmatory: with 57 benchmark questions and judge-based scoring, the evidence we report is best read as descriptive support for or against a hypothesis, not as a definitive test.

All experiments are conducted via OpenRouter using `openai/gpt-5.4-mini`, `x-ai/grok-4.20`, and `qwen/qwen3.6-35b-a3b` with reasoning disabled. The empirical tables (Table 2, Table 3, and Table 4) report all three base models side by side, so the same observations can be cross-checked across them. For readability, our discussion of **RQ1** through **RQ3** concentrates on the `openai/gpt-5.4-mini` columns, and the cross-model comparison for **RQ4** concentrates on the balanced configuration; the corresponding figures focus on these slices accordingly. We made this choice for narrative clarity rather than because the other combinations behave very differently, and the qualitative picture across the remaining configuration and model combinations is broadly similar. The complete set of runs, judged traces, and SNA summaries is available in the project repository (see Sect. 5).

4 Results

We report the results in three parts, addressing the four research questions in turn.

4.1 Effectiveness of Process Mining Configurations

This subsection addresses **RQ1** and the first contribution. The overall picture from Table 2 is that, on these 57 questions, none of the conversational configurations clearly dominates direct answering. The no-agent baseline turns out to be a competitive reference point. On `openai/gpt-5.4-mini`, the balanced and artifact-pipeline configurations sit close to direct answering on average, the unpruned verification-heavy configuration sits slightly below it, and the pruned verification-heavy configuration is the highest scoring of the five with somewhat lower variance. On `x-ai/grok-4.20` and `qwen/qwen3.6-35b-a3b`, direct answering is the strongest reference, and the pruned verification-heavy configuration tracks closely behind it on `x-ai/grok-4.20`. We therefore read the results as suggestive rather than conclusive: adding role-specialized personas does not, by itself, reliably translate into better final answers, but it does not appear to be uniformly harmful either when routing is reasonable. Looking at the per-category columns, the configurations differ more from each other on the harder categories (notably the formal modeling category c03 and the object-centric category c07) than on the easier ones, which is consistent with the idea that conversational deliberation matters most when the underlying task is itself harder for a single call.

Table 2. Final-answer scores by base model, configuration, and benchmark category. Overall columns report the mean and population standard deviation over all 57 questions; category columns report mean final-answer score within each PM-LLM-Benchmark category.

Conf.	openai/gpt-5.4-mini										x-ai/grok-4.20										qwen/qwen3.6-35b-a3b									
	Avg.	SD	c01	c02	c03	c04	c05	c06	c07	c08	Avg.	SD	c01	c02	c03	c04	c05	c06	c07	c08	Avg.	SD	c01	c02	c03	c04	c05	c06	c07	c08
Direct answering	6.72	2.16	7.30	8.17	4.45	7.01	7.93	6.53	4.07	8.14	5.99	2.56	7.30	7.09	3.41	6.00	7.31	5.14	3.90	7.90	6.38	1.99	6.64	7.02	5.05	6.17	6.86	7.23	4.33	7.80
Balanced	6.75	1.81	6.97	7.74	4.61	6.91	7.94	6.80	5.08	8.00	5.32	2.16	4.80	6.08	4.80	5.60	6.33	5.97	3.23	5.46	5.39	1.83	5.33	6.04	4.08	5.79	5.97	6.24	3.42	6.20
Artifact pipeline	6.74	2.13	7.44	8.04	5.76	6.37	7.11	7.40	3.30	8.00	5.24	2.47	5.26	6.07	4.92	4.39	5.86	5.69	4.10	5.28	5.51	1.63	6.14	6.11	4.54	4.53	6.46	6.11	3.78	6.30
Verification heavy	6.55	1.99	7.15	7.70	4.28	7.47	7.00	6.89	3.80	8.06	5.84	2.13	5.96	6.48	5.64	4.70	6.57	6.50	4.55	5.96	5.32	1.74	6.04	4.78	4.49	5.50	6.33	5.51	3.85	6.34
Ver. heavy excl. hand.	6.98	1.70	7.84	8.01	6.06	6.97	7.14	6.90	4.35	8.28	5.88	2.11	5.10	6.93	4.40	6.67	6.27	6.04	4.92	6.84	5.48	2.05	6.12	6.53	4.72	5.14	5.11	5.54	4.10	6.34

4.2 SNA-Based Diagnosis of Roles and Handoffs

This subsection addresses **RQ2** and **RQ3** and illustrates the second contribution. The SNA diagnostics provide two complementary views of a configuration: a node-level view, which asks how individual roles score when they are selected, and an edge-level view, which asks how the next contribution scores after a particular handoff. Both views are informative, and they tend to highlight different kinds of issues.

At the node level (Table 3), reviewer-oriented roles tend to receive comparatively high intermediate scores when they are selected. On `openai/gpt-5.4-mini`, this shows up in the counterexample hunter, the bias and impact auditor, and the data consistency auditor within the verification-heavy configuration, and in the hypothesis generator and query analyst within the balanced configuration. Broader generative roles such as the process modeler are also selected often, but their scores are more variable. This view is useful on its own: it helps locate where a configuration is reliably contributing and where its variance comes from, independently of the final answer. The category usage in Table 4 adds a useful overlay, showing that certain roles are activated mainly for the categories they target (for example, the fairness reviewer concentrates on c06), while others such as the artifact parser and the answer precision auditor appear broadly across the benchmark.

The edge-level view turns out to be more actionable. In the unpruned verification-heavy configuration, a small set of handoffs recurs across questions and consistently produces lower-scoring next contributions, even when the roles on either end are individually reasonable. Treating those transitions as forbidden, while keeping the same role set, produces the pruned configuration; the resulting handoff graph (Fig. 4) is smaller and more focused, and the associated

Table 3. Role-level SNA node scores by configuration and base model. Count is the number of judged role appearances; Avg. and SD are the mean and population standard deviation of the role's intermediate judge scores. For the no-agent baseline, Direct answer denotes the final answer.

Configuration	Role	openai/gpt-5.4-mini			x-ai/grok-4.20			qwen/qwen3.6-35b-a3b		
		Count	Avg.	SD	Count	Avg.	SD	Count	Avg.	SD
Direct answering	Direct answer	57	6.72	2.16	57	5.99	2.56	57	6.38	1.99
Balanced	Event log interpreter	13	6.75	1.64	32	5.15	2.39	14	6.72	1.66
	Process modeler	22	6.24	1.98	58	4.14	2.00	45	4.32	1.66
	Conformance anomaly checker	12	7.22	1.20	52	3.29	2.01	33	5.05	2.20
	Query and sql analyst	7	7.31	1.33	22	4.60	2.06	12	4.74	1.69
	Hypothesis generator	6	7.82	0.83	46	5.11	1.79	20	5.51	1.84
	Fairness and ethics reviewer	7	7.14	1.28	34	3.73	2.29	20	5.17	2.57
	Optimization consultant	10	6.77	1.71	32	3.67	2.15	39	3.86	1.84
Artifact pipeline	Schema and artifact mapper	41	6.58	1.53	49	6.08	1.63	26	5.92	1.19
	Trace and variant reasoner	2	5.35	2.75	33	4.08	1.87	18	4.44	1.62
	Control flow semantics expert	7	7.16	1.97	57	4.41	2.05	28	5.18	2.03
	Performance and resource analyst	0	–	–	20	3.01	1.16	16	4.74	1.38
	Data question designer	0	–	–	24	4.62	1.91	3	4.43	1.70
	Risk fairness and compliance reviewer	5	7.40	1.25	35	4.29	1.97	11	5.89	1.49
	Answer precision auditor	56	7.53	1.71	53	4.74	2.53	28	4.66	1.71
Verification heavy	Artifact parser	55	6.89	1.79	59	6.88	2.14	26	6.19	1.46
	Domain reasoner	28	6.93	1.32	50	5.27	1.68	54	5.10	1.71
	Formal semantics checker	12	7.02	1.43	63	4.30	2.06	12	5.42	2.17
	Counterexample hunter	10	8.16	0.54	49	5.67	2.46	9	5.56	2.84
	Data consistency auditor	7	7.79	0.78	42	5.38	2.30	8	5.48	1.59
	Bias and impact auditor	4	7.83	0.52	37	3.91	1.85	6	5.80	1.76
	Benchmark answer judge	7	6.73	1.00	58	5.69	2.26	26	5.06	2.09
Ver. heavy excl. hand.	Artifact parser	55	6.53	1.99	58	6.81	2.32	22	6.50	1.69
	Domain reasoner	34	6.82	1.68	52	5.09	1.69	57	5.18	1.49
	Formal semantics checker	10	7.64	1.07	71	4.63	2.35	10	3.50	1.82
	Counterexample hunter	13	7.82	1.14	47	5.66	2.73	16	6.48	2.31
	Data consistency auditor	1	5.80	0.00	41	5.45	2.46	10	4.85	1.41
	Bias and impact auditor	3	7.83	0.73	27	4.38	1.73	7	6.37	1.38
	Benchmark answer judge	2	7.15	0.05	56	5.20	2.23	4	5.02	1.04

final-answer mean reported in Table 2 is modestly higher with somewhat lower variance on `openai/gpt-5.4-mini`. The category-level usage of the pruned configuration in Table 4 indicates that the routing changes preserve the original coverage rather than concentrating effort on a smaller part of the benchmark.

Reading the two views together brings out a recurring pattern: a strong intermediate contribution does not automatically translate into a strong final answer, because the synthesis step still has to integrate that contribution and the conversation has to reach synthesis in a useful state. The edge-level diagnostic is helpful precisely here. It points to transitions that tend to leave the conversation in a worse state, rather than to roles that need to be redesigned. We report the pruning result as an illustration of this diagnostic workflow rather than as a causal claim, since the weak handoffs were identified and re-evaluated on the same benchmark.

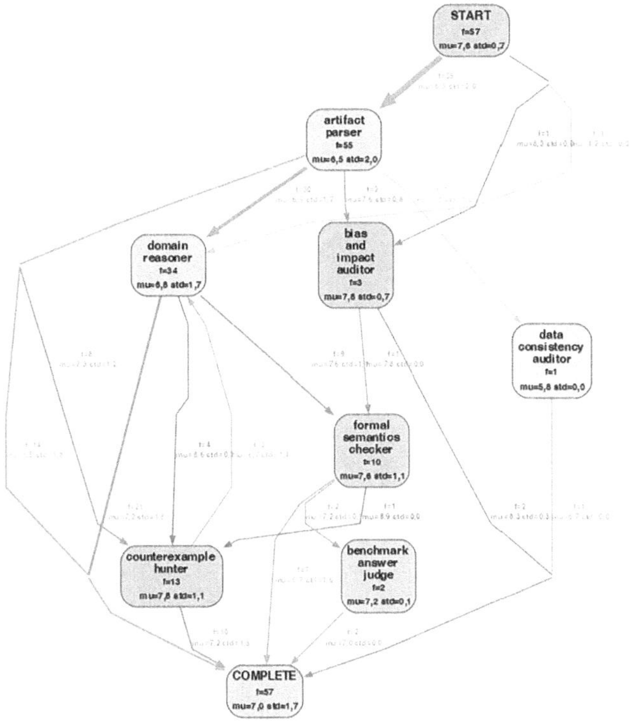

Fig. 4. Social network of the pruned verification-heavy configuration based on `openai/gpt-5.4-mini`. Nodes represent selected roles and final completion; directed edges represent observed handoffs after weak transitions from the unpruned run were excluded.

4.3 Configuration Portability Across Base Models

This subsection addresses **RQ4**. Comparing runs of the same balanced configuration across the three OpenRouter base models (Fig. 5, with supporting counts and role scores in Table 3 and Table 4) suggests that a configuration file is only a partial description of the resulting conversational system. In Fig. 5, node and edge colors encode the average judge score (green for higher means, red for lower means), so the visual texture of each network reflects not just its routing pattern but also the quality of the contributions it elicits. With `openai/gpt-5.4-mini` the selector tends to stop early and yields a sparse network in which most balanced roles are activated only a handful of times across the 57 questions. With `x-ai/grok-4.20` and `qwen/qwen3.6-35b-a3b`, the same roles and the same constraints lead to denser networks with many more role activations and a richer mix of higher and lower scoring transitions. Final-answer quality also varies across models, as can be read off Table 2, but the more interesting observation for our purposes is structural: the social network induced by a configuration is shaped jointly by the role set and by how a particular base model plays the

Table 4. Role usage by benchmark category, configuration, and base model. Each model cell reports counts ordered as c01/c02/c03/c04/c05/c06/c07/c08. For the no-agent baseline, Direct answer counts final answers.

Configuration	Role	openai/gpt-5.4-mini	x-ai/grok-4.20	qwen/qwen3.6-35b-a3b
Direct answering	Direct answer	8/9/8/7/7/7/6/5	8/9/8/7/7/7/6/5	8/9/8/7/7/7/6/5
Balanced	Event log interpreter	7/1/1/0/0/0/2/2	7/4/9/0/2/2/3/5	7/2/0/0/0/0/2/3
	Process modeler	1/3/9/4/0/1/2/2	11/9/17/4/3/2/7/5	8/4/9/5/6/2/7/4
	Conformance anomaly checker	0/5/0/2/3/1/0/1	8/9/13/2/5/4/6/5	6/7/3/4/4/3/2/4
	Query and SQL analyst	0/0/0/3/4/0/0/0	6/1/0/8/1/0/3/3	1/0/0/3/2/4/1/1
	Hypothesis generator	0/0/0/0/6/0/0/0	9/6/1/3/10/6/6/5	0/3/0/0/7/5/4/1
	Fairness and ethics reviewer	0/0/0/0/0/7/0/0	6/4/0/3/2/10/4/5	1/1/0/2/3/12/1/0
	Optimization consultant	0/2/0/1/1/0/1/5	7/4/0/3/4/3/6/5	5/6/2/3/6/5/8/4
Artifact pipeline	Schema and artifact mapper	8/5/7/6/5/2/3/5	8/7/8/6/6/3/6/5	7/1/2/4/2/2/5/3
	Trace and variant reasoner	0/1/1/0/0/0/0/0	6/6/2/4/4/1/5/5	2/4/4/1/2/0/5/0
	Control-flow semantics expert	0/2/2/3/0/0/0/0	6/8/17/7/6/2/6/5	1/6/7/4/4/1/4/1
	Performance and resource analyst	0/0/0/0/0/0/0/0	3/2/0/3/3/0/4/5	0/2/0/2/2/0/6/4
	Data question designer	0/0/0/0/0/0/0/0	2/2/0/4/6/1/4/5	0/0/0/0/2/0/1/0
	Risk fairness and compliance reviewer	0/0/0/0/0/5/0/0	2/3/0/5/6/10/4/5	0/2/0/1/2/5/0/1
	Answer precision auditor	8/9/9/7/6/7/5/5	5/8/9/6/6/7/6/6	1/5/4/6/6/4/1/1
Verification heavy	Artifact parser	8/9/8/7/7/6/5/5	8/9/11/7/7/6/6/5	4/4/1/4/2/3/6/2
	Domain reasoner	2/1/8/6/6/0/0/5	8/8/6/6/6/5/6/5	6/9/9/8/7/2/7/6
	Formal semantics checker	0/1/6/3/0/2/0/0	8/9/18/7/4/6/6/5	1/0/4/3/0/1/3/0
	Counterexample hunter	1/2/1/1/1/2/0/2	5/7/13/4/4/7/5/4	0/2/0/2/1/0/2/2
	Data consistency auditor	2/2/0/0/3/0/0/0	7/6/7/4/4/4/6/4	4/0/1/1/0/1/1/0
	Bias and impact auditor	0/0/0/0/0/4/0/0	0/4/0/2/7/14/6/4	0/0/0/1/0/4/0/1
	Benchmark answer judge	1/1/1/2/1/0/0/1	6/7/11/7/10/7/5/5	4/3/5/3/4/3/3/1
Ver. heavy excl. hand.	Artifact parser	8/9/8/7/7/6/5/5	8/10/9/7/7/5/7/5	5/2/3/3/2/2/5/0
	Domain reasoner	3/4/7/7/5/1/2/5	8/8/7/7/7/3/6/6	8/9/8/7/9/3/7/6
	Formal semantics checker	0/1/5/3/0/1/0/0	10/8/26/7/4/5/6/5	1/1/4/3/1/0/0/0
	Counterexample hunter	2/4/1/1/2/2/0/1	6/6/12/7/4/4/5/3	1/3/3/2/3/0/2/2
	Data consistency auditor	1/0/0/0/0/0/0/0	6/6/5/8/4/4/4/4	2/1/1/1/3/1/1/0
	Bias and impact auditor	0/0/0/0/0/3/0/0	3/3/0/5/4/9/0/3	0/1/0/1/0/5/0/0
	Benchmark answer judge	0/0/0/1/0/1/0/0	7/9/5/8/10/4/8/5	0/1/0/1/1/1/0/0

selector role. This argues for treating configurations as portable templates whose behavior should be re-diagnosed when the underlying model changes, rather than assumed, and it is one reason we report SNA summaries alongside final-answer scores throughout the paper.

5 Implementation and Availability

The framework is publicly available at https://github.com/fit-alessandro-berti/agents-sna under GPL-3.0. It consists of a Python package and a single-file browser tool that share the same configuration format, so a configuration designed in one can be loaded in the other without modification.

Repository Structure. The repository follows the four steps of Fig. 3. The package `src/agents_sna` contains the orchestrator with the selector, agent, and final-synthesis prompts; a CLI entry point `agents-sna`; a `benchmark_runner` that executes a configuration over a local PM-LLM-Benchmark checkout; an `agent_judge` that scores saved benchmark artifacts with an LLM-as-a-judge;

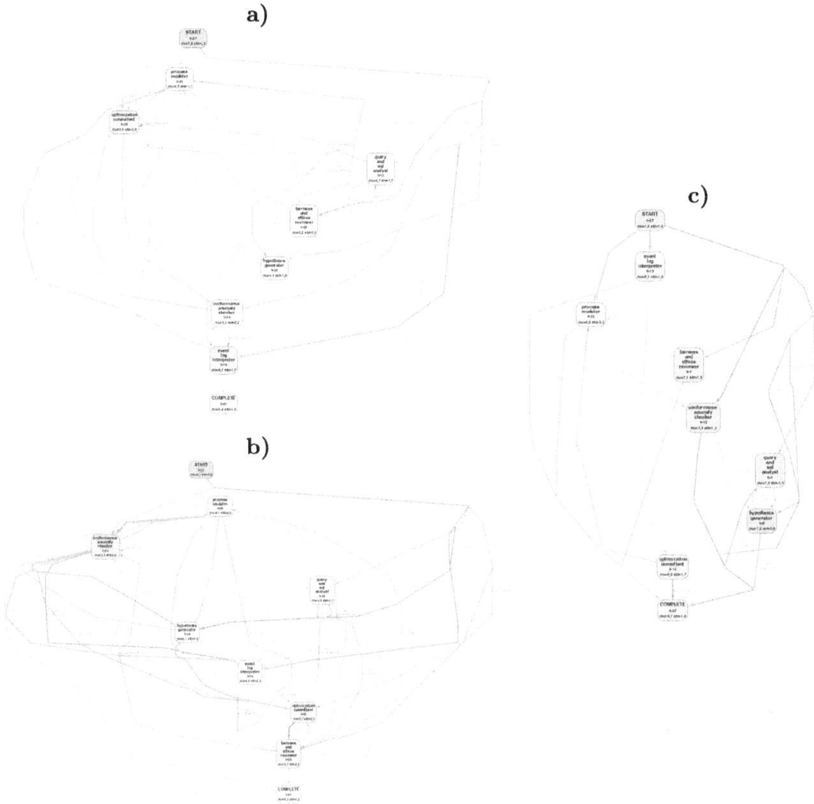

Fig. 5. Balanced-configuration social networks for three OpenRouter base models (reasoning disabled): (a) `qwen/qwen3.6-35b-a3b`, (b) `x-ai/grok-4.20`, and (c) `openai/gpt-5.4-mini`. The same role set leads to different routing behavior: denser deliberation for Grok and Qwen, and a sparser early-stopping pattern for GPT.

and an `evaluation_network` module that aggregates judged traces into the SNA statistics used in this paper. Configurations live in `configs/`, benchmark outputs in `benchmark_runs/`, judged node sequences in `agent_evaluations/`, and unit tests in `tests/`. A shared OpenRouter client lets the base model and any extra payload (e.g., `temperature`, `reasoning`) be set from a single flag.

Configuration Format. A configuration is a self-contained JSON object with four fields. The `agents` list holds role personas, each given by a `name` and a `description` that becomes the system prompt for that role; an empty list reduces the system to direct answering. The integer `max_iterations` bounds deliberation including the final synthesis call, and the boolean `single_agent_per_iteration` switches between one-agent-per-turn and multi-agent regimes. The list `excluded_handoffs` contains pairs {`from, to`} that the selector is forbidden to take, and is the field used to express the SNA-guided routing constraints discussed in Sect. 4.2.

Pipeline Modules. Given a configuration and a prompt, the orchestrator runs the selector-mediated loop of Sect. 3.1: each iteration sends the selector the original prompt, the previous agent answers, and the allowed next agents, and the loop ends when the selector returns `FINAL` or the iteration budget is exhausted. Full request bundles and a compact conversation trace can be exported to JSON. The benchmark runner replays this orchestration over every PM-LLM-Benchmark question, supports concurrent execution and retries on transient OpenRouter failures, and skips already answered questions. The agent judge reads the saved artefacts of a run and produces, for each question, a list of judged nodes with a 1 to 10 score and an explanation; the first node is always `START` and the last `COMPLETE`, which gives the final-answer score. The evaluation network module turns a folder of judged traces into the directed social network used in this paper, computing node and edge counts, mean scores, standard deviations, a category-by-role usage table in LaTeX, and a Graphviz drawing in which colour encodes mean scores and edge thickness encodes handoff frequency.

Browser Tool. Figure 6 shows `agent_chatbot.html`, a self-contained browser tool that exercises the same configuration format interactively. The left panel holds the OpenRouter credentials, the base model, and a JSON editor with a validate button and built-in presets corresponding to the configurations evaluated in this paper, including the pruned verification-heavy variant. The centre panel is a chat composer that shows the prompt, an expandable *run details* block listing each selector decision and agent contribution, and the final answer. The right panel maintains a live handoff network across the session, with counters

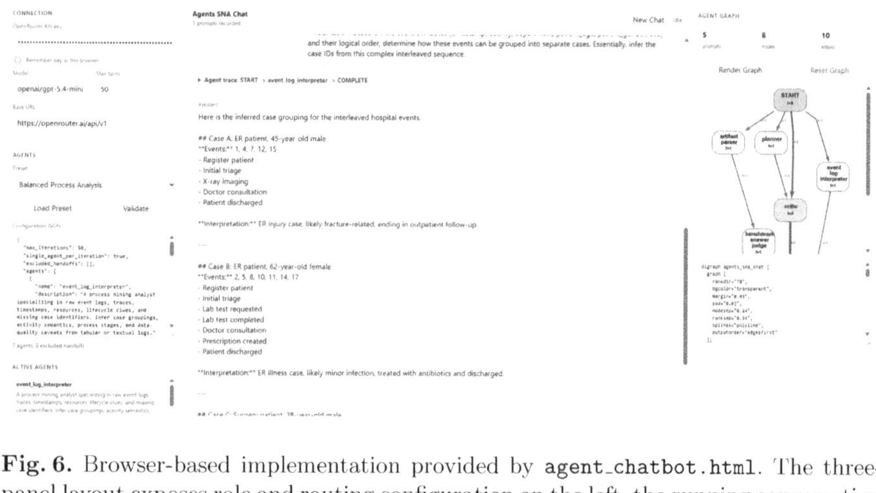

Fig. 6. Browser-based implementation provided by `agent_chatbot.html`. The three-panel layout exposes role and routing configuration on the left, the running conversation in the centre, and a live handoff network on the right.

for prompts, roles, and edges, a *Render graph* button that draws the network in-browser through a Graphviz WebAssembly build, and a read-only DOT-source view that can be copied for offline use. The tool therefore acts both as a usability front end for the configurations and as an inspector for the same SNA diagnostics computed in batch by the Python pipeline.

6 Discussion and Conclusion

We studied selector-mediated conversational agent configurations for process mining and proposed an SNA-based methodology to diagnose them from recorded traces. The agents are role-prompted personas on a shared LLM without external tools, so any value they add must come from how the conversation is structured. We evaluated three role sets on PM-LLM-Benchmark and used node- and edge-level judge scores to identify weak handoffs, which were then removed to obtain a pruned configuration.

Three findings stand out. First, agentic configurations are not automatically better than direct answering: the no-agent baseline is competitive, and only the pruned configuration improved over it on average, suggestively rather than conclusively on 57 questions. Second, role and handoff usefulness are distinct: strong intermediate reviewers do not guarantee strong final answers, so synthesis and routing matter at least as much as role design. Third, the same configuration file behaves differently across base models, producing sparse early-stopping networks under one and denser deliberation under others, so role sets are not portable in isolation from selector behavior.

Two practical points follow. Direct answering should remain a strong baseline, and personas should be added only when they contribute a distinct angle such as formal checking, counterexample search, or fairness review. Conversational agent frameworks should also expose traces, configurations, and handoff statistics as first-class artifacts so that routing can be revised on observed evidence rather than intuition.

Several limitations qualify these results. Judge scores are proxies with possible model-specific biases; the pruning was diagnosed and re-evaluated on the same benchmark and should be re-tested on held-out splits; we did not measure cost, latency, run-to-run variance, or human agreement; and tool-less agents mean that "formal" and "conformance" roles remain LLM personas rather than computational checkers. Future work should combine SNA-based diagnosis with human validation, explore tool-augmented personas (e.g., PM4Py or SQL engines), and study category-specific routing and cost-aware optimization. The broader takeaway is that agent conversations can themselves be treated as event data: both a source of answers and evidence for improving the system that produced them.

Acknowledgments. Funded by the European Union. This work has received funding from the European High Performance Computing Joint Undertaking (JU) and from the German Federal Ministry of Research, Technology and Space (BMFTR), the Ministry of Culture and Science of North Rhine-Westphalia (MKW NRW), and the Hessian Ministry of Science and Research, Arts and Culture (HMWK) under grant agreement No 101250682.

This research has been partially funded by the Federal Ministry of Research, Technology and Space (BMFTR) under grant 16IS23065. The responsibility for the content of this publication lies with the authors.

References

1. Berti, A., Kourani, H.: Evaluating large language models in process mining: capabilities, benchmarks, and evaluation strategies. In: BPMDS/EMMSAD@CAiSE. Lecture Notes in Business Information Processing, pp. 13–21. Springer (2024)
2. Berti, A., Kourani, H., van der Aalst, W.: PM-LLM-benchmark: evaluating large language models on process mining tasks. In: ICPM Workshops. Lecture Notes in Business Information Processing, pp. 610–623. Springer (2024)
3. Berti, A., Qafari, M.: Leveraging Large Language Models (LLMs) for Process Mining. CoRR abs/2307.12701 Technical Report (2023)
4. Hong, S.: MetaGPT: Meta programming for a multi-agent collaborative framework. In: ICLR. OpenReview.net (2024)
5. Li, G.: CAMEL: communicative agents for "mind" exploration of large language model society. In: NeurIPS (2023)
6. Liu, Y.: G-Eval: NLG evaluation using GPT-4 with better human alignment. In: EMNLP, pp. 2511–2522. Association for Computational Linguistics (2023)
7. van der Aalst, W.: Process mining manifesto. In: Business Process Management Workshops (1). Lecture Notes in Business Information Processing, pp. 169–194. Springer (2011)
8. van der Aalst, W.: Process Mining - Data Science in Action, 2nd edn. Springer (2016)
9. van der Aalst, W., Reijers, H., Song, M.: Discovering social networks from event logs. Comput. Support. Cooperative Work. **14**(6), 549–593 (2005)
10. van der Aalst, W.M.P., Song, M.: Mining social networks: uncovering interaction patterns in business processes. In: Desel, J., Pernici, B., Weske, M. (eds.) BPM 2004. LNCS, vol. 3080, pp. 244–260. Springer, Heidelberg (2004). https://doi.org/10.1007/978-3-540-25970-1_16
11. Wasserman, S., Faust, K.: Social Network Analysis: Methods and Applications. Cambridge University Press (1994)
12. Wu, Q.: AutoGen: Enabling Next-Gen LLM Applications via Multi-Agent Conversation Framework. CoRR abs/2308.08155 (2023)
13. Zheng, L.: Judging LLM-as-a-judge with MT-bench and chatbot arena. In: NeurIPS (2023)

Natural Language Querying for Large Declarative Rule Repositories

Max Blaschke[1(✉)], Richard Weber[2], Anjo Seidel[2], Abdur Rehman Anwar Qureshi[2], Adrian Rebmann[3], and Mathias Weske[2]

[1] University of Liechtenstein, Vaduz, Liechtenstein
max.blaschke@uni.li
[2] Hasso Plattner Institute, University of Potsdam, Potsdam, Germany
richard.weber@student.hpi.uni-potsdam.de,
{anjo.seidel,rehman.qureshi,mathias.weske}@hpi.de
[3] SAP Signavio, Berlin, Germany
adrian.rebmann@sap.com

Abstract. In organizations, business processes are constrained by a large number of business rules that can, for example, be represented as process models, declarative constraints, or textual descriptions. When executing, auditing, or managing these processes, users may face specific questions that can only be answered by the declarative business rules that apply in this exact scenario. Non-technical users may only be able to specify their requirements in natural language queries, which need to be answered by specific business rules. However, the efficient matching between the natural language user query and the appropriate declarative business rules is not trivial because rules should be identified based on semantic similarity to the query, which may include synonyms and domain-specific technical terms. In this paper, we propose a system design that combines state-of-the-art technologies, namely Large Language Models, Vector Databases, and Retrieval Augmented Generation, to support cost-efficient natural language querying on large declarative rule repositories to identify appropriate rules. We provide a prototypical implementation of the system design and evaluate it on a large real-world declarative rule repository, showing the technical feasibility of the approach.

Keywords: Declarative Process Models · Process Querying · Large Language Models · Vector Databases · Retrieval Augmented Generation

1 Introduction

Business process management (BPM) has the goal to model and analyze business processes by defining ordering constraints between activities in organizations to realize business goals and create business value [29]. While such processes can be modeled as procedural workflows resulting in standard languages such as

K. Kluza et al. (Eds.): BPM 2026, LNBIP 600, pp. 179–194, 2027.
https://doi.org/10.1007/978-3-032-36710-5_12

BPMN [23], there is also a long tradition in defining processes declaratively through atomic business rules. For instance, DECLARE [1] or DCR [11] define a set of constraints, where each constraint defines a behavioral pattern between a pair of activities. By not capturing all activities at once, this allows to model processes more flexibly. Similarly, business process compliance [16] bridges legal regulations in organizations with process models, by defining prohibited or allowed ordering constraints between activities and with data. Recently, the importance of such declarative business rules has been recognized in industry, for instance, under the term *process atoms* [3,13]. Taking into account a real-world setting, running processes can be constrained by a combination of multiple models [2], resulting in various sources for declarative rules and leading to a large number of rules. Such rules can be stored centrally to represent a unified view of the organizational business rules, for example, envisioned under the term *company memory* [12]. As analyzed in this work, a real-world rule repository may contain hundreds of thousands of rules.

To operationalize business processes in running information systems, process querying [24] interprets business rules and process logic as database queries to evaluate them on the real-world data. This allows process analysts to monitor the performance of their processes, spot violations of business rules, or investigate specific process instances. Still, these insights need to be queried precisely from databases. Declarative business rules can be directly interpreted as process queries, for instance, using PQL [27] or SIGNAL [15]. However, identifying appropriate business rules is especially difficult for non-technical users. We propose the use of a natural language interface for the purpose of fetching declarative business rules from large repositories. However, directly applying large language models (LLMs) is infeasible, motivating to the following requirements:

R1 Resolve natural language questions that use domain terminology and synonyms that do not perfectly match declarative rules.
R2 Employ LLM reasoning cost-efficiently when handling large declarative rule repositories.
R3 Invoke external knowledge to resolve questions that use technical business terms that cannot be resolved without additional context.

In this paper, we propose a system design that is able to answer such natural language user questions to retrieve rules from large repositories. This is achieved by combining state of the art technologies. To support the semantic comparison of activities and ordering constraints (R1), we employ vector databases. To reduce LLM costs (R2), the system design keeps the size of each individual LLM request to a minimum by only considering relevant rules. To resolve technical terms and add additional knowledge (R3), we employ retrieval augmented generation (RAG) as a backup strategy for solving a natural language query. We provide a prototypical implementation of the proposed system design and evaluate it on a real-world large-scale declarative rule repository of 320,972 rules.

In the following, Sect. 2 introduces a motivating example setting for large business rule repositories and natural language queries with requirements R1-3. As the basis of our system design, Sect. 3 provides the technical background on the applied technologies, before Sect. 4 describes the system design and the process to solve natural language queries. The prototypical implementation and its evaluation are presented in Sect. 5, while Sect. 6 discusses our results and related work, before Sect. 7 concludes the paper.

2 Motivating Example

To illustrate our assumed setting of large declarative rule repositories, we use the motivating example of an organization that realizes order-to-cash processes. Once a customer order arrives, the inventory is checked, and the order is packed and shipped. Afterwards an invoice is created and sent to the customer who pays for it. This process can be described and documented through different process models. For instance, a procedural BPMN process diagram describes the intended ordering of activities, a DECLARE model describes ordering constraints more flexibly to allow for deviations, compliance rules define malicious activity orderings that must be avoided, and textual descriptions document the process and best practices in natural language text.

For example, consider the DECLARE model in Fig. 1 that defines different types of ordering constraints between activities. Among others, it defines a *Response(RO, CI)*, such that on every order received (RO), an inventory check must follow (CI). Furthermore, it defines a *Precedence(CI,SO)* enforcing that shipping an order (SO) can only happen if the inventory was previously checked (CI). In total, every arc in the model defines one ordering constraint, resulting in 13 constraints from this model alone.

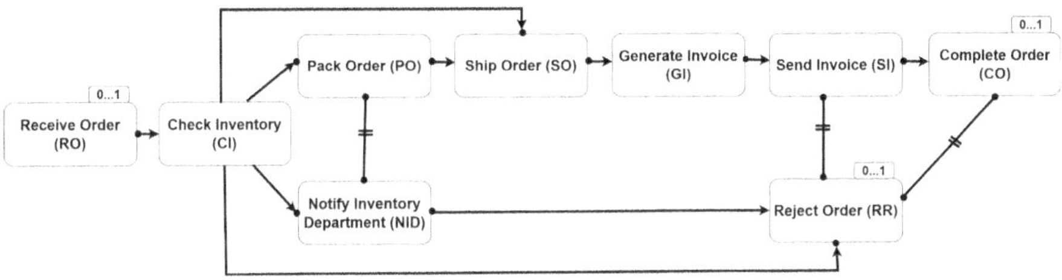

Fig. 1. A DECLARE model for an order-to-cash process.

To access these different sources of process knowledge by process managers, we propose the use of an intermediate representation for storing business rules. As illustrated in Fig. 2, we assume a setting in which all those different sources of process knowledge are brought together in one central declarative rule repository. Such a rule repository combines rules in a well-defined language. Now, we face the challenge that non-technical process managers aim to access such a large rule repository for process analysis and monitoring, who require a natural language interface to query their intended repository rules. Additionally, matching the appropriate declarative rules to their questions is not straightforward. Consider the following two example questions:

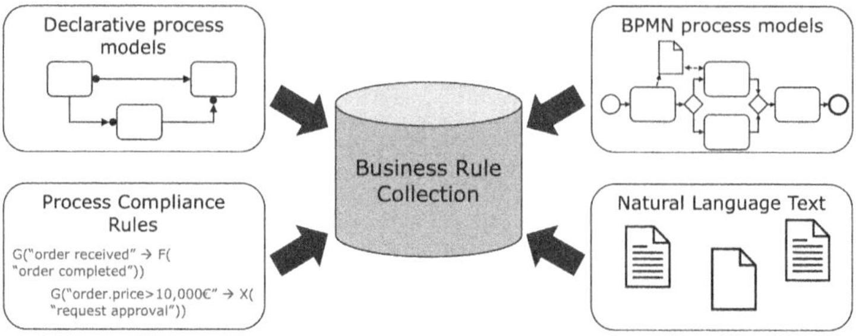

Fig. 2. The setting of large declarative rule repositories that provide a central store for business process rules from different sources.

Q1 "Can I only let the delivery company ship the order if the items ordered are available in our inventory?"
Q2 "Do we enforce Fulfillment Driven Billing in our workflow?"

Even though question Q1 does not directly map to the activity labels and ordering constraints in Fig. 1, it shows a semantic similarity to the rule *Precedence(CI,SO)*, which shall be returned to the user (cf. R1). However, this is even more difficult for question Q2, which refers to the technical term 'Fulfillment Driven Billing'. With the additional domain knowledge that this term refers to triggering the generation of an invoice only after the shipment of an order, one expected rule is *Response(SO,GI)* (cf. R3).

3 Preliminaries

Our approach combines several established technologies. All natural-language interpretation steps of our system are performed by a *Large Language Model* (LLM). LLMs interpret unstructured text flexibly, but large contexts per request drive their computational cost and latency, and can degrade reasoning quality. Reasoning over the whole rule repository is therefore neither cost-efficient nor reliable (cf. R2). Consequently, we keep every LLM call small and targeted and combine LLMs with three further building blocks. For semantic matching of activity labels (R1), we employ vector databases and embeddings. Retrieval-Augmented Generation allows resolving domain-specific terminology (R3), and finally a majority vote improves robustness.

Vector Databases and Embeddings. An *embedding* maps a piece of text x to a d-dimensional real-valued vector $\varphi(x) \in \mathbb{R}^d$ such that semantically related texts lie close to one another in the resulting vector space [21]. Modern embedding models extend this idea from single words to whole phrases and sentences, yielding representations whose geometric proximity reflects semantic similarity and can be compared directly, for instance, through cosine similarity [25]. A *vector database* stores such embeddings and supports retrieval by proximity. Given a query, it returns the stored entries closest, and therefore most semantically related. The *nearest neighbour* of a query text q among a set of stored texts is the entry a whose embedding minimizes the distance $\|\varphi(q) - \varphi(a)\|$. Because exact search over all stored vectors becomes expensive, vector databases rely on approximate nearest-neighbour indices to keep query times low even for large sets.

Retrieval Augmented Generation. The knowledge of an LLM is fixed in its parameters after training, which makes it unreliable for information it has not seen, such as proprietary or domain-specific terminology, and prone to producing plausible but incorrect answers. RAG addresses this by coupling the LLM with an external, non-parametric knowledge source [19]. Before generating, a retriever selects from this source the documents most relevant to the input, which is typically done through a nearest-neighbour query over their embeddings, as described above. Then the retrieved text is added to the LLM's context. Hence, the model can ground its output in information that is not contained in its parameters, which improves factual accuracy and allows the knowledge source to be updated without retraining. RAG can thus serve as a fallback for resolving domain-specific terms that the LLM cannot ground by itself (R3) as unresolved terms can be looked up and be replaced by their retrieved definitions.

Majority Voting. Majority voting is a simple ensemble technique in which several independent predictors each produce a candidate output and the most-supported one is chosen as the final decision; in LLM systems it underlies, for instance, self-consistency over sampled answers [28]. Provided the predictors are reasonably accurate and err somewhat independently, their idiosyncratic mistakes are outvoted by the others, making the combined decision more robust than any single predictor's. Transferred to a vector database, several embedding models each map the phrase to its nearest canonical label, and the label with the most votes is selected. Because different embedding models encode slightly different notions of semantic similarity, a synonym one model maps to a wrong label is likely mapped correctly by the others. Majority voting thereby improves the matching against the lexical ambiguity of synonyms (R1), at the cost of running the search across multiple embedding models.

4 System Design

This section describes the system design for identifying declarative rules from a repository according to a natural language question, which we call the *Filter* in the following. The Filter is designed to meet the three requirements R1–R3 derived in Sect. 1, by using the technologies introduced in Sect. 3. To satisfy R1, activity labels are represented as vector embeddings, and user phrasings are aligned to them through nearest-neighbour search. To satisfy R2, the repository is narrowed to the rules defined over the few matched activities, so that every

LLM call reasons over a small candidate set rather than the full repository. To satisfy R3, unresolved terms are looked up in external knowledge sources and substituted through Retrieval-Augmented Generation (RAG) before the workflow is repeated. We first state the setup and the assumptions our design rests on, including a definition of the repository it operates on, and of the expected natural language questions. Then the workflow of the Filter is described by walking through the two example questions from Sect. 2.

4.1 Setup and Assumptions

The system architecture is displayed in Fig. 3. We assume the *Declarative Rule Repository* to be well defined and provide a glossary of activities as an *Activities (Vector DB)*, consistent with the rules in the *Declarative Rules (SQL DB)*. The user interacts with the Filter through a *Chat Interface*, which receives a natural language question and returns the matching declarative rules. Within the Filter, the question is processed by a chain of components that exchange it along with two intermediate artifacts, the *Activity Candidates* and the *Rule Candidates*. The *Activity Filter* extracts the activities referenced in the question and resolves them against the Activities (Vector DB); the *Rule Filter* retrieves the rules defined over those activities from the Declarative Rules (SQL DB); and the *LLM as a Judge* decides which of the candidate rules answer the question. If no rule is found, the *NLQ Modifier* reformulates the question using *External Knowledge* via RAG, and the process repeats once.

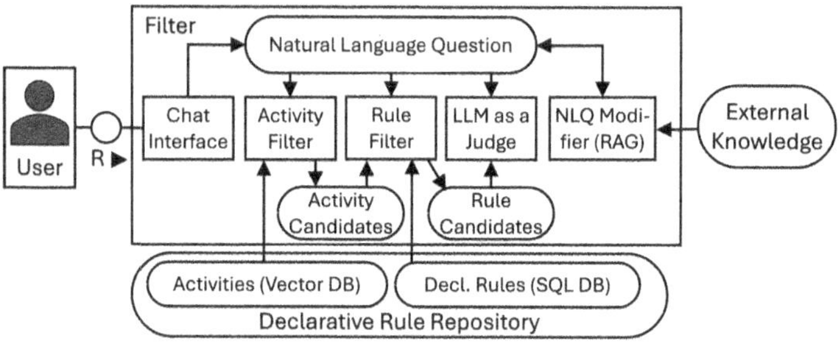

Fig. 3. The system architecture of the Filter for querying declarative rules from large repositories based on natural language questions.

Following Di Ciccio & Montali [6], declarative rules are defined over a fixed set of declarative *templates* Σ_T, i.e., the logical predicates that define the rule types, each of an arity m. The templates in Σ_T are standardized, for example, through declarative modeling languages like DECLARE.

Definition 1 (Declarative Rule Repository, based on [6]). *A declarative rule repository RR is a tuple* $RR = (\Sigma_A, DR)$ *where*

- *Σ_A is a finite non-empty set of activities,*
- *DR is a finite non-empty set of constraints or declarative rules, namely pairs $(\mathrm{T}(x_i, \dots, x_m), t)$ where $\mathrm{T}(x_i, \dots, x_m) \in \Sigma_T$, and t is a mapping that for every $i \in \{1, \dots, m\}$ assigns variable x_i with an activity $t(x_i) = a_i$ where $a_i \in \Sigma_A$, which is compactly denoted as $\mathrm{T}(a_1, \dots, a_m)$.*

We treat the repository *RR* as the unified repository of all business rules in a given context. Although Sect. 1 established that this knowledge originates from heterogeneous sources such as process models, compliance rules, and textual documentation, we assume that these sources have already been mapped into the common declarative representation of Definition 1. The construction of such a repository and the distributed retrieval from its underlying sources are out of scope. We assume a unified access interface that exposes the rules *DR* and their activities Σ_A in this format.

Within the repository, the *Activities (Vector DB)* holds the vector embeddings of all activity labels in Σ_A and supports the nearest-neighbour search (Sect. 3), while *Declarative Rules (SQL DB)* stores the rules *DR*. Beyond the repository, an external knowledge space—*External Knowledge* in Fig. 3—provides documents such as glossaries or domain documentation that the NLQ Modifier consults to resolve domain-specific terminology (Sect. 3). All natural-language processing steps within the Filter are performed by an *LLM*.

Finally, we characterize the questions the system is meant to answer. Note that an *educated* user, i.e., a process manager with at least an approximate idea of what they seek and the ability to paraphrase it as a single natural language question, is assumed to interact with the system.

Definition 2 (Natural Language Question). *Given a declarative rule repository $RR = (\Sigma_A, DR)$, a* natural language question *q is a finite text that describes one or more behavioral constraints over the activities in Σ_A. A question references (i) a set of activities, possibly through synonyms or domain-specific terminology rather than their canonical labels in Σ_A; (ii) one or more ordering or existence constraints relating these activities; and (iii) the* modality *under which they are queried, which is either* existential, *asking whether the described behavior is* possible, *or* universal, *asking whether it is* enforced. *Its answers form the subset of declarative rules $R_q \subseteq DR$ that decide q under this modality. This subset may be empty, for example, when answering the question requires domain knowledge that is not captured in DR.*

To illustrate, consider the first running example of Sect. 2, Q1: *"Can I only let the delivery company ship the order if the items ordered are available in our inventory?"* It references (i) two activities, "Ship the order" and "if [...] available in our inventory" (checking the availability), both given as paraphrases rather than the canonical labels *Ship Order* and *Check Inventory*; (ii) a single ordering constraint relating them, namely that shipping is conditioned on a prior availability check; and (iii) a *universal* modality, as the phrasing "only [. . .] if" asks whether this ordering is *enforced*. Therefore, a possible set of answers is $R_q = \{Precedence(Check\ Inventory, Ship\ Order)\}$.

Consequently, questions that do not follow this convention or ones that do not reference constraints of the repository are not expected to be answered correctly or at all. Definition 2 also explicates the main difficulties. First, the activities and constraints in q are expressed in the user's own words (R1), and may be hidden behind domain-specific terms that first need to be resolved against external knowledge (R3). Together with the size of DR, which prohibits the naive approach of handing the entire repository to the LLM at once (R2), these difficulties motivate the following workflow.

4.2 Workflow for Resolving Natural Language Questions

Figure 4 depicts the workflow of the Filter. It first narrows the question down to a set of relevant activities, then retrieves only the rules defined over those activities, and finally assesses these few candidates individually. Each LLM call, therefore, operates on a small fraction of the repository. The workflow is described by following the two example questions of Sect. 2.

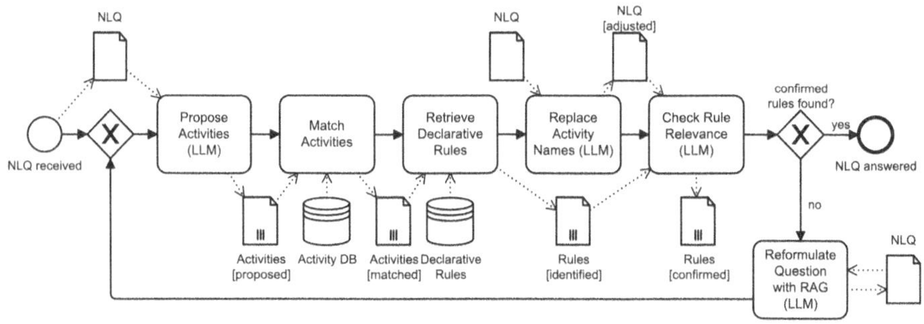

Fig. 4. Workflow for identifying suitable declarative business rules for a given natural language question.

Answering Q1 without RAG. First, consider Q1, *"Can I only let the delivery company ship the order if the items ordered are available in our inventory?"*, which does not contain technical terminology but references the activity labels via synonyms directly. After the question is received, the Activity Filter prompts the LLM to *propose activities*, i.e., to extract the phrases that plausibly denote process activities. For Q1, the LLM returns the two candidates *inform delivery company to ship order* and *items available in inventory*. Note that neither phrase matches a label in Σ_A verbatim.

In the *match activities* step, each candidate phrase is embedded and compared with the embeddings of all activity labels in the activity database, and its nearest neighbour—agreed upon by a majority vote over several embedding models (Sect. 3)—is selected as the canonical label. The two candidates are thereby matched to the existing labels *Ship Order* and *Check Inventory*, which represent the Activity Candidates. This bridges the lexical gap between the user's wording and the repository's labels and realizes the embedding-based matching decision introduced above, by which the system handles synonyms (R1).

Given the matched activities, the Rule Filter can *retrieve declarative rules* by querying DR for every rule defined over *Ship Order* and *Check Inventory*, yielding, among others, *Precedence*(*Check Inventory*, *Ship Order*) as the Rule Candidate(s). Only the rules over these few activities are retrieved, keeping the subsequent reasoning small, and no LLM call is invoked with the full repository DR, which is how the design addresses R2.

Because the canonical labels differ from the user's wording, the LLM then *replaces the activity names* in the original question with the matched labels, producing an adjusted question that refers to *Ship Order* and *Check Inventory* directly. Aligning the question with the repository's vocabulary makes the following assessment by another LLM call more reliable.

In the *check rule relevance* step, the LLM as a Judge is asked, for each retrieved candidate rule in isolation, whether the rule answers the adjusted question. The rule *Precedence*(*Check Inventory*, *Ship Order*) is judged to be relevant, as it expresses precisely that shipping may only occur after the inventory has been checked, while unrelated rules are discarded. Since at least one relevant rule has been confirmed, the gateway routes directly to providing the answer to the question, and the confirmed rule is returned to the user.

Answering Q2 with RAG. The injection of RAG applies for Q2 *"Do we enforce Fulfillment Driven Billing in our workflow?"*. When the LLM proposes activities, it only identifies *Enforce Fulfillment Driven Billing*, a domain-specific term that does not directly correspond to any activity in Σ_A. Its nearest-neighbour search maps to its closest existing label, *Generate Invoice*. As only this single activity is matched, no declarative rule is defined over it, and the retrieval step returns no constraints. The LLM as a judge detects that no rule answers the question properly and the gateway diverts the flow to *reformulate question with RAG* (Sect. 3). The NLQ Modifier looks up the unresolved term in the knowledge space and replaces it with its definition. *Fulfillment Driven Billing*, for instance, is reformulated as the statement that an invoice is generated only after the order has been shipped. The workflow then repeats once, starting again from *propose activities* but now operating on the reformulated question. This time, the LLM extracts *ship order* and *generate invoice*, which are matched to *Ship Order* and *Generate Invoice*. Hence, the retrieved rule *Response*(*Ship Order*, *Generate Invoice*) is confirmed to be relevant and is returned. Resolving domain-specific terminology against external documents via RAG in this way is how the system satisfies R3.

5 Evaluation

To evaluate the proposed Filter, we provide a prototypical implementation and assess it along the three requirements that motivate our approach, i.e., scaling to industry-sized rule repositories, bridging the lexical gap between a user's wording and the activity labels in the repository by resolving synonyms and naming inconsistencies, and resolving domain-specific terminology that the rules themselves do not capture. First, the implementation (Sect. 5.1) is described, then the evaluation setup and the construction of the test data (Sect. 5.2), and finally the results are presented and interpreted (Sect. 5.3). The prototype and technical configurations plus all evaluation data for an artificial non-proprietary dataset are publicly available online[1].

5.1 Prototypical Implementation

The Filter described in Sect. 4 was implemented as a Python application, structured as a pipeline of composable filters, each implementing one step of Fig. 4 over a shared data model, which is orchestrated through LangChain. Because the evaluation relies on proprietary data, all language models are run locally via Ollama. The main reasoning model is `mistral-small3.1:24b`, used for activity extraction, query reformulation, and relevance judgment.

The declarative rule repository is stored in a PostgreSQL database, while the activity database is realized as a pgvector store. To map a proposed activity onto a canonical label, the system embeds it with five embedding models (`granite-embedding`, `nomic-embed-text`, `mxbai-embed-large`, `bge-large`, and `all-MiniLM-L6-v2`) and combines their nearest-neighbour results through a majority vote (Sect. 3). The optional RAG step (Sect. 3) draws on a separate knowledge space of plain-text documents, indexed in a FAISS store, that define domain-specific terms. In the exemplary order-to-cash setting, for instance, it provides the definition of *Fulfillment Driven Billing*. The eleven supported DECLARE templates are each grounded in LTL_f semantics and paired with a natural-language description and a template-specific judge prompt. Execution is traced throughout with Arize Phoenix for monitoring and analysis.

5.2 Evaluation Setup

Evaluation of the filter is done on two declarative business rule repositories. The *Large Rule Collection (LRC)* is a proprietary industry-scale set of 320,972 declarative rules, and serves to assess scalability under realistic conditions. For cost-effective testing, a *Small Rule Collection (SRC)* was developed as a subset of 52 declarative rules on 5 distinct activity labels. These are distinct from the order-to-cash example from Sect. 2, with which the public repository is instantiated.

The test data are derived manually from an initial process model that we treat as ground truth. As characterized in Definition 2, a question references a set of activities and one or more constraints on them, queried under the given modality. Accordingly, each test question is constructed by selecting one or two activities and a relational constraint over them, choosing the ground-truth answer from

[1] https://github.com/DroneArd/Natural-Language-Querying-for-Large-Declarative-Rule-Repositories.

the collection, and phrasing the constraint as a single natural language question. A question may be answered by a single rule or by several rules over the same activities. Its answer may also be empty, for instance, when the queried behavior is not constrained fittingly in the collection. The resulting standard test set comprises 32 questions with straightforward labels and their predetermined answers.

First, we evaluate the activity proposal (Test 1) and the activity replacement (Test 2) in isolation on a separate set of 50 questions, half of which reference a single activity and half of which reference two. The end-to-end capabilities are evaluated in the standard set with straightforward labels (Test 3). In the synonym and naming variants, the phrases referencing activities are replaced by increasingly distant synonyms (Test 4), or are chosen to target activities whose labels are worded inconsistently across rules (Test 5). In these synonym-heavy variants, the majority vote and the activity replacement are additionally toggled to quantify their individual contribution (Test 6). Finally, the queried constraint is expressed through a domain-specific term that does not occur among the activity labels and must first be resolved against the knowledge space (Test 7).

The F_1-score is used as the primary metric. For each question, the rules returned by the Filter are compared against the predetermined ground-truth answer to obtain true positives, false positives, and false negatives, from which precision, recall, and the F_1-score are computed. The two stage-isolation tests for activity replacement and proposal instead report the raw number of correctly handled questions, as they concern a single extracted or replaced label rather than a set of rules. All experiments were run on a Mac Studio (Apple M2 Ultra, 192 GB RAM).

5.3 Evaluation Results

The results of the individual tests are reported in Table 1, building up from the isolated pipeline stages to the complete end-to-end Filter. All F_1-scores in the

Table 1. Summary of test results, reporting the correctness and F_1-score, and a short description. All F_1-scores were measured on the Large Rule Collection (LRC).

ID	Result	Short description
1	48/50 correct	Activity proposal only (50-question set)
2	48/50 correct	Activity replacement only (50-question set)
3	$F_1 = 0.85$	End-to-end Filter, straightforward labels
4	$F_1 = 0.45$	Synonym-heavy NLQs
5	$F_1 = 0.38$	Label naming inconsistencies
6	$F_1 = 0.28/0.30/0.45$	Synonym-heavy NLQs without majority voting/without activity replacement & full
7	$F_1 = 0.39/0.80$	Technical-term-heavy NLQs without/with RAG

table are measured on the LRC. Wherever the SRC yields a substantially different picture, it is reported in the text. The average response time per question was around 60 s on the hardware described in Sect. 5.2.

First is the validation of the two LLM-driven stages in isolation (Tests 1 and 2). On a dedicated set of 50 questions, the system extracts the correct activities in 48 cases and, given the confirmed labels, replaces them correctly in the question in 48 cases. Both stages are thus reliable, which lets us attribute the errors observed below primarily to the matching and judgment.

Building on these stages, Test 3 evaluates the complete Filter on questions with straightforward labels and yields an F_1-score of 0.85 on the LRC. On the SRC, the same test reaches 0.87. The negligible difference between a collection of 52 and the one of 320,972 rules indicates that the approach scales to industry-sized repositories without a notable loss in quality, as long as the wording of the question stays close to the activity labels.

Once the questions diverge lexically from the labels, this changes. With increasingly distant synonyms (Test 4), the F_1-score drops to 0.45 on the LRC, while the SRC still attains 0.65. The widening gap between the two collections shows that the difficulty is not scale alone but its interaction with lexical ambiguity: in a larger collection, a synonym has many more nearby labels to which it can be mismatched. Label naming inconsistencies (Test 5) amplify this further, with the LRC falling to 0.38 against 0.82 on the SRC, as the SRC does not have much ambiguity. Test 6 isolates the contribution of the two features that target this ambiguity. Starting from 0.45 of the full pipeline on synonym-heavy LRC questions, disabling the embedding majority vote lowers the score to 0.28, and disabling the activity replacement in the question lowers it to 0.30. Both mechanisms therefore recover a substantial part of the performance lost to synonyms, although neither closes the gap to the straightforward case.

Finally, Test 7 examines the RAG fallback on technical-term-heavy questions. Without it, such questions cannot be grounded and the Filter reaches only 0.39. Enabling the fallback raises the score to 0.80, as the unresolved terms are replaced by their definitions before the workflow is repeated once. The fallback does not harm questions that need no resolution as rerunning the straightforward questions of Test 3 with RAG enabled left the F_1-score unchanged at 0.85 but increased the average response time from 60 s to 95 s. RAG thus contributes where external knowledge is required and incurs only a runtime cost otherwise.

In terms of the requirements, the evaluation supports scaling to an industry-sized repository (R2, see test 3) and the resolution of domain-specific terminology via RAG (R3, see test 7), while synonyms and naming inconsistencies remain the main limitation (R1), only partially mitigated by the majority vote and activity replacement (see tests 4, 5, and 6).

6 Discussion and Related Work

Our system design aims to reduce the size of LLM calls and therefore also associated token consumption and costs. Following the main objective to enable the querying of large rule repositories in the first place, this paper does not focus on a systematic evaluation of the cost drivers of such a system. The principal cost drivers nonetheless follow from the workflow as the relevance judgment issues one LLM call per candidate rule, so the number of LLM invocations per query scales with the number of rules defined over the matched activities rather than with the repository size $|DR|$. Further drivers are the number of embedding models used in the majority vote (five in our prototype) and the RAG fallback, which adds a retrieval step and one repeated pass, reflected in the increase from 60 s to 95 s per question reported previously. On locally hosted models, these costs would most naturally be expressed in tokens or GPU-seconds, and in API pricing when using proprietary ones. In the future, different configurations of rule repositories at different scales, natural language questions, and different use cases should be analyzed to provide benchmarks on good tradeoffs between quality of results and processing costs. Such benchmarks should also broaden the set of reported metrics by reporting precision and recall separately alongside the F1-score, since retrieval for auditing prioritizes recall whereas interactive querying may favor precision, and the matching stage could be assessed with rank-aware measures such as recall@k to distinguish retrieval failures from judgment failures. Furthermore, our evaluation only employs one LLM instance. In the future, different models should also be compared. In addition, a first, preliminary comparison across the eleven supported DECLARE templates indicated that performance varies by template, but a systematic study of this effect is still required; likewise, the influence of the judge-prompt design and of model size on both quality and cost remains open, as the current evaluation fixes a single prompt configuration.

For the scope of this paper, we assume that a given business rule repository is an input. However, to our knowledge, the construction of such repositories has not yet been studied in detail and is only envisioned by large vendors [12–14]. Therefore, our evaluation restricts itself to a large set of declarative rules that are in line with Definition 1. Still, we show that such a large proprietary rule repository can be efficiently queried. In the future, more research is needed on the characteristics of business rule repositories in real-world settings, and our formalization needs to be extended to capture those features.

In BPM, different works study the effective integration of AI, in general, and LLMs, in particular, with business processes. An overview of AI-augmented BPM to support process managers with AI technologies can be found in [7], while the use cases of LLMs are described by Vidgof et al. [26] and Estrada-Torres et al. [8]. The rise of LLMs and their potential as AI agents results in the vision of agentic BPM [5] with the goal of having AI agents execute business processes autonomously. Our approach is tailored to human users and their questions. However, AI agents may also use our approach to query business rules.

More specifically, LLMs have been studied to produce declarative constraints represented as LTL from prompts [10,20]. Furthermore, LLMs have been studied for the task of process modeling [18] and process mining tasks [4]. Especially conversational process modeling [17], can be a starting point for future research to improve our results by involving the user in solving their questions. In addition, LLMs are used to explain business processes [9]. For this use case, RAG has been shown to improve the results of process explanation [22]. In a similar direction, the intermediate artifacts of the Filter already, partially, lend themselves to explaining its answers. The resolved activity labels reveal how a user's wording was mapped to canonical labels, the adjusted question shows the reformulation, the per-rule judgments record why each candidate rule was kept or discarded, and any definition retrieved via RAG documents how a domain-specific term was resolved. While the prototype already surfaces these artifacts as an explanation trace, evaluating such explanations with users is a promising direction for future work.

Applying RAG in our approach assumes that the required information to answer a question can be retrieved from the organization. In our design, this information is drawn from the External Knowledge space, which we envision to consist of organizational glossaries and domain documentation that define the technical terms in use. It needs to be studied whether real-world organizations hold this information in an accessible way. Where a queried term is neither expressed through the repository's activity labels nor defined in the available documentation, it cannot be grounded, and the corresponding question falls into the empty-answer case already anticipated by Definition 2. How often the necessary definitions can realistically be expected to exist is an empirical question that we leave to future work.

7 Conclusion

Large business rule repositories can capture organizational process knowledge. However, identifying relevant rules for a specific question remains challenging, particularly for non-technical users who express their information needs in natural language. This paper presented a system design for natural language querying of large business rule repositories that combines Large Language Models, vector databases, and Retrieval Augmented Generation. The approach addresses the key challenges of handling synonyms and technical terms, and scaling to large rule repositories. By combining semantic retrieval with targeted LLM reasoning, the proposed architecture enables efficient identification of relevant business rules while limiting the amount of information processed in each reasoning step. A prototypical implementation and evaluation of a small and a large real-world rule repository demonstrate the technical feasibility of the approach.

Acknowledgements. The authors thank Timotheus Kampik and Gregor Berg from SAP Signavio and the Bachelor students Oscar Ulbricht, Victor Lindow, Dennis Sima, Noah Khalil, and Benedikt Zimmermann from the Hasso Plattner Institute, University of Potsdam, for their contributions to this research.

References

1. van der Aalst, W.M.P., Pesic, M., Schonenberg, H.: Declarative workflows: balancing between flexibility and support. Comput. Sci. Res. Dev. **23**(2) (2009). https://doi.org/10.1007/s00450-009-0057-9
2. Alman, A., Maggi, F.M., Rinderle-Ma, S., Rivkin, A., Winter, K.: Towards a multi-model paradigm for business process management. In: Conference on Advanced Information Systems Engineering. Lecture Notes in Computer Science. Springer (2024). https://doi.org/10.1007/978-3-031-61057-8_11
3. Badakhshan, P.: Process atoms: towards AI-powered business process management (2025). https://community.sap.com/t5/technology-blog-posts-by-sap/process-atoms-towards-ai-powered-business-process-management/ba-p/14295617, SAP Community Blog. Accessed 01 June 2026
4. Berti, A., Kourani, H., van der Aalst, W.M.P.: PM-LLM-benchmark: evaluating large language models on process mining tasks. In: Process Mining Workshops - ICPM. Lecture Notes in Business Information Processing. Springer (2024)
5. Calvanese, D., et al.: Agentic business process management: a research manifesto. Inf. Syst. **140**, 102738 (2026)
6. Ciccio, C., Montali, M.: Declarative Process Specifications: Reasoning, Discovery, Monitoring. Lecture Notes in Business Information Processing, vol. 448. Springer (2022). https://doi.org/10.1007/978-3-031-08848-3_4
7. Dumas, M., et al.: AI-augmented business process management systems: a research manifesto. ACM Trans. Manag. Inf. Syst. **14**(1) (2023). https://doi.org/10.1145/3576047
8. Estrada-Torres, B., del-Río-Ortega, A., Resinas, M.: Mapping the landscape: exploring large language model applications in business process management. In: Enterprise, Business-Process and Information Systems Modeling. Lecture Notes in Business Information Processing. Springer (2024)
9. Fahland, D., Fournier, F., Limonad, L., Skarbovsky, I., Swevels, A.J.E.: How well can a large language model explain business processes as perceived by users? Data Knowl. Eng. **157** (2025). https://doi.org/10.1016/J.DATAK.2025.102416
10. Fuggitti, F., Chakraborti, T.: Nl2ltl - a python package for converting natural language (NL) instructions to linear temporal logic (LTL) formulas. In: Proceedings of the AAAI Conference on Artificial Intelligence (2023). https://doi.org/10.1609/aaai.v37i13.27068
11. Hildebrandt, T.T., Mukkamala, R.R., Slaats, T.: Nested dynamic condition response graphs. In: Fundamentals of Software Engineering. Lecture Notes in Computer Science. Springer (2011)
12. Houchen, D.: Why enterprise agents need sap signavio's company memory (2026). https://www.signavio.com/post/why-enterprise-agents-need-sap-signavios-company-memory/, SAP Community Blog. Accessed 01 June 2026
13. Kampik, T.: The role of process atoms in modern business process management (2023). https://community.sap.com/t5/technology-blog-posts-by-sap/the-role-of-process-atoms-in-modern-business-process-management/ba-p/13578285, SAP Community Blog. Accessed 01 June 2026
14. Kampik, T., et al.: Large process models: a vision for business process management in the age of generative AI. KI - Künstliche Intelligenz **39**(2) (2025). https://doi.org/10.1007/s13218-024-00863-8
15. Kampik, T., Lücke, A., Horstmann, J., Wheeler, M., Eickhoff, D.: SIGNAL - the SAP signavio analytics query language. CoRR abs/2304.06811 (2023)

16. Kharbili, M.E., Medeiros, A.K.A.D., Stein, S., van der Aalst, W.M.: Business process compliance checking: current state and future challenges. Modellierung betrieblicher Informationssysteme (MobIS 2008) (2008)
17. Köpke, J., Safan, A.: Efficient LLM-based conversational process modeling. In: Business Process Management Workshops. Lecture Notes in Business Information Processing. Springer (2024). https://doi.org/10.1007/978-3-031-78666-2_20
18. Kourani, H., Berti, A., Schuster, D., van der Aalst, W.M.: Evaluating large language models on business process modeling: framework, benchmark, and self-improvement analysis. Softw. Syst. Model. (2025)
19. Lewis, P., et al.: Retrieval-augmented generation for knowledge-intensive NLP tasks. In: Advances in Neural Information Processing Systems (2020). https://doi.org/10.48550/arXiv.2005.11401
20. Liu, J.X., et al.: Grounding complex natural language commands for temporal tasks in unseen environments. In: Conference on Robot Learning (CoRL). Proceedings of Machine Learning Research, vol. 229 (2023). https://arxiv.org/abs/2302.11649
21. Mikolov, T., Chen, K., Corrado, G., Dean, J.: Efficient estimation of word representations in vector space. arXiv preprint arXiv:1301.3781 (2013)
22. Minor, M., Kaucher, E.: Retrieval augmented generation with LLMs for explaining business process models. In: Case-Based Reasoning Research and Development. Lecture Notes in Computer Science. Springer (2024). https://doi.org/10.1007/978-3-031-63646-2_12
23. OMG: Business Process Model and Notation (BPMN), Version 2.0 (2011). http://www.omg.org/spec/BPMN/2.0
24. Polyvyanyy, A.: Business process querying. In: Encyclopedia of Big Data Technologies. Springer (2019). https://doi.org/10.1007/978-3-319-63962-8_108-1
25. Reimers, N., Gurevych, I.: Sentence-bert: sentence embeddings using siamese bert-networks. arXiv preprint arXiv:1908.10084 (2019). https://arxiv.org/abs/1908.10084
26. Vidgof, M., Bachhofner, S., Mendling, J.: Large language models for business process management: opportunities and challenges. In: Business Process Management Forum. Lecture Notes in Business Information Processing. Springer (2023). https://doi.org/10.1007/978-3-031-41623-1_7
27. Vogelgesang, T., Ambrosy, J., Becher, D., Seilbeck, R., Geyer-Klingeberg, J., Klenk, M.: Celonis PQL: a query language for process mining. In: Process Querying Methods. Springer (2022). https://doi.org/10.1007/978-3-030-92875-9_13
28. Wang, X., et al.: Self-consistency improves chain of thought reasoning in language models. In: International Conference on Learning Representations (2023). https://doi.org/10.48550/arXiv.2203.11171
29. Weske, M.: Business Process Management: Concepts, Languages, Architectures. Springer (2024). https://doi.org/10.1007/978-3-662-69518-0

Representing BPMN Choreographies in OCEL 2.0

Richard Hobeck[1(✉)], Alessandro Marcelletti[2], Andrea Morichetta[2], and Ingo Weber[3,4]

[1] Service-Centric Networking, Technical University of Berlin, Berlin, Germany
richard.hobeck@tu-berlin.de
[2] University of Camerino, Camerino, Italy
{alessand.marcelletti,andrea.morichetta}@unicam.it
[3] School of CIT, Technical University of Munich, Munich, Germany
ingo.weber@tum.de
[4] Fraunhofer Gesellschaft, Munich, Germany

Abstract. Inter-organizational business processes describe the interaction between distributed parties. BPMN choreographies offer a modeling notation for business interactions between participants, setting aside their internal behaviors. In process mining, existing literature views choreographies through a single-case notion lens, attempting to fit interaction data involving several business entities into a flat event log format. In this paper, we investigate how the object-centric event log format OCEL 2.0 can be used to express choreography data. To that end, we present a mapping between the BPMN choreography metamodel and OCEL 2.0 elements. The mapping enforces BPMN choreography semantics by defining data constraints. We demonstrate and evaluate the approach with 19 datasets from different domains. We provide implementations to extract choreography data, validate choreography event logs, and convert them to BPMN choreography diagrams.

Keywords: BPMN · Choreography · OCEL 2.0

1 Introduction

Business processes frequently span multiple organizations, where interactions between process participants play a defining role. Business Process Model and Notation (BPMN) provides a well-established notation for representing inter-organizational processes in models. These include *collaboration models* which show participants' internal behavior and interaction points, and *choreography models* which focus on the interaction of participants, their message exchange, and their hierarchical structure [17].

Prior work has applied *process mining* to inter-organizational processes using both of BPMN's notational bases, collaboration models [1,2] and choreography models [12,18]. The choreography mining initiatives assumed a single-case

K. Kluza et al. (Eds.): BPM 2026, LNBIP 600, pp. 195–210, 2027.
https://doi.org/10.1007/978-3-032-36710-5_13

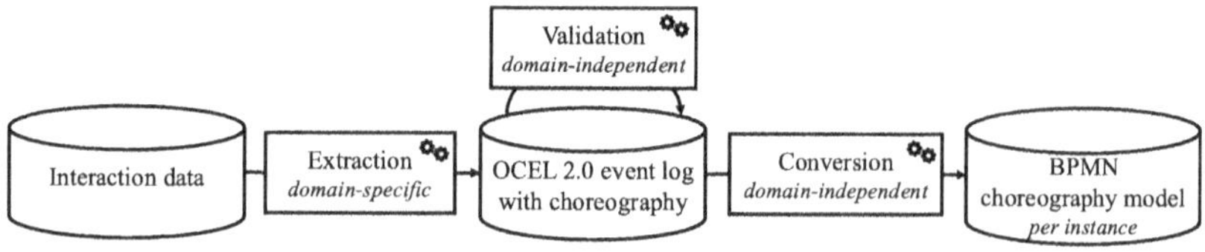

Fig. 1. Data transformation- and tool-chain (cylinders: data; rectangles: tools).

notion for the event logs that form their data basis, which caused complications as choreography data combines and relates multiple entities. To mitigate the issue, choreography-tailored extensions to existing single-case notion event log formats were created [12], but remained limited to capturing interactions as events and associating participants, without reflecting the full expressiveness of BPMN choreographies. Object-centric event log formats, such as OCEL 2.0, overcome the limitations of single-case notion event log formats, and capture processes with multiple interrelated entities [4] and potentially multiple case notions [19]. Thus, the capabilities of OCEL 2.0 make it a natural fit to capture choreographies, yet the format's potential in that domain remains largely unexplored. Challenges lie in representing choreography concepts, such as participants, message exchanges, and sub-choreographies.

The advances object-centric process mining have made in analyzing participant-internal processes have not been extended to analyzing inter-organizational processes. In particular, object-centric process mining on choreographies between participants with unknown internal processes does not have an established formal basis in the literature. A prerequisite for interaction analysis is the representation of interaction data in event logs. Once interaction data can be represented in event logs, several analytical scenarios become feasible, such as supply chain interactions with goods and information flowing between partners, communication between government agencies and the public such as for disaster warning [16], decentralized application interactions on a blockchain [14], or emerging fields like artificial intelligence (AI) agent interactions [7].

In this paper, we make four contributions: (i) a mapping of BPMN choreography elements to OCEL 2.0, (ii) a set of 17 constraints, ensuring valid representation, (iii) an evaluation of the approach on 19 real-world datasets, and (iv) a tool pipeline implementing the mapping and constraint validation (see Fig. 1). Our approach uses the OCEL 2.0 format and refrains from defining extensions to preserve compatibility of existing process mining tools with resulting event logs. Our proposal supports choreography elements we could evaluate with available real-world data. The representation captures choreography tasks, participants, messages, and nested sub-choreographies in point-to-point interactions between single participants. Although limited, this spectrum of supported choreography elements exceeds the supported elements in previous approaches.

Next, Sect. 2 introduces BPMN choreographies and OCEL 2.0. Section 3 discusses related work. Section 4 describes the proposed representation of BPMN choreographies in OCEL 2.0 event logs and the mapping between their elements. Section 5 presents the implementation and evaluates the mapping on real-world datasets. Section 6 discusses relevant outcomes. Section 7 summarizes the work.

2 Preliminaries

BPMN Choreography diagrams are models that represent interactions *between* business participants in terms of message exchanges [17, p.315]. BPMN choreography models have dedicated elements to capture the message exchanges [17, pp.317ff]. *Start* and *end* events mark the beginning and end of a choreography. *Sequence flows* show the execution order of choreography tasks and sub-choreographies. *Choreography tasks* show the interactions between the choreography participants. Within choreography tasks, there is either only a request message sent by the initiator (one-way task), or the request message is followed by a response message sent by the non-initiating participant (two-way task) (pp.323ff). A *message flow* wraps message information about the sending participant (*source*), the receiving participant (*target*), and the related choreography task. *Sub-choreographies* serve as containers for choreography elements; they can be used for scoping details about a process. A sub-choreography does not represent an atomic interaction; it has no messages, but can have more than one non-initiating participant band (pp.328ff).

An object-centric event log consists of **events** from the universe of events $\mathbb{U}_{\mathrm{ev}}$, which have an event type from the universe of event types $\mathbb{U}_{\mathrm{etype}}$, and **objects** from the universe of objects $\mathbb{U}_{\mathrm{obj}}$, which have object types from the universe of object types $\mathbb{U}_{\mathrm{otype}}$. Events and objects can have attributes from the universe of attributes $\mathbb{U}_{\mathrm{attr}}$. Timestamps for events or object attributes are from the universe of timestamps $\mathbb{U}_{\mathrm{time}}$. There are also object-to-object and event-to-object relationships that have qualifiers from the universe of qualifiers $\mathbb{U}_{\mathrm{qual}}$. A widely adopted object-centric event log format is *OCEL 2.0*, which is formally a tuple.

Definition 1 (OCEL 2.0 event log [4]). $L = (E, O, EA, OA, evtype, time, objtype, eatype, oatype, E2O, O2O)$ *where:*

- $E \subseteq \mathbb{U}_{ev}$ *is the set of events.*
- $O \subseteq \mathbb{U}_{obj}$ *is the set of objects.*
- $evtype : E \to \mathbb{U}_{etype}$ *assigns types to events.*
- $time : E \to \mathbb{U}_{time}$ *assigns timestamps to events.*
- $EA \subseteq \mathbb{U}_{attr}$ *is the set of event attributes.*
- $eatype : EA \to \mathbb{U}_{etype}$ *assigns event types to event attributes.*
- $objtype : O \to \mathbb{U}_{otype}$ *assigns object types to objects.*
- $OA \subseteq \mathbb{U}_{attr}$ *is the set of object attributes.*
- $oatype : OA \to \mathbb{U}_{otype}$ *assigns object types to object attributes.*
- $E2O \subseteq E \times \mathbb{U}_{qual} \times O$ *are the qualified event-to-object relationships.*
- $O2O \subseteq O \times \mathbb{U}_{qual} \times O$ *are the qualified object-to-object relationships.*

The central novelties with respect to single-case notion event logs are (i) that no single case notion is forced onto the event data and (ii) the (meta-model level) many-to-many E2O- and O2O-relationships carrying qualifiers to specify the nature of the connection between events and objects.

3 Related Work

Literature on inter-organizational processes has two streams. One stream retains the *internal* perspective of collaborating participants in the event log data [3,9,13,20,21]. The other stream targets the interaction of process participants in the spirit of BPMN choreography models specifically: [10] transforms electronic message exchange data into an XES event log. The message exchanges are represented as events, which are grouped into traces representing process instances; business data accompanying the messages is stored using a custom XES extension. With a similar approach, [12] represent choreography message exchange as events and extend XES to represent collaboration tasks, message events, source and target participants. [18] uses the XES extension proposed by [12] and provides tool support to discover BPMN choreography diagrams from the extended XES logs. While these papers address the representation of inter-organizational processes with BPMN choreographies, they do not represent message content or scoping via sub-choreographies. To capture the supported choreography elements, they rely on the single-case XES standard and propose extensions. The object-centric format OCEL 2.0 overcomes XES-limitation, however, the object-centric capabilities have not yet been applied to logging choreography data. So far, only [15] investigates object-centric process mining and choreographies, proposing an object-centric extension of BPMN choreography semantics. However, the work does not address logging the execution data beyond reachable Petri net markings. We address this gap and define how BPMN choreography elements can be represented in OCEL 2.0 without extensions. Our mapping covers participants, messages, and sub-choreographies, with constraints that enforce choreography semantics directly on the event log.

4 BPMN Choreography Elements in OCEL 2.0

In this section, we describe the proposed representation of BPMN choreographies in OCEL 2.0 event logs. We present the guiding design principle, lay out the conceptual mapping, and a set of constraints that apply.

Our approach uses OCEL 2.0's qualified relationships to express the execution semantics of BPMN choreographies in an OCEL 2.0 event log without prescribing fixed event and object types. The aim is to find an event log representation for the most-widely used BPMN choreography elements to capture the execution of choreography tasks, messages, initiating and non-initiating participants, and scoping in sub-choreographies. Concretely, we represent choreography task executions in **events**, choreography participants and messages in **objects**, and choreography semantics (e.g., initiating participant, message source and target, etc.) via **qualifiers** on E2O- and O2O-relationships. Consequently, by adding relationships, we provide an option to enhance existing OCEL 2.0 logs with choreography-specific information while leaving the original types and attribute structure unchanged. We adopt a fixed object type binding to represent choreography scope and hierarchy. Inspired by [6], we introduce *dedicated* object types for objects that group related events and represent the scope of choreographies and sub-choreographies. E2O-relationships are then used to connect these scoping objects with their events and define their hierarchical structure.

4.1 Data Mapping

In this section, we describe technical details of the mapping between BPMN choreography elements and OCEL 2.0 elements.

Definition 2 (BPMN Choreography to OCEL 2.0 Mapping). *Given an OCEL 2.0 event log L (see Definition 1), a representation of a BPMN choreography consists of the following elements:*
Choreography task: *An* ***event*** $e \in E$ *such that* $(e \times \texttt{choreo:instance} \times o_{inst}) \in E2O$ *for some* $o_{inst} \in O$ *and* $(e \times q \times o) \in E2O$ *for some* $o \in O$ *and for each* $q \in \{\texttt{choreo:initiator}, \texttt{choreo:participant}, \texttt{choreo:message}\}$.
Initiating participant: *An* ***object*** $o_i \in O$ *such that* $(e \times \texttt{choreo:initiator} \times o_i) \in E2O$.
Non-initiating participant: *An* ***object*** $o_p \in O$ *such that* $(e \times \texttt{choreo:partici- pant} \times o_p) \in E2O$.
Message: *An* ***object*** $o_m \in O$ *such that* $(e \times \texttt{choreo:message} \times o_m) \in E2O$.
Message flow: *A message flow is represented* ***implicitly*** *by* o_m, *its source* o_{src} *with* $(o_m \times \texttt{choreo:source} \times o_{src}) \in O2O$, *and its target* o_{tgt} *with* $(o_m \times \texttt{choreo:target} \times o_{tgt}) \in O2O$, *with* $o_{src} \neq o_{tgt}$ *and* $o_{src}, o_{tgt} \in \{o_i, o_p\}$.
Choreography: *An* ***object*** $o_{inst} \in O$ *with* $objtype(o_{inst}) = \texttt{choreographyInstance}$;
Sub-choreography: *An* ***object*** $o_{sub} \in O$ *with* $objtype(o_{sub}) = \texttt{subchoreography- Instance}$; *Sub-choreography containment of an event e is represented by* $(e \times \texttt{choreo:contained-by} \times o_{sub}) \in E2O$; *Sub-choreography nesting of a sub-choreography* o'_{sub} *in* o_{sub} *is represented by* $(o_{sub} \times \texttt{choreo:contains} \times o'_{sub}) \in O2O$.

4.2 Constraints

For an OCEL 2.0 event log to properly encode a BPMN choreography, constraints apply. We formulate them based on Definition 1 (OCEL 2.0 event log). C0–C10 apply on task-level and whenever events are interpreted as choreography task executions. C11–C14 apply when sub-choreography scopes are represented. C15–C16 concern model-level sequencing assumptions that we check assuming a single-threaded model reconstruction from the event logs. In event logs with parallel behavior, violations of C15–C16 should be read as warnings for subsequent choreography discovery rather than as violations of the task-level OCEL 2.0

encoding. All the constraints are rooted in the BPMN 2.0.2 specification [17], the direct reference to which we provide for every constraint.

Definition 3 (Choreography Shorthands). *We define shorthands for events and objects based on their qualified relationships. For events $e \in E$:*

$$
\begin{aligned}
init(e) &= \{\, o_i \in O \mid (e \times \texttt{choreo:initiator} \times o_i) \in E2O \,\} \\
noninit(e) &= \{\, o_p \in O \mid (e \times \texttt{choreo:participant} \times o_p) \in E2O \,\} \\
msg(e) &= \{\, o_m \in O \mid (e \times \texttt{choreo:message} \times o_m) \in E2O \,\} \\
scope(e) &= \{\, o_{sub} \in O \mid (e \times \texttt{choreo:contained-by} \times o_{sub}) \in E2O \,\} \\
inst(e) &= \{\, o_{inst} \in O \mid (e \times \texttt{choreo:instance} \times o_{inst}) \in E2O \,\}
\end{aligned}
$$

For objects $o_m \in O$:

$$
\begin{aligned}
src(o_m) &= \{\, o_{src} \in O \mid (o_m \times \texttt{choreo:source} \times o_{src}) \in O2O \,\} \\
tgt(o_m) &= \{\, o_{tgt} \in O \mid (o_m \times \texttt{choreo:target} \times o_{tgt}) \in O2O \,\}
\end{aligned}
$$

The set of choreography task events *is* $E_T = \{e \in E \mid init(e) \neq \emptyset \vee noninit(e) \neq \emptyset \vee msg(e) \neq \emptyset \vee inst(e) \neq \emptyset\}$. *Events not belonging to* E_T *do not fall under the constraints below.*

Choreography task constraints C0–C10.

C0: Instantiations of choreography tasks happen within the scope of exactly one choreography [17, p. 316].

$$\forall e \in E_T : \ |inst(e)| = 1$$

C1: Messages associated with choreography tasks are sent or received by the initiator or the non-initiating participant of that same task [17, p. 324].

$$\forall e \in E_T, \forall o_m \in msg(e) : \ src(o_m) \cup tgt(o_m) \subseteq init(e) \cup noninit(e)$$

C2–C4: Each choreography task has exactly one initiating participant [17, p. 322] and exactly one non-initiating participant [17, p. 323]; a participant cannot simultaneously be initiating and non-initiating in the same task [17, p. 322&119].

$$
\begin{aligned}
&(\text{C2}) \quad \forall e \in E_T : \ |init(e)| = 1 \\
&(\text{C3}) \quad \forall e \in E_T : \ |noninit(e)| = 1 \\
&(\text{C4}) \quad \forall e \in E_T : \ init(e) \cap noninit(e) = \emptyset
\end{aligned}
$$

C5–C6: A message has exactly one source participant and exactly one target participant [17, p. 122].

$$
\begin{aligned}
&(\text{C5}) \quad \forall e \in E_T, \forall o_m \in msg(e) : \ |src(o_m)| = 1 \\
&(\text{C6}) \quad \forall e \in E_T, \forall o_m \in msg(e) : \ |tgt(o_m)| = 1
\end{aligned}
$$

C7–C8: Each choreography task has exactly one initiating message and at most one non-initiating message [17, pp. 323 ff].

$$\text{(C7)}\quad \forall\, e \in E_T : \ \left|\{\, o_m \in msg(e) \mid src(o_m) \cap init(e) \neq \emptyset\}\right| = 1$$
$$\text{(C8)}\quad \forall\, e \in E_T : \ \left|\{\, o_m \in msg(e) \mid src(o_m) \cap noninit(e) \neq \emptyset\}\right| \leq 1$$

C9–C10: For every choreography task, the initiating message sent by the initiating participant must be received by the non-initiating participant, and vice versa for the non-initiating message [17, pp. 323 ff].

$$\text{(C9)}\quad \forall\, e \in E_T,\ \forall\, o_m \in msg(e) : \ src(o_m) \cap init(e) \neq \emptyset \\ \Rightarrow\ noninit(e) \subseteq tgt(o_m)$$

$$\text{(C10)}\quad \forall\, e \in E_T,\ \forall\, o_m \in msg(e) : \ src(o_m) \cap noninit(e) \neq \emptyset \\ \Rightarrow\ init(e) \subseteq tgt(o_m)$$

Sub-choreography constraints C11–C14:

Definition 4 (Sub-Choreography Relations). *Let $O_S \subseteq O$ be the set of all sub-choreography objects o_{sub}. The events directly contained in a scoping object o_{sub} are*

$$enclosed(o_{sub}) = \{\, e \in E_T \mid o_{sub} \in scope(e)\,\}.$$

Nested scoping objects directly contained in o_{sub} are

$$children(o_{sub}) = \{\, o'_{sub} \in O_S \mid (o_{sub} \times \texttt{choreo:contains} \times o'_{sub}) \in O2O\,\}.$$

The directly enclosing sub-choreography of o_{sub} is

$$parent(o_{sub}) = \{\, o'_{sub} \in O_S \mid (o'_{sub} \times \texttt{choreo:contains} \times o_{sub}) \in O2O\,\}.$$

$descendant(o_{sub})$ is the smallest set such that $children(o_{sub}) \subseteq descendant(o_{sub})$ and $\forall\, o' \in descendant(o_{sub}) : \ children(o') \subseteq descendant(o_{sub})$. All choreography events transitively enclosed in o_{sub} are

$$allevents(o_{sub}) = enclosed(o_{sub}) \cup \bigcup_{o' \in descendant(o_{sub})} enclosed(o').$$

All participants transitively involved in o_{sub} are

$$involved(o_{sub}) = \bigcup_{e \in allevents(o_{sub})} (init(e) \cup noninit(e)).$$

Finally, $\succ (e_1, e_2)$ holds iff $e_1, e_2 \in E_T$, $inst(e_1) = inst(e_2)$, $time(e_1) < time(e_2)$, and there is no $e_3 \in E_T$ with $inst(e_3) = inst(e_1)$ and $time(e_1) < time(e_3) < time(e_2)$.[1]

[1] When two events share a timestamp, their order is determined by additional attributes (e.g., transaction index and call order in blockchain data)..

C11: Choreography tasks and events can be directly contained in at most one sub-choreography [17, pp. 87 & 328].

$$\forall e \in E_T : |scope(e)| \leq 1$$

C12: Every sub-choreography contains at least one choreography task, event, or sub-choreography [17, p. 328].

$$\forall o_{\text{sub}} \in O_S : enclosed(o_{\text{sub}}) \neq \emptyset \ \vee \ children(o_{\text{sub}}) \neq \emptyset$$

C13: All events transitively contained in the same sub-choreography belong to the same choreography instance [17, p. 316].

$$\forall o_{\text{sub}} \in O_S, \forall e_1, e_2 \in allevents(o_{\text{sub}}) : \ inst(e_1) = inst(e_2)$$

C14: Sub-choreographies can be nested; each has at most one directly enclosing sub-choreography, and the containment structure is acyclic [17, pp. 328 ff].

$$\text{(a)} \quad \forall o_{\text{sub}} \in O_S : \ |parent(o_{\text{sub}})| \leq 1$$
$$\text{(b)} \quad \forall o_{\text{sub}} \in O_S : \ o_{\text{sub}} \notin descendant(o_{\text{sub}})$$

Model-level constraints C15–C16 :

C15: "The Initiator of a Choreography Activity MUST have been involved (as Initiator or Receiver) in the previous Choreography Activity." [17, p. 336] For all $e_1, e_2 \in E_T$ with $\succ (e_1, e_2)$, and, whenever non-empty, let $o_1 \in scope(e_1)$, and $o_2 \in scope(e_2)$:

(a) $scope(e_1) \neq \emptyset \ \wedge \ scope(e_2) \neq \emptyset \ \wedge \ o_1 \in descendant(o_2)$
$\Rightarrow \ init(e_2) \subseteq involved(o_c)$ where $o_c \in children(o_2)$
with $o_1 \in descendant(o_c) \cup \{o_c\}$

(b) $scope(e_1) \neq \emptyset \ \wedge \ scope(e_2) = \emptyset \ \Rightarrow \ init(e_2) \subseteq involved(o_r)$ where $o_r \in O_S$ with $o_1 \in descendant(o_r) \cup \{o_r\}$ and $parent(o_r) = \emptyset$

(c) otherwise $\Rightarrow \ init(e_2) \subseteq init(e_1) \cup noninit(e_1)$

C16: "Sequence Flows cannot cross the boundary of a Sub-Choreography" [17, p. 34]; Footnote 1 applies.

$$\forall e_1, e_2, e_3 \in E_T, \ \forall o_{\text{sub}} \in O_S :$$
$$(time(e_1) < time(e_2) < time(e_3) \ \wedge \ inst(e_1) = inst(e_2) \ \wedge \ inst(e_2) = inst(e_3)$$
$$\wedge \ e_1 \in allevents(o_{\text{sub}}) \ \wedge \ e_2 \notin allevents(o_{\text{sub}})) \ \Rightarrow \ e_3 \notin allevents(o_{\text{sub}})$$

Table 1. Data correspondences for evaluation data sources.

Interaction data		OCEL 2.0 choreography schema	BPMN choreography element
Blockchain calls	*XES collab. log*		
Transaction	XES trace	Object, type `choreographyInstance`	Choreography instance
Function call	`send` & `receive` events	Event with `choreo:instance`	Choreography task
Calling account	Sender `org:group`	Object via `choreo:initiator`	Initiating participant
Callee account	Receiver `org:group`	Object via `choreo:participant`	Non-initiating participant
Call input parameters	Message by `msgInstanceId`	Object via `choreo:message`; `choreo:source`→initiator, `choreo:target`→receiver	Request message
Call return value	*n/a* (one-way only)	Object via `choreo:message`; `choreo:source`→receiver, `choreo:target`→initiator	Response message (two-way tasks)
Nested call hierarchy	*n/a* (flat traces)	Object, type `subchoreographyInstance`; `choreo:contained-by`, `choreo:contains`	Sub-choreography containment
Block timestamp and order of internal calls	`send` event timestamp and document order	Event `time`	Sequence flow (instance-level, naïve converter)

5 Implementation and Evaluation

In this section, we apply and evaluate the proposed mapping from Sect. 4. First, we explain the feasibility of the mapping using an example from the *Tornado.Cash* governance contract. Then, we apply the mapping to (1) real-world Ethereum blockchain data provided in a pre-existing repository provided by [8][2], and (2) "realistic" and case study collaboration logs provided by [9][3] and re-used by [18]. We evaluate the event logs against the constraints in Sect. 4.2, based on the BPMN 2.0.2 specification [17]. A successful evaluation shows constraint violations due to source data particularities, not due to the mapping itself.

Applying the mapping requires *source data assumptions*: (i) Each interaction record must be assignable to a choreography instance, (ii) the initiating and non-initiating participants of an interaction must be identifiable, and (iii) message objects must be available. (iv) For two-way choreography tasks, the data must distinguish request and response messages. (v) To represent sub-choreographies, the source data must expose containment or nesting information between interactions; if not, the mapping can still be applied to choreography tasks and messages, but no scoping objects are produced. (vi) Events belonging to the same choreography instance must be orderable, either by timestamps or

[2] https://github.com/AlessandroMarcellettiUnicam1/BlockchainDataset, accessed 02-06-2026.

[3] https://bitbucket.org/proslabteam/colliery_validation/, accessed 02-06-2026.

by an additional ordering attribute. The blockchain source data exhibits (i – vi), [9]'s event logs exhibit (i – iii, and vi).

5.1 Tool Pipeline

From the considered data, we (i) create OCEL 2.0 event logs representing BPMN choreographies, (ii) validate the event logs based on constraints C0–C16, and (iii) convert individual choreography instances in the event logs to BPMN choreography models. For all three steps, we created software artifacts with Python, see Fig. 1. We provide the data (event logs & BPMN choreography files) and additional results of the validation in an online repository[4].

The extractors (domain-specific): *trace2ocelchor* and *xescol2ocelchor* take as input a set of domain-specific interaction records (blockchain transaction traces; XES collaborative event logs). For each record, the extractors (i) create a `choreographyInstance` object representing one choreography execution; (ii) apply Definition 2: for each task execution they create an event, create objects for the associated participants and messages, and link them via E2O-relationships `choreo:initiator`, `choreo:participant`, `choreo:message` and O2O-relationships `choreo:source`, `choreo:target` to reflect the message flow; (iii) order the events in each scope by execution order; (iv) emit the resulting events, objects, and relationships as OCEL 2.0 event logs. The two extractors differ in how they identify the data of their respective domains, see Table 1.

trace2ocelchor (**blockchain calls**). One `choreographyInstance` per blockchain transaction; one event per function call. Links caller and callee as initiator and participant; links input and output data as source and target. Calls without nested sub-calls produce one event carrying both request and response (two-way); calls with nested sub-calls produce two events (Request [...] / Respond to [...], one-way) bracketing the nested scope. For each call with nested sub-calls, the extractor creates a `subchoreographyInstance` scoping object linked via `choreo:contained-by` (E2O) to its enclosed events and via `choreo:contains` (O2O) to its child scopes. For DELEGATECALL calls, the extractor reattributes the issuer of nested sub-calls from caller to callee. Ordering uses the block timestamp combined with the order of internal calls.

xescol2ocelchor (**XES collaboration logs**). One `choreographyInstance` per XES trace; one combined event per `send`-event and matching `receive`-event occurrences (identified by shared message ID). The sender's `org:group` (XES attribute for organization) becomes the initiator, each receiver's `org:group` becomes a non-initiating participant. All choreography tasks are one-way and produce one event each (no marker pairing request with response in the data). The data does not indicate event hierarchy, so no subchoreographyInstance objects are produced. Internal events (without message attribute) are dropped. Ordering uses the send event's timestamp combined with the document order.

[4] https://github.com/23722/ocelchor/tree/v1.1-BPM-2026-PTF, accessed 03-06-2026.

The validator (domain-independent): `ocelchorvalidator` takes as input an OCEL 2.0 event log. From the events, objects, and E2O- and O2O-relationships, the validator builds the variables defined in Definition 3 and Definition 4. The validator then evaluates each constraint C0C16 from Sect. 4. It records a violation if for one element a constraint is false. The output is the validation result as a table and a list to track the violations per constraint.

The converter (domain-independent): `ocelchormodel` takes as input an OCEL 2.0 event log. The converter identifies all choreography instance objects o_{inst}; For each o_{inst} the converter (i) selects the events related to o_{inst} via a `choreo:instance`-relationship and the scoping objects reachable via `choreo:contained-by`- and `choreo:contains`-relationships; (ii) applies Definition 2/Table 1; (iii) orders events inside each scope by their timestamp; (iv) encodes the result as BPMN 2.0 choreography XML (incl. artificial start and end events for each scope). The output is one BPMN choreography diagram per choreography instance (hence converter, not discovery).

5.2 Internal Calls of *Tornado.Cash* as a BPMN Choreography

We illustrate the construction of a choreography from transaction calls with an excerpt from the *Tornado.Cash* application. Figure 2(a) shows the execution sequence. A `Proxy` contract invokes the `Unlock` function of a `Governance` contract. The `Unlock` function lets users withdraw previously locked governance tokens. To execute the `Unlock` function, the `Governance` contract must invoke the `Transfer` function of the `TORN` token contract and process its response. Upon successful completion of the transfer the unlock operation finalizes.

Translating Tornado.Cash. According to the conversion logic above, the transaction calls are translated by the extractor into the events `Request unlock`, `Respond to unlock`, and `Transfer`. Additionally, objects are created to represent: (i) the (initiating) participant (types: `Proxy`, `Governance`, `TORN`), (ii) the messages (types: `unlock call`, `unlock call response`, `transfer call`, `transfer call response`), (iii) the choreography instance, and (iv) the sub-choreography instance. Figure 2(b) shows the final conversion as a property graph representing the data in OCEL 2.0 (for readability without object types or the object/relationships representing the choreography instance). In addition to the events and objects, we insert **relationships** to specify the objects' roles as illustrated in Fig. 2(b). From the OCEL 2.0 log, we derive the BPMN choreography diagram (Fig. 2(c)) with the converter and display the mapped elements (here, messages display their attributes, not their object names or types). From the same event log, we discovered an Object-Centric Directly-Follows Graph (OC-DFG, Fig. 3) using the Python library *PM4Py* [5]; The OC-DFG shows overlapping control-flow with the BPMN choreography, the latter adding role assignments and task hierarchy. We discuss the implications in Sect. 6.

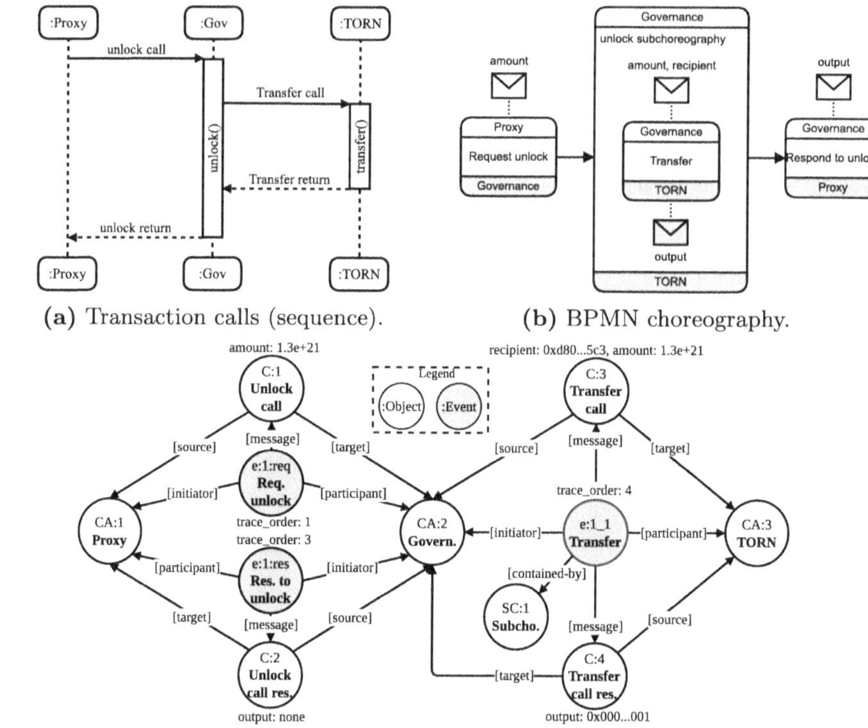

(a) Transaction calls (sequence). **(b)** BPMN choreography.

(c) Property graph representing OCEL 2.0 event log with a selection of elements.

Fig. 2. Tornado.Cash scenario from three perspectives: (a) transaction calls, (b) BPMN choreography, and (c) OCEL 2.0 property graph representation. In (c), relationship qualifiers are abbreviated and stripped from the qualifier prefix *choreo:*.

5.3 Validation with Datasets

To evaluate, we converted the twelve blockchain datasets and the seven collaboration logs by [9] to OCEL 2.0 event logs representing choreographies. We tested

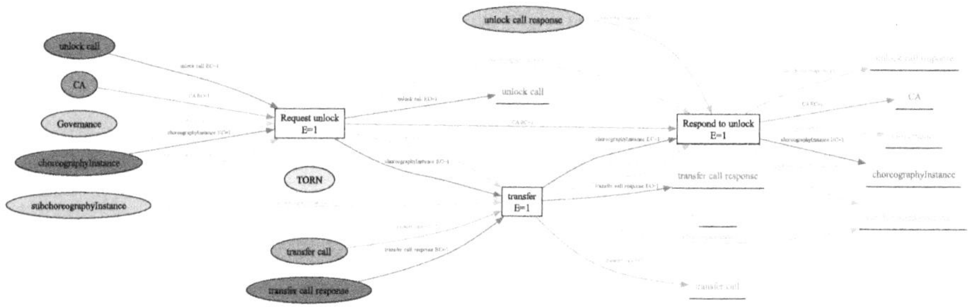

Fig. 3. Tornado.Cash scenario as an OC-DFG.

Table 2. Evaluation data overview

DApp contract/Interaction event log by [9]	$\#_v$	$\#_e$	$\#_m$	$\#_{so}$	$\#_p$	$\#_{E2O}$	$\#_{O2O}$
Consensus Layer: DepositContract	1	9	17	1	3	52	34
ENS: ENSGovernor	9	21	31	6	8	106	63
Tornado.Cash: GovernanceProposalStateUpgrade	11	35	47	17	16	176	100
Tornado.Cash: TornadoRouter	2	85	146	13	22	484	303
PancakeSwap: MasterChefV3	13	241	361	64	45	1312	776
Uniswap: UniswapV2Pair (USDC–WETH)	6	61	88	17	16	326	190
SushiSwap: Router	19	267	415	67	55	1464	880
CryptoKitties: Core (KittyCore)	22	204	302	59	78	1096	646
CryptoKitties: SaleClockAuction	2	6	10	2	4	32	20
Yuga Labs: BoredApeYachtClub	9	18	21	5	13	84	45
Nouns DAO: NounsToken	7	12	13	3	11	54	28
Beanstalk Farms: Attack data	3	489	703	139	59	2656	1542
Real1: Two-robot search	1	2	2	0	2	8	4
Real2: Travel booking	96	842	480	0	2	3368	960
Real3: Smart thermostat	59	472	321	0	3	1880	634
Real4: ZooClub registration	1	4	4	0	3	16	8
Real5: Academic paper review	3	15	15	0	3	60	30
Healthcare: Hospitalization	17	116	116	0	4	455	223
Smart agriculture: Tractor coordination	10	114	100	0	3	452	206

Notes. $\#_v$: choreography variants; $\#_e$: events representing choreography tasks; $\#_m$: message objects; $\#_{so}$: scoping objects; $\#_p$: participants; $\#_{E2O}$: E2O-relationships; $\#_{O2O}$: O2O-relationships.

if the resulting event logs respect C0–C16, to check if the BPMN choreography notion is accurately represented in the data. Each log includes only events and objects associated with choreography instances. We used one representative of each initial blockchain contract function call, and respectively unique interaction trace variants from [9]'s data to avoid including duplicate instances per choreography variant in the evaluation. For the data from [9], our extracted choreography control-flows show the same trace variants as [18]'s independently created XES choreography logs. In sum, the event logs included 291 choreography variants, 3,013 events, 3,192 messages, 393 sub-choreography objects, and 350 participants (see Table 2). C0–C16 were checked between 393 (C12, C13, C14, C16) and 3,192 (C5, C6) times. In the blockchain data, violations of the constraints occurred for the contracts *KittyCore* (C4, 2 violations/204 checks), *PancakeSwap: MasterChefV3* (C4, 10 violations/241 checks), and in the *Beanstalk Farm* attack transaction (C15, 1 violation/486 checks). In [9]'s data, violations occurred for *Real3* (C3 & C6, each 8 violations/472 checks; and C15, 99 violations/413 checks), *Healthcare* (C3 & C6, each 9 violations/116 checks; and C15,

9 violations/99 checks), and *Smart agriculture* (C3 & C6, each 28 violations/114 checks; and C15, 9 violations/104 checks). We discuss those next.

6 Discussion

The comparison between Fig. 3 and Fig. 2(c) illustrates the analytical value in representing choreographies in OCEL 2.0. Both models were created based on the same data. The OC-DFG shows directly-follows relations of events and object involvement. However, it does not show information like: (i) which participant in the event is sending or receiving a message, (ii) which message is a request or response, and (iii) that *Transfer* is contained in the scope of *Unlock*. In contrast, these aspects are shown in the choreography, by interpreting the E2O- and O2O-relationships added with the mapping. Meanwhile, the representation of choreographies in the event log preserves the OCEL 2.0 structure, so that existing OCPM tools (such as the OC-DFG miner) can work with the log.

In Sect. 5, we apply the mapping to validate its applicability to execution data. In the *blockchain data*, we observed violations of the constraints C4 (initiator $\neq$ participant) and C15 (initiator participates in previous activity). The violation of C4 occurred when internal calls were issued from and to the same contract. They occurred due to the peculiarities of blockchain-related programming patterns managing self-executing functions through explicit calls, resulting in the caller being equal to the callee. The violation of C15 occurred when a contract was created and immediately after started to issue calls on its part. In both cases, the blockchain execution semantics were not aligned with the BPMN choreography assumptions. However, our validation approach and the related testing of constraints successfully identified these issues. Moreover, these violations depend on blockchain-specific aspects and not on a fundamental mismatch between the mapping and BPMN choreography semantics. Consequently, they can be solved with a dedicated data transformation step. In [9] *'s event logs*, we observed violations of C3 (one non-initiating participant), C6 (one target participant), and C15. C3 & C6 violations were causally related. They occurred when participant's processes terminated and were no longer available as a message recipient (*Real3*); or due to data inconsistencies, when no matching `receive`-events were provided for `send`-events, causing C15 violations downstream (*Healthcare*); Other C15 violations were caused by repetitive process loops in a parallel sequence flows that interfered with single-threaded execution of the trace assumed in the check of C15 in the data (*Real3*, a warning flag for parallel control flows for future process discovery). A set of related C3, C6, and C15 violations was triggered by the source data exposing message broadcasts that the current mapping and constraint set does not cover (*Smart agriculture*).

The mapping enriches OCEL 2.0 event logs with choreography semantics, while preserving the event log structure. Adding the semantics comes at the cost of additional E2O- and O2O-relationships. Table 2 shows that the number of inserted relationships for choreography representations is more than the total number of events and objects in the event log. The mapping also introduces artificial objects for scoping and representing sub-choreographies. [11]

flag this method as a risk for misinterpretation. We mitigate this by using fixed object types (`choreographyInstance`, `subchoreographyInstance`), for scoping objects. However, an analyst unaware of the object type convention treats the scoping objects as ordinary objects, as illustrated in Fig. 3 and discussed in [19]. For the choreography semantics to be *added to an existing* OCEL 2.0 event log, it has to record one-way interactions as single events. Interactions recorded in separate send and receive events, can still be transformed into a choreography log, but not purely additively. The mapping does not cover the full spectrum of BPMN choreography elements: (i) *BPMN choreography events* were excluded. Their integration would require a distinction between OCEL 2.0 events representing (a) BPMN choreography tasks and (b) BPMN choreography events. Such events would fall under constraints C0, C11–C14, and C16. (ii) *Multi-instance participant* was excluded because the mapping covers point-to-point interaction (per C2 and C3), while multi-instance participant allows one-to-many exchanges; (iii) *call choreography and global choreography task* and *loop and multi-instance sub-choreography* concern reuse and repeated invocation, which is usually not recorded explicitly in an event log making their identification part of process discovery; similarly (iv) *gateways* are not logged in event data but discovered by a miner as control-flow.

7 Conclusion

With this paper, we established the foundation to use the object-centric event format OCEL 2.0 for capturing point-to-point interactions between participants, corresponding messages, and hierarchical decomposition over interactions. To that end, we mapped BPMN choreography elements to OCEL 2.0 elements and formulated a set of constraints to ensure that the data entries reflect well-formed BPMN choreographies. The approach exclusively used existing OCEL 2.0 elements and does not require extending the specification, making it compatible with existing object-centric process mining tools and techniques. We used (real-world) interaction logs to evaluate our contribution. In future work, we plan to extend the mapping to additional BPMN choreography elements and to develop a prototype to discover BPMN choreography models from OCEL 2.0 event logs using our encoding. With the prototype, we intend to investigate choreography-based interaction analysis in case studies.

Acknowledgment. In this paper, artificial intelligence (Claude and ChatGPT with different models) was used to support programming and presentation. We thank Til Drewe for manually drafting process models based on an early version of the data mapping.

References

1. van der Aalst, W.: Intra- and inter-organizational process mining: discovering processes within and between organizations. In: The Practice of Enterprise Modeling, pp. 1–11. Springer, Heidelberg (2011)

2. van der Aalst, W.: Federated process mining: exploiting event data across organizational boundaries. In: SMDS, pp. 1–7 (2021)
3. Benzin, J.V., Rinderle-Ma, S.: Collaboration miner: discovering collaboration petri nets. In: Cooperative Information Systems, pp. 373–381. Springer, Cham (2025)
4. Berti, A., Koren, I., Adams, J.N.: OCEL (object-centric event log) 2.0 specification. Technical report, RWTH Aachen University (2023), v2.0
5. Berti, A., van Zelst, S., Schuster, D.: Pm4py: a process mining library for python. Softw. Impacts **17**, 100556 (2023)
6. Bertrand, Y., Imenkamp, C., Malburg, L.: An object-centric core metamodel for IoT-enhanced event logs. arXiv preprint arXiv:2506.21300 (2025)
7. Calvanese, D., et al.: Agentic business process management: a research manifesto. Inf. Syst. **140**, 102738 (2026)
8. Corradini, F., Luzi, L., Marcelletti, A., Morichetta, A., Re, B.: Extracting and decoding internal transactions in EVM-based blockchains. In: 2026 IEEE International Conference on Pervasive Computing and Communications Workshops and other Affiliated Events (PerCom Workshops), pp. 1–6. IEEE (2026)
9. Corradini, F., Pettinari, S., Re, B., Rossi, L., Tiezzi, F.: A technique for discovering BPMN collaboration diagrams. Softw. Syst. Model. **23**(6), 1323–1343 (2024)
10. Engel, R., et al.: Analyzing inter-organizational business processes. IseB **14**(3), 577–612 (2016)
11. Fahland, D., et al.: Towards a simple and extensible standard for object-centric event data (OCED) – core model, design space, and lessons learned (2024). https://arxiv.org/abs/2410.14495
12. González, L., Delgado, A.: Compliance requirements model for collaborative business process and evaluation with process mining. In: Latin American Computing Conference, pp. 1–10 (2021)
13. Hernandez-Resendiz, J.D., Tello-Leal, E., Sepúlveda, M.: A data-driven approach to discovering process choreography. Algorithms **17**(5) (2024)
14. Hobeck, R., Berti, A., Weber, I., van der Aalst, W.M.P.: Object-centric process mining for blockchain applications: extracting and representing ethereum execution data in OCEL 2.0. Enterp. Model. Inf. Syst. Archit. **20**(2), 1–21 (2025)
15. Lichtenstein, T., Seidel, A., Weske, M.: Detecting conflicting object views in object-centric process choreographies. In: Conceptual Modeling, pp. 204–222 (2025)
16. Murdock, H.J., Rodriguez Castro, D., Dewals, B., Heidenreich, A., Thieken, A.H.: Learning from the past to inform flood risk management: analysis of public survey data in Belgium on flood early warning and response during the July 2021 flood. EGUsphere **2026**, 1–35 (2026)
17. Object Management Group: Business process model and notation (BPMN), version 2.0.2. Spec. formal/2013-12-09, Object Management Group (2013). http://www.omg.org/spec/BPMN
18. Peña, L., Andrade, D., Delgado, A., Calegari, D.: Inter-organizational collaborative BPMN 2.0 business process discovery. J. Intell. Inf. Syst. (2024)
19. Seidel, A., et al.: Object-centric process management: a research manifesto. Inf. Syst. **141**, 102728 (2026)
20. Zeng, Q., Duan, H., Liu, C.: Top-down process mining from multi-source running logs based on refinement of petri nets. IEEE Access **8**, 61355–61369 (2020)
21. Zeng, Q., Sun, S.X., Duan, H., Liu, C., Wang, H.: Cross-organizational collaborative workflow mining from a multi-source log. Decis. Support Syst. **54**(3), 1280–1301 (2013)

Organizational Memory for Agentic Business Process Execution

Lukas Kirchdorfer(✉), Adrian Rebmann, Christian Warmuth, Timotheus Kampik, Theiss Heilker, and Gregor Berg

SAP Signavio, Berlin, Germany
{lukas.kirchdorfer,adrian.rebmann,christian.warmuth,timotheus.kampik, theiss.heilker,gregor.berg}@sap.com

Abstract. LLM-based agents offer new opportunities for automating business process execution beyond the limits of rule-based systems. However, general-purpose LLMs lack the organization-specific knowledge required for reliable execution, which is typically fragmented across human-oriented artifacts such as policies, process models, and standard operating procedures. While such knowledge can technically be encoded in individual prompts or agent-specific retrieval setups, this approach does not scale in enterprises, as it gives rise to knowledge silos and rule duplicates, and makes consistent updates and learning across agents difficult. We argue that this calls for an *organizational memory* for agentic business process execution: a shared, governed, and agent-consumable reference layer of evolving organization-specific procedural knowledge about how work should be executed. We derive requirements for such a memory, propose an architecture for its curation and consumption, and demonstrate its effectiveness in a proof-of-concept based on a procurement scenario.

Keywords: Agentic Process Automation · Large Language Models · Memory · Business Process Management

1 Introduction

Organizations are increasingly adopting LLM-based agents to execute business processes [4,7]. Unlike deterministic automation approaches such as robotic process automation (RPA), which are confined to structured and repetitive settings, these agents can interpret natural language and autonomously handle multi-step tasks including exception handling. Thus, they offer a path to automating processes that were previously difficult to formalize with rule-based systems [26].

However, general-purpose LLMs lack access to organization-specific knowledge by default. They are not inherently aware of company policies, process conventions, role structures, system landscapes, or exception-handling practices. Yet effective process execution in organizational settings depends precisely on such contextual knowledge, which varies across organizations, departments, and

K. Kluza et al. (Eds.): BPM 2026, LNBIP 600, pp. 211–227, 2027.
https://doi.org/10.1007/978-3-032-36710-5_14

individual teams. This knowledge gap poses a fundamental challenge to scaling agentic business process execution in enterprise settings.

A natural response is to encode organization-specific knowledge in prompts or retrieval setups specific to individual agents [16,17]. However, this approach does not scale: in large organizations, thousands of agents are expected to operate simultaneously across evolving processes and heterogeneous knowledge sources. Per-agent knowledge encoding creates new silos, where rules are duplicated, updated inconsistently, and interpreted differently across agents. At the same time, organizations already possess vast amounts of process-relevant knowledge, represented in BPMN models, standard operating procedures (SOPs), policies, knowledge bases, and process analyses based on historical execution data. This knowledge is fragmented across artifacts, many of which were designed for human interpretation, not for LLM agent consumption. The core challenge is, therefore, not knowledge availability, but its transformation into a unified, agent-consumable form and its consistent provision to all agents in the organization.

We argue that what is needed is an *organizational memory* for agentic business process execution: a shared, human-governed, and agent-consumable reference layer of organization-specific procedural knowledge about how work should be executed. Such a memory must integrate distributed knowledge from heterogeneous sources into a unified representation that agents can consume at runtime. It must provide the right context to the right agent at the right time, rather than flooding agents with irrelevant information. Furthermore, it must evolve with organizational change, reflecting policy updates or lessons learned from agent executions. Finally, since organizational memory directly shapes agent behavior across processes and departments, its creation and evolution cannot be fully automated and must instead support human governance.

This paper makes four contributions: (1) we introduce the concept of *organizational memory for agentic business process execution*, defining it as a unified representation of organizational process knowledge that enables LLM-based agents to execute process tasks in a context-aware manner; (2) we derive requirements for such an organizational memory; (3) we propose an organizational memory architecture, describing its components, curation, and consumption; and (4) we demonstrate its effectiveness in a proof-of-concept implementation.

The paper is structured as follows. Section 2 provides background and related work. Section 3 motivates the need for organizational memory, based on which Sect. 4 defines requirements for an organizational memory architecture, which we present in Sect. 5. Then, Sect. 6 showcases the efficacy of our architecture in a proof-of-concept, before Sect. 7 concludes the paper.

2 Background and Related Work

This section provides background and related work on agentic BPM and agentic memory systems.

2.1 Agentic Business Process Management

The concept of agents has periodically attracted attention in BPM over the past three decades. Early work studied autonomous software agents that schedule, execute, and monitor business activities [11]. Later, RPA introduced software robots for automating rule-based tasks [1]. More recently, advances in generative AI and LLMs have renewed interest in agent-oriented process execution. This development has motivated the emerging paradigm of *Agentic Business Process Management* [4,24], which extends traditional BPM by treating autonomous agents as first-class participants in business process execution. Thus, rather than viewing processes primarily as predefined control-flow models, agentic BPM emphasizes process execution as the result of autonomous agents that perceive, reason, act, and interact within organizational constraints. Several recent studies have advanced this agent-centric perspective through data-driven approaches to process discovery, simulation, and optimization [13,14,20,23]. However, these approaches adopt a generic notion of agents rather than specifically instantiating them with LLMs. In that context, several technical contributions have explored the use of LLMs and LLM-based agents for process automation and execution [9,10,18], yet not considering organization-aware execution. More closely related to this work, Kaltenpoth et al. [12] and Skolik et al. [21] show that LLM agents can execute business processes more reliably when guided by semi-structured process rules, which they generate from BPMN models or natural-language process descriptions and retrieve from a rule database during execution. Our work differs in its enterprise-level perspective and the extensiveness of the architecture. While the existing process-rule approaches support process-aware execution for specific agents or processes, our organizational memory treats organization-specific process knowledge as a shared enterprise resource. It integrates many heterogeneous sources, such as policies, SOPs, system documentation, process models, knowledge bases, and execution experience, handles human governance, and addresses lifecycle concerns such as organizational change and reuse across multiple agents and processes. Moreover, unlike strategy-based guardrailing [8], organizational memory does not formally constrain agents; instead, it provides an *informal frame* through additional and useful context.

2.2 Agentic Memory Systems

While traditionally, agentic architectures manage symbolic knowledge in *belief bases* [3], recent works tend to focus on *memory* mechanisms for LLM-based agents. Existing approaches manage memory tiers to extend limited context windows [19], maintain memory streams with dedicated controllers [25], organize memories dynamically through indexing and linking [27], or store long-term conversational memories to improve personalization and coherence across interactions [28]. More broadly, cognitive agent architectures such as CoALA conceptualize memory as a core component of language agents, alongside action

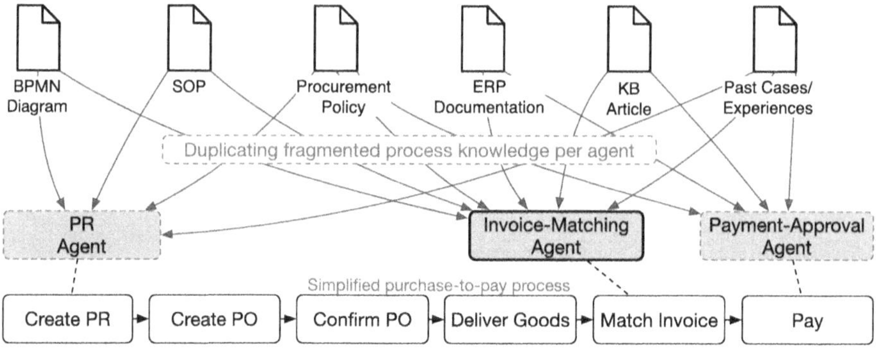

Fig. 1. Running purchase-to-pay example: each agent depends on overlapping knowledge that is fragmented across heterogeneous artifacts.

spaces and decision-making mechanisms [22]. These works highlight the importance of persistent memory for LLM agents, but primarily focus on individual agents rather than the enterprise-level challenge of curating, governing, maintaining, and consistently providing organization-specific process knowledge across many agents and processes. Retrieval-augmented generation (RAG) [16] offers a complementary mechanism by grounding LLM outputs in retrieved passages from external document collections. While useful for accessing organizational documents, RAG alone may be insufficient for enterprise process knowledge that is overlapping, evolving, and often relevant beyond the user's immediate query. Different agents may retrieve different chunks, miss cross-cutting rules, or rely on outdated or conflicting sources. Our notion of organizational memory addresses this gap by treating process knowledge as shared, traceable, governed, and reusable memory elements rather than as query-dependent document fragments.

3 Motivation

In this section, we motivate the need for organizational memory in agentic business process execution by means of a running example, from which we then derive several challenges that arise in this setting.

Problem Illustration. We consider enterprise settings in which LLM-based agents execute business process tasks at runtime, where correct behavior depends on organization-specific procedural knowledge such as policies, responsibilities, systems, and exception-handling practices. Consider a simplified purchase-to-pay process as visualized on the bottom of Fig. 1. The process starts with the creation of a purchase requisition (PR), followed by issuing a purchase order (PO) to a vendor. The vendor confirms the order, delivers the goods, and submits an invoice. Before a payment can be executed, the invoice must be checked against the corresponding purchase order. Suppose an organization wants to automate parts of this process using LLM-based agents. One agent may support employees in creating PRs, while another may handle invoice matching.

Focusing on the invoice-matching agent, assume a received invoice does not exactly match the corresponding purchase order. The agent must decide whether the invoice can be posted automatically, whether it should be routed to a procurement specialist, or whether it requires compliance review. Other agents in the same process, such as a PR agent and a payment-approval agent, depend on overlapping knowledge, as illustrated in Fig. 1. At first glance, this appears to be a straightforward reasoning task: compare the invoice to the purchase order, identify deviations, and decide how to proceed. However, the correct decision depends on organizational knowledge that is not contained in the LLM itself.

For instance, a BPMN model may specify that invoices with mismatches require an "invoice clarification" activity. A standard operating procedure may state that price deviations below EUR 250 can be accepted for preferred suppliers. A procurement policy may define that deviations involving medical equipment always require compliance review, regardless of their amount. ERP documentation may specify where supplier risk classifications are stored, while an internal knowledge-base article may explain that, for a particular subsidiary, the responsible procurement specialist is determined by plant rather than by product category. Relevant knowledge may also emerge from previous executions, for example, when human experts repeatedly resolve a recurring mismatch in the same way or when agents receive corrections for exceptional cases. Each of these sources contains potentially relevant knowledge, but no single source provides the complete execution context required by the agent.

Challenges. Several challenges arise in this setting:

Fragmented and Human-Oriented Knowledge. Process-relevant knowledge is distributed across heterogeneous sources. These sources typically differ in format, terminology, granularity, and quality, and they are primarily created for human interpretation rather than for direct consumption by LLM-based agents. Moreover, such knowledge is often not complete and up-to-date, and sometimes too extensive for humans to review in its entirety. In the purchase-to-pay example, the agent must combine knowledge from a BPMN model, a procurement policy, an SOP, ERP documentation, subsidiary-specific instructions, and potentially past exception handling to make a single invoice-handling decision. The challenge is therefore to transform fragmented organizational knowledge into a unified, agent-consumable representation.

Shared Knowledge Across Agents and Processes. Organization-specific knowledge is rarely relevant to only one agent. The same procurement policy, supplier classification, approval rule, or compliance constraint may be needed by multiple agents operating in different parts of the same process or in related processes. If each agent is equipped with its own prompt, local memory, or retrieval setup, organizations risk creating new knowledge silos: rules are duplicated, updated inconsistently, and interpreted differently across agents. In the purchase-to-pay example, a PR agent, an invoice-matching agent, and a payment approval agent may all depend on overlapping procurement and finance rules. These rules should therefore be maintained once and made available consistently, rather than being reimplemented separately per agent.

Organizational Knowledge Change and Learning. Organizational knowledge is not static. Policies are updated, responsibilities shift, systems change, and new exception-handling practices emerge. This phenomenon is well-known in BPM under *concept drift* [15]. Consequently, the knowledge used by agents must evolve with the organization. For instance, if the company changes the acceptable mismatch threshold from EUR 250 to EUR 100, all affected agents must reliably act according to the updated knowledge. The challenge is to keep the shared knowledge layer current without requiring each agent to be updated individually.

Runtime Accessibility and Efficiency. Agents require organization-specific knowledge at the moment of execution, but only the knowledge that is relevant to the current task and context. Simply providing all potentially relevant documentation is neither scalable nor reliable, as it increases latency, introduces irrelevant information, and may distract the agent from the actual decision. In the example, the invoice agent should receive the applicable mismatch tolerance, the supplier-specific rule, the relevant compliance constraint, and the responsible organizational role, but not unrelated procurement guidelines or ERP manuals.

Human Governance. Since organization-specific knowledge affects how agents behave across processes and departments, its creation and evolution cannot be left entirely to automated mechanisms. Human decision-makers must be involved in deciding which knowledge should be added, modified, or discarded, and how conflicts between sources should be resolved. This raises the question of who is allowed to access, approve, and change which parts of the memory. In the purchase-to-pay example, procurement experts may be responsible for invoice-matching rules, finance for payment policies, compliance for regulatory constraints, and local process owners for subsidiary-specific exceptions. At the same time, not every user who can read a rule should also be able to modify it, and changes made by one organizational unit may affect agents operating in another. Organizational memory therefore requires governance mechanisms for ownership, access control, approval workflows, and accountability.

The challenges show that reliable and scalable agentic process execution requires more than agent-specific prompts or local memories. It calls for a shared organizational memory of procedural knowledge about how work should be executed.

4 Architecture Requirements

Based on the challenges described above, we derive the requirements for an organizational memory architecture.

R1: Integration of heterogeneous organizational knowledge. Integrate process-relevant knowledge from heterogeneous sources. In the example, this includes the

BPMN model, procurement policy, invoice-matching SOP, ERP documentation, and past exception handling.

R2: Agent-consumable representation. Transform fragmented, often human-oriented process knowledge into a unified representation that is decomposed into self-contained units of procedural knowledge, so LLM-based agents can interpret and apply it during process execution. Unification alone (e.g., turning sources to plain text) is insufficient, since without decomposition into self-contained units, individual rules cannot be addressed, compared, updated, or selectively retrieved. For instance, a BPMN model and a textual rule should not only be expressed in a common format but also be broken down to the level of individual rules.

R3: Conflict detection and consistency management. Support the identification and resolution of inconsistencies, redundancies, and conflicts between memory fragments originating from different sources, which is a general problem in AI [6]. For example, if an SOP allows deviations below EUR 250 while a plant-specific policy defines a lower threshold, this conflict must be detected and resolved.

R4: Traceability to source artifacts. Preserve links between memory fragments and their original sources to enable transparency. In the purchase-to-pay example, an agent's decision to route an invoice to compliance should be traceable to the underlying policy document.

R5: Context-specific retrieval. Provide agents with the organization-specific knowledge relevant to the current process task and context. For instance, the invoice-matching agent should receive the applicable mismatch tolerance, supplier-specific rule, compliance constraint, and responsible role, but not unrelated HR guidelines.

R6: Low-latency runtime accessibility. Provide organizational memory efficiently at runtime so agents can access relevant context without prohibitive latency.

R7: Support for adaptive memory. Support the continuous evolution of organizational memory over time, both through deliberate updates to policies, organizational structures, and systems, and through insights gained from process execution experiences. For example, if the allowed mismatch threshold changes from EUR 250 to EUR 100, all affected agents should follow the updated rule after a single change to the memory. Similarly, repeated human corrections for a recurring invoice mismatch may indicate a newly observed exception or successful practice and become a candidate update to the organizational memory.

R8: Human governance and ownership. Support human oversight of the creation and modification of organizational memory. This includes defining ownership of memory fragments, specifying read and write permissions, supporting approval workflows for updates, and ensuring accountability for changes. For instance, procurement department may own invoice-matching rules, whereas finance department may own payment policies.

R9: Scalability across agents and processes. Scale to enterprise settings in which multiple agents operate across different processes, departments, organizational

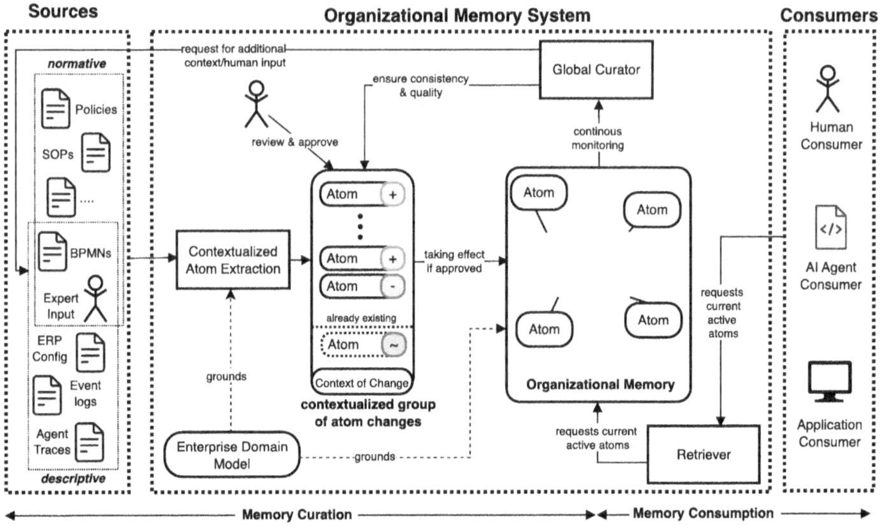

Fig. 2. Organizational memory architecture.

units, and information systems while relying on a shared organizational memory. In the example, a PR agent, invoice-matching agent, and payment approval agent should all be able to access the same maintained procurement and finance rules.

5 Organizational Memory Architecture

This section introduces the proposed organizational memory architecture, addressing the requirements outlined above. Figure 2 provides a high-level overview of the architecture. It shows how heterogeneous sources are transformed into fragments, which we call *process atoms* (*atoms* for short), that collectively constitute the organizational memory. This memory can then be accessed by consumers, such as LLM-based agents, to support organization-specific process execution. In the following, we first make the core design choices underlying the architecture explicit in Sect. 5.1, before describing how organizational memory is curated in Sect. 5.2 and how it is consumed during process execution in Sect. 5.3.

5.1 Design Choices

Before describing the architecture in detail, we make core design choices explicit, justifying them against the requirements.

Atomic Decomposition. Memory fragments could in principle be represented as plain text, document chunks, or larger pieces of structured process descriptions. We instead represent each fragment as a *process atom*: a self-contained unit that captures exactly one rule, constraint, or responsibility, with its applicability conditions separated from its required action and purpose. Concretely, each process atom has the following core attributes:

- *Name*: a short name for the desired process behavior;
- *Source*: a reference to the input source from which the atom was extracted;
- *Content*: a structured textual description with three parts:
  - *Applicability*: the context in which the atom applies (e.g., process, activity, role, organizational unit, business object, or precondition);
  - *Action*: the concrete rule, constraint, or required behavior;
  - *Purpose*: the rationale for the rule, i.e., why it exists.

Atomic decomposition addresses R2 and enables the requirements that operate at the level of single rules and behavioral constraints: conflict detection between candidate and existing knowledge (R3), targeted updates as policies change (R7), per-rule traceability to source artifacts (R4), and retrieval of the smallest sufficient set for a given task (R5). A plain-text representation can be unified and agent-consumable, but it makes it difficult to be addressed, compared, or updated at the rule level without decomposition. Conversely, decomposing further than the described level, would lose the business meaning that domain experts rely on to govern memory (R8), even if a finer logical decomposition would be possible. This idea of representing procedural knowledge as atomic units is, among others, also followed in declarative process modeling and analysis [5]; we provide an integrated structure, i.e., name, source, content, and tags (described below), tailored to organizational memory for LLM-based agents.

Domain-Model Grounding via Tags. Atoms must be grounded in organizational concepts so they can be retrieved in a specific context, for the right process, role, business object, or unit. We ground atoms through categorized *tags* that jointly represent an *Enterprise Domain Model*. A natural alternative is to ground atoms in a formal ontology or a knowledge graph with explicit relations [2]. A formal ontology requires upfront schema design and continuous ontology engineering, though, which is a strong requirement. A tag vocabulary can be assembled incrementally and easily revised as the organization changes (R7). Tags also align well with the way LLM-based retrievers and agents consume context.

Retrieval Pattern. At runtime, we follow a (hybrid) retrieval pattern rather than a specific mechanism. We combine filtering using the domain-model tags with semantic or LLM-based selection within the filtered set. Semantic search over atom content in isolation does not scale well and cannot reliably distinguish, e.g., an invoice-matching rule that applies in subsidiary A from one that applies in subsidiary B, because that distinction may be encoded in the organizational metadata, not the rule text. Pure tag-based filtering cannot resolve fine-grained relevance among many atoms that share the same tags. The architecture therefore features a *Retriever* that takes a context request and returns a relevant atom set, leaving the concrete mechanism to instantiation.[1]

[1] Note that we do not claim or evaluate that a single mechanism is best in this paper.

5.2 Memory Curation

This section describes how organizational memory is created and maintained. As shown in Fig. 2, the curation phase takes multiple input sources, extracts atoms, creates a group of atom changes, and finally updates the memory through a human-governed review and change process.

Input Sources. The curation phase can take different types of organizational knowledge as input, including policies, SOPs, process models, system documentation, knowledge-base articles, and traces from previous process executions. These sources differ in format, structure, terminology, and degree of formality.

Step 1: Create Group of Atom Changes. In the first step, we create a contextualized group of atom changes by deriving a set of candidate process atoms from the input sources and contextualizing these with the existing atoms in the organizational memory. Since different source types encode knowledge differently, source-specific extraction services are required. For instance, a BPMN model must be analyzed differently from a policy document or an execution trace. Note that the design of extraction services for all possible input sources is beyond the scope of this paper. For simplicity, we focus on PDF documents, such as policies and SOPs, which also serve as the input sources in our proof-of-concept evaluation in Sect. 6. In general, the purpose of this step is to identify process-relevant rules, constraints, responsibilities, and exception-handling knowledge and represent them as atoms (see Sect. 5.1 for the atom representation).
Assume a finance control directive states that every PR must by assigned a valid cost center. The extraction service may create the following atom:

- *Name*: Ensure that a cost center is assigned on each PR.
- *Source*: Finance Control Directive, Sect. 3.
- *Content*:
  - *Applicability*: All purchase requisitions, in all company codes.
  - *Action*: Every PR must be assigned to a valid cost center before it can be submitted; if the requester does not provide a cost center, the PR must be rejected.
  - *Purpose*: Ensures that every expenditure is traceable to a responsible organizational unit for budgetary control and audit purposes.

The extraction services also ground each candidate in the *Enterprise Domain Model* through one or more categorized tags, such as `procurement` (corporate function), `purchase requisition` (business object), or `purchase-to-pay` (end-to-end process). These tags provide a shared semantic vocabulary for organizing and later retrieving atoms from the memory.

Candidate atoms are then assessed by the *Global Curator*, an active entity in our architecture that has read access to the existing organizational memory. The Global Curator performs two complementary checks. First, it carries out quality checks on each candidate atom in isolation, such as verifying that the atom has a meaningful name, a clearly defined scope, and a single rule. Second, it relates each candidate to the existing memory, checking whether it duplicates, overlaps with, or conflicts with atoms already stored. For example, if a candidate atom, which is extracted from a policy document, is already present in the organizational memory, it does not need to be added again. Conversely, if a candidate atom conflicts with an existing atom, the conflict must be detected and resolved, for instance by updating the existing atom, replacing it, or flagging the conflict for human review. The outcome of this step is a proposed set of atom changes to the memory, which may include atoms to be added, removed, or modified.

Step 2: Update Organizational Memory. Given the contextualized group of atom changes, a *Human* expert reviews the changes, checks the atoms and their sources, and decides if changes should be applied. They may accept the proposal, modify individual atoms, reject individual changes, or resolve conflicts differently than suggested by the curator. In the purchase-to-pay example, a finance expert may approve changes to cost-center rules, while procurement or compliance experts may be responsible for invoice-matching or regulatory constraints. This human-in-the-loop design ensures that the memory remains governed and that changes to the shared knowledge layer are overseen by experts.

Organizational Memory. The output of the curation phase is the updated organizational memory: a governed collection of validated atoms grounded in the enterprise domain model. Through their tags and source references, atoms can be organized, traced, and retrieved for runtime use. The memory thereby acts as the shared knowledge layer from which agents can obtain task-specific organizational context during process execution. The *Global Curator* can continuously monitor the organizational memory and request additional context or human input.

5.3 Memory Consumption

This section describes how organizational memory is consumed during process execution. As shown in Fig. 2, memory consumption starts with a context request from the runtime agent, continues with a query to the organizational memory, and results in a set of relevant atoms that is provided to the agent as execution context.

Step 1: Create Context Request. At runtime, an agent is assigned to execute parts of a business process. To do so reliably, it may require organization-specific knowledge that is not contained in its prompt or model knowledge. The agent therefore creates a context request for the *Retriever*. This request describes the current execution context, including the process, activity, agent role, organizational unit, input data, and case state. In the purchase-to-pay example, an invoice-matching agent may include information about the supplier, invoice amount, purchase order, material category, subsidiary, and current activity.

Step 2: Query Organizational Memory. The *Retriever* uses the context request to identify atoms in the *Organizational Memory* that are applicable to the current situation. Different retrieval mechanisms may be used for this purpose, such as tag-based filtering, semantic search, rule-based matching, or LLM-based selection. The tags assigned during memory curation are particularly useful because they ground atoms in the enterprise domain model and make them retrievable by process, activity, role, object type, or organizational context. The goal is to retrieve the smallest sufficient set of atoms for the current task, rather than exposing the agent to all potentially related documentation. Reliably avoiding both under-retrieval and over-retrieval is a key challenge for enterprise-level settings and motivates future research.

Step 3: Provide Set of Atoms. The *Retriever* returns the selected atoms to the runtime agent. These atoms are appended to the agent's context and guide its decision-making and activity execution.

In summary, the proposed architecture addresses the requirements froms Sect. 4 through its two complementary phases. The curation phase integrates heterogeneous input sources (R1), transforms them into unified atoms (R2), checks candidate atoms against the existing memory for duplicates and conflicts (R3), preserves source references (R4), and supports governed updates as organizational knowledge changes (R7–R8). The runtime consumption phase enables context-specific retrieval (R5) with low overhead (R6) by providing agents with a compact set of relevant atoms rather than full documents. Finally, the shared organizational memory addresses scalability across agents and processes (R9), since multiple agents can rely on the same maintained knowledge layer instead of separate local memories or retrieval setups.

6 Proof-of-Concept

This section presents a preliminary proof-of-concept of whether the proposed organizational memory architecture can improve the ability of LLM-based agents to execute business process tasks in an organization-aware manner.[2] Specifically, we examine whether access to organizational memory supports agents in making decisions that are in line with desired behavior during process execution. We describe the experimental setup in Sect. 6.1 and the results in Sect. 6.2.

6.1 Experimental Setup

We use the purchase-to-pay process introduced in Sect. 3, where an LLM-based agent acts as a procurement assistant supporting employees in creating purchase requisitions (PRs) in the company's ERP system. Given a natural-language request from a user, the agent must decide whether the requested purchase complies with the organization's rules. If so, the agent should create the corresponding PR with the correct information. If the request violates organizational rules, the agent should refuse the creation and explain the reason for non-compliance.

[2] The source code for our implementation cannot be made publicly available due to proprietary reasons. Yet, we provide a detailed description of the artifact's design and the experimental procedure in the paper, and supply additional evaluation details (incl. data and ground truth): https://github.com/lukaskirchdorfer/org-memory-paper.

Evaluation Data. To create a controlled evaluation environment, we construct a synthetic document collection representing the organizational knowledge base of a hypothetical company. The collection consists of 9 manually created PDF documents, comprising 5 procurement-related documents and 4 distractor documents from other organizational functions, such as HR. The procurement-related documents resemble typical enterprise artifacts, such as policies and SOPs, and encode 11 distinct rules governing the purchase-to-pay process. These rules cover approval thresholds, supplier restrictions, category-specific requirements, and exception-handling procedures. The rules were manually specified and embedded in realistic document contexts rather than provided as isolated rule lists. This design allows us to evaluate whether the proposed architecture can identify and retrieve relevant organizational knowledge from unstructured enterprise documents. The distractor documents are included to reflect a broader enterprise knowledge base and to assess whether the architecture can provide task-relevant context without exposing the agent to excessive or irrelevant information.

Scenarios. We designed 10 scenarios, each representing a specific employee request to the agent. The scenarios cover varying levels of complexity: they range from straightforward compliant requests, in which the employee provides all required information, to requests that violate a single policy, and finally to cases in which the agent must jointly consider multiple policies to determine the correct response. Each scenario is annotated with a binary expected outcome—CREATE or REFUSE—and a corresponding ground-truth specification of PR field values (material, quantity, vendor) that makes evaluation possible in a rule-based manner. The ground truth was created manually by the authors based on the synthetic policy documents, ensuring that each decision and annotated field value can be traced directly to a specific clause in the source documents.

Agent Setups. We compare three setups (*Base*, *RAG*, and *Memory*) using the same LLM and system prompt, yet, differing in organizational knowledge access:

Base: No Access to Organizational Process Knowledge. The agent receives no policy information and must decide whether to create a PR based solely on the information in the employee request and the LLM's parametric knowledge.

RAG: RAG-Based Access to Organizational Process Knowledge. Before each request, the agent retrieves the top-5 most relevant chunks from a vector database that contains the document corpus via dense retrieval embeddings. The ChromaDB vector database has been created in a preprocessing step from the given input documents. Each document is split into overlapping chunks of approximately 600 characters using a paragraph-aware chunker with a 100-character overlap between adjacent chunks, and all chunks are embedded with `text-embedding-3-large` using cosine similarity as the distance metric.

Memory: Access to Organizational Memory (Ours). Before each request, the agent requests the *Retriever* to provide the set of necessary atoms for the specific procurement situation. The atoms have been automatically extracted and created from the given input documents.

Note that although granting the agent direct access to all documents is a conceivable setup, we deliberately exclude it from our evaluation because it is neither realistic nor scalable in enterprise settings.

LLMs. We instantiate all three agents with both OpenAI's GPT-4.1 and Anthropic's Claude Sonnet 4.5 and report results for both models.

Metrics. We evaluate agent behavior using the *Policy Compliance Rate* (PCR), defined as the fraction of scenarios in which the agent produces the policy-compliant outcome. For CREATE scenarios, the agent is considered correct only if it creates a PR with the expected material, quantity, and vendor, which we verify through rule-based comparison against ground-truth fields. For REFUSE scenarios, the agent is considered correct only if it refrains from creating a PR.

6.2 Results

Figure 3 reports the PCR for all three agent configurations, averaged over ten scenarios and four runs per scenario. The organizational memory agent achieves the highest PCR for both models, reaching 88% with GPT-4.1 and 95% with the more advanced Sonnet 4.5. It clearly outperforms both the base agent, which reaches 30% for both models, and the RAG-based agent, which reaches 70% with GPT-4.1 and 80% with Sonnet 4.5. This indicates that structured organizational memory can improve policy-compliant process execution beyond both parametric model knowledge and document-level retrieval. The base agent performs poorly, often creating PRs even when the request violates a policy. Its correct decisions mostly occur in simple scenarios where the relevant information is explicit in the employee request and no organization-specific reasoning is required. RAG improves substantially over the base setup, showing that document retrieval can help when the relevant policy is semantically close to the employee request. However, RAG still fails in scenarios where the required rule is cross-cutting rather than query-specific. For example, when an employee requests notebooks without specifying a cost center, the RAG agent creates the PR, although the organization requires a cost center for every PR. Since this universal rule is stated in a finance policy and is not semantically close to the notebook request, it is not retrieved.

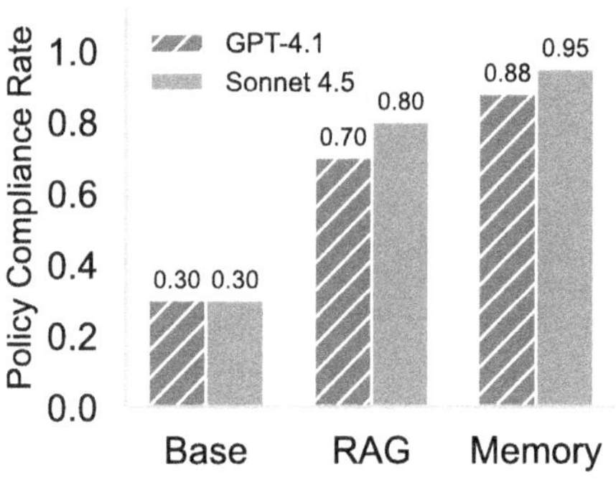

Fig. 3. Results across all 10 scenarios.

The organizational memory agent mitigates this limitation by retrieving task-relevant atoms rather than relying solely on similarity between the request and document chunks. This allows it to consider both request-specific rules and generally applicable constraints, such as the mandatory cost center requirement. The memory agent also handles scenarios that require multiple policies to be considered jointly. Its remaining errors are mainly over-restrictions: in some runs, the agent refuses a request that is actually compliant. Inspection shows that this is caused by an imprecisely extracted atom whose applicability scope is too broad. This highlights that the effectiveness of organizational memory depends not only on runtime retrieval, but also on the quality of memory curation.

Overall, the results provide initial evidence that organizational memory can improve organizational-knowledge-aware process execution, while at the same time, it is not designed to and cannot ensure correct process execution in all cases. Notably, the results should be interpreted with caution, given the preliminary nature and limited scope of the evaluation.

7 Conclusion

LLM-based agents create new opportunities for automating business process execution, but reliable execution in enterprise settings requires access to organization-specific procedural knowledge. In this paper, we introduced *organizational memory* as a shared, governed, and agent-consumable reference layer for such knowledge. We derived requirements for organizational memory, proposed an architecture that covers both memory curation and runtime consumption, and instantiated the architecture through process atoms that can be extracted from organizational artifacts, governed by humans, and retrieved by agents during execution. A proof-of-concept in a procurement scenario indicates that organizational memory can improve policy-compliant agent behavior.

Our work is an initial step and should be interpreted in light of its limitations. The proof-of-concept covers a single synthetic procurement process with a small policy corpus consisting only of PDF documents. Thus, generalization to real enterprise settings with numerous overlapping policies, process models, conflicting sources, and legacy documentation is not yet demonstrated. Future work should therefore evaluate organizational memory in larger and more realistic enterprise environments. Beyond larger-scale evaluation, our work opens several directions for future research at both the engineering and management levels. From an engineering perspective, our experiments indicate that execution quality strongly depends on the precision of atom extraction and curation. Future research should therefore develop robust extraction mechanisms for a broader range of organizational knowledge sources beyond PDF documents. From a management perspective, future work should investigate how organizational memory can be governed effectively and efficiently by human experts in large organizations, including questions of responsibility, accountability, lifecycle management, update propagation, and scalability as organizational knowledge evolves over time.

References

1. van der Aalst, W.M.P., Bichler, M., Heinzl, A.: Robotic process automation. Bus. Inf. Syst. Eng. **60**(4), 269–272 (2018)
2. Bein, L., Pufahl, L.: Karibdis - a knowledge-graph-based business process management system for semantically explainable process execution support. In: Intelligent Information Systems (2026)
3. Boissier, O., Bordini, R.H., Hübner, J.F., Ricci, A., Santi, A.: Multi-agent oriented programming with jacamo. Sci. Comput. Program. **78**(6), 747–761 (2013)
4. Calvanese, D., et al.: Agentic business process management: a research manifesto. IS **140**, 102738 (2026)
5. Ciccio, C.D., Montali, M.: Declarative process specifications: reasoning, discovery, monitoring. In: Process Mining Handbook, pp. 108–152. Springer (2022)
6. Corea, C., Kuhlmann, I., Thimm, M., Grant, J.: Paraconsistent reasoning for inconsistency measurement in declarative process specifications. Inf. Syst. **122**, 102347 (2024)
7. Dumas, M., Milani, F., Chapela-Campa, D.: Agentic business process management systems. CoRR abs/2601.18833 (2026)
8. Giacomo, G.D., Kampik, T., Kirchdorfer, L., Montali, M., Weinhuber, C.: Formal foundations of agentic business process management (2026)
9. Guan, Y., et al.: Intelligent agents with LLM-based process automation. In: KDD (2024)
10. Jain, A., Paliwal, S., Sharma, M., Vig, L., Shroff, G.: Smartflow: robotic process automation using LLMs. CoRR abs/2405.12842 (2024)
11. Jennings, N.R., Faratin, P., Johnson, M.J., Norman, T.J., O'Brien, P.D., Wiegand, M.E.: Agent-based business process management. IJCIS **5**(2&3), 105–130 (1996)
12. Kaltenpoth, S., Skolik, A., Müller, O., Beverungen, D.: A step towards cognitive automation: integrating LLM agents with process rules. In: BPM (2025)
13. Kirchdorfer, L., Blümel, R., Kampik, T., van der Aa, H., Stuckenschmidt, H.: Discovering multi-agent systems for resource-centric business process simulation. Process Sci. **2**(1), 4 (2025)
14. Kirchdorfer, L., Doumeni, A., van der Aa, H., López, H.A.: From global policies to local strategies: multi-objective optimization of resource-specific handover policies. In: BPM. Lecture Notes in Computer Science. Springer (2026)
15. Kraus, A., van der Aa, H.: Comprehensive characterization of concept drifts in process mining. Inf. Syst. **135**, 102584 (2026)
16. Lewis, P., et al.: Retrieval-augmented generation for knowledge-intensive NLP tasks. In: NeurIPS (2020)
17. Liu, P., Yuan, W., Fu, J., Jiang, Z., Hayashi, H., Neubig, G.: Pre-train, prompt, and predict: a systematic survey of prompting methods in natural language processing. ACM Comput. Surv. **55**(9) (2023)
18. Monti, F., Leotta, F., Mangler, J., Mecella, M., Rinderle-Ma, S.: Nl2processops: towards LLM-guided code generation for process execution. BPM (Forum) (2024)
19. Packer, C., Fang, V., Patil, S.G., Lin, K., Wooders, S., Gonzalez, J.E.: Memgpt: towards LLMs as operating systems. CoRR abs/2310.08560 (2023)
20. Shen, Q., Polyvyanyy, A., Lipovetzky, N., Kampik, T.: Applying organizational mining to discover agent systems from event data. Inf. Syst. **138**, 102669 (2026)
21. Skolik, A., Kaltenpoth, S., Beverungen, D., Müller, O.: A modular large language model agent architecture for adaptive and autonomous process-aware execution. Inf. Syst. **141**, 102748 (2026)

22. Sumers, T.R., Yao, S., Narasimhan, K., Griffiths, T.L.: Cognitive architectures for language agents. TMLR **2024** (2024)
23. Tour, A., Polyvyanyy, A., Kalenkova, A.A., Senderovich, A.: Agent miner: an algorithm for discovering agent systems from event data. In: BPM (2023)
24. Vu, H., Klievtsova, N., Leopold, H., Rinderle-Ma, S., Kampik, T.: Agentic business process management: practitioner perspectives on agent governance in business processes. In: BPM (RBPM, PT, and Educators Forum) (2025)
25. Wang, B., et al.: SCM: enhancing large language model with self-controlled memory framework. In: DASFAA (6), pp. 188–203 (2025)
26. Weske, M.: Business Process Management: Concepts, Languages, Architectures. Springer (2024)
27. Xu, W., Liang, Z., Mei, K., Gao, H., Tan, J., Zhang, Y.: A-MEM: agentic memory for LLM agents. In: NeurIPS (2025)
28. Zhong, W., Guo, L., Gao, Q., Ye, H., Wang, Y.: Memorybank: enhancing large language models with long-term memory. In: AAAI (2024)

Object-Centric Process Querying in SQL: Translation Rules and In-Database Execution

Aaron Küsters(✉), Justin Graß, and Wil M. P. van der Aalst

Process and Data Science (PADS), RWTH Aachen University, Aachen, Germany
{kuesters,wvdaalst}@pads.rwth-aachen.de, justin.grass@rwth-aachen.de

Abstract. Object-Centric Process Querying (OCPQ) is a visual query language and accompanying tool for expressing queries and constraints over Object-Centric Event Data (OCED). Such queries span multiple object and event types that case-centric query languages cannot natively address. Its existing execution engine processes data in memory: Dataset size is capped by host memory, and the relational database systems that organizations already operate cannot serve as backends, forcing data export or migration. In this paper, we present a translation from OCPQ to SQL, defined as a recursive mapping over the OCPQ query tree with a rule per construct, emitting SQL for SQLite, DuckDB, and PostgreSQL. An *OCEL 2.0 Virtualization Layer* accommodates source schemas that diverge from the OCEL 2.0 layout via direct rewrite or `VIEW` generation, so the translation can target them without data migration. We evaluate using three types of workloads: Seven queries from prior work, an automatically generated corpus of 6 919 OCPQ trees, and a non-OCEL Enterprise Resource Planning (ERP) database schema queried through the virtualization layer. The resulting query outputs match the in-memory engine across all queries, demonstrating that the translation preserves its semantics. Our evaluation also shows the memory decoupling enabled by the translation: SQL backends scale to support large datasets on which the native OCPQ engine runs out of memory. When data fits in memory, the in-memory OCPQ engine remains fastest, with DuckDB roughly 3–12× slower and PostgreSQL 14–71× slower as drop-in replacements.

Keywords: Object-Centric Process Mining · Process Querying · SQL

1 Introduction

In organizations, process execution data are increasingly stored and queried directly from relational database systems [5,12]. At the same time, the domain of process querying is shifting from case-centric to *Object-Centric Event Data* (OCED), which preserves the multi-object structure of real-life processes [1]. OCED drops the single-case assumption that each event belongs to exactly one defined case. Instead, OCED contains a set of objects and a set of events of specified types, together with relationships between them. This lets OCED represent

K. Kluza et al. (Eds.): BPM 2026, LNBIP 600, pp. 228–244, 2027.
https://doi.org/10.1007/978-3-032-36710-5_15

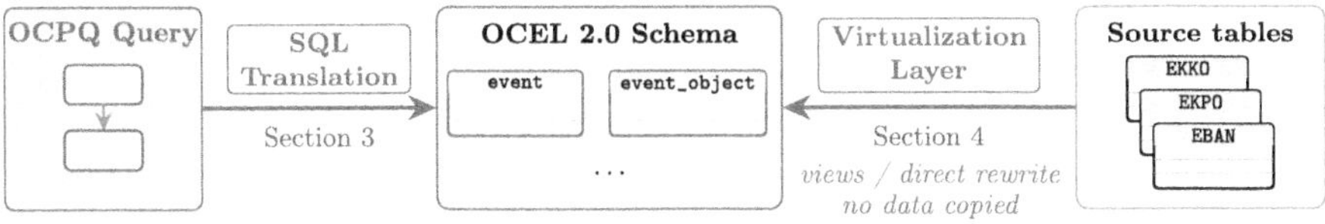

Fig. 1. An OCPQ query is translated to a SQL query that targets the fixed OCEL 2.0 schema. When the data resides in a different operational schema, e.g., the purchasing tables of an SAP system, the virtualization layer presents the OCEL 2.0 schema over the source tables through views or direct rewrite, without copying data.

many real-life processes much more accurately than traditional, flat event logs. For example, in an order management process, the different objects interacting in the process may include `customers`, `orders`, `items`, and `packages`. OCEL 2.0 [4] specifies a standardized relational representation of OCED: One table per event and object type, with junction tables for the relationships between them, as well as general event and object tables. Organizational process data, however, rarely follows this layout. It typically resides in the source tables of operational systems, such as Enterprise Resource Planning (ERP) software, in schemas that OCEL-based tooling cannot use directly. Figure 1 previews the setting addressed in this paper: Queries are translated to SQL against the standardized OCEL 2.0 tables, and a thin virtualization layer maps these tables onto whatever schema the database actually holds, so no data has to be migrated or copied.

To query OCED, recent work has proposed Object-Centric Process Querying (*OCPQ*) [9], a graphical nested querying approach that *natively supports multiple object and event types*. OCPQ is attractive for process analysis, as OCED concepts, such as objects and events, are first-class citizens of the language. An analyst expresses a question spanning several objects and event types directly, rather than through the manual joins and subqueries an equivalent SQL formulation requires (cf. the verbose SQL query in Fig. 2). OCPQ supports a range of tasks, from querying process data and checking constraints to creating feature tables for machine learning applications.

An OCPQ query is structured as a tree of *binding boxes*. Each binding box introduces typed event and object variables and filters them through predicates, most importantly the three basic predicates event-to-object (E2O), object-to-object (O2O), and time-between-events (TBE), which relate variables through OCED relationships and event timestamps. Children of a binding box extend the bindings of their parent and can be combined with size filters (e.g., "orders with at least three items"). In addition to filter predicates, constraints can be used to specify rules that should hold, marking violations in the output. Figure 2 shows an example OCPQ constraint from [9]: *After an Application was accepted, there should be at least one associated Offer accepted afterwards.* The root node queries combinations of application objects and E2O-related accepted events. The child node queries offer objects that are related to the parent application (O2O) with their offer-accepted events occurring afterwards (TBE). Each parent

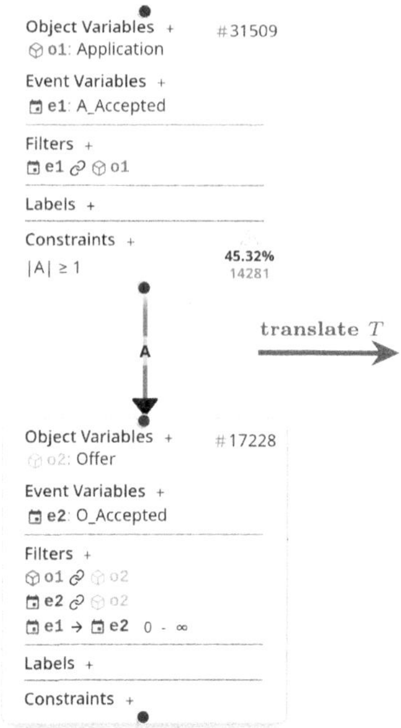


```sql
SELECT O1.ocel_id AS "O1", E1.ocel_id AS "E1", (
      EXISTS (SELECT 1
FROM "object_Offer" AS O2
INNER JOIN "object_object" AS OR1
  ON OR1.ocel_source_id = O1.ocel_id
 AND OR1.ocel_target_id = O2.ocel_id
INNER JOIN "event_object" AS ER2
  ON ER2.ocel_object_id = O2.ocel_id
INNER JOIN "event_O_Accepted" AS E2
  ON ER2.ocel_event_id = E2.ocel_id
WHERE E2.ocel_time >= E1.ocel_time
  AND O2.ocel_changed_field IS NULL
)) AS satisfied
FROM "event_A_Accepted" AS E1
INNER JOIN "event_object" AS ER1 ON ER1.
      ocel_event_id = E1.ocel_id
INNER JOIN "object_Application" AS O1 ON ER1.
      ocel_object_id = O1.ocel_id
WHERE O1.ocel_changed_field IS NULL
```

execute on DB

O1	E1	satisfied
Application_1	A_Accepted_8	TRUE
Application_3	A_Accepted_15	FALSE
Application_5	A_Accepted_24	TRUE

31 509 rows total, 14 281 (45.32%) violated

Fig. 2. *Left:* an OCPQ query tree modeling "*After an Application was accepted, there should be at least one associated Offer accepted afterwards*". The root node binds `o1: Application` objects, and `e1: A_Accepted` events linked by `E2O`. The child node binds `o2: Offer`, `e2: O_Accepted` linked by `O2O` and `E2O`, and specifies a time filter (`e2` after `e1`). The parent's size constraint $|A| \geq 1$ marks each output binding as satisfied if and only if the child set is non-empty. *Right:* the SQL query Q produced by T (Sect. 3) based on the OCPQ query tree. *Bottom:* schematic of the result rows on BPIC 2017.

binding (application and application-accepted event) is marked satisfied if and only if the child binding set (offer and offer-accepted event combinations) is non-empty, as per the $|A| \geq 1$ Child-Binding-Set (CBS) size constraint.

The OCPQ tool has a high-performance execution engine that achieves fast runtime, even on datasets with millions of events [9]. However, the existing engine relies on in-memory processing of OCEL files using custom data structures. For organizations that store their process data in a database, this design is restrictive in several ways: The data must first be exported to an OCEL 2.0 [4] file before queries can be evaluated, query results then go stale as soon as the underlying data in the database changes, and the mature infrastructure that database systems already offer (indexing, distributed processing, etc.) cannot be leveraged. Moreover, in-memory processing has natural scalability limits: When the data exceeds host memory, the in-memory engine cannot produce results.

This paper investigates the following question: *How can OCPQ queries be translated to SQL so they execute directly on the relational databases that hold process data?* We answer it with a translation from OCPQ to SQL, defined as a recursive mapping over the binding-box tree with one rule per construct. The structural predicates (the basic relations E2O, O2O, TBE, together with attribute filters, child-binding-set (CBS) predicates, and child-composition constraints) become SQL joins, `WHERE` clauses, correlated count or `EXISTS` subqueries, and a `satisfied` column. The same rules apply to SQLite, DuckDB, and PostgreSQL with small variations (e.g., for timestamp arithmetic). The right side of Fig. 2 shows the SQL query produced for this OCPQ tree, and a schematic of the resulting rows after execution on a database.

OCPQ returns a set of output bindings for every tree node, as a table with one column per variable. SQL, in contrast, returns a single table per query. Our translation therefore targets the output bindings of the root node. Child results can be used in the parent output, for example through CBS predicates, but are not returned themselves. Retrieving child outputs on demand follows the same scheme and is left for future work.

Most OCPQ queries translate directly to SQL that can run on the database. However, OCPQ also allows for scripting using the Common Expression Language (CEL) (https://cel.dev), which has no direct SQL equivalent. We support these as an extension (cf. Sect. 3.2). When used, as much work as possible is pushed to the database, and only CEL predicates are post-processed host-side as residuals.

The translation targets the relational schema defined by OCEL 2.0. On such a schema, the emitted SQL runs directly without exporting or migrating data. To support scenarios where process data resides in a different schema, we add an optional *OCEL 2.0 Virtualization Layer* (Sect. 4) that mediates between this OCEL 2.0 surface and the actual source schema. It offers two paths: a *direct rewrite* that maps OCEL 2.0 table and column names to source names, so the translation targets the source schema with no view indirection; or, when the schema diverges further, generated `VIEW`s over the source tables that reshape the data into an OCEL 2.0-like layout.

We integrate the translation into the open-source OCPQ tool [9] and evaluate it on three relational backends: SQLite, DuckDB, and PostgreSQL. We evaluate and verify the translation on three workloads: Seven BPIC 2017 queries reused from prior work, an automatically generated corpus of 6 919 OCPQ trees, and an ERP-like database schema that differs from the OCEL 2.0 layout and is queried through the virtualization layer.

Contributions. This paper contributes a SQL backend for OCPQ that enables queries to execute directly on the relational systems already deployed in organizations. It produces the same root output bindings as the in-memory engine, so downstream uses that consume root bindings are unaffected. Concretely, we make three contributions. First, we define a recursive translation T from OCPQ to SQL whose per-construct rules cover all binding-box predicates, with dialect-aware emission for SQLite, DuckDB, and PostgreSQL. As an extension, CEL filters, label functions, and AdvancedCEL quantifiers (CEL expressions that access the child binding set, e.g., for aggregation) are evaluated host-side. Second, we add support for diverging database schemas through a declarative OCEL 2.0 Virtualization Layer with direct-rewrite and generated-view options. Third, we integrate this into the OCPQ backend as an open-source extension and evaluate it for correctness, runtime, and memory across different workloads.

2 Background

Object-Centric Event Data (OCED) models a process as events and objects of different types, related through qualifiers [1]. Objects may relate to other objects, and object attributes are time-varying. The OCEL 2.0 standard [4] specifies a SQL schema for OCED. The junction tables `event_object` and `object_object` carry the E2O and O2O relationships with an `ocel_qualifier` column. Per-type tables `event_<T>` and `object_<T>` hold attributes, with an `ocel_time` timestamp for events as well as object attribute history. The translator targets this schema as its logical surface. When the source schema differs, the OCEL 2.0 Virtualization Layer (Sect. 4) mediates the difference.

Object-Centric Process Querying (OCPQ) structures a query as a tree of *binding boxes* [9]. Each box declares typed event and object variables (V_E, V_O) and constrains them with predicates. The three *basic relational predicates* reference OCED structure directly: $\text{E2O}(e, o, q)$ and $\text{O2O}(o_1, o_2, q)$ require an event-to-object or object-to-object relationship between their arguments (with optional qualifier q), and $\text{TBE}(e_1, e_2, t_{\min}, t_{\max})$ bounds elapsed time between events. *Attribute filters* test an event or object attribute against a constant or range, where object attribute filters carry a time qualifier: `Always` and `Sometime` quantify over the object's attribute history, while `AtEvent` evaluates at a specified event's timestamp. *CEL filters* carry an expression in the Common Expression Language (CEL) evaluated per candidate binding, e.g., `e1.attr("amount") > 1000`. *Label functions* attach computed CEL values to surviving bindings under a named tag. Child edges are named (`A`, `B`, ...) so a parent can reference its children by edge label. A *binding* of box B is a mapping that assigns concrete events and objects to all variables in $V_E \cup V_O$ such that every predicate of B holds. Binding multiple events and objects of arbitrary types at once lets OCPQ express joint conditions across the process graph that case-centric query languages cannot. Evaluation is bottom-up: A binding c *extends* a parent binding b, written $b \sqsubseteq c$, if the two agree on every variable that b assigns. For a node N and an ancestor binding b, we write $out_b(N)$ for the set of all bindings that satisfy N's box and extend b; this is the *child binding set* produced for b on N's edge, and at the root, where b is empty, $out(N)$ is the full result of N. We follow the binding formalization of [9]. The *child-binding-set predicate* $\text{CBS}(A, n_{\min}, n_{\max})$ requires the child binding set $out_b(M)$ on edge A (to child M) to have cardinality in $[n_{\min}, n_{\max}]$. The shorthand $|A| \geq 1$ used in Fig. 2 stands for $\text{CBS}(A, 1, \infty)$. Each predicate above can serve two roles: As a *filter*, which drops bindings that violate it, or as a *constraint*, which keeps the binding but labels it as satisfied or violated. Constraint results of child nodes can be used with logical compositions (e.g., OR). For more details, we refer to [9] or https://ocpq.aarkue.eu.

Figure 2 demonstrates these definitions. The root box binds an `Application` object and its `A_Accepted` event through E2O, the child binds an `Offer` and its `O_Accepted` event (linked by O2O, E2O, and a TBE ordering), and the root constraint $\mathrm{CBS}(\mathtt{A}, 1, \infty)$ marks each parent satisfied when its child set is non-empty. Next, in Sect. 3 we sketch how such a query is translated to SQL.

3 Translating OCPQ to SQL

We define translation as a recursive mapping $T(N) = (Q_N, R_N)$ over each binding-box node N of the OCPQ tree. Q_N is the SQL fragment for N's subtree and R_N is a list of *residuals*: Host-side predicates that N (or one of its descendants) contributes but that have no SQL counterpart. The root node's Q is the query sent to the database. When a query carries no CEL ($R = \emptyset$), the returned rows are the final binding set, so a single SQL query fully captures the OCPQ semantics. The mapping recurses into children, and per-construct rules (Sect. 3.1) define how each predicate contributes to Q. Figure 3 shows the architecture. In the common case where no CEL predicates are involved, the translation maps the OCPQ tree to a single SQL query Q, which is then adapted by the virtualization layer to the source schema and executed on the database, with the returned rows as the final result.

When CEL predicates are used, the SQL query is relaxed to over-approximate OCPQ semantics. Thus, as much of the filtering and work as possible is pushed to SQL and the database, leaving only the CEL predicates as residuals to post-process on top of that. These residuals are then evaluated host-side over the returned rows, dropping or labeling each binding accordingly.

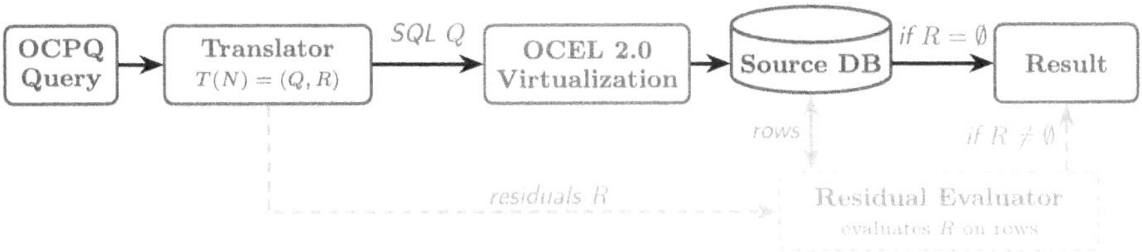

Fig. 3. Translation architecture. The translator T emits a SQL query Q and a residual list R. The query Q is executed against the source database via the Virtualization Layer. When $R = \emptyset$, the database rows are the final result. Otherwise, the CEL extension (Sect. 3.2) invokes an optional residual evaluator host-side.

3.1 Per-Construct Rules

Each variable in N gets a stable alias in the SQL query: `E`m for the m-th event variable and `O`k for the k-th object. Junction joins use `ER`n on `event_object` and ORn on `object_object`. Children inherit their parent's alias set so cross-node variable references resolve outward. Table 1 states each rule by the SQL it emits. A full per-rule formalization is beyond the scope of this paper; we state the semantics T preserves and its soundness below, and refer to the released implementation (Sect. 7) for the remaining details. As an example, E2O(e, o, q) contributes an `INNER JOIN` on `event_object` with a `WHERE` on the qualifier q if specified. An event attribute filter on e contributes a `WHERE` on the per-type attribute column of e. Constraints *label* bindings rather than filtering them out. Every node, root or non-root, exposes a boolean `satisfied` column built from its constraint list via a `CASE` expression. Empty-child-set rules (SAT vacuously true, ANY false) match the in-memory engine.

The recursion composes SQL as follows: For each child edge ℓ of N, the child's query is embedded into Q_N as the body of the correlated subqueries that the CBS and composition rules referencing ℓ require (e.g., inside `EXISTS` or a `COUNT`), with parent variables resolved through the inherited aliases and the child's residuals appended to R_N.

Correctness. We argue that T preserves OCPQ semantics, referring to the notation and definitions of Sect. 2 and [9]. Let N be a node with $T(N) = (Q_N, R_N)$ and b an output binding of N's parent (empty at the root). Assume N's subtree contains no CEL predicate, so $R_N = \emptyset$. Then the rows Q_N returns for b, projected onto the columns of $V_E^N \cup V_O^N$, are exactly $out_b(N)$. Constraints do not filter out rows, but indicate violations, so each returned row additionally carries a `satisfied` value built using a `CASE` expression and slightly modified SQL fragments from Table 1. The invariant follows by induction over the tree:

Leaf Nodes. A leaf declares variables and filters only. Its query joins the per-type table of each variable through the `event_object` and `object_object` junctions, so one row is one type-correct assignment to $V_E^N \cup V_O^N$. Every filter is mainly encoded by the clause in the corresponding row of Table 1, and each such clause is simply a relational implementation of the predicate's definition. For example, E2O(e, o, q) holds iff the qualified `event_object` join has a matching row, and the `Always`/`Sometime`/`AtEvent` scopes select exactly the attribute rows their definition quantifies over. A row survives Q_N iff its assignment satisfies every filter, i.e. iff it lies in $out_b(N)$, where identical assignments arising from several witnessing junction rows are coalesced by the `DISTINCT` the emitter applies.

Inner Nodes. Assume the invariant for every child. The query of a child node M is embedded in Q_N as a subquery whose outer reference is a parent binding b (the current row), so by hypothesis it enumerates only child bindings $out_b(M)$ of the parent binding b. The CBS rule reads this subquery as a `COUNT` or `EXISTS`, so the test $n_{\min} \leq |out_c(M)| \leq n_{\max}$ is decided exactly. Other compositions (e.g., SAT) combine the children's `satisfied` values similarly. Basic and attribute predicates of the inner node itself are handled as in the leaf case.

CEL Predicates. A subtree with CEL predicates has no exact clause in Table 1. Instead, T widens Q_N to a superset of $out_b(N)$, and the residual evaluator removes the extra output rows host-side (Sect. 3.2).

Table 1. Per-construct translation rules. Each row shows the SQL fragment a construct contributes to Q. Used as a filter, the fragment restricts Q. Used as a constraint, the same fragment forms a branch of the parent's satisfied CASE. Predicates without a SQL counterpart (CEL, AdvancedCEL) are covered in Sect. 3.2.

	Construct	**Contribution to Q**
Basic relations	E2O(e, o, q), O2O(o_1, o_2, q) (event-/object-to-object)	INNER JOIN on event_object / object_object, and q adds an ocel_qualifier WHERE
	TBE$(e_1, e_2, t_{\min}, t_{\max})$ (time-between-events)	WHERE on the e_1, e_2 timestamps; $e_2 \geq e_1$ shortcut for $t_{\min} = 0$, otherwise dialect-specific epoch arithmetic (see *Dialect Handling*)
	Variable disequality $x \neq y$	WHERE on ocel_id inequality
Attributes	Event attribute filter	WHERE on the per-type attribute column of e
	Object attribute filter (Always/Sometime/AtEvent)	Scoped EXISTS on object_<T>; the time qualifier selects the historical rows considered (all, any, or the row valid at the named event, respectively)
Child-binding-set	CBS$(A, n_{\min}, n_{\max})$ (child binding count)	Correlated COUNT(DISTINCT...) over edge A (defaulting to 0 via COALESCE), checked against $[n_{\min}, n_{\max}]$; NOT EXISTS/EXISTS shortcuts for $[0, 0]/[1, \infty)$
	CBS over projection variable v	As above, with COUNT(DISTINCT) over v instead of the full binding key
	Binding-set equality $A = B$	Two EXCEPT subqueries checking $A \setminus B$ and $B \setminus A$; both must return no rows
Composition	SAT ℓ (all children satisfied)	NOT EXISTS child row on edge ℓ with satisfied = FALSE; vacuously true on the empty set
	ANY ℓ (some child satisfied)	Correlated count of satisfied = TRUE rows on edge ℓ must be ≥ 1; false on the empty set
	AND, OR, NOT over $\{\ell_i\}$	Boolean composition of the per-edge SAT forms

Example. Figure 2 shows an example mapping T. At the root, the E2O(e_1, o_1) rule aliases the per-type tables E1/O1 and adds the event_object junction ER1, and the constraint CBS$(A, 1, \infty)$ takes the EXISTS shortcut into the root's satisfied CASE, with the child's Q as its body. In the child, O2O and E2O emit the object_object (OR1) and event_object (ER2) joins to tables O2/E2, and the zero-bound TBE reduces to E2.ocel_time >= E1.ocel_time. With $R = \emptyset$, Q yields 31 509 root bindings when executed on BPIC 2017.

Dialect Handling. Generally, the same rules apply across SQLite, DuckDB, and PostgreSQL. The main difference is timestamp arithmetic for general TBE bounds: Writing t for the event timestamp column, SQLite emits `(julianday(t) - 2440587.5) * 86400`, DuckDB `EPOCH(t)`, and PostgreSQL `EXTRACT(EPOCH FROM t)`. When either bound is zero, the translation compares the timestamp columns directly (`E2.ocel_time >= E1.ocel_time`) on all three dialects.

3.2 Extension: Predicates Without a SQL Counterpart

The rules in Sect. 3.1 cover every OCPQ predicate that maps to a relational operator, so a CEL-free query translates to a single self-contained SQL statement ($R = \emptyset$). CEL filters and label functions are extensions appended to R and evaluated host-side over the rows returned by Q. Simple CEL expressions could be pushed into a `WHERE` clause, which we leave as future work (Sect. 7). Other operators cannot be translated to SQL. For example, a regular-expression match in `attr("name").matches("iPhone \d+")` has no portable SQL form (SQLite, for instance, lacks regex support by default), so it is incompatible with the uniform, multi-dialect translation. More fundamentally, CEL is a general expression language that OCPQ extends with custom host-evaluated extensions, allowing for any predicate or functionality implementable as Rust functions.

Residuals are handled in two steps. First, T relaxes the SQL query so that Q over-approximates OCPQ semantics and never wrongly excludes a valid binding. Second, the residual evaluator post-processes the rows Q returns. It evaluates the CEL filters and labels and re-checks the structural clauses that Q could not enforce, such as upper-bounded child-binding-set (CBS) counts, then drops or labels each binding accordingly.

Relaxation targets any parent clause that constrains a child whose subtree carries residuals (i.e., CEL predicates). In such cases, the child count seen in SQL is only an upper bound and not yet final, because the evaluator may still drop child rows, so the parent rule cannot commit to it. The SQL form is allowed to over-include but never under-include. Clauses for which over-inclusion is safe stay in Q exactly: ANY, a CBS with no upper bound, and SAT or AND over CEL-free children. The remaining clauses, NOT, OR, an upper-bounded CBS, and binding-set equality, are relaxed from Q and re-checked host-side. For example, CBS$(A, 0, 2)$ over a CEL-filtered child cannot be enforced in Q, since CEL may later drop child rows and lower the count. Thus, Q keeps every parent binding and the evaluator applies the exact bound afterward.

Post-processing the CEL predicates in R needs referenced attribute values, so T widens the relevant `SELECT` to project them if possible. Attributes unreachable through the parent's projection, e.g., across a CBS predicate, are fetched by a dedicated query against the referenced IDs, so the SQL path never materializes the full linked OCEL graph. CEL based on child bindings additionally needs the parent's bindings in scope for its child query, which the emitter provides with a correlated join where the dialect supports it, and otherwise by substituting the parent bindings into a batched subquery.

4 OCEL 2.0 Virtualization Layer

The translation so far targets the OCEL 2.0 database schema (e.g., assuming per-type event and object tables `event_<T>`, `object_<T>`, and `event_object` junction tables). Organizations rarely store process data this way. Source schemas diverge from this layout to varying degrees. Some differ only in naming, where a table maps to an OCEL type by renaming alone. Others differ more substantially, for example when an object's identifier spans several columns. The OCEL 2.0 Virtualization Layer bridges this gap. Based on a declarative description of the source schema, it presents the standard OCEL 2.0 SQL surface over an arbitrary source layout, so the translation runs against an organization's existing database without first migrating the data. The layer realizes the description in two ways.

Direct rewrite handles sources that differ only in their table and column names. The emitter substitutes the source names at emission time, producing plain SQL over the source tables with no view indirection and no overhead of its own. Defaults reproduce the standard OCEL 2.0 SQL exporter convention, so stock SQLite, DuckDB, and PostgreSQL exports work without configuration.

Generated views handle anything beyond renaming. The description carries `CREATE OR REPLACE VIEW` statements, installed once before queries run, that the queries then read from. The views hold the reshaping (row splits, type discrimination, deduplication, computed identifiers) and expose the OCEL 2.0 surface as named relations that any OCEL 2.0 consumer can query, not only our translator. Because the view definition can contain arbitrary SQL, this path is the more general of the two, able to persist, index, and reuse the reshaped surface in ways inline substitution cannot.

We evaluate the generated-view path as RQ3 in Sect. 5 on a synthetic ERP-shape source with deliberately stressful reshaping. The OCEL 2.0 Virtualization Layer is a first step towards executing object-centric queries directly on the databases where process data already lives. Federation across sources, automatic schema inference, and query optimization remain future work (Sect. 7).

5 Evaluation

We organize the evaluation around three research questions:

RQ1. How do the runtimes of the SQL backends compare to the existing in-memory engine? We investigate this on the query set from [9] and across a generated corpus of OCPQ queries, both on BPIC 2017.

RQ2. Do SQL backends decouple memory from data size and scale out to support larger datasets, and at what runtime cost? We evaluate this using memory stress on BPIC 2017 in two modes: Tight memory limits and scale-out through replication.

RQ3. What is the runtime overhead of running OCPQ over a non-OCEL SQL source schema via generated views?

Table 2. Per-query wall-clock on BPIC 2017 (ms, mean with standard deviation subscript over 10 warm runs, darker cells faster). Construct tags follow Sect. 3.1, and Q4 is the running example of Fig. 2.

Query	Features	In-mem	DuckDB	Postgres	SQLite
Q1	E2O, CBS	$8.7_{\pm 0.8}$	$40.2_{\pm 4.3}$	$615_{\pm 22.1}$	$365_{\pm 24.3}$
Q2	2×E2O, TBE, CBS	$19.4_{\pm 1.2}$	$80.9_{\pm 3.4}$	$713_{\pm 57.0}$	$1\,580_{\pm 734}$
Q3	O2O, CBS	$6.5_{\pm 0.7}$	$33.0_{\pm 1.5}$	$187_{\pm 23.1}$	$92.0_{\pm 7.3}$
Q4	2×E2O, O2O, TBE, CBS	$18.3_{\pm 0.9}$	$73.5_{\pm 3.5}$	$870_{\pm 75.8}$	$2\,529_{\pm 1\,321}$
Q5	4×E2O, O2O, SAT	$25.5_{\pm 2.5}$	$160_{\pm 6.7}$	$1\,215_{\pm 45.6}$	$2\,469_{\pm 1\,160}$
Q6	2×E2O, CEL aggregate	$37.0_{\pm 1.2}$	$448_{\pm 31.8}$	$598_{\pm 16.3}$	>60 s
Q7	2×E2O, 2×O2O	$27.5_{\pm 4.2}$	$226_{\pm 10.1}$	$397_{\pm 6.1}$	$4\,433_{\pm 113}$

Correctness is a precondition for every result we report. For each query we require the resulting binding set to match that of the in-memory engine. All cells reported below pass the gate, including all 6 919 trees of the generated corpus. These checks reinforce the translation correctness argued in Sect. 3.1.

Setup and Coverage. We use the BPIC 2017 dataset (1 202 267 events, 106 162 objects) because of the pre-existing query corpus [9] and as it is among the largest publicly available real-life object-centric datasets. Each SQL backend opens the DB connection once, and all four engines report mean ± standard deviation over 10 warm runs (1 untimed warmup). We time the whole path from a query to its output bindings in OCPQ, including translation, execution, rebuilding bindings from the result rows, and any residual evaluation on the host. All experiments ran on an Intel Core Ultra 7 155U with 32 GiB RAM and NVMe SSD storage, running PostgreSQL 18.3, SQLite 3.52, and DuckDB 1.4. All three SQL backends share a uniform index set (primary keys on the four OCEL tables, indexes on both sides of each junction, and an `ocel_id` index on each per-type table). The query-set runtime and the virtualization study use all four engines. The corpus and the scale-out drop SQLite because it is not competitive at corpus and replication scale, and the virtualization study drops the in-memory engine because the layer targets SQL backends.

Runtime (RQ1). Table 2 reports per-query mean durations across all four engines on the seven-query BPIC 2017 set (Q1–Q7, reused verbatim from the OCPQ paper [9]). Construct coverage beyond these seven queries is provided by the generated corpus below. DuckDB stays within roughly 4 to 12× of in-memory. The only time it crosses the 10× factor is for Q6, where CEL is evaluated as residual. PostgreSQL trails by roughly 14 to 71×. SQLite is competitive on a few cheaper queries but drops to the bottom of the benchmark on Q2, Q4, Q5, and Q7. The Q6 outlier is SQLite-specific, where the CEL aggregate over child bindings does not complete within the 60 s timeout.

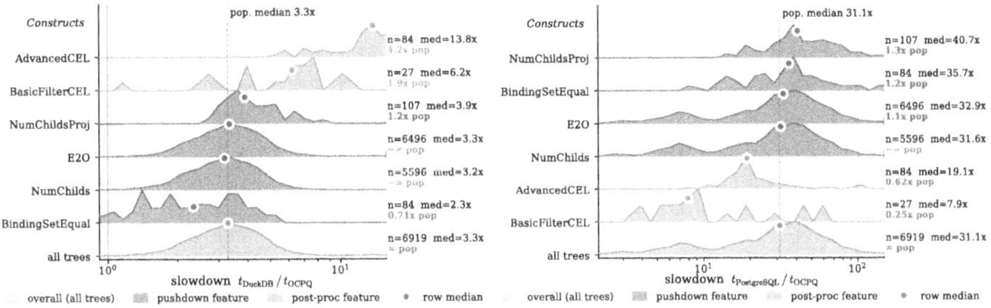

Fig. 4. Slowdown over the in-memory engine by OCPQ construct across the 6 919-tree corpus. Left: DuckDB. Right: PostgreSQL. Each row shows the $\log_{10}$ distribution of slowdowns over trees carrying the construct. The dashed lines indicate each engine's population median (DuckDB 3.28×, PostgreSQL 31.1×).

The in-memory engine wins where the OCEL fits in RAM. The SQL backends matter when it does not, avoiding export, staying fresh against database mutation, reusing existing indexes, and (per RQ2) lifting the RAM ceiling.

To test correctness beyond BPIC 2017, we additionally create seven advanced OCPQ queries over three simulated OCEL 2.0 logs (order-management, container-logistics, purchase-to-pay), spanning qualified E2O, more nesting, TBE windows, and advanced CEL labels. All three backends return the same output on all these queries.

Corpus Runtime Distribution (RQ1). Beyond Q1–Q7, we automatically generate a corpus of OCPQ trees on BPIC 2017 over five event types and two object types. Each tree shape satisfies $|V_E^N| + |V_O^N| \leq 3$ per node, depth ≤ 2, and at most two structurally identical sibling children; every variable must participate in at least one E2O or O2O predicate. On top of each connected shape, the generator enumerates every filter, every CBS filter variant, every AdvancedCEL configuration, and every composition operator (SAT, ANY, AND, OR, NOT). Labels are layered at the root only. Four construct families fall outside the corpus and are instead covered manually by tests: Constraint-mode CBS, E2O/O2O qualifier variation, non-zero TBE windows, and the `AtEvent`/`Always` object-attribute time qualifiers.

The generator produces **6 919 distinct trees** and the translator accepts every one. We run each on three engines (in-memory, DuckDB, PostgreSQL) under a 120 s per-tree timeout. All 6 919 trees yield exactly the same output binding sets (variable, `ocel_id` tuples, label values, the per-binding satisfied flag), with zero SQL errors and zero timeouts. Across the corpus, 99.78% of per-tree predicates land in SQL (51 130 of 51 241). The remaining 0.22% are residuals. Translation takes a median 0.11 ms per tree, under 1% of any backend's runtime.

In Fig. 4, we break down the slowdown of each backend over the in-memory engine by query construct, showing the distribution of slowdowns for trees carrying each construct. On DuckDB (median 3.28×) the structural constructs (like NumChilds and E2O) are close to the median, while CEL-bearing constructs

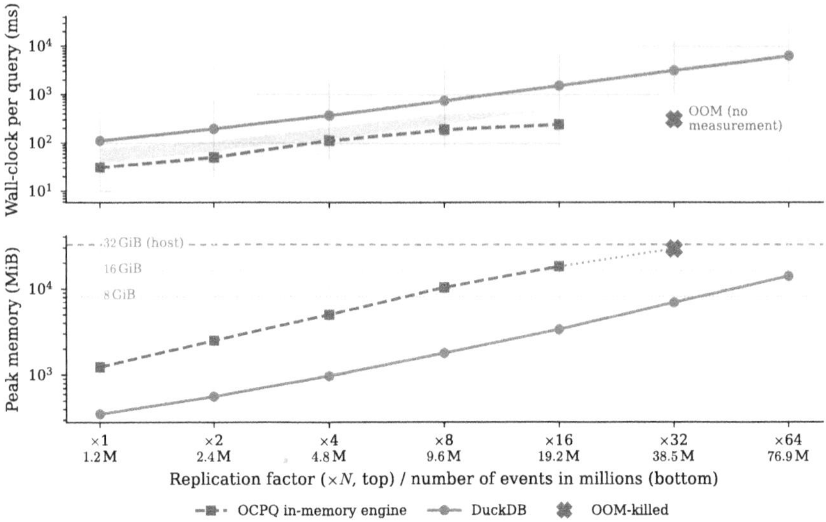

Fig. 5. BPIC 2017 replication ladder, in-memory engine vs. DuckDB. Runtime (top) is the per-engine Q1–Q7 geometric mean with min–max band. Memory (bottom): the heavy dashed line is the 32 GiB host, and the ×32 red "X" is the last observed in-memory value (28.6 GiB) before it ran out of memory. The runtime "X" is a log-log extrapolation of the in-memory trend (no measurement).

sit above it because their predicate is re-evaluated host-side. PostgreSQL is uniformly slower (median 31.1×). There, CEL-bearing trees sit *below* the median, likely reflecting structural corpus features (e.g., simpler query shape when CEL is generated) rather than a real speedup. Such residual predicates occur in only 2.2% of trees, so excluding them barely shifts either median.

Takeaway. When the data fits in RAM the in-memory engine is fastest, and running the same queries inside the database costs a small constant factor, a corpus median of 3.28× for DuckDB. Across the full 6 919-tree corpus every backend returns the same bindings, so the translation is correct over the construct combinations the corpus enumerates.

Memory Decoupling (RQ2). We run two memory stress tests: A *scale-out* replication ladder and a *tight* host-memory *cap*. *Scale-out.* We replicate BPIC 2017 by ID-suffixing from ×1 up to ×64 (76.9 M events, 153.9 M E2O at ×64), evaluating the Q1–Q7 queries from [9] at each step. Figure 5 reports the per-engine runtime (geometric mean) and peak memory usage across Q1–Q7. The in-memory engine runs out of memory at ×32 (terminated during OCEL load), while DuckDB completes the full ladder (13.8 GiB at ×64), a peak-memory gap of roughly 3.5 to 5.8× across the ladder. Notably, DuckDB also supports data larger than memory [11], so the scalability argument is not only a constant-factor memory reduc-

Table 3. Virtualization cost on the synthetic ERP schema (ms, mean ± standard deviation over 10 warm runs). Native is the standard OCEL 2.0 tables, and Virt. is generated VIEWs over a non-OCEL source (darker is faster, and Virt./Native ≥ 10× is shown in bold).

Rows	Query	DuckDB		PostgreSQL		SQLite	
		Native	Virt.	Native	Virt.	Native	Virt.
10 k	V1 (E2O)	$13.2_{\pm 2.6}$	$18.1_{\pm 2.5}$	$20.4_{\pm 2.6}$	$25.5_{\pm 1.2}$	$39.9_{\pm 9.9}$	$67.1_{\pm 3.4}$
10 k	V2 (CEL)	$21.4_{\pm 0.8}$	$25.5_{\pm 1.2}$	$28.8_{\pm 1.0}$	$38.0_{\pm 3.0}$	$53.2_{\pm 10.9}$	$73.8_{\pm 2.2}$
10 k	V3 (size filter)	$14.9_{\pm 2.7}$	$13.9_{\pm 1.7}$	$119_{\pm 11.6}$	**18 179**$_{\pm 1\,140}$	$39.0_{\pm 4.1}$	**72 027**$_{\pm 3\,461}$
100 k	V1 (E2O)	$136_{\pm 7.2}$	$174_{\pm 6.6}$	$225_{\pm 12.1}$	$355_{\pm 14.0}$	$671_{\pm 267}$	$970_{\pm 47.5}$
100 k	V2 (CEL)	$248_{\pm 7.7}$	$292_{\pm 24.2}$	$355_{\pm 12.5}$	$470_{\pm 7.4}$	$789_{\pm 272}$	$1\,095_{\pm 37.9}$
100 k	V3 (size filter)	$99.5_{\pm 4.8}$	$97.0_{\pm 4.0}$	$1\,098_{\pm 73.8}$	**>180 s**	$493_{\pm 32.8}$	**>180 s**

tion. On warm RAM-resident data the in-memory engine is 3.6 to 6.4× faster than DuckDB, and the gap does not close as data grows.

Tight Cap. We re-run Q1–Q7 on BPIC 2017 under a 400 MiB cap enforced as a hard per-process memory limit. The in-memory engine runs out of memory before the warmup completes. All three SQL backends complete the query set under the cap. SQLite peaks at 162 MiB and DuckDB at 350 MiB. PostgreSQL saturates the cap with default settings, and its client process holding query results adds 37 to 134 MiB outside the cap. DuckDB's higher footprint reflects its default buffer-pool target (80% of system RAM).

Takeaway. Memory is decoupled from data size. The in-memory engine runs out of memory at ×32 replication and under a 400 MiB cap, while the SQL backends complete every query in both settings.

Virtualization Layer (RQ3). We evaluate the generated-view path (as direct rewrite only renames) on a synthetic ERP source with three reshapes from Sect. 4: A type-discriminated event table (i.e., multiple event types in one table), a deduplicated `Customer` object (i.e., an object type without a dedicated table), and a composite `ocel_id` via concatenation, at 10 000 and 100 000 source rows. Three queries V1–V3 run across SQLite, DuckDB, and PostgreSQL.

View indirection on V1 and V2 is a constant overhead across both scales (DuckDB 1.2 to 1.4×, PostgreSQL 1.3 to 1.6×, SQLite 1.4 to 1.7×). V3 exposes a cliff on the row-oriented SQLite and PostgreSQL: the CBS's correlated count re-evaluates the view's per-row composite-identifier concatenation, which no underlying index satisfies, while DuckDB's columnar engine processes it in a single pass. A stored generated column on the composite identifier could close the gap. Overall, querying non-OCEL source schemas through generated views is feasible, with performance depending on the engine and the schema's shape (Table 3).

Limitations. Soundness is argued by construction (Sect. 3.1) and checked through our evaluation query corpus but not established by a full formal proof. Under the tight cap the PostgreSQL client process is not subject to the cap (only the server), so we report client and server peaks separately. Scale-out scales by ID-suffixed replication and so stresses size, not graph complexity. RQ3 uses one synthetic ERP schema with worst-case reshape, and other production shapes remain untested. Host-side residuals carry a transfer cost when an unselective SQL fragment meets a selective CEL filter. Runtime figures use default engine configurations and one uniform SQL shape, not per-engine tuning, so cross-engine factors likely do not reflect the engines' full potential.

6 Related Work

Process Query Languages. Prior process query languages (BP-QL [3], APQL [8], PIQL [10], Celonis PQL [15]) target case-centric settings and do not natively span multiple object types or multiple instances of the same object type. OCPQ [9] closes that gap at the language level. We contribute an orthogonal execution backend that maps OCPQ to SQL.

Process Mining on Relational Databases. Keeping process data inside the database during execution predates this work in the case-centric setting: DB-XES [14] computes intermediate structures via SQL rather than over an in-memory log, and Dijkman et al. [6] realize a discovery operator running entirely inside the database engine. Schönig et al. [13] and Riva et al. [12] translate DECLARE-style templates to SQL for constraint checking in the single-case setting. The object-centric setting adds multi-typed variable bindings joined through junction tables and correlated child-binding-set counts, neither of which arises in single-case DECLARE templates. We additionally handle non-translatable predicates as host residuals, a boundary those works do not encounter.

Object-Centric Process Mining. Object-centric behavioral constraint models [2] introduce object-centric predicates at the modeling level by grounding behavior in an ER-style schema of typed entities and relationships, anticipating the relational execution target we adopt for OCPQ. Esser and Fahland [7] query OCED in Neo4j via hand-written Cypher; graph databases fit OCED's structure naturally, but organizations rarely store process data in them, so querying one requires a data export that our approach avoids. Berti et al. [5] and Xiong et al. [16] extract OCEL from ERP and via virtual knowledge graphs. We contribute the executable OCPQ-to-SQL query layer over OCED already in relational storage, and assume a mapping from the raw data to an OCEL 2.0 surface to be given as input [1,4].

7 Conclusion

We presented a translation from OCPQ to SQL: A recursive mapping $T(N) = (Q, R)$ over the binding-box tree, with per-construct rules emitting SQL for SQLite, DuckDB, and PostgreSQL and routing CEL filters, labels, and AdvancedCEL quantifiers to host-side residuals. An OCEL 2.0 Virtualization Layer extends it to source schemas that diverge from the standard layout. Memory decouples from dataset size: The SQL backends complete BPIC 2017 under a tight memory cap and DuckDB scales to 76.9 M events, both of which exhaust the in-memory engine, while binding sets match it across all 6 919 corpus trees.

Practitioner Guidance. Use the in-memory engine when the log fits comfortably in RAM. Switch to a SQL backend when data approaches or exceeds the memory ceiling, when results must stay fresh against a live database, or when the data already resides in one. Among the SQL backends, DuckDB is the default for analytical and oversized workloads, PostgreSQL is suited to operational-database integration, and SQLite suits small embedded deployments.

Future Work. Simple CEL shapes such as `attr("X") == const` could be pushed into the SQL query instead of evaluated host-side. The generated-view path could be exercised on a real production schema. More database backends could be supported, and the emitter could pick between `EXISTS`, `COUNT`, and `LEFT JOIN` shapes based on index statistics. Moreover, the translation can be improved by optimizing the simplicity and performance of the emitted SQL. Finally, similar translation rules could target a second query language such as Cypher.

Reproducibility. The translation code is available and integrated in OCPQ (see https://github.com/aarkue/OCPQ/). The evaluation setup, datasets, and scripts are available at https://github.com/aarkue/ocpq-sql-eval.

Acknowledgments. The authors gratefully acknowledge the German Federal Ministry of Research, Technology and Space (BMFTR) and the state government of North Rhine-Westphalia for supporting this work as part of the NHR funding.

References

1. van der Aalst, W.M.P.: Object-centric process mining: unraveling the fabric of real processes. Mathematics **11**(12) (2023)
2. van der Aalst, W.M.P., Artale, A., Montali, M., Tritini, S.: Object-centric behavioral constraints: Integrating data and declarative process modelling. In: Description Logics. CEUR Workshop Proceedings. CEUR-WS.org (2017)
3. Beeri, C., Eyal, A., Kamenkovich, S., Milo, T.: Querying business processes with BP-QL. In: VLDB, pp. 1255–1258. ACM (2005)
4. Berti, A., et al.: OCEL (object-centric event log) 2.0 specification. CoRR abs/2403.01975 (2024)
5. Berti, A., Park, G., Rafiei, M., van der Aalst, W.M.P.: A generic approach to extract object-centric event data from databases supporting SAP ERP. J. Intell. Inf. Syst. **61**(3), 835–857 (2023)

6. Dijkman, R.M., Gao, J., Syamsiyah, A., van Dongen, B.F., Grefen, P., ter Hofstede, A.H.M.: Enabling efficient process mining on large data sets: realizing an in-database process mining operator. Distrib. Parallel Databases **38**(1), 227–253 (2020)
7. Esser, S., Fahland, D.: Multi-dimensional event data in graph databases. J. Data Semant. **10**(1–2), 109–141 (2021)
8. ter Hofstede, A.H.M., Ouyang, C., Rosa, M.L., Song, L., Wang, J., Polyvyanyy, A.: APQL: a process-model query language. In: AP-BPM. LNBIP, pp. 23–38. Springer (2013)
9. Küsters, A., van der Aalst, W.M.P.: OCPQ: object-centric process querying and constraints. In: RCIS (1). LNBIP, pp. 383–400. Springer (2025)
10. Pérez-Álvarez, J.M., Díaz, A.C., Parody, L., Quintero, A.M.R., Gómez-López, M.T.: Process instance query language and the process querying framework. In: Process Querying Methods, pp. 85–111. Springer (2022)
11. Raasveldt, M., Mühleisen, H.: DuckDB: an embeddable analytical database. In: SIGMOD Conference, pp. 1981–1984. ACM (2019)
12. Riva, F., Benvenuti, D., Maggi, F.M., Marrella, A., Montali, M.: An SQL-based declarative process mining framework for analyzing process data stored in relational databases. In: BPM Forum. LNBIP, vol. 490, pp. 214–231. Springer (2023)
13. Schönig, S., Rogge-Solti, A., Cabanillas, C., Jablonski, S., Mendling, J.: Efficient and customisable declarative process mining with SQL. In: CAiSE. LNCS, vol. 9694, pp. 290–305. Springer (2016)
14. Syamsiyah, A., van Dongen, B.F., van der Aalst, W.M.P.: DB-XES: enabling process discovery in the large. In: SIMPDA, pp. 63–77. CEUR Workshop Proceedings. CEUR-WS.org (2016)
15. Vogelgesang, T., Ambrosy, J., Becher, D., Seilbeck, R., Geyer-Klingeberg, J., Klenk, M.: Celonis PQL: a query language for process mining. In: Process Querying Methods, pp. 377–408. Springer (2022)
16. Xiong, J., Xiao, G., Kalayci, T.E., Montali, M., Gu, Z., Calvanese, D.: A virtual knowledge graph based approach for object-centric event logs extraction. In: ICPM Workshops. LNBIP, pp. 466–478. Springer (2022)

LLM-Based Orchestration of Business Process Management Toolchains

Oscar Ulbricht[1(✉)], Max Blaschke[2], Anjo Seidel[1], and Mathias Weske[1]

[1] Hasso Plattner Institute, University of Potsdam, Potsdam, Germany
oscar.ulbricht@student.hpi.de, {anjo.seidel,mathias.weske}@hpi.de
[2] University of Liechtenstein, Vaduz, Liechtenstein
max.blaschke@uni.li

Abstract. There is a tradition in business process management research not only to develop theoretical concepts and powerful algorithms, but also to engineer research prototypes to showcase the contributions. These tools are typically developed in isolation and are rarely designed to interoperate. As a result, researchers and practitioners need to manually combine multiple tools to construct toolchains that address their specific needs, limiting the practical usability and reuse of existing prototypes. In this paper, we propose the PRAXIS (Process Research Artifact eXecution and Integration System) framework that leverages large language models as a natural language interface for practitioners to compose and execute process management toolchains. For a specific user question, the system automatically orchestrates and executes the respective BPM research tools. The framework is prototypically implemented based on state-of-the-art techniques to integrate several existing BPM research tools. We provide an evaluation methodology that generates test cases for a given set of tools to test the framework. The application of the evaluation methodology shows the usefulness of our framework.

Keywords: Business Process Management · Large-Language-Models · Research Prototypes · Orchestration

1 Introduction

Business Process Management (BPM) has emerged as a multidisciplinary field of study, bridging the gap between management science and information technology [19]. The BPM community has a strong record of designing and implementing research prototypes. These can be associated to the different phases of the BPM lifecycle [8,23], which consists of four phases: design and analysis, configuration, enactment, and evaluation. Many of the available tools address the design and analysis phase [16], but research prototypes exist for other phases as well. The combination of such tools is challenging due to the independent development of the individual tools and incompatible interfaces [16].

The manual orchestration of toolchains poses a high entry barrier and requires significant effort, especially for non-technical users. Such a planning of

K. Kluza et al. (Eds.): BPM 2026, LNBIP 600, pp. 245–259, 2027.
https://doi.org/10.1007/978-3-032-36710-5_16

sequences of tools, which originally requires human intelligence, can be reduced to a token prediction problem with a sufficiently large model [20]. This makes LLMs a natural candidate technology for orchestrating BPM tools and making the BPM tool space more accessible to human users through a natural language interface. Although the probabilistic nature of LLMs allows for very reasonable outputs in many cases, they are known to hallucinate, leading to factually wrong sequences [11]. However, so-called reasoning systems can improve the quality of LLM answers [15]. Many modern LLM-based systems or conversational agents, e.g. ChatGPT or Claude Desktop, combine reasoning systems with the ability to use other pieces of software, which is often referred to as *tool calling*. Still, as pointed out in [21], it is important to integrate LLMs into process-centric information systems with caution to not obscure deterministic process logics.

Dedicated to the vision of integrating LLMs into BPM, we propose the PRAXIS framework (Process Research Artifact eXecution and Integration System), which combines toolsuites with a natural language interface and an LLM-based system to solve user queries. It hides the technical complexities of BPM tools, whose individual reliability remains unaffected by the usage of a probabilistic orchestration approach. Following a problem-centered design-science research approach [18], PRAXIS aims to satisfy the following requirements.

R1 Orchestration Effectiveness: The artifact shall be able to choose a fitting solution strategy for a given query by selecting the appropriate tools, ordering them correctly, and orchestrating them.

R2 Reproducibility: Additionally, the artifact must be reproducible in finding and executing its solution strategy in order to perform a rigorous evaluation.

R3 Separation of Concerns: To combine the capabilities of LLMs carefully into the existing and growing BPM landscape, the artifact shall separate tasks of natural language processing, reasoning, and deterministic BPM tools into dedicated components by design. This also supports the extensibility with new tools and the exchangeability of system components.

We propose a general architecture for the PRAXIS framework which satisfies the requirements R2 and R3 by design, and provide a publicly available prototypical implementation[1] with 14 distinct BPM tools covering each phase of the BPM lifecycle, showing the technical feasibility and meeting the requirements R2 and R3. To evaluate the orchestration effectiveness (R1) of such an implementation, we provide an evaluation methodology to automatically generate test cases for a given toolsuite. Following this methodology, our prototype evaluation generates 118 test cases, which are simulated and show that, in most cases, the correct tools are combined.

The remainder of this paper is structured as follows. The context of this study is discussed in Sect. 2, which situates the presented work in the current state of research. Section 3 details the PRAXIS framework, and Sect. 4 presents the prototypical implementation. The following Sect. 5 describes the evaluation methodology and its application to the prototype. Finally, Sect. 6 discusses the findings and concludes the article.

[1] https://github.com/ockapunktnet/PRAXIS.

2 Related Work

Integrating LLMs into BPM and its tool sets has emerged with the broader adoption of LLMs in general. A multitude of works analyze all phases of the BPM lifecycle to identify application areas for LLMs [22]. The focus mainly lies on the conversational capabilities of LLMs to improve process understanding or assist decision making [22]. Furthermore, there is a strong tendency for current research and academic prototypes to emphasize single phases of the BPM lifecycle (e.g. discovery, analysis, or monitoring) only [10]. Additionally, current approaches put the responsibility for final decisions or for independent, goal-based actions on the human [22]. PRAXIS extends this state-of-the-art by shifting the focus toward a holistic and tool-agnostic orchestration spanning the entire lifecycle.

In more visionary approaches, LLMs are used beyond their conversational capabilities with a limited degree of autonomy. The vision of AI-Augmented BPM Systems (ABPMS) [7] utilizes a diverse set of LLM-based methodologies to support process execution, analysis, and optimization and is constrained to a controlled environment. The focus is on user support rather than full automation of complete workflows. In contrast, PRAXIS gives greater agency to an LLM as the orchestrator. Large Process Models (LPMs) [12] envision a system that learns based on process data, recognizes patterns, and follows defined rules, which allows for the definition and analysis of context-dependent processes [12,22]. Agentic Business Process Management (APM) [5] extends traditional BPM by treating autonomous process-aware agents as first-class entities and by providing governance mechanisms that frame and align their behavior with organizational process goals. While our work does not fully realize the visions of LPMs or APM, it represents a significant step toward them by enabling the transparent orchestration of existing BPM artifacts through LLM-based reasoning.

The work of Buss et al. [4] examines how fine-tuning LLMs based on BPMN process models and BPM-specific knowledge can enhance the results from an LLM without invoking specific tools. Similarly to our approach, Kourani et al. [14] have developed a system for automatic generation of BPMN process models based on a natural language description via automatic generation of Python code from a set of predefined functions. Likewise, PRAXIS is provided with predetermined functionalities by the tools and is only allowed to execute those.

To allow users without technical knowledge to ask questions about process mining and receive understandable answers, Barbieri et al. [1] built a system that translates a natural language question into an intermediary representation and from there into executable PM4Py [2] functionalities. Outside of BPM, similar tool orchestration patterns have been shown to improve LLM performance on deterministic tasks such as mathematical problem solving [9].

3 PRAXIS Framework

In this section, we describe the general functionality of the PRAXIS framework. First, we provide a motivating example, before describing its architecture and the process of orchestrating a toolchain. Throughout this paper, the term *tool* is used as a reference to a single functionality with well-defined inputs and outputs that can be automatically invoked.

3.1 Motivating Example

As a motivating example, consider a toolsuite that comprises different BPM tools. Amongst others, a tool to analyze event logs, a translator of BPMN process diagrams into Petri nets, a Petri net analysis tool, and a conversational process modeling tool. Now, a user of this toolsuite has the natural language prompt: *"Provide general information about the event log and check whether the BPMN model is sound by translating it into a Petri net. If not, fix it"*. The aim of the framework is to take this prompt as input and to construct a BPM toolchain that allows to solve the prompt.

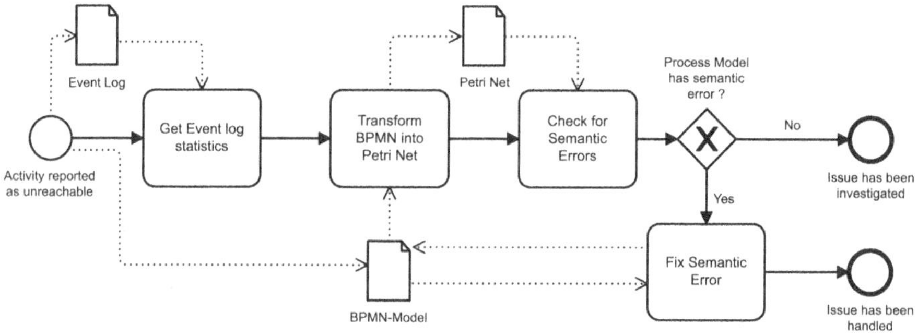

Fig. 1. Sample BPM toolchain, expressed as a BPMN process diagram.

For the given prompt, the framework shall orchestrate an effective toolchain. For the example prompt, we illustrate the desired toolchain in Fig. 1 as a BPMN process diagram. The toolchain starts with a deployed BPMN process diagram of a process, where, e.g., a process participant reports an activity as unreachable. A traditional analysis pipeline for the corresponding process owner could be similar to the following. First, the event log from the deployed model is examined for general information. Afterward, the deployed BPMN process diagram is translated into a Petri net, which makes checking certain syntactic properties easier. This check is then performed on the Petri net, leading to either a positive or negative outcome. If the outcome is positive, meaning no semantic errors can be found, no further action has to be taken regarding the deployed BPMN process diagram. However, if there is an error a new version of the BPMN shall be created to fix the error.

3.2 Architecture

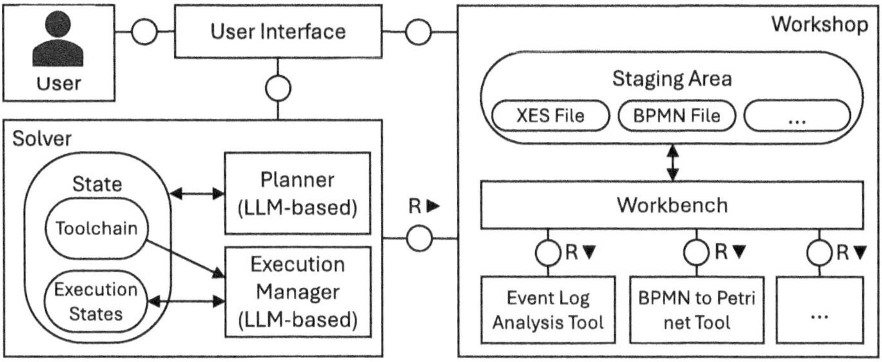

Fig. 2. The system architecture, where rectangles refer to software components, ellipses to data stores, circles to communication channels, and R to request/response communication.

The PRAXIS framework aims to combine a set of BPM tools with the functionalities of LLMs. Given a user query, the framework plans a toolchain and executes it step by step using existing tools. The architecture, illustrated in Fig. 2, consists of three main components, each with clearly defined responsibilities: the *User Interface*, the *Solver*, and the *Workshop*. In the following, we outline the architecture design which aims to satisfy the three requirements of orchestration effectiveness (R1), reproducibility (R2), and separation of concerns (R3).

The user interface mediates between the user and the core logic components of the system. It accepts natural language queries, forwards them to the solver, and presents the results back to the user. It supports file upload and display, enabling multimodal input and output. The user interface holds no task logic or state of its own; it serves purely as the interaction layer, allowing users to input natural language questions like the one in the motivating example above.

The solver is the central reasoning component. Its responsibility is twofold: planning how to solve a given task and supervising that plan's execution by coordinating the available tools. It is the only component that maintains awareness of the overall task context. Given a user request, the LLM-based *Planner* first derives a plan in the form of a graph that orders execution steps for a given query, inspired by a Graph of Thought [3]. These steps may rely on intermediary results from previous steps. For the motivating example, the planner may construct the toolchain of using an event log analyzer first, then a translator from BPMN into a Petri net, and later a soundness check on the Petri net.

Once this plan is established, the *Execution Manager*, also based on an LLM, iterates the steps in the plan, delegates its execution to the workshop by invoking the appropriate tools, and records the result. If execution reveals that the current plan is flawed, the execution manager can trigger replanning by the planner, submitting feedback from the failed attempt. This feedback loop between planning and execution constitutes a multi-layered realization of the ReAct paradigm [24], fulfilling the design principle of orchestration effectiveness (R1): the system reasons about its own progress and adapts its plan if necessary. For example, if the soundness check on the Petri net has a negative result, the execution manager can prompt the planner again to remodel the model to be sound.

Throughout this process, the solver maintains an internal *State* that records both the planned toolchain and an execution protocol capturing the inputs and outputs of each completed step, i.e., execution states. Operationally, this state provides the context required to determine the next action and to detect when replanning is necessary. Beyond that, it ensures Reproducibility (R2): every decision and intermediate result is logged in a structured manner, making the system's reasoning process transparent and enabling meaningful error communication to the user. In the example, the state is simply the plan consisting of the three and later four actions and their corresponding input and output, e.g. an event log or a Petri net.

The workshop is responsible for invoking the external tools that the solver relies on during execution. While the solver decides *which* tool to invoke and *why*, the workshop is responsible for *how* that invocation is carried out. Each tool is integrated through a standardized interface into the *Workbench*, so that adding a new tool requires only an adapter to this interface. This directly addresses the requirement of extensibility in (R3): the set of capabilities available to the system grows with each newly registered tool, without requiring changes to any of the components in the architecture. For example, the workshop would hold a registry containing the tools for analyzing event logs, translating BPMN models to Petri nets, and others. Those tools can be accessed by the workbench.

Beyond tool registration, the workshop is responsible for data management during execution. Its *Staging Area* stores input and output files, such as process models or analysis results, with unique identifiers. This allows the solver to reference data without embedding file contents in the LLM's context window, avoiding the risk of unintended modification or truncation during LLM interaction. This supports the strict separation between reasoning in the solver and data operations in the staging area. The framework allows the combination of diverse tools and file types to support a variety of modeling languages and multimodal outputs. In the beginning of the motivating example, the staging area would hold an XES event log and a BPMN process diagram, and later add a Petri net and possibly an updated BPMN process diagram.

4 Prototypical Implementation

In this section, we present a prototypical implementation of a system following the PRAXIS framework. In addition, we demonstrate integration into an existing BPM platform. This shows that the requirements of reproducibility (R2) and separation of concerns (R3) are feasible. Furthermore, the prototype serves as the foundation for evaluation in Sect. 5 to assess its effectiveness (R1).

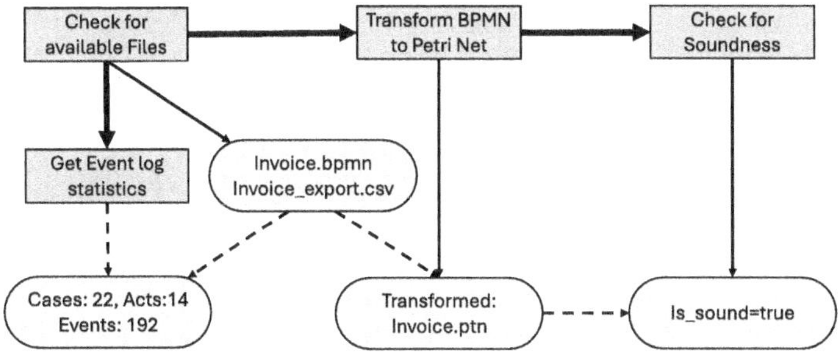

Fig. 3. Illustration of the state planning and execution for the motivating example. Boxes represent tools that are ordered via arcs and results in execution states represented as ovals.

4.1 System Implementation

In the following, we elaborate how state-of-the-art techniques can be applied to implement the framework in Sect. 3. The prototype is implemented using LangChain and LangGraph for LLM interaction and state handling, OpenAI models for reasoning, and a graph-based state structure inspired by Graph of Thought [3]. For testing purposes, the user interface can be realized through the debugging web applications of LangChain and MCP.

Building on this design primitive, a graph-based data structure, inspired by Graph of Thoughts [3], implements the execution states of the solver. In this structure, tasks and (sub)steps required to fulfill a given task are represented as nodes, while dependencies are represented as edges. Results and the dependencies of results on other steps are captured through a separate class of nodes and edges. This graph-based data structure is managed by the solver and documents both the plan and the execution of tasks, implementing the states of PRAXIS. This design allows both the planning and the execution of the solving process to be monitored and logged at any time, which directly fulfills the reproducibility requirement. Figure 3 shows the graph structure for the example from Fig. 1. The first check of the available files yields both the event log as well as the BPMN process diagram of a deployed process. Then, the event log can be analyzed ('Get Event log statistics') and the BPMN file can be translated ('Transform BPMN to Petri net'). The translation step is then followed by the soundness check. Note that since the translation and the event log analysis have no precedence relation, they can be executed in any order or in parallel.

The planner and the execution manager are implemented according to the paradigms of Graph of Thoughts [3] and ReAct [24]. Each component reads the state, performs an LLM call, and then either makes a workflow decision or prepares a tool call. In addition, they evaluate progress and determine subsequent actions. For the implementation, OpenAI models (GPT-5.2 and GPT-5-mini) were used. This choice is motivated both by the variety of models that OpenAI provides and by the high rankings these models achieved at design time on LLM Arena, an LLM evaluation platform based on the work of Chiang et al. [6].

The workshop is implemented as a Model Context Protocol (MCP) server that has access to a file directory and provides basic data management functions to the solver as tools. MCP is an open protocol published and established by Anthropic at the end of 2024 to standardize the interface between AI systems and resources, in our case the solver, the workbench, and the tools. To integrate existing BPM tools, an adapter needs to be implemented according to the MCP. Therefore, we implement most tools as standalone, stateless MCP servers.

When a tool is requested by the solver, the workbench resolves file references and forwards the function call to the corresponding tool, including the file content. The return value of the tool is, if applicable, stored by the workbench as a file, and a reference is passed back to the solver, keeping the LLM context free of large file contents. In total, four MCP servers were implemented, providing 14 tools for carrying out the activities depicted in Fig. 1 and others. Eight tools implement distinct process mining functionalities from the PM4Py library [2] like event log statistics, process discovery, and model visualizations. Furthermore, tools implement modeling functionalities from bpmn-io/bpmn-lint and deployment and monitoring functions from Camunda 7 Community Edition. To support formal analysis, tools include a translator of BPMN models into a Petri net, and formal model checking for Petri nets. A mocked tool implementation simulates the remodeling of a BPMN process diagram, which can be replaced by tools like [13]. The use of MCP allows an easy extension with new tools which implement different functionalities for BPM.

4.2 Platform Integration with OpenBPT

In addition to the prototype, an adapter for a local development instance of OpenBPT [16] was implemented, enabling its backend to be used as a workbench in conjunction with the standalone solver implemented for the prototype. OpenBPT is a platform for the conceptual modeling and analysis of processes [16]. Its backend is conceptually similar in structure to the workbench, and some existing tool integrations are already available for use. Due to the limited availability of services and the absence of an automated evaluation mechanism, the evaluation was performed manually to test the overall usability.

Five prompts covering the current model inventory, an identifier lookup, BPMN-to-Petri-net conversion, deadlock detection and property output were executed four times each, with two authors judging the success. In addition to the initial prompt, a single follow-up input was permitted if the output had not yet been assessed as successful; the time required for this follow-up was excluded from the recorded processing time. In the tests conducted with OpenBPT [16], the mean response time was 96.7 s and the median response time was 52.5 s. Overall, 95% of responses were correct, while 85% required no further specification of the initial request. This integration demonstrates that single components can be exchanged, fulfilling the exchangeability requirement.

5 Evaluation

In this section, we show how to evaluate the orchestration effectiveness of a system that implements the PRAXIS framework. First, we present a general methodology that generates test cases semi-automatically for a given toolsuite. Then, we provide an exemplary evaluation setup for the prototype provided in Sect. 4. Finally, we discuss the obtained results. We note that this evaluation purposefully does not include a comparison against an external baseline. The closest related work either targets a single phase of the lifecycle [1,13] or is the simple use of an LLM [4,14]. Instead, our methodology assesses whether PRAXIS, given some toolsuite, achieves high orchestration accuracy in absolute terms, against a ground truth derived from the tool suite itself.

5.1 Evaluation Methodology

The goal of our evaluation methodology is to support the assessment of how well an implemented PRAXIS system is able to orchestrate a given set of BPM tools. We do not consider the quality of the individual tools but only how well a natural language prompt can be mapped to a useful sequence of tools.

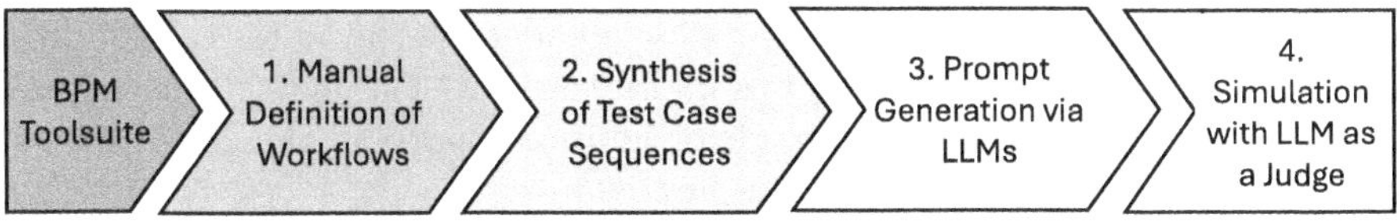

Fig. 4. The methodology for evaluating PRAXIS systems consisting of four steps.

To achieve this goal, we propose the four steps depicted in Fig. 4. First, a workflow model of the existing tools is created to define useful ordering relations between the tools and to provide a ground truth of useful toolchains. This definition may require manual input, and it allows one to only focus on directly following relations between tasks instead of defining all possible sequences manually. Second, this workflow model is analyzed for all possible paths in the workflow. Each path represents a unique test case for one sequence of tools. In the third step, each test case, which is a ground truth sequence of tools, is complemented with a natural language prompt that shall serve as the input for testing the framework. This input prompt is generated by an LLM that takes the expected resulting sequence and is asked to generate a natural language prompt for it. Additionally, each sequence can be labeled with the information whether it belongs to the intRA-phase Test set (RAT) of tools within one phase of the BPM lifecycle, or the intER-phase Test set (ERT) of tools that span multiple phases. Finally, all test cases are automatically simulated by using the natural language prompt as input to the PRAXIS system. The sequence chosen by the system is then compared against the ground truth sequence to assess orchestration quality.

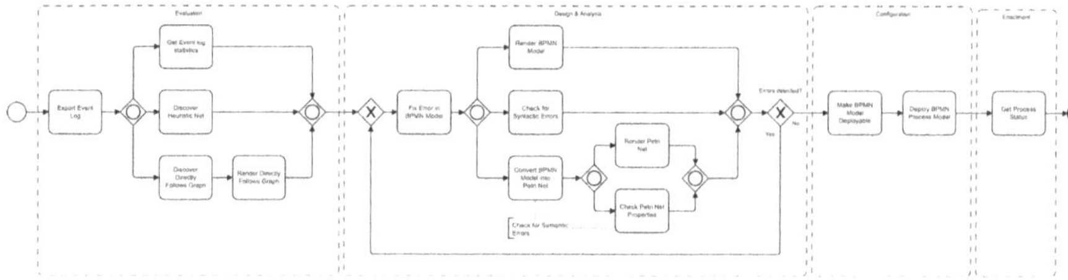

Fig. 5. The extended example for BPM tool workflows along the four lifecycle phases.

5.2 Prototypical Evaluation Setup

1. Manual Definition of Use Cases. Figure 5 extends the motivating example depicted in Fig. 1 into a process that spans all four phases of the lifecycle, with 14 activities each mapped to exactly one implemented tool. These tools are representative by covering all BPM lifecycle phases and different modeling and analysis tasks. Starting from event log export, the workflow allows parallel execution of analysis, heuristics-net discovery, and DFG discovery in the evaluation phase, followed by BPMN remodeling, linting, rendering, and Petri-net conversion in the design phase, then downgrading and deployment in the configuration phase, and finally process-status retrieval in the enactment phase. A use case can now consist of activities from a single or multiple phases of the BPM lifecycle, comprising the inter-phase and intra-phase use cases.

2. Synthesis of Test Case Sequences. For the workflows in the implemented toolsuite, we automatically synthesized the test case sequences as follows. For each tool, a clear natural language prompt as well as the input and output file(s) were manually crafted once. An exemplary prompt reads: *"Run the export_event_log tool with the process definition key 'invoice' "*. The corresponding activity has no input file but produces the output file `export_event_log_result.csv`. The input and output files are aligned such that dependencies between activities are accounted for in the use cases. Accordingly, for the subsequent activity 'Discover a Directly Follows Graph', the input file would be `export_event_log_result.csv`. The expanded BPMN model described above was transformed into a directed graph where nodes represent activities, and an edge indicates that one activity can be executed after another. Through traversal of the graph, all valid, loop-free subsequences of length 1 to 14 were derived and stored as sets. Identical sets were removed because the corresponding use cases differ only in the execution order of parallel activities. Testing these variations would significantly increase the size of the test set without yielding substantial additional insight.

Two test sets are derived. The RAT test set evaluates the orchestration effectivity within one phase of the BPM lifecycle, and the ERT test set evaluates the effective tool orchestration over at least three phases. The RAT test set comprises all tools within the design phase of the BPM lifecycle. The length of those test cases is up to 6 tools. Meanwhile, the ERT test set contains all use cases that span at least three phases of the BPM lifecycle. These inter-phase use cases consist of 12 to 14 tools. Assuming that the capability to execute more complex cases in the ERT test set can be extrapolated to simpler cases, we excluded use cases that span only two phases for computational and cost efficiency.

3. Prompt Generation via LLMs. For each set, a test case was created by concatenating the prompts and expected output files. The minimal number of required input files was determined through topological sorting. For instance, if the activities 'Export Event Log' and 'Discover Directly Follows Graph' are part of the same test case, the previously defined input file 'export_event_log_result.csv' for 'Discover Directly Follows Graph' is not required, as it can be generated by the 'Export Event Log' activity. To make the clear and explicit natural language prompts more realistic, each prompt was transformed by an LLM. This involved removing numbering, generalizing function names and file names, and making the text more prose-like and removing structure. For example, *"1. Run the export_event_log tool with the process definition key invoice. 2. Run the discover_dfg tool with the file export_event_log_result.csv as input"* was transformed into: *"Export the event log for the invoice process and compute the Directly Follows Graph from it"*. By applying this approach, 70 test cases were generated for the RAT and 48 test cases for the ERT.

4. Simulation with LLM as a Judge. To examine the effectiveness in orchestrating tools, the prototype was tested in an automated fashion. To test a given use case, the solver initially receives the natural language prompt and access to the corresponding files required to successfully execute the activities of that use case. Once the solver has completed the use case, an LLM is employed as a judge to evaluate whether the output aligns with the intended toolchain by the testcase. The evaluation is based on the state of the solver, the initial prompt, all generated files, and a grading rubric. The grading rubric is a framework which is provided as context to the LLM as a judge to evaluate whether an orchestration was successful or not. The corresponding test for each use case contains the prompt, input file, output file, grading rubric, and metadata. For automated testing, the test runner creates an isolated instance with the input files for each test case, invokes the solver, evaluates the execution using an LLM as a judge, and documents the test run in LangSmith[2], a free-to-use, yet closed source web application, which is compatible with the other Langchain-AI libraries.

5.3 Evaluation Results

To avoid relying on a single execution, each test case was executed 3 times, resulting in 354 total runs. In the RAT, the success rate was 87%, with average costs of 0.09 \$ and a median response time of 150.31 s. In the ERT set, approximately 87 % of all use cases were orchestrated correctly, while the average costs

[2] https://smith.langchain.com/.

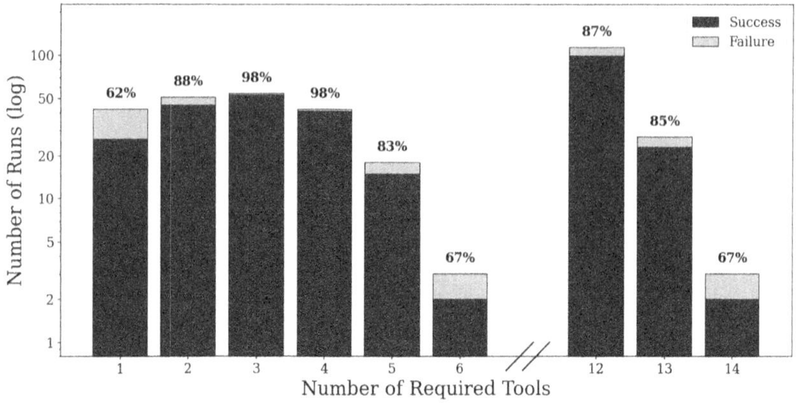

Fig. 6. Orchestration success rate of the RAT (1–6) and ERT (12–14) test cases. Each test case was run three times.

and median response time of 0.20 $ and 449.92 s were considerably higher. Considering both test sets, PRAXIS requires an average of 0.02 $ and 41 s per tool call to process a use case. The graph in Fig. 6 shows that in both the RAT and the ERT the test cases with the most activities posed a greater challenge than others. While this can be attributed to the growing complexity of use cases with larger activity count, test cases containing only a single activity actually perform the worst. Here, the LLM plans more tool usage than intended. An example is the prompt *"deploy the BPMN model"* with the target being the 'deploy_bpmn' tool. However, the solver 'over-elaborates' and hallucinates additional tools, such as invoking the linting service before trying to deploy a model. This relates to the inverse scaling paradox [17] which suggests that bigger models tend to add additional complexity in easy tasks or use cases.

6 Discussion and Conclusion

In this paper, we propose the PRAXIS framework, which assists process managers throughout the BPM lifecycle by combining LLMs as an interface to orchestrate useful BPM toolchains. In the following, we summarize the approach before discussing limitations and future work.

PRAXIS combines LLMs with deterministic BPM tools to orchestrate toolchains on the basis of a prompt in natural language. The proposed architecture satisfies reproducibility (R2) and separation of concerns (R3) by design. The implemented prototype with 14 tools demonstrates the technical feasibility. The satisfaction of orchestration effectiveness (R1) is shown by the applied evaluation methodology.

6.1 Limitations

The proposed evaluation of the PRAXIS framework is highly dependent on our prototypical implementation. Therefore, we need to discuss the limitations of the prototype and the evaluation methodology.

The prototype is implemented using a specific reasoning approach and does not consider different architectures that could increase the quality of the results. Furthermore, it uses a fixed set of publicly available BPM tools. To facilitate the evaluation, some tools were only mocked. The combination with other tools and in real-world implementations should be studied in the future. While the architecture simplifies tool integration, the prototype still requires an MCP adapter for every tool. Additionally, the prototype assumes that all files and tools for solving a prompt are present, which might not always be the case for real world prompts. Also, proprietary OpenAI models were chosen as LLMs, which may introduce a bias to the results. Other vendors and open models should be compared to improve privacy and reduce costs. To show the exchangeability, we integrated the framework with only one process modeling platform. The integration into other BPM platforms remains a future endeavor.

The reliability of the test case evaluation depends on the above limitations of the prototype, and on the implementation of the evaluation itself, which relies on an LLM-as-a-judge approach. The evaluated test cases comprise a set of questions within the design phase of the BPM lifecycle, and questions spanning multiple phases. The focus on the design phase may introduce a bias that should be studied with other tool sets. The prototype integrates tools that have only one input and one output file, while more complex tools may complicate the creation of a toolchain and should be considered in the future. Also, more file types, and additional data formats should be considered. The evaluation setup assumes that all files and tools are present for a given prompt. As this might lead to overthinking and producing wrong results, the handling of missing tools or data should be evaluated in the future as well. We implemented the evaluation of each test prompt as a zero-shot prompt, so without any additional example or context. In practice, repeated interactions and user feedback can improve the system's results.

6.2 Future Work

As shown by the novel PRAXIS framework, it is possible to combine existing BPM tools with an LLM interface to orchestrate useful toolchains. This allows us to combine the merits of LLMs with traditional and deterministic tools to stay in control in business environments. However, our evaluation is just a starting point. As discussed above, it should be applied to more tools, file types, and real world scenarios. More empirical research on the usage of PRAXIS is also needed to investigate how users interact with such a system and what additional requirements they have. Furthermore, the costs of PRAXIS in action need to be discussed. With more complex settings, calls to proprietary LLMs become more costly. Costs can be optimized by adjusting the prompts, reducing the

LangGraph nodes, or using local LLM instances. In addition, costs need to be set in relation to the actual business value of PRAXIS.

In summary, the PRAXIS framework advances the adoption of LLMs in BPM, while preserving the best of two worlds: existing tools for deterministic results and LLMs to enable natural language interfaces.

References

1. Barbieri, L., Stroeh, K., Madeira, E.R.M., van der Aalst, W.M.P.: An LLM-based Q&A natural language interface to process mining. In: Process Mining Workshops - ICPM 2024 International Workshops, Lyngby, Denmark, 14–18 October 2024, Revised Selected Papers. Lecture Notes in Business Information Processing, vol. 533. Springer (2024). https://doi.org/10.1007/978-3-031-82225-4_1
2. Berti, A., van Zelst, S.J., Schuster, D.: Pm4py: a process mining library for python. Softw. Impacts **17** (2023). https://doi.org/10.1016/J.SIMPA.2023.100556
3. Besta, M., et al.: Graph of thoughts: solving elaborate problems with large language models. In: Proceedings of the Thirty-Eighth AAAI Conference on Artificial Intelligence and Thirty-Sixth Conference on Innovative Applications of Artificial Intelligence and Fourteenth Symposium on Educational Advances in Artificial Intelligence. AAAI'24/IAAI'24/EAAI'24. AAAI Press (2024). https://doi.org/10.1609/aaai.v38i16.29720
4. Buss, A., Kratsch, W., Schmid, S.J., Wang, H.: ProcessLLM: a large language model specialized in the interpretation, analysis, and optimization of business processes. In: Business Process Management Workshops - BPM 2024 International Workshops. Springer (2024). https://doi.org/10.1007/978-3-031-78666-2_17
5. Calvanese, D., et al.: Agentic business process management: a research manifesto. Inf. Syst. **140**, 102738 (2026). https://doi.org/10.1016/j.is.2026.102738. https://www.sciencedirect.com/science/article/pii/S0306437926000529
6. Chiang, W.L., et al.: Chatbot arena: an open platform for evaluating LLMs by human preference. In: Proceedings of the 41st International Conference on Machine Learning. ICML 2024. JMLR.org (2024)
7. Dumas, M., et al.: AI-augmented business process management systems: a research manifesto. ACM Trans. Manag. Inf. Syst. **14**(1) (2023). https://doi.org/10.1145/3576047
8. Dumas, M., La Rosa, M., Mendling, J., Reijers, H.A.: Fundamentals of Business Process Management, 2nd edn. Springer, Berlin (2018). https://doi.org/10.1007/978-3-662-56509-4
9. Dunn, C.W., Hashemi Tonekaboni, N.: Transforming generative large language models' limitations into strengths using gestalt: a synergetic approach to mathematical problem-solving with computational engines. In: Proceedings of the 57th Hawaii International Conference on System Sciences (HICSS) (2024). https://doi.org/10.24251/HICSS.2024.622
10. Estrada-Torres, B., del-Río-Ortega, A., Resinas, M.: Mapping the landscape: exploring large language model applications in business process management. In: Enterprise, Business-Process and Information Systems Modeling - 25th International Conference, BPMDS 2024, and 29th International Conference, EMMSAD 2024. Lecture Notes in Business Information Processing, vol. 511. Springer (2024). https://doi.org/10.1007/978-3-031-61007-3_3

11. Kalai, A.T., Vempala, S.S.: Calibrated language models must hallucinate. In: Proceedings of the 56th Annual ACM Symposium on Theory of Computing (2024). https://doi.org/10.1145/3618260.3649742
12. Kampik, T., et al.: Large process models: a vision for business process management in the age of generative AI. KI - Künstliche Intelligenz **39**(2) (2024). https://doi.org/10.1007/s13218-024-00863-8
13. Klievtsova, N., Ehrendorfer, M., Mangler, J., Rinderle-Ma, S.: AutoBPMN.AI: conversational process modeling and automation. In: Joint Proceedings of the Best BPM Dissertation Award, Doctoral Consortium, and Demonstration & Resources Forum co-located with 23rd International Conference on Business Process Management (BPM 2025). CEUR Workshop Proceedings, vol. 4032. CEUR-WS.org (2025). https://ceur-ws.org/Vol-4032/paper38.pdf
14. Kourani, H., Berti, A., Schuster, D., van der Aalst, W.M.P.: Process modeling with large language models. In: Enterprise, Business-Process and Information Systems Modeling - 25th International Conference, BPMDS 2024, and 29th International Conference, EMMSAD 2024. Lecture Notes in Business Information Processing, vol. 511. Springer (2024). https://doi.org/10.1007/978-3-031-61007-3_18
15. Li, Y.: A survey of RAG-reasoning systems in large language models. In: Findings of the Association for Computational Linguistics: EMNLP 2025, Association for Computational Linguistics, Suzhou, China (2025). https://doi.org/10.18653/v1/2025.findings-emnlp.648
16. Lichtenstein, T., König, M., Körner, M., Seidel, A., Ghasemi, A., Weske, M.: OpenBPT: an extensible platform for conceptual modeling and analysis. In: Intelligent Information Systems. Springer, Cham (2025). https://doi.org/10.1007/978-3-031-94590-8_23
17. McKenzie, I.R., et al.: Inverse scaling: when bigger isn't better. Trans. Mach. Learn. Res. (2023). https://doi.org/10.48550/arXiv.2306.09479. https://openreview.net/forum?id=DwgRm72GQF, featured Certification
18. Peffers, K., Tuunanen, T., Rothenberger, M., Chatterjee, S.: A design science research methodology for information systems research. J. Manag. Inf. Syst. **24** (2007)
19. Rosemann, M.: Foreword. In: Fundamentals of Business Process Management, 2nd edn. Springer, Berlin (2018). https://doi.org/10.1007/978-3-662-56509-4
20. Shanahan, M.: Talking about large language models. Commun. ACM **67**(2) (2024). https://doi.org/10.1145/3624724
21. van der Aalst, W.M.P., Reijers, H., Hinz, O., Pfeiffer, J.: Process mindlessness: when we lose sight of what AI is supposed to improve. Bus. Inf. Syst. Eng. **67**(6) (2025). https://doi.org/10.1007/S12599-025-00972-0
22. Vu, H., Klievtsova, N., Leopold, H., Rinderle-Ma, S., Kampik, T.: Agentic business process management: practitioner perspectives on agent governance in business processes. In: Business Process Management: Responsible BPM Forum, Process Technology Forum, Educators Forum. Springer, Cham (2026). https://doi.org/10.1007/978-3-032-02936-2_3
23. Weske, M.: Business Process Management: Concepts, Languages, Architectures, 4th edn. Springer, Heidelberg (2024). https://doi.org/10.1007/978-3-662-69518-0
24. Yao, S., et al.: React: synergizing reasoning and acting in language models. In: The Eleventh International Conference on Learning Representations, ICLR 2023, Kigali, Rwanda, 1–5 May 2023. OpenReview.net (2023). https://openreview.net/forum?id=WE_vluYUL-X

Educators Forum

From Classroom to Practice: Bridging BPM Education and Industry Expectations Through a Process Mining Tool Comparison Project

Najah Mary El-Gharib(✉), Daniel Amyot, and William Van Woensel

University of Ottawa, Ottawa, Canada
{nelgh031,damyot,wvanwoen}@uottawa.ca

Abstract. This paper presents an innovative assignment approach implemented in the Business Process Modelling, Mining, and Automation course for Master's students at the *University of Ottawa*. The course integrates Active Learning principles to train industry-ready business analysts, capable of supporting modern business process management, mining, optimization, and automation initiatives. The assignment approach involves a project-based process mining tool comparison project, where students evaluate industry-relevant tools using real event logs to address a concrete business problem. The assignment design integrates experiential learning, formative feedback, and peer review activities to support both technical and professional skill development. The assignment is structured around a set of guiding principles. First, it emphasizes the development of real-world skills aligned with the evolving demands of digital business processes. Second, it encourages students to critically challenge their assumptions, tool limitations, and existing practices through a data-driven, evidence-based mindset. Finally, it promotes continuous self-reflection and the active exchange of constructive feedback among peers to reinforce deeper learning and professional growth. These guiding principles offer a practical framework that can support other educators in the BPM community in designing effective and practical assignment strategies that help bridge the gap between academic learning and industry expectations.

Keywords: Business Process Management · Process Mining · Active Learning · Academia · Industry · Process Mining Tools

1 Introduction

In the context of digital transformation, organizations are increasingly relying on data-driven approaches to understand, monitor, and improve their business processes. Among these approaches, *Process Mining* (PM) has emerged as a key analytical capability, enabling organizations to extract actionable insights

K. Kluza et al. (Eds.): BPM 2026, LNBIP 600, pp. 263–278, 2027.
https://doi.org/10.1007/978-3-032-36710-5_17

from event logs and support operational decision-making [19]. Industry adoption of PM has accelerated across sectors such as finance, healthcare, and public services, driven by the need for transparency, efficiency, and automation. As a result, there is a growing demand for graduates who not only understand core PM concepts, but are also able to apply them effectively in real-world settings. In parallel with this industry demand, universities have begun to incorporate PM into Business Process Management (BPM) education by offering dedicated courses or integrating PM modules within BPM, data analytics, information systems, and digital transformation curricula. In response, PM tool vendors such as Celonis [5] and Apromore [3] started adding Academic Alliance options. Notable examples include the courses developed by van der Aalst at RWTH Aachen University and the online course "PM: From Theory to Execution", jointly offered by PADS and Celonis [1, 17]. ProcessMining.org maintains a repository of academic and professional PM training opportunities [2].

These initiatives primarily focus on teaching core PM concepts, methods, and tool usage, which are essential to build a solid foundation of knowledge and skills. However, Grzybowska and Hoffa-Dąbrowska [12] identified a major gap in BPM education between such conceptual knowledge and practical application, pointing out the need for developing practical competencies through active engagement, critical thinking, and experiential learning. Their proposed framework emphasizes the integration of theoretical concepts with real-world BPM challenges, encouraging students to acquire knowledge through problem-solving, collaboration, reflection, and application-oriented activities. In doing so, they gain familiarity with PM techniques and tools [12].

Nevertheless, there are opportunities to develop other salient, industry-critical skills, such as evaluating tool capabilities, interpreting outputs in a business context, and making evidence-informed and data-driven decisions. Indeed, in industry settings, analysts must be able to select appropriate tools (and justify their choices), adapt to varying data and organizational contexts, and present evidence-driven answers to concrete business problems.

This paper presents an innovative assignment approach, which involves a *PM tool comparison project*. The assignment was implemented in a graduate-level course on Business Process Modelling, Mining, and Automation at the *University of Ottawa*. Rather than focusing on technical functionality per se, students are asked to utilize the tools to address business problems within an organizational context, and, while doing so, critically analyze their suitability in this task. The assignment requires students to analyze the same event log dataset using multiple tools (they must select at least two relevant PM tools among Apromore [3], Celonis [5], Mindzie [15], Disco [9], ProM [7] or others), and to critically assess their capabilities across key PM perspectives, including process discovery, conformance checking, and performance analysis. This assignment thus goes beyond traditional tool-based exercises, by not only asking to solve a realistic business problem of their choice using various tool features, but also by emphasizing a critical reflection on the tools employed. The assignment design is grounded in *Active Learning* [4] principles and incorporates multiple formative and summative assessment elements throughout the semester. Students progressively develop their final project through a proposal, milestone submissions, peer discussions, and feedback activities, enabling continuous reflection and iterative improvement. In addition, peer review is integrated into the assessment structure to strengthen the student's ability to critically evaluate professional reports.

The contribution of this paper is threefold. First, it provides a detailed description of a practical and transferable BPM assignment that aligns closely with current industry expectations for PM competencies. Second, it demonstrates how active learning strategies can be integrated into BPM education to improve student engagement and practical skill development. Third, it offers insights and recommendations for educators seeking to design industry-oriented BPM assignments that include critical thinking, active learning, and professional readiness.

The remainder of the paper is structured as follows. Section 2 discusses the background and related work. Section 3 details the assignment design and its different parts. Section 4 highlights the guidelines for educators that are interested in adopting this assessment approach. Section 5 discusses challenges in conducting such an assignment and outlines future directions. Finally, Sect. 6 concludes the paper.

2 Background and Related Work

The *Business Process Modelling, Mining, and Automation* course from the University of Ottawa explores BPM languages and technologies, with a focus on modelling, analyzing, and improving business processes using BPMN and process mining/automation. Students learn key PM concepts, including event log preprocessing, algorithms, and the use of industry tools such as Apromore to perform process discovery, conformance checking, performance analysis, and process improvement. The course also introduces process simulation techniques to assess improvement opportunities. In addition, students are given an overview of Robotic Process Automation (RPA) and broader process automation approaches using Business Process Management Systems (BPMS) and low-code platforms. Key challenges and opportunities related to automation – including data quality, system integration, governance, and organizational readiness – are examined throughout the course. Finally, the course explores the evolving relationship between BPM and Artificial Intelligence, including Generative AI and emerging agentic AI applications. Concepts are reinforced through practical exercises and real-world examples drawn from a variety of industries and application domains. As for Course Learning Outcomes (CLO), students will be able to:

- **CLO1:** explain important process modelling concepts and notations;
- **CLO2:** read, create, and analyze business process models;
- **CLO3:** apply pre-processing techniques to business process event logs;
- **CLO4:** analyze business process event logs using process mining techniques;
- **CLO5:** select a process mining tool and use it at an intermediate level;
- **CLO6:** understand process automation tools, opportunities, and challenges.

As mentioned, universities and professional training providers have started introducing specialized PM courses and/or modules covering more specific topics. Many courses emphasize hands-on learning through the use of industry-relevant tools (such as those mentioned in the introduction), often incorporating real or synthetic event logs to provide experiential learning opportunities. However, we found that many PM courses continue to focus primarily on technical tool usage and analytical techniques, with limited emphasis on structured assignment designs that integrate formative activities, peer reviews, and industry-oriented tool comparison projects to prepare students for modern process intelligence initiatives. At the same time, multiple studies have compared PM tools to support practitioners and educators in selecting appropriate technologies [6,14]. Indeed, these studies demonstrate that tool suitability depends on the specific context, objectives, and user requirements. To the best of our knowledge, however, limited attention has been given to the pedagogical value of such tool comparison activities in PM education.

Below, we elaborate on aspects from the relevant literature that motivate our proposed assessment approach.

2.1 Tool Comparison and Selection

Panis [16] highlighted that selecting enterprise software tools is a complex decision-making process, requiring organizations to evaluate usability, functionality, integration capabilities, stakeholder requirements, and long-term organizational fit. In their case study, the author discusses the practical and organizational challenges involved in comparing and selecting a requirements management platform. These observations reinforce the importance of preparing students with critical tool evaluation and comparison skills within BPM education. Indeed, the presented PM tool comparison project asks students to assess multiple platforms against structured criteria, justify trade-offs, and provide evidence-based recommendations.

2.2 Active Learning

Active learning strategies in BPM education have involved *project-based and collaborative activities* that encourage students to apply BPM concepts in realistic contexts. In doing so, they engage in Active Learning as they go beyond passive knowledge dissemination. Gdowska et al. [10] discuss how collaborative BPM projects can enhance student engagement, interdisciplinary problem solving, and applied learning experiences. The authors showed that project-based learning approaches improved student participation, critical thinking, and practical understanding of BPM methodologies. Similarly, we propose a project-based and collaborative assignment approach.

Other examples of Active Learning have emphasized the *importance of experiential learning* to better prepare students for real-world BPM initiatives. A study by Lippe et al. [13] engaged students in practical BPM consulting activities, which was shown to not only enhance technical BPM competencies, but also professional skills such as stakeholder communication, teamwork, problem-solving, and critical thinking. The authors argue that traditional BPM teaching methods often focus heavily on theoretical concepts and isolated modelling exercises, which may not fully reflect the complexity of industry practice. Their findings highlight the value of tailored assignments in bridging the gap between academic BPM education and organizational needs.

Grzybowska and Hoffa-Dąbrowska [12] also argue that traditional lecture-based teaching methods are often insufficient for developing practical BPM competencies. The authors propose a pedagogical framework that emphasizes active participation, active learning, and competency development. The framework focuses on improving how students acquire, apply, and reflect on BPM knowledge through interactive and practice-oriented educational activities. Both studies highlight the value of combining theoretical BPM foundations with experiential learning. We aim to bridge the gap between academia and industry by having students use industry-relevant tools to address concrete business problems within an organizational context, and critically analyze the tools' suitability while doing so.

2.3 Use of Realistic Teaching Cases

Recent BPM education research has emphasized the importance of using realistic teaching cases to expose students to practical process analysis challenges. Grangel et al. [11] encourage students to apply BPM and PM concepts in a real-world context. In addition, case-based and hands-on learning approaches are increasingly adopted in BPM education to improve engagement and practical skill development. Grzybowska and Hoffa-Dąbrowska [12] similarly emphasized the role of real-world scenarios in supporting deeper student engagement and learning outcomes. In our proposed assignment method, students are encouraged to define a realistic business context, including the organization in question and business problem to be solved, and subsequently solve them within the defined real-world context.

2.4 Peer-Review Feedback

Winstone and Carless [20] position peer-review feedback not only as a mechanism for receiving suggestions for improvement, but also as an opportunity for students to learn by reviewing the work of others. In doing so, it encourages critical evaluation, reflection and communication skills; exposes them to different ways of structuring and presenting analysis results; and pushes for a deeper engagement with PM concepts—as they are now "grading" others on *their* conceptual understanding. Peer review contributes to active learning by positioning students as active participants in the assessment process rather than passive recipients of instructor feedback; thus further reinforcing Active Learning principles. Our proposed assignment requires students to critically analyze peer reports using a structured rubric and provide constructive, actionable, and evidence-based feedback. To support meaningful participation, the above-mentioned rubric scaffolded the feedback process by providing clear evaluation criteria and promoting constructive inputs.

3 Assignment Design: Tool Comparison Project

3.1 Assignment Overview

We introduce a project-based PM assignment centered on the comparative evaluation of PM tools within a concrete business context. The assignment is implemented in a graduate-level Business Process Modelling, Mining, and Automation course. It is designed to incorporate multiple formative and summative elements throughout the semester, as students develop their project through a proposal, milestone submissions, peer discussions, and (peer) feedback activities.

The assignment requires students to perform a hands-on evaluation of two PM tools (e.g., Celonis, Apromore, Disco, ProM), using their relevant PM features (e.g., process discovery, conformance checking, performance analysis) to address the business problem in a concrete organizational context[1]. Students evaluate the tools across multiple PM perspectives, including the technical capabilities of relevant PM features, their usability and practical applicability. The students are expected to submit a project proposal, which outlines the concrete organizational context, chosen PM tools, and project plan; after receiving the instructors' feedback, they proceed with the project and ultimately submit a final report with their findings and recommendations on tool use.

The concrete artefacts associated with the project can be found in our online supplemental material [8].

3.2 Learning Objectives and BPM Competency Alignment

The project is explicitly designed to support the development of key BPM competencies, including both technical proficiency and analytical reasoning skills. Upon completion, students are expected to:

- **LO1**: Understand event log structures, including case identifiers, activities, and timestamps (CLO3, CLO4);
- **LO2**: Perform process discovery, conformance checking, and performance analysis to address concrete business problems within an organizational context (CLO4, CLO5);
- **LO3**: Critically evaluate the capabilities and limitations of PM tools in a data-driven and evidence-based way (CLO5);
- **LO4**: Translate analytical findings (from LO2 and LO3) into actionable and business-oriented recommendations on tool suitability (CLO5).

The assignment thus goes beyond tool training by emphasizing interpretation of results, critical reflection, evaluation, and analytical decision-making skills (i.e., choosing one tool over another), which are essential in professional BPM practice.

[1] The same event log is used for each tool.

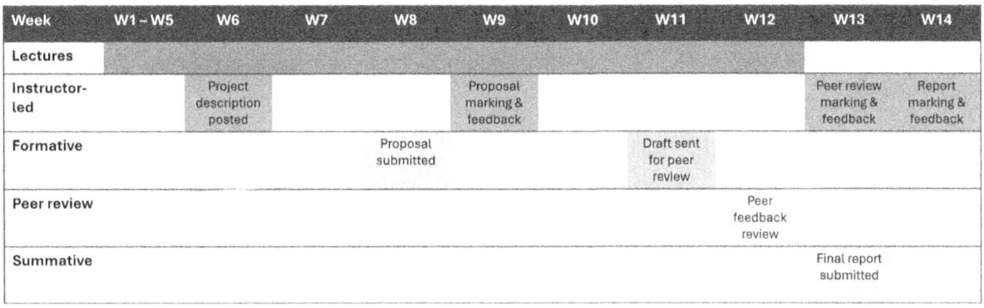

Fig. 1. Project Timeline

3.3 Assignment Design and Structure

The assignment is structured as a three-stage project, consisting of a proposal phase, followed by a draft report that is sent for peer-review, and a final report. This staged approach enables formative feedback while maintaining a strong summative evaluation component. Figure 1 presents the timeline of assignment activities associated with the project deliverables throughout the course. The project begins in Week 6 with the release of the project description, allowing students to start planning their work. Students submit a project proposal in Week 8 and receive instructor feedback within a week (Week 9) to help refine their approach before proceeding with the analysis and final report development. In Week 11, teams submit a draft version of their report for peer review. During Week 12, students review another team's work and provide structured feedback. Finally, students incorporate both peer and instructor feedback before submitting the final report in Week 13. This sequence integrates formative and summative assessment activities, providing multiple opportunities for feedback, reflection, and iterative improvement throughout the project lifecycle.

Project Proposal. Students start by defining the evaluation setting, including the target organization, the business problem(s) to be addressed, the intended users of any chosen tool, the types of process analysis required, other possible uses of the tool, and any tool deployment constraints (e.g., on-premise). Then, they select two PM tools for evaluation that are deemed suitable for the identified context, and justify their selection based on the organization's needs and the tools' capabilities (currently, gleaned from documentation or superficial use).

Next, students identify a set of key features to compare across the selected tools that are aligned with the business problem and analytical requirements of the context. For each feature, students provide a rationale explaining its importance. Finally, students develop an evaluation plan describing how the tools and selected features will be used and tested to address the business problem, how the findings will support a recommendation, and which event logs or datasets will be used during the assignment.

Project Final Report. Students are required to conduct a structured comparison based on predefined and student-selected evaluation criteria. The analysis focuses on key functionalities, including:

1. Process discovery and model generation;
2. Conformance checking capabilities;
3. Performance analysis (e.g., cycle time, bottlenecks, rework);
4. Usability and user experience;
5. Data handling and preprocessing requirements;
6. Advanced features (e.g., simulation, automation insights, AI support).

Analytical Phases. To guide students through the assignment, the project is organized into six analytical phases, linked to the prior learning objectives:

1. **Problem Framing and Context Definition:** Students begin by defining the organizational and business context for which the comparison is conducted. They identify the business problem to be addressed, the intended users of the tool, and the expected analytical outcomes. Students are also encouraged to consider practical constraints such as data privacy requirements, deployment limitations, scalability expectations, and organizational objectives. This phase emphasizes that the critical evaluation of PM tools (LO3) always depends on the organizational context.
2. **Event Log Understanding and Preparation:** Students explore the structure and quality of the event log used for analysis. They examine key attributes such as case identifiers, activities, timestamps, and resources while identifying potential data quality issues including missing values, inconsistencies, duplicate events, or incomplete traces. Students perform preprocessing tasks required by the selected tools, e.g., attribute enrichment or data cleaning. This phase meets LO1, and, similar to above, reinforces the understanding that analytical outcomes heavily depend on data quality.
3. **Tool-Based Analysis:** Students use the selected PM tools for conducting PM analysis tasks to address the business problem, using the same dataset to ensure consistency and comparability. Depending on the business problem, the analysis may include process discovery, conformance checking, bottleneck identification, performance analysis, and visualization. Students are expected to explore how the tools generate and present insights, as well as how configuration choices influence the outputs. Here, students gain practical experience with previously theoretical PM concepts (LO2); the hands-on exploration step further encourages students to interpret and critically assess analytical results rather than simply accepting and reporting system-generated outputs.
4. **Evaluation Strategy Development:** Students design a structured and transparent evaluation framework for comparing the tools. This includes defining evaluation criteria aligned with the business context and analytical objectives, such as usability, visualization quality, analytical capabilities, scalability, learning curve, integration support, collaboration features, security, and reporting functionalities. Students may also develop scoring models or weighting schemes to prioritize criteria based on organizational needs.

Students are encouraged to justify their evaluation choices and reflect on how evaluation strategies can influence decision-making outcomes. This phase helps meet LO3 by having students not just check off a predefined feature list, but rather having them reflect on the evaluation strategy itself.

5. **Comparative Analysis:** Using the evaluation framework (phase 4), and in tandem with the tool-based analysis (phase 3), students conduct a systematic comparison of the selected PM tools across technical, analytical, and user-oriented dimensions identified earlier (phase 1). They analyze similarities and differences in discovered process models, insights generated, usability experiences, feature availability, and interpretation support. The phase thus completes the project's realization of LO3. Unlike traditional tool comparison reports and exercises that focus primarily on listing and scoring features, the proposed assignment emphasizes fit-for-purpose evaluation. Students are required to consider the organizational context, business objectives, intended users, and operational constraints before assessing tool capabilities. Rather than assuming that more features imply a better tool, students learn that a feature is only valuable if it addresses the needs of the given context. Consequently, the same pair of PM tools may lead to different recommendations depending on the evaluation scenario, reflecting the context-dependent nature of technology selection in practice. An overview of sample projects (each including its business context, PM tools compared, evaluation goal, recommended tool, and rationale) is provided with the supporting material online [8].
6. **Recommendation and Reflection:** In the final phase, students synthesize their findings into a justified recommendation regarding tool suitability for the defined business context. Recommendations must be supported by evidence (gathered in phase 5) and aligned with organizational priorities (identified in phase 1). Students also reflect on lessons learned during the comparison process, including challenges encountered, assumptions made, limitations of the tools, and broader conclusions for real-world PM initiatives. This reflective component encourages students to connect technical analysis with professional judgment, and meets LO4.

Peer Review. Peer review was integrated as a structured formative assessment activity to support students in refining their PM tool comparison reports before final submission. The activity was designed not only to improve the quality of the deliverables, but also to develop students' critical evaluation, reflection and communication skills, and encourage deeper engagement with PM techniques (Sect. 2, *Peer-review feedback*). Peer review commenced after students submitted a draft version of their project report. Each group received another group's report and was asked to evaluate it using a structured review rubric aligned with the assignment learning objectives and grading criteria. The review criteria included clarity of the definition of the business problem, the appropriateness of the comparison context, the quality of the PM analysis, justification of the evaluation criteria and scoring models, interpretation of the findings, and the strength of the final recommendations. Students were further encouraged to suggest alternative analytical approaches.

3.4 Assessment Rubric Design

The assignment emphasized analytical reasoning, critical evaluation, and evidence-based decision making. A structured rubric (Table 1) was used to assess the project's learning objectives, including event log preparation (LO1), process analysis (LO2), comparative evaluation of tool capabilities (LO3), and the quality of recommendations and reporting (LO4).

Table 1. Summary of Rubric Dimensions

Criterion	Evaluation Focus
Problem Framing and Introduction	Assesses how clearly students define the business problem, project objectives, process context, stakeholders, and scope of the analysis.
Dataset Understanding and Preparation	Evaluates students' understanding of the event log, including the key attributes (e.g., Case ID, Activity, Timestamp) and the quality of data preprocessing performed before analysis.
PM Analysis	Evaluates the accuracy of process discovery and the quality of insights derived from the analysis.
Methodology and Tool Usage	Measures the appropriateness and rigor of applying the selected PM tools, as well as the students' ability to explain their analytical methods and demonstrate technical competence.
Tool Comparison and Evaluation	Focuses on the quality of the comparative analysis, including critical evaluation of tool capabilities, justification of findings, and consideration of practical and industry-related implications.
Reflection	Assesses the student's ability to critically reflect on the analysis, discuss limitations and challenges, and identify key lessons learned.
Clarity, Structure, and Presentation	Assesses the organization, readability, professionalism, and effective use of visual elements such as figures and tables to communicate results.

Alignment of Assignment Milestones with Targeted Skill Development. Figure 2 illustrates how the major assignment milestones (Fig. 1) are designed to progressively develop technical, analytical, and professional competencies throughout the PM tool comparison project. The peer feedback review activity (Week 12) shifts the focus toward analytical and professional competencies by engaging students in critical evaluation of another team's work, providing constructive feedback, and reflecting on strengths and areas for improvement. Finally, the submission of the final project report (Week 13) is the primary summative assessment, integrating all targeted competencies. Students are required to demonstrate technical proficiency in applying PM techniques and tools, analytical skills through tool comparison, critical evaluation, and recommendation development, as well as professional skills related to communication and reflective practice.

Assessment Task	Technical skills			Analytical skills			Professional skills		
Week #	Event log	PM	PM tools	Critical evaluation	Tool comparison	Decision making	Peer review	Communi-cation	Reflection
Proposal submitted Week 8	±		±					±	
Peer feedback review Week 12				±			✓	✓	±
Final report submitted Week 13	✓	✓	✓	✓	✓	✓		✓	✓

✓ Major contribution ± Minor, indirect contribution

Fig. 2. Contributions of project deliverables to desired skills

4 Practical Guidelines for Educators

A key strength of the proposed assignment lies in its practical applicability and ease of adoption across diverse educational contexts. As indicated below, the assignment design prioritizes flexibility, enabling instructors to implement or adapt the assignment with minimal effort while maintaining alignment with BPM learning objectives.

Guideline 1: Adapt the project to the teaching modality

The course has been offered in both fully online and in-person formats. This necessitated an assignment structure that can be adapted to different teaching modalities while retaining student engagement, collaboration, and practical skill development. Given our successful experience with both modalities, we expect educators to be able to adopt this assessment approach in either modality as well.

Guideline 2: Adapt the project to the student population

The assignment is adaptable to various educational levels, including undergraduate, graduate, and executive education programs. For introductory courses, similar to above, instructors may limit the scope to process discovery features and basic tool comparison strategies; while more advanced courses can incorporate more advanced features (e.g., conformance checking, performance analysis) and structured evaluation strategies with more in-depth analysis. This flexibility ensures that the assignment remains pedagogically relevant across different levels of student expertise.

Guideline 3: Adapt to class size, time constraints, and course depth

The assignment structure allows instructors to scale the scope of analysis, depending on time constraints and course depth; for instance, varying the required breadth of the business problem, number and types of features to be compared, and depth of the tool comparison strategy. We note that the project can further be implemented in both small and large classes by varying the team sizes (minimum 2 students per team).

Guideline 4: Leverage publicly available datasets

The assignment leverages publicly available and easily accessible datasets, such as event logs from open repositories or instructor-generated synthetic data. This reduces barriers for students and allows instructors to tailor the assignment to specific domains. Moreover, an instructor's choice for real-world event logs keeps the assignment relevant for realistic organizational settings.

Guideline 5: Clearly define deliverables and expectations

The assignment design includes clearly defined deliverables, structured phases, and an explicit evaluation rubric, which enhances transparency for students and simplifies implementation for instructors. These elements make the assignment readily transferable, allowing educators to reuse or adapt it without extensive redesign. As mentioned, the assignment artefacts are available as online supplementary material [8].

5 Limitations and Future Work

The implementation of a PM tool comparison project within a BPM course presents several pedagogical and practical challenges that are highlighted in this section. Several challenges related to teaching PM were already discussed by Tentina et al. [18], such as aligning goals with diverse learner profiles, finding suitable data and event logs, implementation and evaluation of courses, selecting an appropriate PM tool, and lack of reusable teaching material.

Challenge 1: Diverse backgrounds of students

Students often possess varying levels of technical expertise, analytical capabilities, and familiarity with BPM and PM concepts. As a result, designing an assignment that is sufficiently challenging for advanced learners while still accessible to students with limited technical experience can be difficult. In particular, activities related to event log preprocessing, tool configuration, advanced PM features, and interpretation of PM outputs may introduce steep learning curves for some students.

Challenge 2: Access to PM tools

This challenge is related to the accessibility and availability of PM tools. Commercial tools differ significantly in licensing models, deployment requirements, and educational support. For instance, some tools offer only limited (or no) academic access, while others require complex installation and configuration processes. These differences can affect the consistency of student experiences and create barriers to equitable learning opportunities. In addition, PM platforms continuously evolve, resulting in changing functionalities and interfaces that require ongoing updates to course materials and assignments.

Challenge 3: Use of real-world event logs

The use of real-world event logs, meant to support realistic organizational contexts, also introduces challenges. Industrial datasets are often incomplete, noisy, inconsistent, or poorly structured, requiring extensive preprocessing before meaningful analysis is possible. Although this may reflect industry conditions and contributes to experiential learning, students may spend substantial time addressing data quality issues rather than focusing on analytical reasoning and business interpretation, the chief goals of the assignment.

Challenge 4: Assignment consistency

Student teams may select different business scenarios, evaluation dimensions, datasets, and PM tools, leading to variability in project scope and complexity. To mitigate this issue, structured rubrics, peer-review mechanisms, and formative assessment activities were incorporated into the course design to improve fairness and provide clearer evaluation expectations.

Challenge 5: Continuous instructor involvement

The project-based and feedback-oriented nature of the assessment requires continuous instructor involvement. Providing formative feedback (e.g., project proposal), facilitating peer-review activities, and supporting students throughout the technical and analytical aspects of the assignment can become difficult to scale in larger class settings without additional teaching support.

The proposed assignment provides pedagogical value beyond teaching students how to use PM tools. By comparing alternative tools, students develop higher-order skills such as critical evaluation, evidence-based decision-making, and reflective analysis. The assignment also prepares students for industry practice, where technology selection requires balancing technical capabilities, business objectives, and organizational needs and constraints. As a result, students gain transferable competencies in technology evaluation that are applicable beyond PM.

A key distinction of the assignment is that students do not evaluate tools based on the number of available features. Instead, they learn to identify which capabilities create value within a specific organizational setting. Throughout the projects, students recognized that features irrelevant to the evaluation context

Table 2. Challenges and successes reported informally by students

Challenges	Successes
Tool learning	Opportunity to explore existing tools
Tools can be overwhelming	Lots of support and guidance
Lack of practical experience	Clear instructions
Total cases limit with free licenses	The examples provided during class are really helpful
Determine event logs	Amazing team to brainstorm with
Learning curve and data understanding	Very structured way to learn different tools
Source data file compatibility	Project proposal feedback was great too
Apple MacBook support	Lecture materials helped with potential characteristics to compare
Technical difficulties	

should not influence the selection decision. This shift from generic feature-based comparison to fit-for-purpose reasoning mirrors professional technology selection practices, where organizational objectives, user requirements, and constraints determine the most appropriate solution.

Future studies could investigate how activities such as scoring-model development exercises, peer-review workshops, and iterative feedback mechanisms influence students' analytical depth, report quality, confidence, and engagement levels. It is also a challenge to fit all these activities within the time frame of the course in addition to the lecture material and other assignments.

Moreover, the assignment approach could be validated across different educational contexts, including undergraduate programs, interdisciplinary courses, and institutions outside *Canada*. Such cross-institutional evaluations would help assess the generalizability, adaptability, and scalability of the proposed active learning framework within BPM education.

Additionally, we would like to collect students' feedback through a survey and analyze the collected results so that we can continue to improve the PM tool comparison project-based assignment similar to the approach that was used by Grangel et al. [11]. Using *Wooclap*[2], we have collected some informal and anonymous comments on challenges and successes related to this assignment (Table 2), but a more quantitative and repeatable approach would be useful here.

Future work includes evaluating the long-term impact of this assignment approach on student learning outcomes, industry readiness, and engagement across different teaching formats and class sizes. Additional research may also explore extending the framework to include emerging AI-driven process analysis capabilities and broader automation-oriented BPM scenarios.

[2] See https://www.wooclap.com/.

6 Conclusion

This paper presented an Active Learning-oriented assignment approach designed for a graduate course on Business Process Modelling, Mining, and Automation. The proposed assignment engages students in a realistic PM tool comparison project in which they analyze a given business problem and event log using different PM platforms. In doing so, the assignment aims to bridge the gap between BPM (in particular PM) education and industry expectations. Moreover, by integrating hands-on experience, structured tool comparisons, peer review, and reflective learning activities, the approach aims to leverage Active Learning principles to improve student engagement.

The assignment design emphasizes the development of both technical and professional skills, including process analysis, critical evaluation, data-driven decision making, communication, and collaborative problem solving. Through the use of real business problem scenarios and comparative analysis tasks, students are encouraged to go beyond tool usage towards critically understanding the strengths, limitations, and capabilities of different PM technologies in organizational contexts. The integration of formative activities in class and peer feedback further supports continuous learning and helps scaffold the development of analytical and evaluative skills for the final project deliverable.

Overall, the work contributes to the growing body of BPM education research advocating for more innovative, learner-centered, and practice-oriented teaching strategies to bridge the gap between academic knowledge and industry expectations.

Acknowledgments. This work was supported by the Natural Sciences and Engineering Research Council of Canada (Discovery Grant), the Ontario Graduate Scholarship (OGS) Program, and the University of Ottawa.

References

1. van der Aalst, W.: Process mining: from theory to execution (2021). https://www.tf-pm.org/newsletter/newsletter-stream-8-12-2021/new-course-process-mining-from-theory-to-execution, online course jointly developed by RWTH Aachen University and Celonis
2. van der Aalst, W.: Courses (2026). https://www.vdaalst.com/courses/courses.html. Accessed 29 May 2026
3. Apromore: Full spectrum process intelligence platform (2026). https://apromore.com
4. Barkley, E., Major, C.: Interactive Lecturing: A Handbook for College Faculty. Wiley (2018). https://books.google.ca/books?id=0q5JDwAAQBAJ
5. Celonis SE: Enterprise AI powered by Celonis (2026). https://www.celonis.com
6. Drakoulogkonas, P., Apostolou, D.: On the selection of process mining tools. Electronics **10**(4) (2021). https://doi.org/10.3390/electronics10040451
7. Eindhoven University of Technology: ProM tools (2026). https://www.promtools.org/

8. El-Gharib, N.M., Amyot, D., Van Woensel, W.: From classroom to practice: bridging BPM education and industry expectations through a process mining tool comparison project: supporting material (2026). https://doi.org/10.5281/zenodo.20560879
9. Fluxicon: Discover your processes (2024). https://fluxicon.com/disco/
10. Gdowska, K., Brzychczy, E., Kluza, K.: Project-based learning in BPM for interdisciplinary groups — teachers' experience. In: Business Process Management: Responsible BPM Forum, Process Technology Forum, Educators Forum, pp. 315–329. Springer (2026). https://doi.org/10.1007/978-3-032-02936-2_22
11. Grangel, R., Palomares, N., Sánchez-Galán, M., Luna-Aleixos, D., Llagostera-Reverter, I.: Teaching case: process mining for predicting failure to rescue. In: Business Process Management: Responsible BPM Forum, Process Technology Forum, Educators Forum, pp. 417–426. Springer (2026). https://doi.org/10.1007/978-3-032-02936-2_29
12. Grzybowska, K., Hoffa-Dąbrowska, P.: Improving the practice of acquiring knowledge in the field of BPM – a pedagogical framework. In: Business Process Management: Responsible BPM Forum, Process Technology Forum, Educators Forum, pp. 390–400. Springer (2026). https://doi.org/10.1007/978-3-032-02936-2_27
13. Lippe, S., Bärtsch, F., Brönimann, C., Mader, C.: Enabling BPM competencies through real-life student consulting projects. In: Business Process Management: Blockchain, Robotic Process Automation, Central and Eastern European, Educators and Industry Forum, pp. 344–353. Springer (2024). https://doi.org/10.1007/978-3-031-70445-1_23
14. Loyola-González, O.: Process mining: software comparison, trends, and challenges. Int. J. Data Sci. Anal. **15**(4), 407–420 (2023). https://doi.org/10.1007/s41060-022-00379-0
15. Mindzie: AI driven process mining and intelligence (2026). https://mindzie.com/
16. Panis, M.: Reflections on choosing a new requirements management tool. IEEE Softw. **39**(1), 7–10 (2022). https://doi.org/10.1109/MS.2021.3119506
17. Process Mining Organization: Process mining courses and educational resources (2026). https://processmining.org/courses.html. Accessed 29 May 2026. Repository of academic and professional process mining courses including MOOCs and industry training programs
18. Tentina, I., Beerepoot, I., Lu, X., Reijers, H.A., van Dongen, B., Dani, V.S.: Challenges in teaching process mining: insights from process mining educators. In: Proceedings of the International Workshop on Education Meets Process Mining (EduPM 2025). Process Mining Workshops, LNBIP, vol. 575 (2026)
19. van der Aalst, W.M.P.: Process Mining: Discovery, Conformance and Enhancement of Business Processes. Springer, Germany (2011). https://doi.org/10.1007/978-3-642-19345-3
20. Winstone, N., Carless, D.: Implementing peer feedback. In: Designing Effective Feedback Processes in Higher Education: A Learning-Focused Approach, chap. 9. Routledge (2020). https://doi.org/10.4324/9781351115940-9

Simulating the Domain Expert: An Empirical Evaluation of LLM-Based Interview Training for Business Process Modeling

Bedilia Estrada-Torres(✉), José Antonio Parejo, Pablo Fernandez, Manuel Resinas, and Adela del-Río-Ortega

SCORE Lab, I3US, Universidad de Sevilla, Seville, Spain
{iestrada,japarejo,pablofm,resinas,adeladelrio}@us.es

Abstract. Business process discovery relies heavily on stakeholder interviews, yet practicing this skill in educational settings is constrained by limited access to domain experts. This paper empirically evaluates whether LLM-based conversational agents, configured to simulate non-technical domain experts, can effectively support the training of business process elicitation and BPMN modeling skills by providing opportunities for structured practice before engaging with human experts. We conducted a counterbalanced within-subjects pilot experiment with eight Computer Science students, who each interviewed both an LLM-based domain expert and a human expert on two different business process scenarios and subsequently produced a BPMN model. Results show that LLM-based sessions were substantially faster (mean 33.4 vs. 59.5 min), perceived as equally informative, and produced models of comparable quality. Both conditions generated distinct but condition-specific error patterns, which varied across process complexity and gateway type. These findings suggest that LLM-simulated domain experts are a useful training tool for BPM education, supporting repeated practice where expert availability is limited.

Keywords: BPM Education · Simulated Domain Expert · Interview Training · Business Process Modeling · Large Language Models

1 Introduction

Interviews with domain experts are a widely used technique for process discovery in Business Process Management (BPM), providing analysts with insights into how business processes are executed in practice and enabling the generation of accurate process models [2,7,16]. Learning to conduct such interviews effectively requires repeated practice in realistic settings, yet educational contexts rarely provide sufficient opportunities for this. Organizing interviews with domain experts is logistically demanding: class groups are typically large, expert

K. Kluza et al. (Eds.): BPM 2026, LNBIP 600, pp. 279–295, 2027.
https://doi.org/10.1007/978-3-032-36710-5_18

availability is limited, and opportunities for repeated practice are scarce in most academic programs. As a result, students often enter professional environments with limited hands-on experience in process elicitation from human sources.

Recent advances in generative artificial intelligence (AI), particularly in Large Language Models (LLMs), have opened new opportunities across the BPM lifecycle, including process discovery, analysis, and monitoring [8]. In the context of process discovery, most efforts have focused on extracting process elements from textual descriptions or documents (e.g. [3] and [15]). Less attention has been paid to the educational dimension of process discovery, specifically whether LLMs can simulate a domain process expert to provide students with realistic interview practice. Whether such simulations produce learning outcomes comparable to those of human expert interviews remains an open empirical question.

This paper investigates whether LLM-simulated domain experts can provide scalable, repeatable, and pedagogically useful interview practice for BPM education, and how the resulting modeling outcomes compare with those obtained through human-expert chat interviews. To this end, we conducted a counterbalanced within-subjects pilot experiment in which eight Computer Science students interviewed both an LLM-based domain expert and a human expert on two different business process scenarios via text-based chat, subsequently producing a BPMN model based on the elicited information. The LLM-based domain expert was implemented using a custom configurable tool in which instructors can define both the target process and the simulated persona. The agent was prompted to behave as a non-technical domain expert with a defined personality profile, responding only to direct questions, avoiding process modeling terminology, and presenting itself as a human domain expert. Results show that LLM-based interviews were faster, perceived as equally informative, and produced models of comparable quality to those obtained through human expert interviews, and generated distinct error patterns across conditions.

The remainder of the paper is structured as follows. Section 2 reviews background and related work. Section 3 describes the study design. Section 4 presents the results. Section 5 discusses the findings and their pedagogical implications. Section 6 addresses threats to validity, and Sect. 7 concludes the paper.

2 Background and Related Work

Process discovery interviews are a central technique in Business Process Management (BPM), as they allow analysts to elicit how work is actually performed from stakeholders and domain experts. Analysts may proceed forward from triggers to outcomes or backward from results to causes, clarifying inputs, decisions, outputs, and exceptional cases along the way [7]. However, interviews are costly to plan and execute, require active participation from domain experts, and are difficult to repeat in educational settings [4,5]. Similar challenges arise in requirements elicitation practice, where stakeholder availability, communication gaps, and changing needs remain recurrent obstacles [13,22]. Business process models have also been used to support requirements elicitation, reinforcing the close relationship between elicitation competence and process modeling quality [24].

In BPM education, prior work has mainly focused on improving students' modeling performance through pedagogical interventions such as gamification, formative assessment, feedback, and collaborative or individual modeling exercises [11,20,25]. These studies show that model quality is influenced by task complexity, feedback design, and learner characteristics, but give less attention to the elicitation activity through which students gather the information needed to build a model. Interview-based training has been proposed to approximate realistic process modeling conditions [4], yet such activities remain difficult to scale when human experts must repeatedly engage with many students.

Advances in Large Language Models (LLMs) have opened new opportunities across the BPM lifecycle, including process discovery, analysis, and monitoring [8]. Discovery-oriented work has largely focused on extracting process elements from text and supporting conversational process modeling [3,15]. LLMs have also shown potential in broader business modeling tasks such as model creation [12].

Beyond BPM, LLM-based conversational agents are increasingly used to simulate interaction with humans in learning and design contexts. ChEdBot shows how educators can instantiate domain-specific conversational agents for simulation-based learning [18] but they do not explore mechanisms for encoding structured models (such as BPMN) into the agent's conditioning; Conversate uses LLM-supported job interview simulation and dialogic feedback to support reflective interview practice [1]; and medical training studies compare LLM-driven and human-driven virtual conversations in realistic simulations [10]. Recent work on interview-informed generative agents further indicates that LLM simulations may approximate population-level responses [26].

3 Study Design

This study investigates whether LLM-simulated domain expert interviews can support the training of business process discovery and BPMN modeling as a complement to human expert interaction, addressing the following research questions:

- **RQ1.** How does the efficiency of LLM-simulated interviews compare with human expert interviews in terms of session duration and turn frequency?
- **RQ2.** How do students perceive the quality, usefulness, and engagement of LLM-simulated interviews compared with human expert interviews?
- **RQ3.** How does the quality of BPMN models produced after LLM-simulated interviews compare with those produced after human expert interviews?
- **RQ4.** What types of modeling errors emerge under each interview condition?

Efficiency (RQ1) is a measure of the scalability of repeated practice, a central motivation given limited expert availability. Perceived quality (RQ2) captures students' assessment of the tool as a learning resource. Model quality (RQ3) is the closest indicator of elicitation effectiveness. Finally, the identification of condition-specific errors (RQ4) informs how each interview source should be used pedagogically.

We conducted a counterbalanced within-subject experiment in which participants completed two exercises: in one, they interviewed an LLM-based domain expert (BPM-LEIA, *BPM Learning-Enabling Intelligent Assistant* [9]); in the other, they interviewed a human expert via *Microsoft Teams* chat[1] (HUMAN-TEAMS). Each exercise was conducted on a different process scenario (STANDARD and EXTENDED, see Sect. 3.2), and participants produced a BPMN model based on the elicited information. To reduce order effects, the sequence of conditions was counterbalanced across participants.

The study followed a 2×2 counterbalanced within-subjects design, varying both interview condition order and scenario order. Each participant completed both interview conditions, but with different scenarios, thereby reducing systematic order and learning effects while avoiding repeated exposure to the same process scenario. Participants were contacted by email introducing the study and providing access instructions for their first exercise. Upon completion, instructions for the second exercise were provided. For each exercise, they completed a *Google Forms* questionnaire recording start time, exercise order, educational background, and self-reported BPMN knowledge and experience. The questionnaire also included a brief introduction to BPMN notation, modeling best practices, and basic process discovery interview guidelines including identifying task order, sequential and parallel execution, and optional activities. Participants scheduled a one-hour *Microsoft Teams* session with the domain expert to conduct the interview, regardless of whether the human expert condition was completed as the first or second exercise; those assigned to the BPM-LEIA condition accessed the conversational interface directly through a provided link.

After each interview, participants recorded their end time and submitted their BPMN model anonymously as an image file to a shared folder. They then completed a post-exercise questionnaire using a 5-point Likert scale (1 = Totally Disagree, 5 = Totally Agree) covering four dimensions: exercise adequacy (time and difficulty), perceived quality of the domain expert, perceived effect on elicitation and analytical skills, and engagement. Both questionnaires shared equivalent items across the exercise adequacy, perceived quality and elicitation skills dimensions to allow direct comparison, with the BPM-LEIA questionnaire including one additional item on sustained engagement with the tool. Open-ended fields were also provided for participants to report erratic behaviors and general feedback. All study materials are publicly available[2].

3.1 Participants

Ten out of 16 invited students agreed to participate (response rate: 62.5%), comprising eight master's and bachelor's students. Two were subsequently excluded due to incomplete data (one did not submit the models, and another only completed one interview), yielding a final sample of eight valid responses (effective participation rate: 50%). All participants reported prior BPMN knowledge, with

[1] https://www.microsoft.com/es-es/microsoft-teams.
[2] Supplementary material: https://github.com/isa-group/SimulatingDomainExpert.

two indicating high, three intermediate, and three low levels. Practical experience was predominantly limited (three intermediate, five low), making the sample suitable for process modeling training exercises. Table 1 summarizes the eight valid participants (S1–S8, excluding non-respondents and incomplete responses), including degree level, self-reported BPMN knowledge and experience, interview condition order (Human or LLM first), and scenario sequence (STANDARD or EXTENDED first).

Table 1. Participant assignment to conditions and process scenarios

ID	Degree	Knowledge	Experience	Condition Order	First Scenario	Second Scenario
(No resp.)	Master	-	-	Human-BPMLEIA	EXTENDED	STANDARD
S1	Master	High	Intermediate	Human-BPMLEIA	EXTENDED	STANDARD
S2	Master	Low	Low	Human-BPMLEIA	EXTENDED	STANDARD
S3	Master	Low	Low	Human-BPMLEIA	EXTENDED	STANDARD
S4	Master	Intermediate	Intermediate	Human-BPMLEIA	STANDARD	EXTENDED
(Incomplete)	Bachelor	-	-	Human-BPMLEIA	STANDARD	EXTENDED
(No resp.)	Bachelor	-	-	Human-BPMLEIA	STANDARD	EXTENDED
(No resp.)	Bachelor	-	-	Human-BPMLEIA	STANDARD	EXTENDED
S*(Incomplete)	Master	Intermediate	Low	BPMLEIA-Human	EXTENDED	STANDARD
S5	Master	Low	Low	BPMLEIA-Human	EXTENDED	STANDARD
S6	Master	High	Low	BPMLEIA-Human	EXTENDED	STANDARD
(No resp.)	Bachelor	-	-	BPMLEIA-Human	EXTENDED	STANDARD
(No resp.)	Master	-	-	BPMLEIA-Human	STANDARD	EXTENDED
S7	Master	Intermediate	Intermediate	BPMLEIA-Human	STANDARD	EXTENDED
S8	Bachelor	Intermediate	Low	BPMLEIA-Human	STANDARD	EXTENDED
(No resp.)	Bachelor	-	-	BPMLEIA-Human	STANDARD	EXTENDED

3.2 Business Process Scenarios

Two business process scenarios were designed for the study, each representing a realistic operational context from which a BPMN model could be elicited through a stakeholder interview. The processes that informed both the configuration of the BPM-LEIA and the domain expert's knowledge base are illustrated in Figs. 1a and 1b. The complete and detailed descriptions used for each process scenario are provided in the supplementary material.

Hardware Order Fulfillment (STANDARD): encompasses the activities triggered upon receipt of a customer order, whereby the appropriate shipping method is determined based on order volume, third-party logistics providers are engaged as needed, goods are packaged by warehouse staff, and the shipment is dispatched to the carrier for delivery. The reference BPMN model contains 7 activities, 6 gateways (2 exclusive, 2 inclusive, and 2 parallel), and 2 events (start and end).

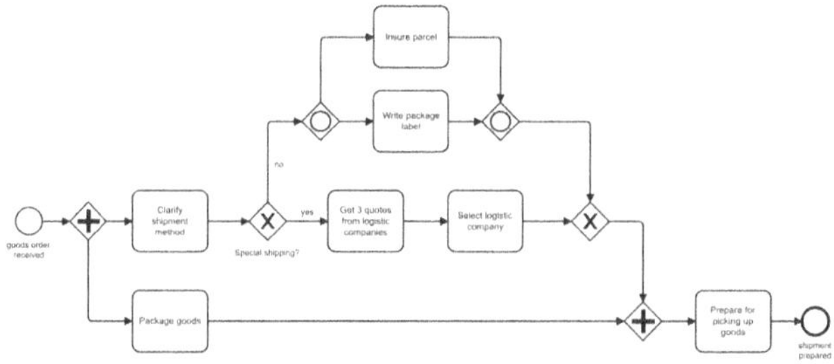

(a) Hardware Order Fulfillment (Standard)

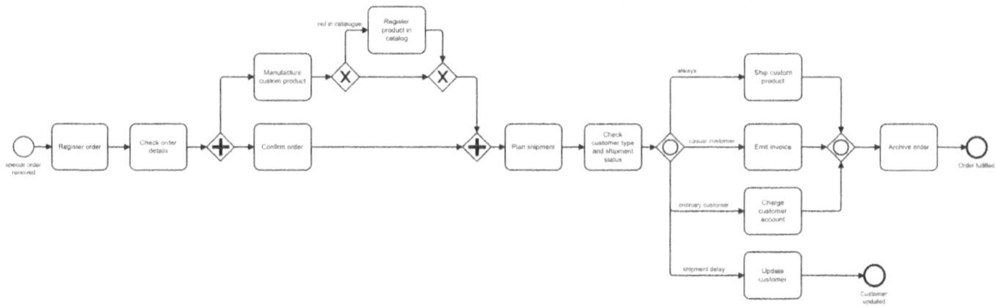

(b) Custom Furniture Order Fulfillment (Extended)

Fig. 1. Processes on the configuration of the BPM-LEIA.

Custom Furniture Order Fulfillment (Extended): encompasses the set of activities triggered upon receipt of a special order, whereby the order is registered and verified, the custom product is manufactured and cataloged if not previously listed, and the shipment is planned accordingly. Billing is handled based on customer type, with ad hoc invoices issued for casual customers and account charges applied for ordinary ones. Shipment delays are communicated to the customer as required. Concurrently, the custom product is shipped, and the process concludes with the archival of the order. The reference BPMN model contains 12 activities, 6 gateways (2 exclusive, 2 inclusive, and 2 parallel), and 2 events (start and end). This is the more complex scenario, with concurrent activities and conditional flows that require careful elicitation to model correctly.

The difference in complexity between the two scenarios is intentional: it allows us to examine whether any observed differences between conditions are consistent across process types or are amplified by complexity.

3.3 Experimental Design

To develop the counterbalanced design, participants were assigned to one of two ordering groups along two dimensions: interview condition order and scenario order. Regarding condition (interview type) order, in the Human-BPMLEIA group participants first interviewed the human expert (HUMAN-teams) and then BPM-LEIA; in the BPMLEIA-Human group, the sequence was reversed.

Regarding scenario order, half of each group started with the *Hardware Order Fulfillment scenario* (STANDARD) and the other half with the *Custom Furniture Order Fulfillment scenario* (EXTENDED). This two-dimensional counterbalancing ensures that neither the interview condition nor the scenario appears systematically first across participants. The assignments were made after the students had agreed to participate, although some ultimately did not submit their responses. All HUMAN-TEAMS interviews were conducted by the same researcher.

3.4 Experimental Conditions

In the *BPM-LEIA condition*, the domain expert was implemented as a conversational agent powered by a large language model (the full architectural details are described in [17]). The BPM-LEIA interface also provided students with a brief exercise context, including a business scenario, a localized setting, and a simulated interviewee profile with motivations and areas of responsibility. Students interacted with BPM-LEIA through a dedicated web-based chat interface with no time limit. Specifically, the BPM-LEIA agent behavior was governed by a set of strict constraints: (i) A passive and reactive stance was mandated in the system prompt, ensuring that information was only disclosed when direct and appropriate questions were posed by the student. This approach was justified by the need to place the full responsibility of the discovery process on the interviewer; by preventing the agent from being proactive, the student's ability to lead a professional inquiry was effectively tested. (ii) In accordance with the experimental design, the prompt was enriched by the inclusion of the BPMN solution (reflecting the reference models described in Sect. 3.2) as an explicit component of the agent's conditioning. This technical reference was supplemented by detailed textual descriptions regarding the functional and non-functional rationale underlying the process model to allow for a deeper level of inquiry. It was intended that if a student were to probe beyond the *what* of a task and ask about the *why*, such as the necessity for specific traceability or the reasons for certain temporal constraints in the hotel industry, the agent would be equipped to provide substantive, context-aware justifications. This setup was established to simulate the depth of knowledge held by a true domain expert, who not only knows the steps of a process but also understands the business objectives and operational limitations that define them. In such a context, a set of guardrails was established to prevent any inappropriate behavior; e.g. any use of technical modeling terminology or BPMN-specific language was strictly prohibited. This restriction was established to force a conceptual translation, where the student was required to map everyday business descriptions onto formal technical structures. Since real-world stakeholders rarely possess knowledge of modeling notations, this gap was intentionally recreated to simulate an authentic engineering environment. (iii) Finally, the agent was instructed to respond using natural, sometimes messy, human prose. This was mandated so that the student would be forced to parse through conversational noise and filler words to identify the underlying business rules. By creating a stakeholder who was characterized as *busy* and *pragmatic*, a sense of professional pressure was introduced, ensuring that only clear and well-prepared questions were rewarded with useful information.

Conversely, in the *HUMAN-TEAMS condition*, one of the authors played the same role via *Microsoft Teams* chat, using the same process knowledge and behavioral guidelines provided to BPM-LEIA. Each session began with the same contextual information available in the BPM-LEIA interface: the business scenario, a localized setting, and the interviewee profile. Interviews were conducted entirely in writing to maintain consistency with the BPM-LEIA condition and were aimed to last under one hour.

3.5 Data Collection and Evaluation

BPMN models were evaluated against the formal reference models, following established approaches to process model quality [19,23], along two dimensions: *element count accuracy*, measured as deviation from the reference in number of activities, total gateways, and gateway type distribution; and *control flow correctness*, assessed qualitatively based on overall flow structure, activity positioning, and loop and gateway usage. A total of 17 models were submitted; one was discarded as the participant did not complete both exercises (see Sect. 3.1), yielding 16 evaluable models (8 participants $\times$ 2 conditions).

Post-exercise questionnaire responses were analyzed descriptively, computing mean scores per item and condition. Interview transcripts from both conditions were retained as supplementary qualitative evidence and are used selectively to illustrate linguistic patterns that may account for observed differences in modeling outcomes. No systematic discourse analysis was performed.

4 Results

This section reports the results of the comparative study on interview efficiency, student perceptions, BPMN model quality, and error patterns, structured by research question and compared across the BPM-LEIA and HUMAN-TEAMS conditions for the STANDARD and EXTENDED scenarios.

4.1 Interview Efficiency

This section addresses RQ1: *How does the efficiency of LLM-simulated interviews compare with human expert interviews in terms of session duration and turn frequency?* BPM-LEIA-based interview sessions were consistently shorter than HUMAN-TEAMS sessions. The mean conversation duration was 33.4 min for the BPM-LEIA condition (range: 4–61 min) and 59.5 min for the human expert condition (range: 38–70 min), a difference of approximately 26 min. The median duration further confirms this pattern: 26.5 min for BPM-LEIA versus 62 min for HUMAN-TEAMS, a difference of 35.5 min. Although one of the sessions lasted only 4 min, the student reports having covered all the elements needed for process modeling.

Despite the shorter duration, BPM-LEIA sessions involved a slightly higher number of student turns on average (13.3 vs. 10.1), with comparable medians (12.5 vs. 10), suggesting that students asked more questions in less time. This is consistent with the almost immediate responses of the LLM compared to the inherent delays of a human chat exchange. The conversation data used to compute these values is available in the supplementary material.

4.2 Student Perceptions

This section addresses RQ2: *How do students perceive the quality, usefulness, and engagement of LLM-simulated interviews compared with human expert interviews?* Fig. 2 shows the response distributions across all items and conditions; mean scores are reported in the text. Detailed descriptive statistics are provided in the supplementary materials.

Within the *information source* block, perception of the domain expert was high and comparable across conditions, with identical scores for 'enough information provided' (mean 4.62 in both conditions) and small differences for 'provided useful information' (4.62 vs. 4.88) and 'no BPMN jargon used' (4.50 vs. 4.88). The jargon item shows the largest divergence and the highest response variability in the BPM-LEIA condition (min = 1, max = 5), driven by a single participant who strongly disagreed and suggested that BPM-LEIA should 'specify parallelism or conditionals' more explicitly (S6). This is consistent with BPM-LEIA's intended configuration, which deliberately avoids process modeling terminology. Another participant reported that BPM-LEIA initially described two tasks as sequential and later corrected this to parallel when pressed (S4), suggesting occasional coherence failures that may account for lower jargon scores.

Within the *elicitation skills* block, scores were high across all five items in both conditions (range 4.25–5.00). The HUMAN-TEAMS condition scored higher on most items, with the largest gap on 'improved open-ended questioning' (4.62 vs. 5.00), the only item where all HUMAN-TEAMS participants responded *Totally Agree*, followed by 'improved communication precision' (4.50 vs. 4.75) and 'improved closed-ended questioning' (4.62 vs. 4.88). 'Improved active listening' was identical across conditions (4.75 vs. 4.75), and 'improved assumption questioning' showed the smallest difference (4.25 vs. 4.38), with both conditions sharing the same minimum score of 3. One participant explicitly identified a potential limitation of BPM-LEIA regarding open-ended questioning: 'it provides a lot of information, even though questions were open ended. This could hinder elicitation abilities of the business process analyst' (S1).

Within the *engagement* block, BPM-LEIA outperformed the HUMAN-TEAMS condition on both comparable items: 'felt engaged' (4.88 vs. 4.62) and 'interaction was motivating' (4.88 vs. 4.62), with a minimum score of 4 across both conditions and both items. The BPM-LEIA questionnaire additionally included 'tool kept interest', which scored 4.88 with no responses below 4, though this item is not directly comparable to the human expert condition. Two participants described the interaction as 'realistic' (S8) and 'more fluid' than the

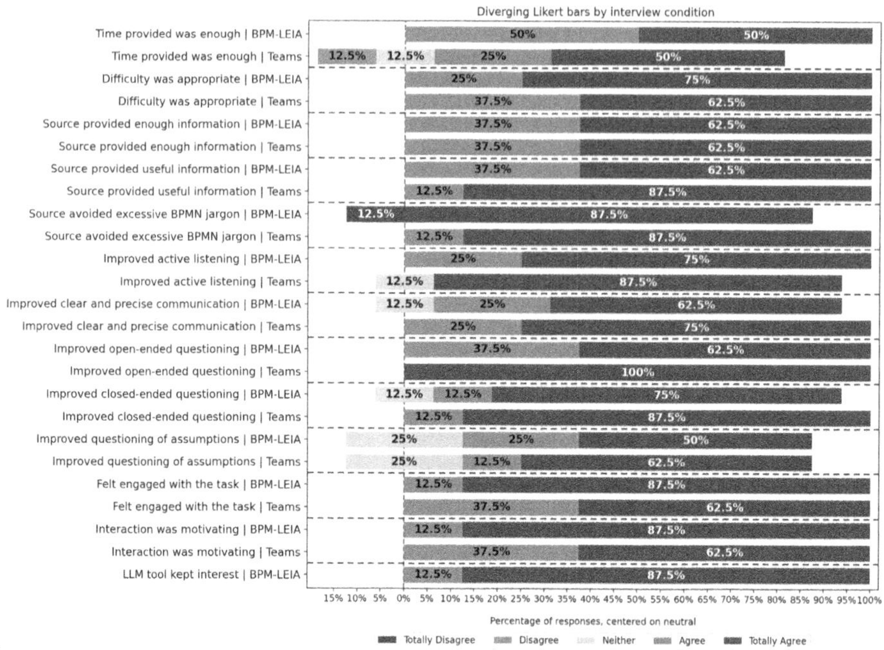

Fig. 2. Diverging Likert bars by condition.

human expert (S3), though S8 suggested the agent could be 'a bit more clumsy' and S5 noted that immediate answers may be 'less informative'.

Cronbach's alpha was computed for each block to assess the internal consistency of the items [6], where values above 0.70 are generally considered acceptable [21]. For the HUMAN-TEAMS condition, alpha was acceptable for exercise adequacy ($\alpha = 0.77$), information source quality ($\alpha = 0.81$), and elicitation skills ($\alpha = 0.75$), and lower for engagement ($\alpha = 0.64$). For the BPM-LEIA condition, exercise adequacy was also acceptable ($\alpha = 0.73$), elicitation skills was moderate ($\alpha = 0.54$), and engagement was perfect ($\alpha = 1.00$), which likely reflects very low response variability rather than strong evidence of a high engagement capability of the BPM-LEIA. The information-source block for BPM-LEIA produced a negative alpha ($\alpha = -0.47$), indicating poor internal consistency, largely due to the outlier response on the jargon item (S6: Totally Disagree vs. all others: Agree or Totally Agree). These reliability estimates should be interpreted cautiously because the sample is very small and several items have restricted variance. To compare paired Likert responses between conditions, Wilcoxon signed-rank tests were used as the primary statistical test, as they are appropriate for ordinal paired data without assuming normality [27]. Paired t-tests and sign tests were included as sensitivity checks. After Holm correction for multiple comparisons [14], no item-level or block-level comparison reached statistical significance, which is consistent with the small sample size.

Overall, student perceptions were high and broadly comparable across both conditions, with differences small in absolute terms and none reaching statistical significance. The HUMAN-TEAMS condition scoring marginally higher on elicitation skills and BPM-LEIA scoring slightly higher on engagement. The raw and clean datasets and the Python analysis scripts are available in the supplementary material.

4.3 BPMN Model Quality

This section addresses RQ3: *How does the quality of BPMN models produced after LLM-simulated interviews compare with those produced after human expert interviews?*. Table 2 summarizes the element counts of the 16 submitted models against the ground truth for both scenarios and conditions, and Fig. 3 shows the corresponding boxplot distributions.

Regarding activities, BPM-LEIA models deviated more from the ground truth in the STANDARD scenario (mean $\Delta = +2.00$) than HUMAN-TEAMS models (mean $\Delta = +1.25$), while in the EXTENDED scenario both conditions were comparable (BPM-LEIA $\Delta = +0.50$, HUMAN-TEAMS $\Delta = +1.00$). Individual models ranged from exact matches (S8-BPM-LEIA-STANDARD, $\Delta = 0$) to substantial overestimation (S2-BPM-LEIA-STANDARD, $\Delta = +4$). Regarding total gateway count, HUMAN-TEAMS models were notably more accurate in the STANDARD scenario (mean $\Delta = -0.25$, nearly exact) compared to BPM-LEIA (mean $\Delta = +0.75$). In the EXTENDED scenario, both conditions overestimated gateways, though more severely in the BPM-LEIA condition (mean $\Delta = +4.25$ vs. $\Delta = +2.33$ for HUMAN-TEAMS), suggesting that process complexity is a primary driver of gateway inflation regardless of interview condition.

Gateway type distribution reveals two contrasting patterns. First, exclusive gateways (X-GW) were overestimated more strongly in HUMAN-TEAMS-EXTENDED (mean $\Delta = +3.33$) than in BPM-LEIA-EXTENDED (mean $\Delta = +0.50$), suggesting that human expert interviews may lead students to over-structure conditional branching in complex processes. Second, the inclusive gateway (O-GW) count in EXTENDED shows a clear difference: HUMAN-TEAMS-EXTENDED models produced a mean of 0.33 O-GW (values: 0, 1, 0), while BPM-LEIA-EXTENDED produced a mean of 3.75. However, this mean is mainly caused by a single extreme case: S4-BPM-LEIA-EXTENDED produced 14 inclusive gateways against a ground truth of 2, while the remaining three BPM-LEIA-EXTENDED models produced 0, 1, and 0 respectively. Excluding S4, the BPM-LEIA-EXTENDED mean drops to 0.33, identical to HUMAN-TEAMS-EXTENDED. The outlier in the O-GW distribution for BPM-LEIA-EXTENDED visible in Fig. 3 corresponds to S4, whose model is discussed further in Sect. 5.

4.4 Error Patterns

This section addresses RQ4: *What types of modeling errors emerge under each interview condition?* A qualitative assessment of each submitted model was con-

Table 2. BPMN model element counts by participant, condition, and scenario. Ground truth: STANDARD = 7 activities, 6 gateways (2 X, 0 O, 2 +), 1 start, 1 end; EXTENDED = 12 activities, 6 gateways (2 X, 2 O, 2 +), 1 start, 2 ends.

ID	Cond.	Scen.	Order	Act	ΔAct	Tot. GW	ΔGW	X-GW	O-GW	+-GW	Start	End
HOF STANDARD scenario—ground truth: 7 act, 6 GW (2X, 0O, 2+), 1 start, 1 end												
S8	BPM-LEIA	STANDARD	LH	7	0	6	0	4	0	2	1	1
S1	BPM-LEIA	STANDARD	HL	10	+3	10	+4	4	0	6	1	1
S2	BPM-LEIA	STANDARD	HL	11	+4	5	-1	3	0	2	1	1
S3	BPM-LEIA	STANDARD	HL	8	+1	6	0	4	2	2	1	1
S4	HUMAN-TEAMS	STANDARD	HL	7	0	6	0	4	0	2	1	1
S5	HUMAN-TEAMS	STANDARD	LH	9	+2	6	0	2	2	2	1	1
S6	HUMAN-TEAMS	STANDARD	LH	8	+1	5	-1	3	0	2	1	1
S7	HUMAN-TEAMS	STANDARD	LH	9	+2	6	0	5	0	1	1	1
Mean BPM-LEIA-STANDARD				9.00	+2.00	6.75	+0.75	3.75	0.50	3.00	1.00	1.00
Mean HUMAN-TEAMS-STANDARD				8.25	+1.25	5.75	-0.25	3.50	0.50	1.75	1.00	1.00
CFOF EXTENDED scenario—ground truth: 12 act, 6 GW (2X, 2O, 2+), 1 start, 2 ends												
S4	BPM-LEIA	EXTENDED	HL	16	+4	18	+12	0	14	4	1	2
S5	BPM-LEIA	EXTENDED	LH	12	0	10	+4	6	0	4	1	1
S6	BPM-LEIA	EXTENDED	LH	12	0	6	0	1	1	4	1	1
S7	BPM-LEIA	EXTENDED	LH	10	-2	7	+1	3	0	4	1	1
S8	HUMAN-TEAMS	EXTENDED	LH	13	+1	10	+4	6	0	4	1	1
S1	HUMAN-TEAMS	EXTENDED	HL	13	+1	7	+1	4	1	2	1	1
S2	HUMAN-TEAMS	EXTENDED	HL	14	+2	8	+2	6	0	2	1	1
S3	HUMAN-TEAMS	EXTENDED	HL	12	0	—	—	—	—	—	—	—
Mean BPM-LEIA-EXTENDED				12.50	+0.50	10.25	+4.25	2.50	3.75*	4.00	1.00	1.25
Mean HUMAN-TEAMS-EXTENDED				13.00	+1.00	8.33	+2.33	5.33	0.33	2.67	1.00	1.00

LH = BPM-LEIA first, then HUMAN-TEAMS; HL = HUMAN-TEAMS first, then BPM-LEIA

Δ values indicate deviation from ground truth (+ = overestimation, - = underestimation).

S3 HUMAN-TEAMS-EXTENDED: gateway and event data unavailable from submitted model.

* Mean O-GW for BPM-LEIA-EXTENDED driven by a single extreme case (S4: 14 O-GW); excluding S4, mean = 0.33.

ducted across three dimensions: activity quality, loop and flow structure, and notable errors. Full details are provided in the supplementary materials.

Four error patterns emerge from the assessment. First, confusion between activities and start events appeared in multiple models regardless of condition (S8, S1, S2 in BPM-LEIA-STANDARD; S8, S1 in HUMAN-TEAMS-EXTENDED), suggesting a BPMN notation gap rather than an effect of the interview source. Second, incorrect activity positioning was the most frequent error in the STANDARD scenario across both conditions, with *Package goods* and *Generate label* consistently misplaced, indicating that sequential ordering is the most difficult aspect to elicit correctly in this process. Third, several participants in both conditions introduced unnecessary BPMN elements such as lanes, pools, intermediate events, typed activities, and timers, further suggesting general gaps in BPMN training rather than condition-specific effects. Fourth, flow correctness in the EXTENDED scenario differed between conditions: S2 and S3-HUMAN-TEAMS-EXTENDED showed multiple non-conforming sections and poorly structured loops, while most BPM-LEIA-EXTENDED models captured the overall process meaning despite structural irregularities. Conversely, activity quality in

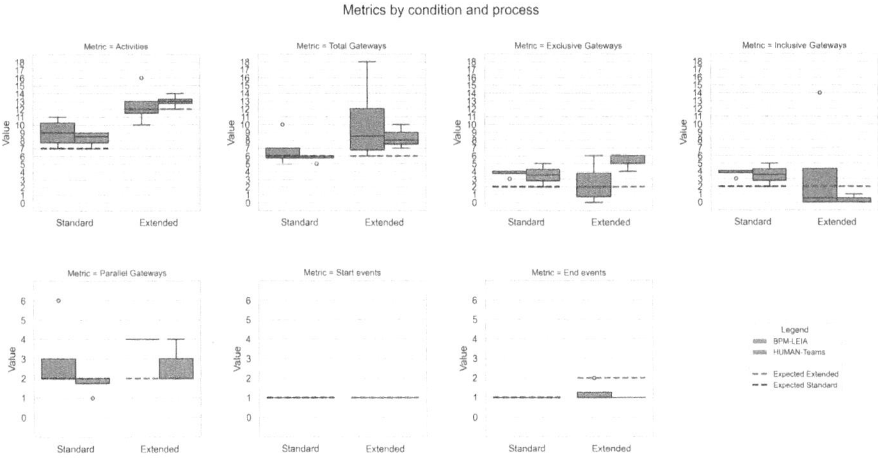

Fig. 3. Distribution of BPMN element counts by condition and scenario

the STANDARD scenario was better in the HUMAN-TEAMS condition, where two of four models were rated Good, compared to none in the BPM-LEIA condition.

5 Discussion

Session duration was substantially shorter in the BPM-LEIA condition (mean 33.4 vs. 59.5 min), a difference participants attributed to response speed rather than information quality (S1, S3, S5). Perceived information quality was comparable across conditions, and overall model quality showed no consistent advantage for either condition. Given the small sample size, these findings should be interpreted as exploratory trends rather than definitive conclusions.

Both conditions produced distinct error patterns. In the BPM-LEIA condition, gateway count deviation was higher in the EXTENDED scenario (mean $\Delta = +4.25$), driven largely by a single model with an extreme overuse of inclusive gateways (S4), possibly linked to BPM-LEIA's use of explicit concurrency markers such as "simultaneously" and "concomitantly", though participant experience cannot be ruled out. Conversely, the HUMAN-TEAMS condition produced stronger overestimation of exclusive gateways in the EXTENDED scenario (mean $\Delta = +3.33$ vs. $+0.50$ for BPM-LEIA) and more severe control flow errors in the same scenario, suggesting that human expert interviews do not necessarily yield more accurate models in the EXTENDED scenario. Errors independent of condition, such as confusion between activities and start events, point to gaps in BPMN training rather than interview source effects.

Participants in the BPM-LEIA condition perceived slightly less improvement in open-ended questioning compared to the HUMAN-TEAMS condition (4.62 vs. 5.00), suggesting that BPM-LEIA's comprehensive responses may reduce the need to refine questioning strategies. BPM-LEIA's behavioral profile is configurable: the degree of helpfulness, response verbosity, and persona characteristics can be adjusted to simulate a wider range of domain expert styles. Varying BPM-LEIA's configuration to introduce less helpful personas represents a promising direction for better replicating real-world elicitation challenges in future training scenarios. This trade-off is also reflected in student feedback (see Sect. 4.2).

In summary, BPM-LEIA appears to offer a faster and at least equally engaging training interaction, while HUMAN-TEAMS may better support the development of open-ended questioning skills. Perceived information quality is broadly comparable, although BPM-LEIA shows more variability in the jargon and coherence-related items.

6 Threats to Validity

Several limitations should be considered when interpreting the results.

Regarding *internal validity*, the small sample size (n = 8) limits the conclusiveness of the findings, the counterbalanced design mitigates but does not fully eliminate order effects, and ground truth models were defined by the researchers, introducing potential evaluation bias. All HUMAN-TEAMS interviews were conducted by a single researcher, which ensured consistency in domain knowledge and behavioral guidelines but introduces a potential interviewer effect, including communication style, rapport, and response latency, that cannot be separated from the condition itself. BPMN model quality was assessed by a single instructor, which may introduce evaluator bias, particularly in the qualitative assessment of control-flow correctness, activity positioning, and gateway usage, which involved expert judgement. This threat was mitigated through a structured assessment protocol combining objective and predefined evaluation criteria. Quantitative element counts were derived directly from the submitted models and objectively compared against the reference models, reducing reliance on subjective interpretation for part of the assessment. In addition, qualitative evaluation followed predefined reference models and explicit dimensions, including activity correctness, gateway use, loop and flow structure, and notable modeling errors. These measures established a consistent evaluation framework and constrained subjective interpretation by grounding assessments in predefined modeling expectations rather than ad hoc judgment. Nevertheless, future replications should include blind assessment by at least two independent evaluators and report inter-rater agreement (e.g., Cohen's κ).

Regarding *external validity*, participants were Computer Science students from a single institution, and both scenarios belong to service and manufacturing domains, limiting generalizability to other populations, curricula, and process types. Finally, regarding *construct validity*, model quality was assessed through element counts and a qualitative review covering activity correctness, flow structure, and notation use; dimensions such as model readability and elicitation completeness were not systematically evaluated. Furthermore, the study measures single-session performance per condition, not learning gains over time.

7 Conclusions and Future Work

This paper presented an exploratory empirical evaluation of LLM-simulated domain experts for training process-discovery interviews and BPMN modeling. In a counterbalanced within-subjects study with eight students, LLM-based interviews were notably faster than human-expert chat interviews and were perceived as similarly informative and engaging. The resulting BPMN models showed broadly comparable quality, although both conditions produced distinct error patterns in gateway selection, activity positioning, and control-flow structuring.

These findings suggest that LLM-simulated domain experts can complement human expert interaction in BPM education by enabling scalable and repeatable practice. They should not be interpreted as a replacement for authentic stakeholder interaction, but as a configurable training environment for repeated elicitation practice. Future work should replicate the study with larger and more diverse cohorts, evaluate learning gains through pre/post-test designs and longitudinal follow-up, and examine how different simulated expert personas affect students' questioning and modeling outcomes.

Acknowledgments. This paper is part of the projects PID2024-155693NB-C44, PDC2025-166101-C21, PDC2025-166101-C22, PID2024-156482NB-I00 and PID2022-140221NB-I00 funded by MICIU/AEI/10.13039/501100011033 and by ERDF/EU, and by the Junta de Andalucía and co-funded by the EU through the FEDER Andalucía 2021–2027 Programme under project DGP_PIDI_2024_01609 and by the Horizon Europe programme under grant agreement No. 101292597, project EOSC-ARENA.

References

1. Daryanto, T., et al.: Conversate: supporting reflective learning in interview practice through interactive simulation and dialogic feedback. ACM Hum. Comput. Interact. **9**(1), 1–32 (2025). https://doi.org/10.1145/3701188
2. Bano, M., Zowghi, D., Ferrari, A., Spoletini, P., Donati, B.: Teaching requirements elicitation interviews: an empirical study of learning from mistakes. Requirements Eng. **24**(3), 259–289 (2019). https://doi.org/10.1007/S00766-019-00313-0
3. Bellan, P., Dragoni, M., Ghidini, C.: Extracting business process entities and relations from text using pre-trained language models and in-context learning. In: Proceedings of EDOC 2022. LNCS, vol. 13585, pp. 182–199 (2022). https://doi.org/10.1007/978-3-031-17604-3_11
4. Candrlic, S., Pavlic, M., Katic, M.A.: Interviewing model to enhance process modelling education. In: EDULEARN20 Proceedings, pp. 6605–6613. IATED (2020)
5. de Castro, D.S., Fantinato, M., Costa, M.B.: Business process design support with automated interviews. In: Proceedings of ICEIS 2025, vol. 2, pp. 734–745. SCITEPRESS (2025). https://doi.org/10.5220/0013362000003929
6. Cronbach, L.J.: Coefficient alpha and the internal structure of tests. Psychometrika **16**(3), 297–334 (1951)
7. Dumas, M., Rosa, M.L., Mendling, J., Reijers, H.A.: Fundamentals of Business Process Management, 2nd edn. Springer (2018)

8. Estrada-Torres, B., del-Río-Ortega, A., Resinas, M.: Mapping the landscape: exploring large language model applications in business process management. In: Enterprise, Business-Process and Information Systems Modeling BPMDS-EMMSAD 2024. LNBIP, vol. 511, pp. 22–31 (2024). https://doi.org/10.1007/978-3-031-61007-3_3
9. Estrada-Torres, B., Parejo, J.A., Fox, A., Fernández, P.: A training framework for BPM modeling interviews based on LLM assistants. In: Joint Proceedings of the Best Dissertation Award, Doctoral Consortium, and Demonstration & Resources Forum at BPM 2025 co-located with 23rd International Conference BPM (2025), vol. 4032, pp. 280–287. CEUR-WS.org (2025). https://ceur-ws.org/Vol-4032/paper-37.pdf
10. Ferreira, W., Negrão, M.D., Maciel, A., Torchelsen, R.P., Nedel, L.: LLM vs. human driven conversation: a user study in virtual reality medical simulation. In: Proceedings of 27th Symposium on Virtual and Augmented Reality, SVR 2025, pp. 527–535. IEEE (2025). https://doi.org/10.1109/SVR67689.2025.00073
11. Garaccione, G., Coppola, R., Ardito, L., Torchiano, M.: Gamification of business process modeling education: an experimental analysis. Softw. Syst. Model. **23**, 1569–1594 (2024). https://doi.org/10.1007/s10270-024-01171-3
12. Görgen, L., Müller, E., Triller, M., Nast, B., Sandkuhl, K.: Large language models in enterprise modeling: case study and experiences. In: Proceedings of MODELSWARD 2024, pp. 74–85. SCITEPRESS (2024). https://doi.org/10.5220/0012387000003645
13. Hidalgo, M., Yanine, F., Paredes, R., Frez, J., Solar, M.: What is the process? A metamodel of the requirements elicitation process derived from a systematic literature review. Processes **13**(1), 20 (2025). https://doi.org/10.3390/pr13010020
14. Holm, S.: A simple sequentially rejective multiple test procedure. Scand. J. Stat. **6**(2), 65–70 (1979)
15. Klievtsova, N., Benzin, J.V., Kampik, T., Mangler, J., Rinderle-Ma, S.: Conversational process modelling: state of the art, applications, and implications in practice. In: BPM Forum, vol. 490, pp. 319–336 (2023). https://doi.org/10.1007/978-3-031-41623-1_19
16. Leyh, C., Bley, K., Seek, S.: Elicitation of processes in business process management in the era of digitization - the same techniques as decades ago? In: Innovations in Enterprise Information Systems Management and Engineering. Proceedings of 5th International Conference on ERP Future 2016. LNBIP, vol. 285, pp. 42–56 (2017). https://doi.org/10.1007/978-3-319-58801-8_4
17. Lojo, N., et al.: C-LEIA: an interactive LLM-based simulator for requirements interviews. In: 33rd International Requirements Engineering Conference Workshops (REW), pp. 59–61. IEEE (2025)
18. Martin, A., Pande, C., Witschel, H.F., Mathez, J.: ChEdBot: designing a domain-specific conversational agent in a simulational learning environment using LLMs. In: Proceedings of AAAI Spring Symposium Series, vol. 3, pp. 180–187 (2024). https://doi.org/10.1609/AAAISS.V3I1.31198
19. Mendling, J.: Metrics for Process Models: Empirical Foundations of Verification, Error Prediction, and Guidelines for Correctness. LNBIP, vol. 6. Springer (2008). https://doi.org/10.1007/978-3-540-89224-3
20. Nobre, M.R.B., Vilela, J., de Barros, L.M.: Model quality and student satisfaction in BPMN education: a quasi-experiment. In: Proceedings of 17th International Conference on Computer Supported Education, vol. 1, pp. 620–627. SCITEPRESS (2025). https://doi.org/10.5220/0013483100003932

21. Nunnally, J.C., Bernstein, I.H.: Psychometric Theory, 3rd edn. McGraw-Hill, New York (1994)
22. Palomares, C., Franch, X., Quer, C., Chatzipetrou, P., López, L., Gorschek, T.: The state-of-practice in requirements elicitation: an extended interview study at 12 companies. Requirements Eng. **26**(2), 273–299 (2021). https://doi.org/10.1007/s00766-020-00345-x
23. Rolón, E., Ruiz, F., García, F., Piattini, M.: Applying software metrics to evaluate business process models. CLEI Electron. J. **9**(1), 1–14 (2006)
24. Valvas, S., Milani, F.: Requirement elicitation using business process models. In: Matulevičius, R., Dumas, M. (eds.) BIR 2015. LNBIP, vol. 229, pp. 67–81. Springer, Cham (2015). https://doi.org/10.1007/978-3-319-21915-8_5
25. Vemuri, P., Poelmans, S., Compagnucci, I., Snoeck, M.: Using formative assessment and feedback to train novice modelers in business process modeling. In: Proceedings of MODELS 2023 Companion, pp. 130–137. MODELS-C 2023, IEEE (2023). https://doi.org/10.1109/MODELS-C59198.2023.00037
26. Wang, Z., Siu, A.: Interview-informed generative agents for product discovery: a validation study. In: Proceedings of Conference on Human Factors in Computing Systems, CHI 2026, pp. 1612:1–1612:27 (2026). https://doi.org/10.1145/3772318.3791918
27. Wilcoxon, F.: Individual comparisons by ranking methods. Biometrics Bull. **1**(6), 80–83 (1945)

Does AI Improve BPMN Modelling? Evidence from Iterative Assignments and Critical Verification Practice

Marzena Grzesiak[1](✉), Piotr Sliż[2], Leszek Szała[3], and Marek Moszyński[1]

[1] Gdańsk University of Technology, 80-233 Gdańsk, Poland
{marzena.grzesiak,marek.moszynski}@pg.edu.pl
[2] University of Gdańsk, Gdańsk, Poland
piotr.sliz@ug.edu.pl
[3] University of Chemistry and Technology, Praha, Czechia
leszek.szala@vscht.cz

Abstract. This paper investigates whether and under what conditions AI assistance improves BPMN modelling in higher education. Building on research in BPM education, active learning, and LLM-assisted modelling, authors designed an iterative two-stage assignment in which students first submitted an initial BPMN model (V1) and then refined it using a rubric-aligned verification checklist, scenario-based test cases, and a short change log (V2). A volunteer subgroup additionally produced an AI-assisted alternative final model (V2–AI), accompanied by an AI Usage Statement capturing intentions, verification behaviors, and reasons for accepting or rejecting AI suggestions. Results show substantial learning gains between V1 and V2, particularly in logic-intensive areas such as gateways and events. Error frequencies decreased across most categories, demonstrating the effectiveness of structured scaffolding for guiding iterative improvement. The comparison of V2 and V2–AI reveals heterogeneous outcomes: while some students benefited from AI support, others introduced new inconsistencies when relying uncritically on AI-generated suggestions. Qualitative analysis of AI Usage Statements indicates that critical verification, rather than intensity of AI use, is the key determinant of high-quality AI-assisted modelling. The primary contribution is empirical evidence on learning gains through iterative revision and structured verification. As secondary, exploratory contributions, the study provides: (2) a fine-grained error taxonomy for analysing student process models, and (3) the first paired comparison of human-refined and AI-assisted BPMN models in higher education. The proposed instructional design offers a scalable, responsible approach for integrating critical AI use into BPM education.

Keywords: BPMN · Process Modelling · Business Process Management Education · AI · BPM · Education · LLM · Learning Gains · Active learning · Iterative assignment design

K. Kluza et al. (Eds.): BPM 2026, LNBIP 600, pp. 296–310, 2027.
https://doi.org/10.1007/978-3-032-36710-5_19

1 Introduction

Business Process Model and Notation (BPMN) is one of the core modelling notations used in Business Process Management (BPM), providing a standard graphical language for representing and communicating business processes [11]. This educational importance is further reinforced by evidence from the BPM education context, where standardizing the teaching of business process modelling in BPMN and introducing certification of BPMN modelling skills have been proposed as key measures to reduce the gap between university curricula and labour market expectations [12]. Despite its widespread inclusion in BPM curricula, the authors' teaching experience in BPM indicates that novice modellers often struggle with key BPMN concepts, including gateway logic, event semantics, message flows, and end-to-end process consistency. These difficulties often persist even in well-structured courses, suggesting that BPMN education should move beyond syntactic familiarity and place greater emphasis on conceptual understanding and reasoning about process logic.

In parallel to these longstanding educational challenges, recent advances in artificial intelligence (AI), particularly large language models (LLMs), have rapidly entered the BPM domain [13]. Consequently, AI is increasingly viewed as a promising way to address common difficulties encountered by students in process modelling. Yet, while AI-based tools may speed up diagram creation and help maintain syntactic correctness, the authors contend that they can also generate subtle semantic inconsistencies, especially in logic-intensive aspects of BPMN modelling, such as gateway logic, exception handling, and collaboration semantics. This leads to a critical educational question that this paper seeks to explore: does AI genuinely enhance students' BPMN modelling competence, or does it mainly transfer the cognitive effort from model construction to the *post hoc* correction of AI-generated outputs? This question reflects broader concerns about the responsible use of AI in education, as emphasized in international policy discussions [10].

Despite growing interest in AI-supported BPMN modelling, several gaps remain insufficiently addressed. First, there are limited empirical studies quantifying learning gains in BPMN modelling achieved through short, iterative instructional cycles supported by structured verification scaffolds. Second, existing studies rarely compare human-refined and AI-assisted BPMN models produced by the same students, making it difficult to isolate the educational contribution of AI. Third, little is known about how students' verification behaviours, particularly their ability to accept, adapt, or reject AI suggestions, relate to observed changes in model quality. To address these gaps, this paper investigates an iterative BPMN modelling assignment implemented in a higher education BPM course.

The main aim of this study is to examine whether AI contributes to improved BPMN model quality beyond gains achieved through iterative modelling and structured verification, and to identify the conditions under which AI support is educationally beneficial. By combining iterative assignments, fine-grained error analysis, and explicit attention to critical AI use, the study contributes empirical evidence and practical insights relevant to BPM educators seeking to integrate AI responsibly into BPMN teaching.

The present study follows a design-based research logic, a well-established approach in educational research for investigating and refining instructional interventions in authentic learning settings [8,9]. Rather than testing narrowly defined causal hypotheses, this approach focuses on understanding learning mechanisms, evaluating instructional designs in authentic settings, and deriving transferable design principles. Building on empirical research in BPMN modelling education [15,16] and recent advances in AI-assisted process modelling [13,14], the study investigates an iterative BPMN assignment design: an initial student-created model (V1), a revised model supported by structured verification scaffolds (V2), and an optional AI-assisted alternative final version (V2–AI). This structure enables the separation of learning effects resulting from iteration and verification from potential additional effects associated with AI assistance.

To capture this dimension, the study includes a lightweight AI Usage Statement that documents students' intentions, verification steps, and decisions to accept or reject AI suggestions.

Based on the identified research gaps and the instructional design described above, the study is guided by the following four research questions (RQ):

- **RQ1**: *Does AI improve BPMN modelling quality beyond the learning gains achieved through iterative modelling and structured verification in BPM higher education?*
- **RQ2**: *To what extent does a two-stage iterative BPMN assignment (V1–V2), supported by a verification checklist and scenario-based test cases, improve students' BPMN model quality?*
- **RQ3**: *How does the quality of AI-assisted BPMN models (V2–AI) compare to human-refined models (V2) produced by the same students?*
- **RQ4**: *How are students' verification behaviours and declared approaches to AI use related to observed changes in BPMN model quality and error patterns?*

Together, these questions enable an integrated examination of learning gains, AI-related effects, and verification practices, providing both empirical evidence and pedagogically relevant insights for BPM educators seeking to integrate AI responsibly into BPMN teaching.

2 Background: AI-Supported BPMN Modelling

Recent AI-based BPMN modelling tools increasingly offer features such as natural language-driven model generation, automated compliance checks, and governance-oriented constraints that improve layout consistency and syntactic validity [2,7]. AI-supported BPMN modelling indicates a shift from *one-shot* process model generation toward interactive tools that support the creation, modification, and validation of BPMN diagrams through natural language. *BPMN Assistant*, for example, uses LLMs to bridge textual process descriptions and formal BPMN representations, relying on a structured intermediate representation to improve reliability, editing efficiency, and semantic consistency. Its evaluation suggests that such tools can reduce the technical barrier to BPMN modelling and support more responsive *human–AI* interaction. However, the reported limitations also show that current AI-based BPMN tools remain constrained by the supported subset of BPMN elements, the absence of full collaboration modelling with pools and lanes, dependence on clear and unambiguous natural language input, and considerable variation in performance across LLMs. These limitations are particularly relevant for BPMN education, where the goal is not only to generate diagrams efficiently, but also to develop students' ability to understand, verify, and critically refine AI-generated process models [6].

Research on LLM-based process modelling shows that generative AI can support the generation and iterative refinement of process models from textual descriptions. Kourani et al. [13] propose a framework that combines prompting strategies, a safety- and quality-oriented model generation protocol, error handling, and a user feedback mechanism to refine generated process models. Their evaluation indicates that GPT-4 generated process models efficiently, resolved adjustable errors, and incorporated user feedback effectively, whereas Gemini produced less reliable results and encountered difficulties in error resolution and feedback integration. Importantly, the authors show that the use of partially ordered workflow language (POWL) as an intermediate process representation helps ensure that the generated models remain sound and executable before being exported to standard notations such as BPMN or Petri nets. At the same time, they explicitly acknowledge that manual effort remains necessary for validating generated models and providing effective feedback. From an educational perspective, this suggests that AI should not be treated as a replacement for BPMN modelling competence, but rather as a tool that increases the importance of critical verification, refinement, and semantic validation of AI-generated process models [13].

Licardo, Tanković, and Etinger [14] propose an NLP- and LLM-based method for extracting BPMN models from textual descriptions, combining a fine-tuned BERT model, GPT-3.5-Turbo, GPT-4, and Graphviz-based visualisation. Their evaluation of 31 textual descriptions showed higher accuracy for GPT-4 than for GPT-3.5-Turbo, 96% and 80% respectively. However, the authors also report limitations that are important when considering educational use, including occasional inaccuracies, dependence on well-formed input text, and support for only selected BPMN elements. Importantly, they indicate that such tools may also serve as educational aids for students learning BPMN, which reinforces the need to study not only AI-generated model quality, but also students' ability to verify and understand these models [14].

These studies show that AI-supported BPMN modelling tools are becoming increasingly capable of generating, editing, and validating process models. However, existing research focuses on technical performance, model generation accuracy, editing reliability, or system efficiency, rather than on educational outcomes. In particular, limited attention has been paid to how students use AI-generated BPMN models, how they verify their correctness, and whether AI support improves modelling competence or merely shifts the cognitive effort from model construction to *post hoc* correction. This educational gap motivates the present study, which examines AI-assisted BPMN modelling in a higher education BPM course.

3 Methodology

3.1 Course Context and Participants

The study was conducted as part of an undergraduate Business Process Management (BPM) course at a technical university in 2026. BPMN modelling constituted a central component of the course, and all participants had previously been introduced to core BPMN elements, including activities, sequence flows, message flows, gateways, and events. In total, 56 students were enrolled in the course. For the analysis, only cases in which at least both the initial ($V1$) and revised ($V2$) model were submitted were included ($N = 50$ students). Participation in the AI-assisted condition ($V2-AI$) was voluntary. Submitting an AI-assisted version did not replace the graded $V2$ submission and was intended as an exploratory learning activity rather than a performance-driven task.

3.2 Assignment Design

Students completed an iterative BPMN modelling assignment over a 2–week period. The assignment consisted of three versions:

- **V1 (Initial model):** an individual BPMN model created after the initial instructional session ($N = 50$).
- **V2 (Revised model):** an improved version produced after structured self-verification using a checklist and scenario-based test cases ($N = 50$).
- **V2–AI (Optional AI-assisted version):** a voluntary alternative final model created with the support of a large language model ($N = 29$).

The iterative structure was designed to promote reflection, error detection, and procedural reasoning, consistent with active learning principles in BPM education [15,19].

3.3 Verification and Feedback

To support deliberate practice and minimise instructor-driven corrections, all students received the same standardised minimal feedback after submitting $V1$. This feedback consisted of three non-prescriptive prompts directing attention to common modelling pitfalls: (1) continuity and completeness of private process flows, (2) correct use of message flows versus sequence flows, and (3) proper closure of gateway splits and merges.

In addition, students used two structured verification scaffolds when revising their models:

- a rubric-aligned BPMN verification checklist, and
- three mandatory scenario-based test cases (happy path, rejection/error scenario, exception or boundary case).

These scaffolds operationalise verification-centred learning and support systematic self-assessment rather than *ad hoc* correction.

3.4 AI-Assisted Modelling Condition

For the optional $V2-AI$ submission, students were allowed to use any LLM-based assistant of their choice. Each AI-assisted submission was accompanied by a short AI Usage Statement documenting the purpose of AI use, the modelling aspects addressed, and examples of accepted and rejected AI suggestions. This instrument provides insight into students' verification behaviour and supports transparency in AI-supported learning [1,15]. Students were free to choose any LLM-based assistant; the most frequently reported tools were Gemini and ChatGPT. No standardised prompt was required, and the starting point for AI-assisted revision (V1 or V2) was not explicitly specified in the protocol. The V2–AI submission counted toward the optional homework component (up to 15% of the lab grade), meaning non-participation reflected a conscious grade trade-off rather than disengagement. Both the unspecified starting point and this self-selection pattern are acknowledged as limitations of the exploratory condition.

3.5 Evaluation and Data Analysis

All submitted models ($V1$, $V2$, $V2-AI$) were evaluated using a rubric covering four BPMN dimensions: diagram structure and readability, message flows and collaboration, events, and gateways. To enable fine-grained analysis, models were additionally coded using an error taxonomy derived from common BPMN modelling errors reported in the literature [5,7].

The analysis combined descriptive statistics, paired comparisons between model versions, and exploratory analyses linking AI usage declarations to observed changes in model quality. Given the ordinal nature of rubric scores and unequal group sizes, non-parametric statistical methods were applied. The evaluation rubric is provided as online supplementary material. The normality of score differences was assessed using the Shapiro–Wilk test. Since the assumptions for parametric paired tests were not consistently met and the rubric scores were ordinal in nature, paired comparisons between model versions were conducted using the Wilcoxon signed-rank test.

3.6 Supplementary Material

To support transparency, replicability, and practical reuse of the proposed instructional design, additional materials are provided as a free online supplementary material. These materials include the verification scaffolds, assignment artefacts, and evaluation instruments used in the study, as well as additional methodological details [20].

4 Results

4.1 Iterative Learning Effects (from V1 to V2)

This subsection reports learning effects associated with the iterative modelling cycle supported by structured verification scaffolds. The comparison between the initial BPMN models ($V1$) and the revised models ($V2$) includes all students who submitted both versions and focuses on overall model quality as well as rubric-aligned modelling dimensions. Overall, students demonstrated a clear improvement in BPMN model quality between $V1$ and $V2$. Median total rubric scores increased substantially, indicating that the iterative revision process led to measurable learning gains. Improvements were observed across all evaluated dimensions, with particularly strong gains in logic-intensive areas such as gateways and events, which are commonly reported as challenging for novice modellers.

Table 1. Summary statistics of rubric-based BPMN model quality for $V1$ and $V2$ ($N = 50$).

Dimension	$V1$ Mean	$V2$ Mean	$V1$ Median	$V2$ Median
Total light-score (0–5)	2.873	3.814	2.900	3.925
Diagram/Readability	1.564	1.775	1.575	1.800
Message Flows	0.637	1.011	0.650	1.000
Events	0.586	0.817	0.600	0.900
Gateways	0.086	0.211	0.000	0.225

As shown in Table 1, the largest relative improvements occurred in the gateway and event dimensions, suggesting that structured verification helped students detect and correct logical inconsistencies that were not apparent during initial model creation. Improvements were also visible in diagram structure and message flows, reflecting increased attention to completeness and collaboration semantics during revision. The descriptive improvements reported in Table 1 were confirmed by non-parametric paired comparisons. The Wilcoxon signed-rank tests showed statistically significant improvements between $V1$ and $V2$ across all rubric categories ($p < 0.001$), with large effect sizes ranging from $r = 0.59$ to $r = 0.84$. This confirms that the observed learning gains were not limited to descriptive score increases, but reflected statistically robust improvements associated with the iterative revision process and verification scaffolds.

Figure 1 illustrates that learning gains were not limited to a small subset of high-performing students but were distributed across the cohort. Most students improved their scores between $V1$ and $V2$, indicating that the observed effects reflect a general learning trend rather than isolated cases. Taken together, these results demonstrate that iterative modelling combined with verification scaffolds leads to substantial improvements in BPMN model quality.

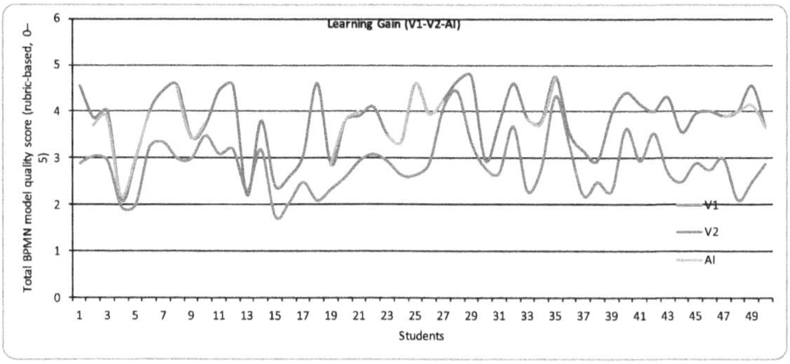

Fig. 1. Slope chart illustrating individual learning gains between $V1$, $V2$, and (optionally) $V2-AI$ total rubric scores.

4.2 AI-Assisted Modelling Effects (from V2 to V2-AI)

This subsection examines whether AI assistance led to additional improvements beyond those achieved through iterative revision. Since the $V2-AI$ submission was voluntary, this comparison should be interpreted as exploratory rather than causal. The analysis is based on paired comparisons between the human-refined models ($V2$) and the AI-assisted alternative versions ($V2-AI$) submitted by a voluntary subgroup of students ($N = 29$). At the aggregate level, no systematic improvement was observed when comparing $V2-AI$ models with their corresponding $V2$ versions. Median total rubric scores remained comparable across both versions, and in some cases slight decreases were observed. This indicates that AI assistance did not consistently enhance BPMN model quality beyond the gains already achieved through verification-centred iteration. Table 2 presents mean and median rubric scores for both versions across the overall score and four BPMN quality dimensions.

Table 2. Paired comparison of rubric-based BPMN model quality for $V2$ and $V2-AI$ ($N = 29$).

Dimension	$V2$ Mean	$V2-AI$ Mean	$V2$ Median	$V2-AI$ Median	Median Diff.
Total light-score (0–5)	3.814	3.763	3.925	3.825	-0.100
Diagram/Readability	1.775	1.771	1.800	1.800	0.000
Message Flows	1.011	0.986	1.000	1.000	0.000
Events	0.817	0.841	0.900	0.900	0.000
Gateways	0.211	0.165	0.225	0.150	-0.075

As shown in Table 2, AI-assisted models did not outperform the student-refined $V2$ models at the aggregate level. The median total rubric score was slightly lower for $V2-AI$ than for $V2$, while median scores for diagram readability, message flows, and events remained unchanged. The most visible decrease concerned the gateway dimension, suggesting that AI assistance may be particularly fragile in logic-intensive aspects of BPMN modelling. Overall, these results indicate that AI assistance did not provide systematic additional improvement beyond structured human revision.

4.3 Verification and AI Use (Exploratory)

Statements were coded inductively into three behavioural patterns: passive acceptance, targeted verification, and broad delegation, with two coders resolving disagreements through discussion. This analysis focused on how students described their goals, verification strategies, and decisions to accept or reject AI suggestions. The results indicate that model quality improvements in $V2-AI$ were associated with evidence of critical AI use rather than with the intensity of AI use.

Students who treated AI as a verification or idea-testing partner, explicitly checking gateway logic, event semantics, and end-to-end consistency, were more likely to maintain or slightly improve model quality. In contrast, students who relied on AI to "complete" or substantially restructure their models without systematic verification rather introduced new logical inconsistencies. Across submissions, AI was primarily effective in supporting technical and rule-based aspects of BPMN modelling, such as identifying missing events or correcting syntactic patterns. However, AI showed limited effectiveness in addressing deeper semantic challenges, particularly those related to decision logic, exception handling, and collaboration semantics. These findings suggest that AI functions best as a technical support tool, while responsibility for logical correctness and semantic appropriateness remains with the student.

5 Discussion and Lessons Learned

5.1 Discussion

The results of this study allow us to address the four research questions and derive substantive interpretations about the educational role of AI in BPMN modelling.

Regarding RQ2, the magnitude of learning gains between V1 and V2 warrants interpretation beyond the descriptive level. The particularly pronounced improvements in gateway and event handling—dimensions that are notoriously difficult for novice modellers [15]— suggest that these competences respond well to structured, reflective revision rather than to repeated exposure alone. This baseline is essential for interpreting the V2–AI comparison: only by demonstrating that structured human revision already yields substantial gains can the study meaningfully assess AI's marginal contribution. This finding is consistent with active learning research in BPM education [4,19], which argues that one-off modelling tasks do not sufficiently develop procedural reasoning. What the verification scaffold appears to achieve is forcing students to adopt a testing-oriented perspective toward their own models—one in which process paths must be explicitly traced and gateway logic justified against concrete scenarios. The breadth of improvement across the cohort further indicates that the observed gains reflect a general pedagogical effect of the intervention design rather than isolated progress among high-achieving students.

Regarding RQ3, the absence of systematic improvement in V2-AI models relative to V2—and the slight regression in the gateway dimension—points to a mismatch between what current LLMs can reliably support and what BPMN modelling at higher quality levels actually demands. Research on LLM-based process modelling has demonstrated that AI tools perform well on syntactic generation tasks but struggle with the semantic reasoning required for decision logic, exception handling, and collaboration-level coherence [6,13]. The present study provides educational evidence consistent with this pattern: AI assistance proved adequate in structurally straightforward aspects of model revision, while tending to introduce inconsistencies precisely where domain-specific process logic is most consequential. This points to a limitation that BPM educators should be aware of: AI-generated models can appear visually complete while concealing semantic flaws that are difficult to detect without sufficient domain knowledge—a risk that resonates with findings on BPMN guideline application, where visual conformance does not imply semantic adequacy [17]. The gateway dimension is particularly telling in this respect: it is precisely the area where formal correctness and business intent must be reconciled, and where AI-generated suggestions proved least trustworthy.

Regarding RQ1, the comparison between V2 and V2-AI provides a natural experiment for assessing the marginal contribution of AI beyond structured human revision. The results indicate that this marginal contribution is, at the aggregate level, negligible or slightly negative—a finding that challenges the assumption that access to AI tools straightforwardly improves the quality of produced artefacts. In the BPMN context, quality improvement depends on the quality of the verification process rather than the sophistication of the generation tool. This aligns with recent empirical work on AI in BPM instructional design [16], which found that generative AI requires careful integration to avoid reinforcing surface-level correctness at the expense of deeper conceptual understanding, and connects to broader concerns about whether AI-assisted production substitutes for or merely defers genuine competence development [10]. If AI accelerates model generation while reducing the cognitive engagement required to evaluate semantic correctness, the educational value of the task may be eroded rather than enhanced.

Regarding RQ4, the qualitative patterns in AI Usage Statements illuminate why aggregate effects were heterogeneous. The critical distinction lies not in how extensively students used AI, but in whether they treated AI outputs as hypotheses to be tested or as solutions to be accepted. Students who structured their AI interaction around targeted verification—testing gateway configurations, checking end-to-end path completeness, querying event semantics—were able to use AI productively. Students who delegated broader modelling decisions to AI without subsequent verification exposed their models to compounded errors, since AI-introduced inconsistencies in one part of a model can propagate and conflict with existing elements elsewhere. This behavioural pattern is consistent with research on critical AI use in higher education [1]: the educational value of AI interaction depends heavily on the metacognitive orientation the student brings to it. The AI Usage Statements proved useful as a lightweight instrument for making this orientation visible—and, arguably, for encouraging it through the act of structured reflection itself.

Taken together, these findings suggest a reframing of the question posed in the title. Rather than asking whether AI improves BPMN modelling in general, the more productive pedagogical question is: under what conditions can AI interaction be structured to support the development of genuine process reasoning? The present study offers preliminary evidence that the answer depends less on the AI tool itself and more on the surrounding instructional design—specifically, on whether verification is embedded as an explicit and assessable learning activity.

5.2 Implications for BPM Educators

The findings translate into several concrete recommendations for educators designing BPMN courses in contexts where generative AI is available to students.

First, the results suggest that iterative assignment designs with structured verification scaffolds can achieve substantial pedagogical impact even without extensive instructor-driven feedback. Rubric-aligned checklists and scenario-based test cases proved effective in orienting students toward the logic-intensive aspects of BPMN where novice modellers most commonly struggle [15]. Verification competence should therefore be treated as a first-order learning objective, not as a secondary quality-control step appended to model creation.

Second, the heterogeneous outcomes in the AI-assisted condition indicate that teaching students how to interact with AI critically is as important as teaching BPMN notation itself. Educators should assume that variation in verification behaviour—rather than access to AI per se—will be the primary driver of quality differences in AI-assisted work. Structured exercises in which students are explicitly asked to identify, challenge, and correct AI-generated modelling decisions may be more educationally valuable than open-ended AI use without accountability requirements.

Third, the AI Usage Statement demonstrated that lightweight transparency instruments can surface verification behaviour without substantially increasing grading burden. Requiring students to document their modelling intentions, the aspects they verified, and AI suggestions they accepted or rejected serves a dual function: it creates a reflective pause that may itself improve the quality of AI-assisted work, and it gives educators insight into students' cognitive engagement with the task. Educators are encouraged to adapt and integrate such instruments into assessment frameworks where AI-assisted submissions are permitted.

Fourth, educators who wish to ensure that students develop genuine process reasoning should deliberately foreground tasks that require semantic judgement: modelling decision logic under ambiguous conditions, handling exceptions that span multiple pools, or justifying gateway configurations against specified business rules. These are precisely the areas where AI assistance proved least reliable, and where student competence therefore remains most consequential. Research on BPMN guideline application similarly indicates that syntactic conformance and semantic adequacy do not co-vary reliably [17], reinforcing the need for assessment instruments that evaluate process logic independently of surface-level diagram quality.

5.3 Implications for Research, Practice, and Lessons Learned

From a research perspective, the results point to a need to extend existing BPM modelling quality frameworks to explicitly account for AI-generated artefacts. Future research should investigate how traditional BPMN quality dimensions—syntactic, semantic, and pragmatic—manifest when process models are co-created by humans and LLMs, and how prompting strategies can be systematically designed to mitigate semantic degradation. Controlled experiments varying the degree of student verification guidance, the type of AI tool used, and the complexity of the modelling scenario would help establish more transferable design principles for AI-integrated BPM education.

From a practical perspective, organisations should use LLMs primarily for supportive tasks such as draft model generation and syntactic error detection, while establishing formal review procedures to ensure semantic correctness.

A key lesson from this study is that the value of AI assistance in process modelling is strongly contingent on prompt quality: the clarity and completeness of the user's input shapes the reliability of AI output, which means that effective AI use requires not only BPMN knowledge but also the ability to formulate precise, context-rich prompts—a competence not yet routinely addressed in BPM training programmes.

At the same time, the risk of false confidence deserves explicit attention: LLM-generated BPMN models may present as formally complete while embedding semantic errors that are difficult to detect without domain expertise, creating conditions under which users—particularly novices—may overestimate model quality. Formal review and validation procedures are therefore not optional safeguards but necessary components of any workflow that incorporates LLM-generated process artefacts. As Maslov et al. [16] have argued, the challenge is not to resist AI integration but to ensure that human judgement remains structurally embedded in the modelling process.

6 Threats to Validity

As with any empirical study, several threats to validity must be considered when interpreting the results. Internal validity may have been affected by the way prompts were formulated and presented to participants, as well as by possible learning effects occurring during the study. Although efforts were made to standardize the instructions and tasks, differences in participants' prior familiarity with BPMN notation or AI-based tools may have influenced their performance and, consequently, the observed outcomes. Construct validity is also an important consideration, particularly with regard to the operationalization of key concepts such as model quality and usefulness. A further threat concerns the nested structure of the AI-assisted condition: since students generated V2–AI starting from their already-revised V2 baseline, the absence of improvement may partly reflect an anchoring effect—aspects correctly resolved in V2 were unlikely to be degraded by AI, limiting the AI's opportunity to demonstrate value in those areas. This means the finding that V2–AI does not outperform V2 is best interpreted as specific to this post-revision setting rather than as a general limitation of AI for BPMN modelling. Additionally, the heterogeneity of LLM tools and prompting strategies used by students introduces uncontrolled variance in the V2–AI condition that cannot be retrospectively disaggregated. Although the study relied on established BPMN quality criteria and evaluation instruments, some degree of subjectivity in the assessment of generated models cannot be fully excluded. External validity is limited by the specific LLMs, modeling tasks, and participant groups included in the study. Therefore, the findings should not be generalized uncritically to all AI-based modeling tools, more complex process scenarios, or highly specialized industry domains. The results provide meaningful insights into the analyzed setting, but their transferability to other technological and organizational contexts requires further empirical verification. Future studies should therefore examine a broader range of AI tools, process types, user groups, and domain-specific modeling situations.

Finally, the modest sample size and predominantly descriptive character of the analysis mean that the findings should be understood as empirically grounded indications rather than definitive proof of general effectiveness across all modelling contexts.

7 Conclusion

This paper examined AI-supported BPMN modelling in higher education through an iterative assignment design combining human revision, structured verification, and optional AI assistance. The findings show that the strongest improvement occurred between the initial student-created models ($V1$) and the revised models ($V2$), confirming the educational value of verification-centred iteration. In contrast, AI assistance did not produce systematic additional gains beyond human refinement, and its effects varied depending on how critically students verified AI-generated suggestions.

The study suggests that AI should not be treated as a replacement for BPMN modelling competence, but as a reflective support tool that can help students compare alternatives, detect errors, and develop modelling judgement. For BPM educators, the results highlight the importance of embedding AI use within structured verification activities, such as checklists, scenario-based test cases, change logs, and AI Usage Statements. Future research should examine similar instructional designs across different courses, process scenarios, AI tools, and student groups.

Acknowledgments. The authors would like to sincerely thank all students who participated in the study for their time, engagement, and valuable contribution to the empirical research. Their participation was essential for the completion of this study.

References

1. de Borba, D., Brinkhues, R.: Artificial intelligence in business process management: challenges, opportunities and strategies for alignment with ISO 42001. Rev. Artif. Intell. Educ. **6** (2025). https://doi.org/10.37497/rev.artif.intell.educ.v6ii.60
2. BPMN.AI: AI-powered business process modeling (2026). https://www.bpmnai.com/. Accessed 2026
3. Grzesiak, M.: Implementation of active learning in BPM education. Decis. Making Manufact. Serv. 117–125 (2024). https://doi.org/10.7494/dmms.2024.si.6696
4. Grzesiak, M., Moszyński, M.: Application of the flipped learning methodology at a business process modelling course – a case study. In: Business Process Management: Blockchain, Robotic Process Automation, Central and Eastern European, Educators and Industry Forum. BPM 2024. Lecture Notes in Business Information Processing, vol. 527. Springer (2024). https://doi.org/10.1007/978-3-031-70445-1_19
5. Kluza, K., et al.: Revisiting BPMN assignments with AI in mind: insights from experiments with large language models in process modeling education. In: Business Process Management: Responsible BPM Forum, Process Technology Forum, Educators Forum. BPM 2025. Lecture Notes in Business Information Processing, vol. 565. Springer (2026). https://doi.org/10.1007/978-3-032-02936-2_25
6. Licardo, J.T., Tanković, N., Etinger, D.: BPMN Assistant: An LLM-based approach to business process modeling. arXiv preprint arXiv:2509.24592 (2026)
7. PwC: BPMN AI: Automating business process model creation with GenAI (2025). https://www.pwc.ch/en/services/consulting/bpmn-ai.html. Accessed 2026
8. Reeves, T.C.: Design research from a technology perspective. In: van den Akker, J., Gravemeijer, K., McKenney, S., Nieveen, N. (eds.) Educational Design Research, pp. 52–66. Routledge (2006)
9. McKenney, S., Reeves, T.C.: Conducting Educational Design Research. Routledge (2012). https://doi.org/10.4324/9780203818183
10. OECD: AI in education: Challenges and opportunities for sustainable development. OECD Publishing (2021). https://doi.org/10.1787/9f5ee6d6-en
11. von Rosing, M., White, S., Cummins, F., Man, H.: Business Process Model and Notation - BPMN, pp. 429–453 (2014). https://doi.org/10.1016/b978-0-12-799959-3.00021-5
12. Senkus, P., et al.: Bridging the gap: an evaluation of business process management education and industry expectations – the case of Poland. In: International Conference on Business Process Management, pp. 250–262. Springer (2023). https://doi.org/10.1007/978-3-031-43433-4
13. Kourani, H., Berti, A., Schuster, D., van der Aalst, W.M.P.: Process Modeling With Large Language Models, pp. 229–244 (2024). https://doi.org/10.1007/978-3-031-61007-3_18

14. Licardo, J., Tanković, N., Etinger, D.: A method for extracting BPMN models from textual descriptions using natural language processing. Procedia Comput. Sci. 483–490 (2024). https://doi.org/10.1016/j.procs.2024.06.196
15. Maslov, I.: Towards empirically validated process modelling education using a BPMN formalism. In: Guizzardi, R., Ralyté, J., Franch, X. (eds.) Research Challenges in Information Science. RCIS 2022. Lecture Notes in Business Information Processing, vol. 446, pp. 1–8. Springer (2022). https://doi.org/10.1007/978-3-031-05760-1_58
16. Maslov, I., Poelmans, S., Wautelet, Y.: Generative AI's aid in business process modeling instructional design: a case study. In: Er, M., et al. (eds.) Business Process Management: Responsible BPM Forum, Process Technology Forum, Educators Forum. BPM 2025. Lecture Notes in Business Information Processing, vol. 565. Springer (2025). https://doi.org/10.1007/978-3-032-02936-2_24
17. Fantinato, M., Thom, L.H.: Enhancing business process clarity: enabling the development of more understandable BPMN models through prioritized guidelines. Bus. Process Manag. J. **31**(3) (2025). https://doi.org/10.1108/bpmj-10-2024-0989
18. Rockembach, G.R., Thom, L.H.: Investigating the use of intelligent tutors based on large language models: automated generation of business process management questions using the revised bloom's taxonomy. In: Anais do Simpósio Brasileiro de Informática na Educação (SBIE), vol. 35. SBC (2024). https://sol.sbc.org.br/index.php/sbie/article/view/31343
19. Delgado, A., Calegari, D.: Teaching BPM fundamentals: a project-based hands-on course for process-driven systems modeling and development. In: Köpke, J., et al. (eds.) Business Process Management: Blockchain, Robotic Process Automation and Educators Forum. BPM 2023. Lecture Notes in Business Information Processing, vol. 491. Springer (2023). https://doi.org/10.1007/978-3-031-43433-4_14
20. Online Supplementary Material (Repository Link) (2026). https://marmoszy.github.io/EduForum2026/FreeSupplement.pdf. Accessed 2026

AI-Decision Checkpoints for AI-Augmented Business Process Management: Framework and Educational Instantiation

Amin Jalali(✉)

Department of Computer and Systems Sciences, Stockholm University, Stockholm, Sweden
aj@dsv.su.se

Abstract. Large Language Models (LLMs), Retrieval-Augmented Generation (RAG), and AI agents are increasingly embedded in operational business processes. Yet Business Process Management (BPM) curricula and frameworks still largely treat artificial intelligence (AI) as an add-on technology, leaving graduates (as potential future process developers) unprepared to reason about AI as a first-class design element of end-to-end processes. This paper addresses that gap by proposing *AI-decision checkpoints*: explicit moments in a process development trajectory where process developers identify AI-candidate sub-processes, assess expected effects on time, cost, quality, and flexibility, consider legal and organisational constraints, and document a reasoned decision to adopt, constrain, or reject specific AI components. The checkpoints are instantiated through a fictitious customer onboarding process as a *BPM Teaching Case*, embedded in a lifecycle-driven framework spanning six modules that combine process modeling, simulation, workflow execution with AI agents, and process mining, with each module's output serving as the next module's input. A preliminary formative reflection draws on instructor observations, submitted artifacts, and discovered process maps from learning-management-system logs. These exploratory observations suggest that the approach supported clearer distinctions between task-level automation and process-level value.

Keywords: BPM education · BPM lifecycle · curriculum design · AI-augmented BPM · teaching case

1 Introduction

Business Process Management (BPM) is a mature information-systems discipline that provides concepts, notations, and methods for identifying, modeling, analyzing, redesigning, executing, and monitoring organizational processes [1,2]. Core pedagogical content has remained remarkably stable for decades: process discovery and modeling, control-flow analysis, simulation-based performance analysis, workflow enactment, and process mining [3].

K. Kluza et al. (Eds.): BPM 2026, LNBIP 600, pp. 311–327, 2027.
https://doi.org/10.1007/978-3-032-36710-5_20

This stability is now challenged by advances in Large Language Models (LLMs), Retrieval-Augmented Generation (RAG), and AI agents. It has become technically feasible to introduce AI agents as process participants in knowledge-intensive tasks [4], such as extracting data from unstructured documents, conducting Know-Your-Customer (KYC) or fraud checks, generating customer communication, and lead generation for claim investigation [5].

Industry adoption of such components currently outpaces their curricular integration, leaving graduates at risk of two complementary deficits. Classically trained BPM analysts can model and analyze processes, but struggle to reason about when and how AI components add value at the process level. Conversely, AI engineers may be able to build proof-of-concept agents and deploy them in production but lack the methodological grounding to assess their impact on end-to-end performance and value creation.

Existing curricular responses tend to follow one of two patterns: appending a stand-alone module on AI or LLMs to an otherwise unchanged course, or adding AI-related examples throughout traditional content without altering its structure. In both cases, AI is treated as a technology topic rather than as a design concern to be evaluated systematically across the BPM lifecycle.

We contend that the discipline already possesses a suitable backbone for structuring AI-related reasoning: the BPM lifecycle [1], which has long served as a textbook spine and didactic organizer. Our contribution is not to propose a new lifecycle, but to turn the existing one into an active driver of AI-related decision-making through what we call *AI-decision checkpoints*: pre-specified points in the learning trajectory at which students must identify, justify, configure, or reject an AI component using the analytical instruments of the corresponding lifecycle phase, transforming the lifecycle from a passive organiser into a sequence of grounded decision situations. Therefore, this paper makes three contributions:

(C1) It presents a fictitious *Banking case* designed to sustain multi-module engagement across the BPM lifecycle,
(C2) It introduces *AI-decision checkpoints* as the organizing pedagogical artifact of the case, with one checkpoint per lifecycle phase, embedded in a lifecycle-driven curriculum framework in which each module's output becomes the next module's input across modeling, simulation, redesign, implementation, and mining,
(C3) It presents a *formative reflection* on a Master-level instantiation of the case, combining instructor observations with insights discovered from the learning-management-system event logs showing how groups progressed across modules.

The remainder of this paper is structured as follows. Section 2 positions the work in relation to previous related literature. Section 3 introduces the AI-decision checkpoint framework and positions each checkpoint within the BPM lifecycle. Section 4 describes its instantiation in a Master-level course built around the Banking teaching case. Section 5 reflects on course delivery and student performance based on instructor observations and learning-management-system data. Section 6 concludes and outlines implications for BPM educators and directions for future work.

2 Related Work

Despite BPM's maturity as a discipline, its integration into higher-education curricula has remained narrower than the scope of professional practice. Moormann and Bandara [6] issued an early call for action, noting that graduates consistently lack the breadth of skills employers require. To identify related work, a Scopus search was conducted using the query shown in Table 1, which returned 45 documents at 2026-05-31. After title-and-abstract screening for papers addressing the intersection of BPM education and AI, three papers were identified as indirectly related.

On curriculum innovation, Dallavalle et al. [7] propose a PBL-Process Mapping Blueprint for undergraduate BPM courses at the University of São Paulo, structured around three next-generation BPM drifts: Conversationality, Autonomization, and Sophistication. The blueprint was developed in three phases, culminating in prospective validation through two focus groups with eight alumni. Participants identified autonomization as the most professionally consequential drift and called for a brief AI-agent primer to be introduced early in the course. The blueprint treats AI as one of several emerging competencies to be woven into existing activities. Our work shares the view that AI must be embedded as a design concern throughout the lifecycle, but operationalizes it differently: rather than a primer, we provide students with a structured decision instrument at every lifecycle phase, requiring them to justify, configure, or reject AI components using phase-appropriate evidence.

On teaching cases, Grangel et al. [8] present a teaching case aimed at postgraduate students in AI, BPM, or Medical Informatics, in which real pseudoanonymised clinical data from a Spanish hospital are used to analyse Failure To Rescue events through process mining and predictive process monitoring. The case covers process discovery, conformance checking, performance analysis, and process redesign, and is accompanied by online instructor-facing teaching

Table 1. Scopus query used for literature identification.

Scopus search query (45 results):

```
TITLE-ABS-KEY(
("business process management" OR BPM)
AND ("artificial intelligence" OR AI OR "generative AI"
OR "large language model*" OR LLM*
OR "AI agent*" OR "retrieval-augmented generation" OR RAG)
AND (teaching OR education OR pedagogy OR curriculum
OR "course design" OR "higher education")
)
```

notes. Like our Banking case, it grounds BPM competencies in a domain-specific scenario with real complexity. The key difference is scope: the Grangel et al. case focuses on applying mining techniques to an existing process, whereas our case requires students to design, implement, and evaluate AI-augmented process variants across the full lifecycle, treating governance concerns such as human-in-the-loop routing and fallback handling as explicit learning objectives.

On LLMs in BPM education, Kluza et al. [9] conduct two experiments in which ChatGPT-4o and Gemini Flash are tested on typical BPMN assignment tasks: answering comprehension questions based on five BPMN models (in PNG and XML format) and detecting errors in 30 flawed BPMN diagrams. The authors conclude that BPMN assignments are not inherently AI-resistant and call for AI-aware pedagogy in which students learn to critically assess AI-generated output. In our course, LLMs are not evaluated as assignment solvers but configured and deployed as process participants. The control-flow reasoning limitations that Kluza et al. document are precisely what motivate the AI-decision checkpoint for implementation, where students must specify schema-constrained outputs, confidence thresholds, and human escalation paths to compensate for unreliable AI behavior.

Collectively, these works show that AI is entering BPM curricula through multiple routes, yet none provides a systematic method for reasoning about AI adoption across all lifecycle phases using the analytical instruments of each phase. The AI-decision checkpoint framework addresses this gap.

3 AI-Decision Checkpoints

The proposed framework uses the BPM lifecycle to structure a sequence of explicit, phase-related decisions about AI adoption. At each checkpoint, students contribute to determining how AI can enhance processes by identifying AI-candidate subprocesses, establishing the baseline, evaluating expected benefits and risks using the analytical instruments of the corresponding lifecycle phase, and documenting a reasoned decision to adopt, adapt, postpone, or reject specific AI components. This approach decomposes AI adoption from a monolithic yes/no judgment into a chain of locally grounded assessments that distribute AI-related reasoning across all phases, anchor decisions in established BPM instruments, and produce a traceable decision trail that can be revisited.

The framework does not prescribe a correct answer at any checkpoint; it foregrounds reasoning quality, requiring students to articulate assumptions, acknowledge trade-offs, and justify decisions. One checkpoint aligns with each lifecycle phase: identification, discovery, analysis, redesign, implementation, and monitoring [1]; individual checkpoints can be weighted or combined without disrupting the logic. Figure 1 positions them relative to the six course modules (M1-M6).

3.1 CP1 - Process Identification: Scoping AI Opportunities

CP1 is located in the process-identification phase. Its purpose is to determine in which processes and subprocesses AI augmentation is sufficiently promising

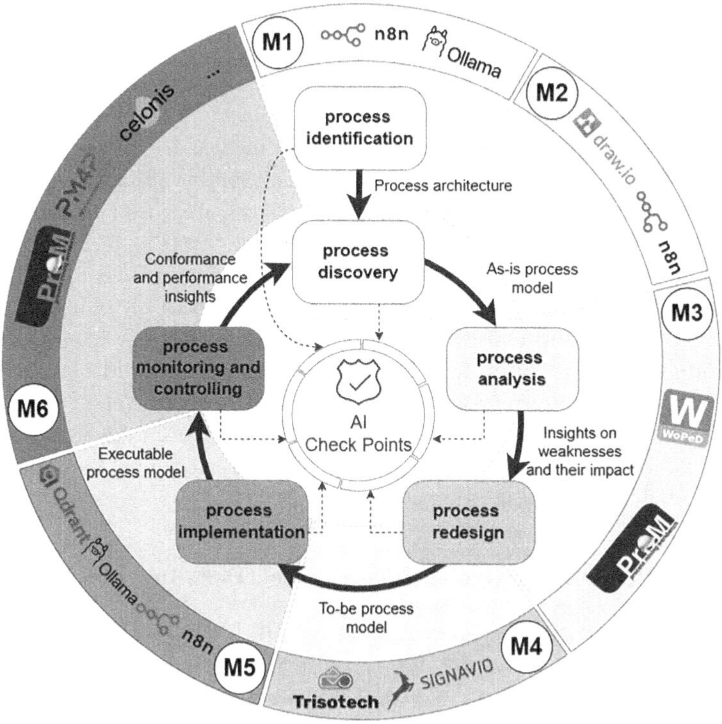

Fig. 1. The BPM lifecycle with six AI-decision checkpoints and their instantiation in course modules (M1-M6) with associated tool support. We operationalized CP4 in two modules in our course: an initial simulation-informed redesign in M3 (workflow-net level) and a final exception-handling redesign in M4 (BPMN level).

to warrant further investigation. The guiding question is: *which elements of the process landscape are plausible AI-candidate subprocesses, and what effects are they expected to have at the level of entire processes?*

The process architecture is examined for subprocesses amenable to AI, specifically those exhibiting high knowledge intensity, data-processing demands, repetitiveness, or dependence on scarce expertise. Each candidate's expected impact is assessed along the *devil's quadrangle* (time, cost, quality, flexibility) [10], and risks are cataloged covering legal and regulatory requirements (e.g. GDPR), ethical considerations, data availability, and organizational readiness. The outcome is an AI opportunity register that combines benefit assessments with initial technical feasibility in a prioritization matrix, determining which candidates proceed to later phases.

3.2 CP2 - Process Discovery: Positioning AI in the As-Is Process

CP2 is located in the process-discovery phase. The focus shifts from the process landscape to a concrete as-is process, and from abstract opportunity identification to a precise understanding of how candidates from CP1 relate to existing organizational work. The guiding question is: *given a description of the selected process, where would the AI candidates conceptually attach, and which human and system actors would they interact with or affect?*

The as-is process in this phase shall be specified as a normative model and enriched with an explicit actor perspective. Candidates prioritized in CP1 are positioned with respect to this model rather than integrated into it. For each candidate, the analysis specifies (i) the segment of the as-is model it would conceptually replace, support, or augment, (ii) the human roles and IT systems currently involved at that segment and how they would be affected, and (iii) the inputs, outputs, and control transfers currently observed. Escalation paths, approval steps, and exception-handling mechanisms that would become relevant under AI augmentation are also sketched at a conceptual level.

The artifact is the as-is process model together with an AI-positioning note that records, for each candidate, its intended placement, the actors affected, and prominent structural obstacles or dependencies.

3.3 CP3 - Process Analysis: Performance Baseline

CP3 is located in the process-analysis phase. While CP1 and CP2 establish where AI augmentation is promising and how it relates to the as-is process, CP3 quantifies the current performance without altering the process structure. The guiding question is: *given an executable representation of the current process, what is the baseline performance against which any redesign (AI-based or otherwise) must be assessed?*

The baseline can be identified in different ways such as extracting the current state of the process using process mining, or transforming the model from CP2 into an executable model and simulating the process to identify the baseline. In courses without prior process data, the latter approach can be feasible; while in practical settings, other approaches can be followed. Regardless of the method, the analysis should yield baseline indicators such as average throughput times, waiting times, resource utilization, and sensitivity to workload variation.

The artifact produced at CP3 is a baseline performance report documenting the chosen method, underlying assumptions, and key performance indicators, with particular attention to structural bottlenecks and systematic delays. The importance is to define a baseline against which upcoming changes can be evaluated (whether AI-driven or otherwise). Crucially, where a simulation model is established as part of this baseline, it serves as a reusable analytical asset: as redesign proposals emerge in later phases, the model can be refined to reflect the proposed changes and re-simulated to assess their expected impact before any real implementation takes place.

3.4 CP4 - Process Redesign: Value-Aware Business Process Redesign

CP4 is located in the process-redesign phase. Building on the baseline from CP3 and the conceptual positioning from CP1 and CP2, it treats AI augmentation as one design option within a broader redesign effort. The guiding question is: *when redesign practices are applied, under which conditions does the inclusion of AI-candidate subprocesses produce to-be process models that outperform the baseline?*

Starting from the executable as-is model and diagnosed weaknesses, the redesign uses recognized principles (task elimination, resequencing, parallelization, integration of responsibilities). To-be scenarios include both classical control-flow and resource changes and variants that embed AI candidates, integrating AI with redesign moves rather than bolting it on. Redesign variants are distinguished by the AI technology embedded: (i) LLM-based agents for email intake and structured data extraction from documents, (ii) RAG-equipped agents for fraud screening that retrieve and reason over AML rule documents, and (iii) classical BPM redesign variants without AI components, which serve as the comparison baseline. All scenarios are simulated under consistent assumptions and compared to the baseline.

The artifact is a comparative redesign-and-simulation dossier in which each candidate is assigned a redesign decision, justified by the redesign rationale and quantitative evidence, including trade-offs such as bottleneck relocation or increased process complexity.

3.5 CP5 - Process Implementation: Realisability and Governance

CP5 is located in the implementation phase and addresses the gap between conceptual redesign and technical realization. The guiding question is: *are the AI-augmented redesign decisions technically realizable within the selected workflow and AI infrastructure?*

The to-be design is mapped onto a concrete workflow platform and integrated with specific AI services, such as LLM APIs or RAG pipelines. For each AI component, required data flows, performance characteristics, error-handling mechanisms, and security and privacy constraints are matched against what the infrastructure can provide. Where the platform does not natively support the required control-flow patterns or error-handling logic, compensating structures are introduced, for instance, additional validation activities, schema-constrained prompts, explicit human approval tasks, or fallback paths triggered on AI failure or low confidence.

The artifact is an implementation feasibility and governance dossier per AI component. It records the extent to which the component could be realized as designed, the workarounds introduced, the governance measures put in place (logging, thresholds, escalation rules, human-in-the-loop points), and any residual technical or organizational risks.

3.6 CP6 - Process Monitoring: Mining-Based Insights

CP6 is located in the monitoring-and-controlling phase and concerns the ex-post evaluation of AI-augmented subprocesses on the basis of execution data. The guiding question is: *given execution data of the deployed process, do the AI-augmented subprocesses behave as intended and deliver net value over time, or should the process be redesigned and reconfigured?*

Execution data are collected as event logs once the deployed process has run sufficiently; in a course setting, pre-generated logs are typically provided. Process discovery is applied and tested against the normative to-be model from CP4/CP5; conformance analysis surfaces deviations such as bypassed AI-enabled subprocesses; and performance analysis quantifies indicators against the CP3 baseline and CP4 simulated expectations. The artifact is a mining-based evaluation summary consolidating execution patterns, conformance deviations, and realized performance. Its insights feed back into CP1, completing the evidence-based governance cycle.

4 Course Instantiation

The framework was instantiated in a Master 's-level course for students in the final year of a Business Information Systems or Computer Science program. The course has been coordinated by the author for more than 15 years and is designed based on experience-based learning [11]. No prior knowledge of BPM or AI was assumed. A total of 152 students enrolled in the course. Of these, 29 had passed the group assignments in a previous year and were required to participate only in the individual exam; the remaining 123 were new students who formed 25 project groups, each with a maximum of five members. The course mandates group work, with results delivered at each module milestone. A key structural principle is the *evolving-artifact* organization: each module's output becomes the next module's input, so students revisit AI-related decisions repeatedly with accumulating evidence.

4.1 The Banking Case

The primary deliverable of this paper is the Banking case. The six modules (M1-M6) are built around this case, where students focus on designing a customer onboarding process comprising customer registration, fraud assessment, and approval, at a fictitious retail bank facing a scalability crisis. The case is intentionally rich in regulatory constraints, knowledge-intensive work, and heterogeneous data sources, which together make it suitable for realistic AI augmentation scenarios. It is structured around two main documents alongside other instructions distributed to students at different points in the course.

Document A (released at M1, needed for M2) describes the current as-is process. Seven roles are involved: a clerk who registers scanned customer requests, a credit investigator who checks customer credits via an external service, two CRM officers who manage customer communications and profile activation, an accounting assistant who enables transactions, and two fraud investigators who manually review each case against the bank's anti-money-laundering (AML) guidebooks. The normal scenario traces a case from registration through credit check, parallel profile and account activation, customer notification, fraud review, and either full access or external fraud investigation. A cancellation scenario (needed and can be released at M4) introduces a regulatory constraint on when a customer can withdraw the case, which is before an external fraud investigation returns a positive result, requiring a carefully bounded cancel region.

Document B (released at M3) describes the redesign brief: the bank cannot scale with the current manual process and proposes a combination of AI agents, where students are guided to prioritize email-based registration extraction and RAG-based fraud screening, as well as web-service automation (credit checks, profile and transaction management, police reporting). The fraud-screening agent is deliberately constrained by a known class-imbalance problem (the data science team reports high recall but imperfect precision), so students must reason about when the agent's output can be trusted autonomously and when a human must remain in the loop.

The hook of the case is established in M1: students build a working prototype of the email-intake AI agent before they have modeled the process, then are asked whether deploying it would actually help the bank. The tension between "the agent works" and "we don't yet know if it adds value" motivates the entire lifecycle journey that follows. Case materials, module instructions, and datasets are openly available in a GitHub repository[1]. Teaching notes, discussion guidance, and sample solutions are available to instructors upon request.

4.2 Module 1 - Process Identification, BPM and AI Foundations, and CP1

The first module (M1) introduces the conceptual foundations required to apply the framework: a concise overview of BPM and its evolution, the BPM lifecycle, and core notions such as processes, subprocesses, and actors, together with an introduction to the AI technologies most relevant to BPM: Large Language Models (LLMs), generative AI, and AI agents.

On this basis, the banking case is presented, and students construct a process architecture for the bank. Within this architecture, they identify AI-candidate subprocesses, documenting their assumptions as artifacts introduced in CP1. In the onboarding context, two prominent candidates are expected to emerge from the given case description: a subprocess handling incoming customer emails and registration requests (including data extraction from attached documents), and a subprocess for assessing potential fraud against the bank's internal guidelines and regulatory documents.

[1] https://github.com/jalaliamin/AI-decision-checkpoints/tree/main.

A complementary lab makes the building blocks tangible: students configure an AI agent in `n8n` that processes customer emails, extracts information from attached PDF forms using a local LLM (served by `Ollama` to meet on-premises regulatory constraints), and registers the result into a mock system. The lab surfaces the distinction between task-level automation and process-level value.

CP1 is discharged by applying the AI opportunity register to the two candidates, yielding a prioritization matrix that determines which proceed to M2.

4.3 Module 2 - Normative As-Is Modeling and CP2

M2 focuses on normative as-is modeling in `draw.io` using workflow and presentation patterns. CP2 is discharged by positioning the CP1 AI candidates (email-intake and fraud assessment) with respect to the as-is model. Students also evaluate `n8n`'s workflow-pattern support from a chief-IT-officer perspective, producing the AI-positioning note that feeds into M3.

4.4 Module 3 - Process Analysis and Redesign, and CP3 and CP4

M3 covers process analysis and an initial redesign pass. Students transform the as-is model into a workflow net (`WoPeD`), verify soundness, configure simulation parameters, and obtain baseline indicators; performance spectrum analysis in `ProM` [12] surfaces bottlenecks. Building on this baseline, students construct and simulate redesign variants (from classical changes such as task elimination, resequencing, and parallelisation [10] to AI-embedding designs), producing a preliminary CP4 report to be finalized in M4. Splitting CP4 across M3 and M4 is intentional: the workflow-net redesign grounds design choices analytically before the BPMN exception-handling redesign in M4.

4.5 Module 4 - BPMN-Centric Process Redesign and CP4

M4 refines the best M3 scenario in BPMN (using `Signavio` and `Trisotech`), shifting from happy-path to exception-handling redesign. Students apply established modeling guidelines to clarify control flow, make subprocess boundaries explicit, and model human-system-AI interactions and cancellation scenarios. This enables a second pass on CP4, revisiting AI-augmented subprocess choices in light of representational clarity and implementability.

The artifact is a consolidated BPMN-based redesign model with clearly identified AI-augmented subprocesses and associated redesign principles. Together with the analytical evidence from M3, this model constitutes the final outcome of CP4 and provides the basis for the feasibility and governance analysis in M5.

4.6 Module 5 - Implementation in Workflow Platforms and CP5

The fifth module (M5) addresses the implementation phase and operationalizes CP5. Building on the selected and refined BPMN-based to-be model from M4, students map the preferred redesign onto a concrete AI-augmented workflow in `n8n`. The emphasis is on realizing the control-flow structures and subprocess boundaries defined in BPMN and on technically integrating the AI-augmented subprocesses retained after CP4.

Students decompose the BPMN model into an executable workflow and realize AI-augmented subprocesses as concrete agents: an email-intake agent for document extraction and a RAG-based fraud-assessment agent that retrieves and reasons over AML rules embedded in `Qdrant`, with a human-in-the-loop step for flagged cases. The fraud agent introduces students to configuring semantic memory within a workflow platform.

CP5 is discharged by assessing, for each AI-augmented subprocess, the extent to which the BPMN-level design from M4 can be realized and governed in the chosen technical environment. The resulting dossier documents (i) alignment between design and implementation, (ii) workarounds and additional control structures introduced (validation and enrichment steps, schema-constrained prompts, fallback paths on AI failure), and (iii) governance and oversight mechanisms. On this basis, the redesign-level decisions from CP4 are refined into implementation decisions.

4.7 Module 6 - Process Mining and CP6

M6 instantiates CP6 via process mining. Students receive a multi-year synthetic event log (constructed to exhibit concept drift). They segment the log, characterize drifts, discover period-specific models, and evaluate them on fitness, precision, simplicity, and generalization. The artifact is a mining-based evaluation summary consolidating execution patterns, drift evidence, conformance deviations, and performance indicators that feed back into CP1.

5 Reflection on Course Delivery and Student Performance

The course's core has been offered by the instructor since 2011; its core scaffolding was therefore pedagogically validated prior to the present intervention, which restructured it around the AI-decision checkpoints and adopted `n8n` as the implementation platform, representing a substantial redesign effort of approximately 3 months. The course follows a project-based learning (PBL) design [13]. At the time of writing, the exam has not yet been administered, so the evaluation focuses on group performance in labs and assignments.

Given the single-course, single-instructor design with no comparison group and no final exam data at the time of writing, all findings below should be read as a preliminary, exploratory reflection on experience rather than as evidence of learning effectiveness. The evaluation addresses three questions: (EQ1) Do students engage with AI adoption as expected phase-specific decision? (EQ2) How is submission timing distributed across modules, and what does this imply for course design? (EQ3) How do students perceive the learning value of each

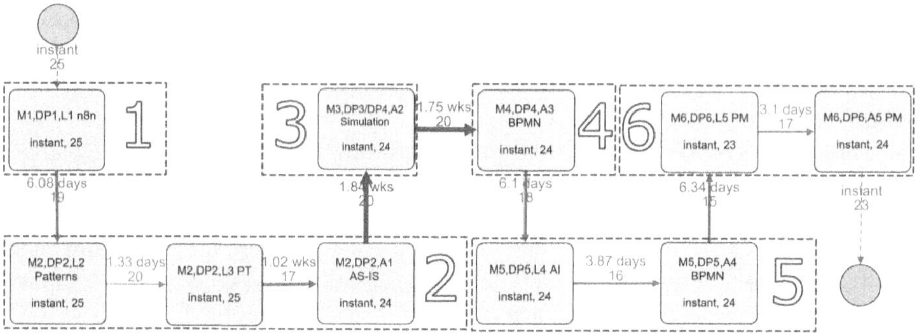

Fig. 2. Discovered process map of student-group progression through labs and milestones, annotated with activity frequencies and throughput times.

module's primary activity? Evidence triangulates instructor observations with insights discovered from LMS event logs extracted by PM4Moodle [14], a voluntary end-of-course group survey (conducted per groups), which capture group interactions at checkpoints. Given the single-instantiation design, findings are presented as preliminary indications rather than generalizable results.

Figure 2 shows a process map discovered with Apromore, using each group as a case identifier and retaining only the main paths; activity durations are instantaneous as the learning management system only captures the completion timestamps for submission boxes, but module durations can be estimated by the sum of actions taken after the previous module until the last touchpoint in each module. Several findings emerge.

Groups and Progression Paths: Of 25 registered groups, 24 were active at the time of analysis; one discontinued after M1. All 24 continuing groups followed the expected sequential progression, suggesting that the checkpoint artifacts were sufficient to proceed.

Submission Timing and Effort Distribution: Because the LMS captures only submission timestamps, module durations are approximated as the interval from the previous module's last touchpoint to the current module's submission. With this caveat, groups appeared to spent between 9 days and 2 weeks on each module after M1. This pattern is consistent with student feedback, suggesting that certain modules should be lightened to reduce overall course density.

Evidence on EQ1 (Revised AI Assessments): Checkpoint artifacts were reviewed by the instructor against criteria assessing (i) completeness of the AI opportunity register or redesign dossier, (ii) quality of justification for AI adoption decisions, and (iii) consistency of reasoning with evidence accumulated in prior modules. This review was qualitative and based on instructor judgment; detailed rubrics will be shared in the public repository. To illustrate how checkpoints encouraged students to evaluate AI adoption during process development, we describe a pattern observed in the CP4 artifacts. Most groups identified the email-intake and fraud-assessment subprocesses as AI candidates in

CP1 after receiving feedback, and the majority initially expected substantial reductions in throughput time; in several cases, they estimated that AI-based intake would halve overall case cycle time. After completing the baseline simulation in CP3, these groups found that the bottleneck lay not in registration but in the two-person fraud-investigation queue, which operated near full utilization. Their CP4 comparative redesign dossiers explicitly recorded revised assessments: the email-intake agent, while technically functional, was projected to shift the bottleneck downstream rather than reduce end-to-end cycle time unless the fraud-investigation subprocess was redesigned with the help of a second AI agent equipped with RAG. These groups consequently reclassified the intake agent from a high-priority candidate to a medium-priority one, conditional on redesigning the fraud investigation with the help of an AI agent and applying a re-sequence pattern to do it earlier in the process. This pattern of initial overestimation followed by evidence-driven revision was evident in the written justifications attached to CP4 artifacts and was confirmed through instructor observations during group supervision sessions. It illustrates the intended function of the checkpoint chain: not to discourage AI adoption, but to ground it in process-level evidence accumulated across phases. A comparable pattern of evidence-driven revision appeared at CP5: several groups that had modeled a fraud-screening path in BPMN (M4) introduced an explicit human-approval step in their `n8n` implementation once they encountered the structured-output reliability limitations of the local LLM, converting a conceptual governance requirement into a concrete workflow control.

Process Identification Phase (M1): No dedicated submission point was designed for students to submit their process architecture and deliverables related to CP1, as communication for this phase occurred primarily through in-class presentations. The absence of such a point limits data capture, and we therefore plan to introduce a dedicated submission box in the next course offering.

Process Discovery Phase (M2): Took longer than expected; the cancel-region pattern proved demanding and will be deferred to M4 in the next offering. Most groups also postponed formal submissions until the deadline despite completing labs early; future offerings will require submission immediately after group presentations.

Process Analysis and Redesign (M3): Took 1.8 weeks, in line with expectations. Applying redesign patterns consumed most effort; re-simulation after model changes proceeded quickly. `ProM` performance spectrum analysis required more time than `WoPeD`'s built-in KPIs.

Process Redesign Using BPMN (M4): Took 1.7 weeks, as expected. Students needed to recap cancellation patterns introduced in M2; deferring those to M4 (planned for next offering) should reduce this overhead. Despite this, students engaged more purposefully with exception-handling patterns once their practical role in redesign was apparent.

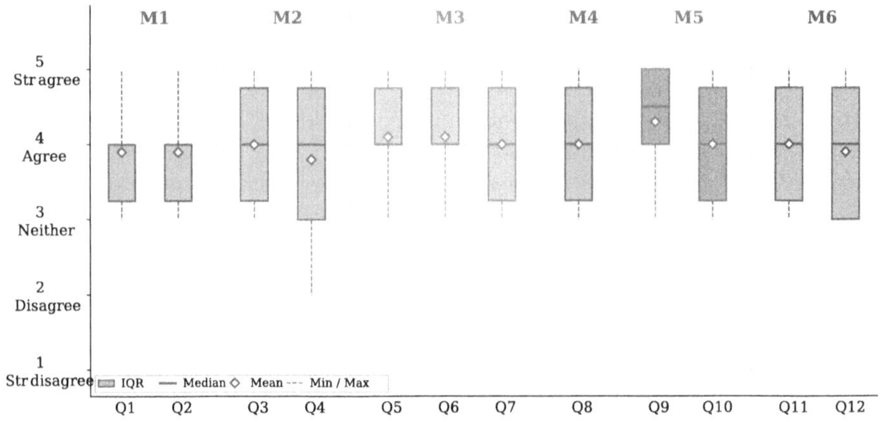

Fig. 3. Perceived learning value per module activity (10 group responses, 5-point Likert scale). Boxes show the IQR; horizontal lines indicate medians; diamonds indicate means; dashed whiskers extend to min/max.

Implementation Phase (M5): Students completed M5 faster than the expected two weeks, attributed to cumulative scaffolding: they had already built an AI agent (M1), evaluated `n8n`'s pattern support (M2), and modeled agent roles and exception paths (M3–M4). Prior checkpoints appeared to pay off at implementation time.

Process Monitoring and Mining Phase (M6): This module proceeded as expected; on average, students submitted both the assignment and the lab within 9 days.

Evidence on EQ3 (Student Perceptions): A voluntary end-of-course survey (10 of 24 groups responded) had each group rate 12 Likert statements (1–5) on the perceived learning value of each module's primary activity: Q1–Q2 (M1), Q3–Q4 (M2), Q5–Q7 (M3), Q8 (M4), Q9–Q10 (M5), Q11–Q12 (M6). Figure 3 summarizes the responses; full survey items are available in the public repository[2].

Overall ratings are positive: all 12 items have a median of 4 (Agree), and no item falls below 3. The M5 implementation items show the highest ratings (Q9: mean = 4.3, median = 4.5), consistent with the LMS-based analysis finding that students completed M5 faster than expected, corroborating the interpretation that earlier checkpoints provided effective scaffolding for implementation. Q4 (workflow-pattern analysis in `n8n`) shows the widest spread (min = 2, mean = 3.8), reflecting variation in groups' prior experience with workflow engines. The survey is limited by a 42 % group response rate and single-instantiation scope; results should be treated as indicative rather than conclusive.

[2] https://github.com/jalaliamin/AI-decision-checkpoints/blob/main/Survey_Questions.md.

5.1 Threats to Validity and Limitations

Several limitations bound the interpretation of the findings above and warrant explicit discussion.

Internal Validity: The reflections rely on a single instantiation of the framework delivered by the authors. Researcher-instructor overlap may have shaped both the design and the interpretation of student behavior, even though triangulation with LMS event logs partially mitigates this risk. Furthermore, the course is still in progress, and individual learning outcomes via the final exam have not yet been assessed.

Construct Validity: The LMS captures only those interactions that occur at the predefined touchpoints. Cognitive work conducted outside the LMS (peer discussion, off-platform prototyping, or iterative refinement of models) is not directly observable and is approximated indirectly through artifact submissions and teaching-team observations [15]. The artifacts produced at each checkpoint also vary in granularity, which constrains comparability across phases.

External Validity: The framework was instantiated in a single Master-level program with a banking onboarding case. While the case is intentionally rich in regulatory constraints and heterogeneous data, generalization to other domains, study levels, or institutional contexts requires further evidence. The choice of supporting tools (`draw.io`, `WoPeD`, `ProM`, `Celonis`, `Signavio`, `Trisotech`, `n8n`, `Ollama`, `Qdrant`) is opinionated and may affect transferability.

6 Conclusion

This paper presented the Banking case and the AI-decision checkpoint framework that structures it. Six checkpoints, one per BPM lifecycle phase, turn the lifecycle from a passive organizer into a chain of evidence-grounded decisions about AI adoption, instantiated in a Master-level course with a continuously evolving banking onboarding artifact.

Formative evidence suggests three preliminary conclusions. First, checkpoints appeared to encourage students to treat AI as conditional on process-level value: groups revised initial over-estimates once simulation evidence accumulated, and incorporated governance structures in their BPMN and implementation artifacts. Second, the discovered process map from LMS's log characterizes submission-timing patterns and highlights two actionable adjustments: deferring cancellation patterns to M4 and adding a CP1 submission touchpoint for systematic data capture. Third, a voluntary group survey (10 of 24 groups) indicates that students perceived positive learning value across all modules (all medians ≥ 4), with M5 implementation activities receiving the highest ratings, consistent with the LMS evidence that prior checkpoints appeared to have effectively scaffolded that phase.

For educators, the framework integrates AI into existing BPM courses without replacing the lifecycle backbone or mandating specific tools. The key transferable principle is the *evolving-artifact* structure, which ensures students revisit

AI decisions with accumulating evidence rather than treating each module as an independent exercise. Future work will extend LMS instrumentation for richer data capture per checkpoint and, after several course runs, codify recurrent reasoning patterns into structured checkpoint templates to support adoption in other institutional contexts.

References

1. Dumas, M., La Rosa, M., Mendling, J., Reijers, H.A.: Fundamentals of Business Process Management. Springer (2013)
2. Weske, M.: Business process management architectures. In: Business Process Management: Concepts, Languages, Architectures, pp. 333–371. Springer (2012)
3. Van Der Aalst, W.: Data science in action. In: Process Mining: Data Science in Action, pp. 3–23. Springer (2016)
4. Lewis, P., et al.: Retrieval-augmented generation for knowledge-intensive NLP tasks. Adv. Neural. Inf. Process. Syst. **33**, 9459–9474 (2020)
5. Khayatbashi, S., Sjölind, V., Granåker, A., Jalali, A.: AI-enhanced business process automation: a case study in the insurance domain using object-centric process mining. In: International Conference on Business Process Modeling, Development and Support, pp. 3–18. Springer (2025)
6. Moormann, J., Bandara, W.: Where are we with bpm education: a call for action. BPTrends (2012)
7. Dallavalle, S., Ribeiro, L.R.B., Moreira, S.A.S.: Educating for next-generation bpm: a PBL-process mapping blueprint anchored in the drifts of the digital era. In: International Conference on Business Process Management: Educators Forum, pp. 299–314. Springer (2025)
8. Grangel, R., Palomares, N., Sánchez-Galán, M., Luna-Aleixos, D., Llagostera-Reverter, I.: Teaching case: process mining for predicting failure to rescue. In: International Conference on Business Process Management: Educators Forum, pp. 417–426. Springer (2025)
9. Kluza, K., et al.: Revisiting BPMN assignments with ai in mind: insights from experiments with large language models in process modeling education. In: International Conference on Business Process Management: Educators Forum, pp. 358–372. Springer (2025)
10. Reijers, H.A., Mansar, S.L.: Best practices in business process redesign: an overview and qualitative evaluation of successful redesign heuristics. Omega **33**(4), 283–306 (2005)
11. Jalali, A.: Teaching business process development through experience-based learning and agile principle. In: International Conference on Business Informatics Research, pp. 250–265. Springer (2018)
12. Denisov, V., Belkina, E., Fahland, D., van der Aalst, W.M.: The performance spectrum miner: visual analytics for fine-grained performance analysis of processes. In: International Conference on Business Process Management: Demo Track, pp. 96–100. CEUR-WS. org (2018)
13. Gdowska, K., Brzychczy, E., Kluza, K.: Project-based learning in BPM for interdisciplinary groups—teachers' experience. In: International Conference on Business Process Management: Educators Forum, pp. 315–329. Springer (2025)

14. Miri, N., Khayatbashi, S., Zdravkovic, J., Jalali, A.: An object-centric approach to multi-entity educational process mining. Decis. Anal. J. **19**, 100724 (2026)
15. Karunaratne, T., Jalali, A., Assom, L.: What, why, and how of datafication in education–a scoping review. In: International Conference on Education and New Learning Technologies, pp. 3745–3756 (2025)

Teaching Declarative BPM: Where, When and to Whom it Matters

Hugo A. López[1(✉)], Claudio Di Ciccio[2], Thomas Hildebrandt[3], and Johannes De Smedt[4]

[1] Technical University of Denmark, Kongens Lyngby, Denmark
hulo@dtu.dk
[2] University of Utrecht, Utrecht, The Netherlands
c.diciccio@uu.nl
[3] University of Copenhagen, Copenhagen, Denmark
hilde@di.ku.dk
[4] KU Leuven, Leuven, Belgium
johannes.desmedt@kuleuven.be

Abstract. The theory of BPM has, for a long time, discussed the interaction between flows and business rules, in what some call the *declarative vs imperative business process spectrum.* While imperative models constitute the standard in the curricula, with standards such as BPMN being imparted in most educations, declarative BPM has received a much more specialized scope. This paper collects and discusses insights coming from experiences teaching declarative process models in academia, in courses ranging from computer science to business informatics across universities in Austria, Belgium, Denmark, and the Netherlands, with languages such as Declare, DCR graphs, and DMN. We report considerations in curriculum development, considerations for diverse backgrounds, accessibility of tools and materials, and overall student perceptions of declarative modelling. Moreover, it reports some of the challenges of teaching declarative models in the classroom and outlines possible mitigation aspects that could support an easier adoption by the students.

Keywords: Declarative BPM · Declare · DCR graphs · DMN · Education in BPM

1 Introduction

The introduction of the declarative paradigm in Business Process Management (BPM) has given rise to various modeling languages and approaches for constructing, deriving, and improving them to facilitate the various stages of the BPM lifecycle [21]. Declarative Business Process Management (DBPM) is an approach within BPM for the modeling, execution, and mining of business processes that focuses on describing the constraints that processes must fulfill, rather than prescribing how activities should be ordered. Unlike classical imperative process modeling languages (e.g., workflow nets, BPMN), which define

K. Kluza et al. (Eds.): BPM 2026, LNBIP 600, pp. 328–347, 2027.
https://doi.org/10.1007/978-3-032-36710-5_21

explicit control-flow sequences, declarative notations specify activities and a set of constraints governing the causal dependencies between them. These rules can impose control-flow as well as temporal and data-driven requirements. Any execution that satisfies all constraints is considered valid. This paradigm is particularly suitable for domains characterized by high variability, knowledge-intensive tasks, and frequent exceptions, such as healthcare, compliance, and case management [18,53]. However, the languages, approaches, and tools supporting DBPM require a different mindset, which also translates into the need for adapted declarative BPM education.

Over the years, the declarative paradigm has found its way into the BPM curriculum at various institutions, from teaching Dynamic Condition Response (DCR) Graphs and DECLARE for control flow and case-based modeling, to mining Decision Model and Notation (DMN) models to teach process redesign. Yet, they face the struggle of factors such as a maturing, often scattered technological landscape, as well as the inherent cognitive complexities that declarative modeling induces in learners. An initial set of challenges and research agenda was outlined in [54]. Now, more than a decade later, we would like to provide an account of how the educational aspects of DBPM have matured over the years and the remaining aspects to be supported in the future. In this way, this paper provides an overview of some of the efforts to integrate DBPM into BPM education. We collected experiences and answers regarding the ways DBPM has been implemented by instructors over the years from four European countries, in different educational contexts and adopting diverse declarative languages. We use these experiences to identify common themes and trends, which were agreed upon by the author team. As a result, we outline the main challenges in educating DBPM and propose a set of educational and research challenges to better integrate declarative approaches into BPM education.

The rest of this paper is structured as follows: Sect. 2 introduces the declarative paradigm for BPM; Sect. 3 outlines experience gained from teaching DBPM and its role in BPM; Sect. 4 discusses extant challenges in education related to DBPM; finally, Sect. 5 concludes the paper with an agenda for the further development of DBPM curricula.

2 Declarative BPM

To illustrate the declarative approaches in BPM, we will briefly describe some of their main exponents: the DECLARE language, the Dynamic Conditions Response (DCR) Graphs, and the Decision Modelling Notation. The following example illustrates a typical process description formulated through behavioral rules in a declarative way. It serves as the basis for the DECLARE map, the DCR Graph, and the DMN table in Figs. 1, 2 and 3, respectively.

Example. A product ordering process specification consists of the following constraints:

- The process starts whenever a customer submits an order,
- Whenever a customer submits an order, it needs to be finalized by the seller. Conversely, a seller cannot finalize orders not submitted by the customer.
- Once an order is finalized, it must be confirmed or rejected.
- A confirmed order may be rejected later on (due to "late" problems), but not vice versa: a rejection cannot be reverted.
- An order can be confirmed only if it has been shipped before.
- An order may be rejected autonomously by the seller. However, if the warehouse notifies the seller of a shipment issue, then the seller has to reject the order.
- The warehouse decision is firm: either the order is shipped, or a shipment issue is notified.
- An order can be finalized at most once.

A valued customer is eligible for express delivery with expensive orders (based on the DMN table in Fig. 3).

Declare and LTL_f Semantics. One of the declarative process modeling languages is Declare [19]. Declare specifications consist of a set of constraints over activities, each typically expressed using templates grounded in Linear Temporal Logic on Finite Traces (LTL_f). LTL_f provides a formal semantics for specifying temporal properties of sequences of events, making it well-suited for reasoning about process executions. The repertoire of standard Declare templates aims to cover the most common patterns used in behavioral process verification (based on a work surveying those in the context of industrial applications [22]) while hiding the complexity of formal logics for their expression. For example, a Response (a, b) constraint between activities a and b dictates that whenever a occurs, b must eventually follow. In LTL_f, this can be expressed as $\square(a \rightarrow \lozenge b)$, where $\square$ denotes "globally" (always), and $\lozenge$ denotes "eventually". The templates are associated with a graphical notation. Response (a, b), e.g., is diagrammatically represented as

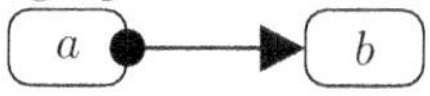

.The example of an order handling process mapped in Declare is depicted in Fig. 1.

Dynamic Condition Response (DCR) Graphs. Another formalism in Declarative BPM is the Dynamic Condition Response (DCR) graphs [29]. While DCR graphs and Declare share some principles, i.e. activities are represented as nodes, and constraints as relations between nodes, other aspects separate the two notations: first, the DCR notation introduces a minimal set of constraints (four in its original presentation) compared to 18+ standard Declare patterns; second, its underlying semantics is given in terms of labeled transition systems; third, each activity node is labelled with an activity name and a finite set of roles,

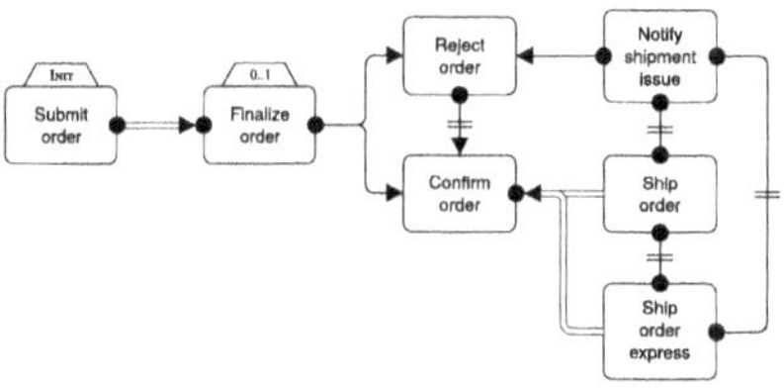

Fig. 1. A Declare map

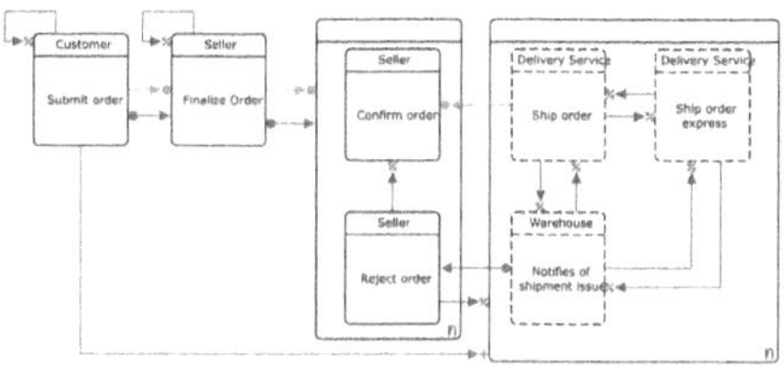

Fig. 2. A DCR Graph

declaring which role can execute the activity; finally, DCR graphs support *defeasible* constraints, which cannot be directly mapped by DECLARE constraints. In short, the state of a DCR graph is given as a *marking* of each activity, recording if it has been executed in the past, whether it is pending, and whether it is currently excluded/irrelevant – depicted graphically as a dashed box (cf. Fig. 2). A relation from an activity a to another activity b either constrains the execution of the activity b (e.g., a condition relation from a to b declares that b can only be executed if 1) it is included, and 2) if a is included, then a must have been executed in the past) or defines a change of state of activity b, when activity a is executed. The notation has been extended with a range of additional features, including data, computations and guarded rules, time, subprocesses, and choreographies. In this paper, we focus on its classical definitions, using the nesting of activities to denote that a constraint is applied to each of its constituents.

Decision Model and Notation (DMN). While DECLARE and DCR graphs focus on temporal constraints, the Decision Model and Notation (DMN) complements the two notations by providing a standard for modeling decision logic. DMN separates decision-making from process flow, typically using decision tables to specify how outputs are derived from inputs (cf. Fig. 3). In a declarative setting, DMN can be used as a qualifier for activities, simplifying otherwise complex logic that would otherwise be directly embedded into process models. DMN has been integrated for both DECLARE and DCR graphs [30, 47].

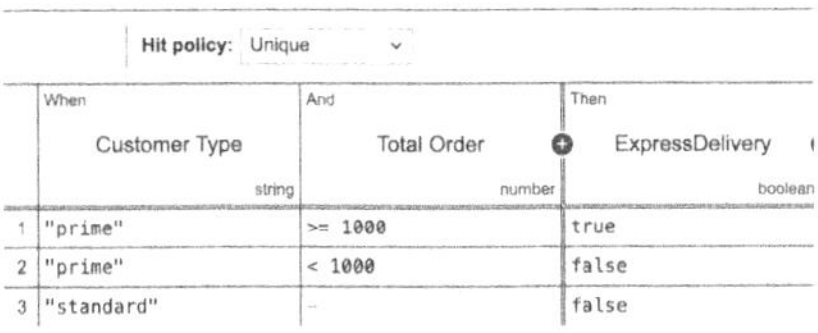
Hit policy: Unique

	When Customer Type string	And Total Order number	Then ExpressDelivery boolean
1	"prime"	>= 1000	true
2	"prime"	< 1000	false
3	"standard"	-	false

Fig. 3. A DMN table (excerpt)

Table 1. Courses teaching DBPM material (a non-exhaustive sample).

Course	Eds.	Level	Audience	University	Country
Knowledge Management and Business Intelligence	16	MSc.	Business Engineering, Information Systems Engineering, Information Management, Computer Science	KU Leuven	BE
Business analysis	16	MSc.	Business Engineering, Information Systems Engineering, Information Management	KU Leuven	BE
Reactive and Event-Based System	6	BSc.	Computer Science	U. Copenhagen	DK
Development of Cloud-Based Health Apps	5	BSc.	Health Science, Global Health and Health Informatics	U. Copenhagen	DK
Process Mining	5	MSc.	Business Informatics, Data Science, Artificial Intelligence	U. Utrecht	NL
Health Data and Interoperability	4	BSc.	Public Health Science, Global Health and Health Informatics	U. Copenhagen	DK
IT and Business Process Modeling	3	BSc.	Business Informatics	IT University Copenhagen	DK
System Integration	3	MSc.	Computer Science	Technical University of Denmark	DK
Process Modeling	3	BSc.	Information Science, Computing Science	U. Utrecht	NL
Formal Foundations of Business Process Modeling	3	MSc.	Information Systems	WU Vienna	AT
Process Mining Applications of Data Science	2	BSc.	Business, Economics and Social Science	WU Vienna	AT
Process Oriented and Event-Based Systems	2	MSc.	Computer Science	Technical University of Denmark	DK
Process Mining	1	MSc.	Computer Science	Technical University of Denmark	DK

3 Development of DBPM Curricula

The development of our educational practice insights takes as a departure point the experience of the author team teaching declarative and decision process models across four countries in twelve courses across bachelor and master programs in Europe. Table 1 documents each of the courses, as well as their main intended audience (Bachelor or Master level), and their provenance (e.g., Information Systems or Computer Science). While the sample is not exhaustive, it aims at balancing geographical distribution and level of expertise, allowing us to compare different strategies when imparting DBPM material.

3.1 Stakeholders, Their Expected Backgrounds, and Coping with Diversity

Overall, the courses where DBPM materials are addressed are in three different fields of education: computer science, business informatics, and health informatics. While some understanding of software development and IT sits at the intersection, the educational background and expected competencies are diverse. Students from business and healthcare informatics often have strong competencies in information systems, including performing fieldwork, organizational management, and enterprise architectures. This comes at the expense of foundational studies in computer science, and they have not been exposed to background material such as modal or temporal logics. On the other hand, the courses in computer science education require maths and experience in software development as prerequisites, but students have rarely worked on projects at the intersection with stakeholders (which is an ability required for process mapping/elicitation tasks). Between the two educations, the health informatics students have specialized knowledge in the healthcare domain and a few introductory courses in software engineering and databases.

The motivators for students in the three types of education differ. Feedback from health and business informatics students emphasizes *the value of modeling*, as in how models help stakeholders in the decision-making process (for instance, improving quality, automation, or optimization). In contrast, computer science students tend to appreciate the *technical challenges* associated with the declarative mindset. Often, computer science students would engage in the programming of technical artifacts such as declarative process engines, verifiers, or extensions for process mining, making a transition from the use of foundational material to the creation of software artifacts.

Another differentiator sits at the core of the courses, where some specialize in the modeling and integration of models, while others focus on the implementation via process-aware information systems and process mining techniques. The inclusion of DBPM tools provides alternatives to students. Declarative process mining, for instance, may provide representations that differ from overly fitting procedural spaghetti models coming from direct-follows representations, and an easier description of business rules used for conformance checking vis-a-vis process alignments.

The teaching approach for DBPM material changes depending on the assumed background of the educational programs. For instance, in computer science education, teaching DECLARE based on its underlying LTL_f formulas aids understanding based on previous knowledge. For business and health informatics students, it may be counter-productive. In those cases, teaching using natural language patterns or positive/negative examples of satisfied templates is preferable.

In most of the considered courses, DBPM techniques cover a part of the overall content, alongside procedural approaches in BPM. The latter are typically supported by standardized and more mature modeling notations such as BPMN or workflow nets [1,21]. Thus, educators need to explain the position of declarative models as a tool part of the modeling repertoire. The emphasis given is mostly on explaining when declarative representations can be more succinct, easier to use, and explain to the business in terms of control flow, and how DMN can offload complex reasoning and fragmentation in procedural systems.

3.2 Integration of Declarative Models Into the BPM Lifecycle

In the next paragraphs, we will discuss the current role of declarative and decision models across the BPM lifecycle [21], as well as some opportunities for further support.

Identification. Typically, process scoping is one of the most difficult tasks for our students, especially for those who are more technically-minded. Not all courses consider scoping as a core activity in the curriculum, given the multiple types of learners' profiles. This phase, being concerned with high-level aspects such as the process landscape and value-chain analysis, does not usually define a specific process modeling notation.

Discovery. While the identification phase provides the list of candidate processes, this phase requires our students to lower the abstraction level by mapping their processes using multiple techniques. Depending on the course, this will involve ethnographies, interviews, document analysis, process mining, or user interface testing. In terms of curriculum development, this is the phase where students are introduced to the different process model notations. Declarative process notations are taught together with imperative modeling notations (e.g., BPMN, workflow nets, or UML activity diagrams).

Teaching the process discovery phase requires two main learning outcomes: first, to be able to elicit process information from potential stakeholders, and second, to familiarize students with the syntax and semantics of formal process modeling languages. These tend to be parallel skills, and given the short time spans, sometimes the courses compromise one of the aspects. For computer science students, this tends to be the process elicitation part. Instead, students are handed process descriptions for their modeling tasks. The nature of descriptions may influence the resulting models. If the descriptions have been written with a specific notation in mind (e.g., a baseline BPMN model), their textual descriptions may be biased towards sequential flows [23]. Using the same description style to map declarative models will create *cognitive dissonance*, as students will need to map concepts foreign to the notations, such as flows and loops, resulting in over-constrained models. The second challenge here is the inherent ambiguity of natural language, which may contain unclear relations, underspecifications, unclear references, or inconsistencies in the text [42]. A final challenge comes from the *ontological misalignment* between imperative and declarative process notations. While essentially both imperative and declarative notions of processes describe traces, their metamodels clash with terms that are named differently but represent the same information (e.g., "effects" in some tools for DCR graphs correspond to "service tasks" in BPMN), or where the same term represents different information (e.g., BPMN *events* do not correspond to DCR *events*, as the latter ones encompass tasks in BPMN, too).

Analysis. Modeling is not a one-off phase, and students iteratively refine their models. They typically follow the process model quality assurance layers defined in [21], including syntactic quality (i.e., *does it conform to a valid model?* and *is it a sound model?*), semantic quality (i.e., *does it represent the real-world process?*), and pragmatic qualities (i.e., *does it conform to domain-specific policies?*). We will describe each of these steps below.

Some aspects in *syntactic quality* can be directly enforced by modeling editors for DCR [16,59], DECLARE [11,49], and DMN [37]. One aspect in this category involves the verification of models. While the students apply standard soundness definitions [21] ported to their declarative setting, particular extensions of declarative formalisms may render unsound models. For instance, declarative models may lead to processes with hidden constraints [15], inconsistencies [17], or that are time-locked [9]. Also, just as for imperative choreographies, one may model declarative choreographic models that result in distributed endpoints whose communications are unrealizable. This can be characterised formally for DCR Choreographies as *end-point projectability* [31]. Soundness itself may be difficult to identify with the naked eye, firstly because realistic declarative process models bear a high degree of interrelated constraints, and secondly because the interplay of multiple dimensions, such as time, data, and decisions may hide implicit dependencies. Currently, the concepts of soundness for declarative models are imparted in some of the courses, and the verification of soundness properties via model-checking techniques [39,40,55] is an active area of research, where students can learn the theory but face difficulties applying it, given the lack of tools supporting these aspects.

Multiple techniques contribute to the validation of semantic quality aspects of declarative process models. The most preferred option is simply a discrete-event simulation. This technique not only supports the model's validity but is also the most widely used for reinforcing understanding of the different types of constraints in a declarative model. Simulations also become the guardrails for subsequent model changes. Using test-driven development techniques [64], students define positive and negative control traces to verify whether changes to the constraint set affect model satisfiability. *Open tests* [12] are used whenever tests should be considered as partial specifications, where only the most important ordering relations between events are specified. Checking for process *completeness* is challenging, as it can only be addressed via validation with domain specialists, and does not pertain to specific problems originating from the use of declarative notations. For instance, initial models based on textual descriptions may be incomplete when students only map causal relations between activities, forgetting the produce-consume relations between artifacts generated by the multiple events. In addition, combinations of multiple constraints might yield the same set of accepting traces, complicating comparisons with baseline solutions. As mitigation actions, students can make use of dual-coding tools to align process documentation, such as process descriptions, SOPs, and business rules, and their corresponding declarative process models using semantic annotations [43].

Pragmatic qualities relate to the usability of process models, and while multiple factors affect the comprehension of declarative process models [3,23,24], a salient aspect corresponds to the creation of modeling guidelines for declarative processes. For instance, an important characterization is that the declarative notations used consider events as first-class citizens, and while this notion is naturally refined into system, computational, and data-driven events [44], their ontological misalignment with imperative notions renders common guidelines for activity readability in imperative process models [46] unusable for declarative models. Another aspect concerns model layout: while imperative models have clearly defined starting points, their declarative counterparts may have multiple enabled events at the start of an execution. Tools like DCR-js [59] implement automatic layout techniques to reduce the number of salient constraints, minimize overlapping edges, and impose left-right alignments, but further research is needed to identify the most effective layout techniques for declarative models with full expressiveness.

Part of the analysis phase includes quantitative and qualitative analysis, and while some qualitative analysis techniques can easily be applied to a declarative mindset (e.g., value-added analysis), important analyses, such as waste analysis and cycle-time analysis, have not yet been described for Declare or DCR graphs. Some of the notions needed are either implicit (e.g., handovers) or not first-class primitives in existing declarative simulators, such as queues and resources on hold. Thus, teaching these aspects has so far been done using imperative specifications and simulation tools.

Redesign. The teaching of process redesign is normally focused on the methods rather than on the tools supporting them. At the same time, we are not aware of a process redesign tool specifically designed for declarative process models. Redesign takes place both as a natural step of refinement in the modeling phase, a predecessor to the implementation of processes, and the resulting step of value-based analysis. Notation change may be a useful opportunity in the redesign of processes. First, because strict process flows can be decomposed into precedence and obligatory constraints, adding flexibility in the orchestration of activities, but also because business adopters of the declarative mindset note flexibility, cost, and modeling time as some of the main advantages over traditional imperative approaches [36].

Two alternative workflows can support redesign initiatives. First, it is possible to depart from an imperative model and generate either a Declare or a DCR graph [8,62]. Such a model directly corresponds to the flows accepted in the

BPMN model, but renders explicit the large number of constraints needed to do so. This is an opportunity to apply redesign patterns to offer more flexibility in the models. Direct encodings of BPMN models map to declarative models with numerous constraints, and students can change these dependencies to add more scenarios. The second possibility is to apply test-driven modeling techniques [57, 63]: departing from an unconstrained model, generate representative traces, and gradually add constraints to understand their impact. While these strategies help outline the boundaries accepted by the process, a lack of functionality is the analysis of time and costs, which are covered in imperative models.

Implementation. Mostly catered to computer science students, this phase transitions from process models intended for communication to executable process models. This step requires extending models with additional dimensions and refined constraints, including data and time, service activities (called *effects* in the tools for DCR graphs by DCR Solutions at DCRSolutions.net), and datatypes. This is also the phase where students need to define an architecture that enables them to interface with outside applications, for instance, by developing a presentation layer, defining the sources of events, and linking to a PAIS via existing APIs. A crucial prerequisite for this step is the involvement of industrial partners who implement APIs, thereby demonstrating their practicality in real-world settings. At the end of the course, students will have experimented and implemented both imperative and declarative models in exemplary PAISs, fostering critical thinking.

Process Mining/Monitoring. Material covered in declarative process mining lectures includes process discovery and conformance checking in terms of rule checking and alignments, for which specific process mining tools have been developed [13,28,33]. The selection of logs to evidence the applicability of declarative process mining techniques plays a key role here. If process variants are not properly distributed, the resulting models will be highly constrained and not benefit from a declarative model. Using models with high diversity (proper for the execution of knowledge-intensive processes [18]) may help to illustrate cases where declarative approaches benefit in the comprehension of the processes. Some examples of logs bearing these characteristics are the Sepsis and Traffic Fine Management logs [38,45]. Scalability of declarative process mining tools is also of consideration when students are building projects with their own logs. In these cases, tools like DCR4Py [28], Declare4Py [20], and MINERful [33] may provide an easy access link to their development workflow since they offer APIs and can be used as libraries. A clear teaching opportunity is the integration of PAIS implementations using imperative models and their monitoring using conformance checking techniques.

3.3 Instructional Approaches and Evaluation Techniques

This section describes the learnings from the evaluation techniques applied to the different courses, including Multiple Choice Questionnaires (MCQs), Project-

Based Learning (PBL), flipped classrooms, and gamification. We proceed to discuss each of them below.

Multiple Choice Questionnaires. Being one of the most scalable alternatives, MCQs have been used to evaluate knowledge in DBPM when dealing with cohorts larger than 200 students. An exemplary exam question would require verifying whether LTL_f formulas and DECLARE constraints are satisfied by a set of traces, or complied with by workflow nets and finite-state automata. Our challenge lies in crafting questions that capture high-level ILOs as defined in the Bloom Taxonomy, as high-level objectives such as synthesis, evaluation, and analysis tend to be better mapped to other evaluation methods. Similarly, evaluating the skills in modeling declarative processes (an active skill) cannot easily be mapped to an MCQ test. In those cases, we resort to additional assignments during the course.

Project-Based Learning. PBL has been the preferred evaluation technique for mid-size courses (up to 100 students). In these cases, students implement an open-ended project engaging with an existing organization in a journey from process identification to recommendations or automation. The students are incentivized to apply modeling techniques from both imperative and declarative approaches, and reflect on the best notations to support their processes. Students assess how the different modeling techniques support the process, and this incentivizes them to explore imperative, declarative, and decision-based models, sometimes even complementary. Indeed, the resulting process landscapes may include models from multiple notations if the students deem it appropriate. This evaluation technique supports high-level intended learning outcomes such as modeling, communicating, and implementing process-aware solutions, but it poses some challenges in the implementation of the coursework. First, it requires engagement with external companies, who sometimes lose interest over the course of the project. Second, it requires intuitive and robust tools for declarative tools at all levels of the lifecycle. Third, it benefits from a mixed student cohort of business informatics (who have experience in stakeholder management) and computer science (who have experience in the development of applications). Finally, these courses become difficult to scale given the amount of follow-up required for projects. An alternative to reduce the risks associated with external partners used is to provide closed projects. For instance, one of the courses resorts to LTL_f and DECLARE for the specification of a process-aware system in the education domain itself, encompassing the processes regarding enrollment, the organization and attendance of courses, and the release of diplomas. LTL_f and declarative constraints are used to specify the inter- and intra-dependencies of those processes, focusing on the control-flow dimension.

Flipped Classrooms. In flipped classrooms, the instructional approach reverses the traditional learning model by shifting direct instruction from the classroom to the individual learning space. We have explored flipped classroom education for DPMs, where videos and lecture material are provided beforehand. The challenge with this approach is the lack of accessible self-study material developed for Declarative models. Except for the 2022 Process Mining Summer School's lecture notes [19], most online material resorts to research papers, PhD theses, or company manuals for specific features, limiting the possibilities for students to acquire knowledge independently. Moreover, concepts explored in research papers may be modified once the declarative language increases in expressiveness (e.g., do the same criteria for soundness apply when we move to declarative models with timing constraints?). While advantageous in theory, student evaluations indicate that students still prefer the clarity of a lecture explanation followed by exercise sessions over the current implementation of flipped classroom education.

Gamification. To counter the heavily identified cognitive load that declarative models pose on modelers and users of DBPM, gamification offers a way to bring relief in the form of visual or salient features that render the modeling process into a more tangible or playful environment. [14] introduced extra visualization allowing users to create extra graphs and annotations explaining dependencies in DECLARE models while executing and simulating behavior, fostering an inquisitive approach to understanding the models. [35,61] introduced virtual 3D environments to visualize the context and real-life execution of processes modeled in DCR Graphs, providing an immersive experience that helps users grasp intricate relations but also fosters a more natural relation with the modeling context.

3.4 Overall Student Perception

Typically, high-level process mappings tend to show sequential operations that are better mapped to imperative notations such as BPMN. However, when students work on projects involving a degree of reactive behavior (e.g., mapping user interfaces, games, and healthcare processes) or explicit constraints (e.g., legal or safety regulations), they begin to appreciate the flexibility and explicit modelling of rules offered by DBPM notations.

The DBPM-related topics within the broader courses in which they are taught spark interest in the learners, who often select those as the themes for their BSc. or MSc. theses and research topics. To briefly give an account of it, we take as a sample the academic year 2025-2026 in a Dutch University where Process Mining and Process Modeling are taught. At the time of writing, one Bachelor's and two Master's honors paths (research projects held by students with outstanding curricula), and five Master's and one Bachelor's theses are specifically centered around declarative process mining and modeling. The students who selected these topics tend to have a strong technical or formal background. The novelty of the approach also calls for innovation projects, and some of our students have created startups based on declarative modelling techniques (integrated with imperative models).This aspect can be seen as a sign of the relative novelty of the approach, more at the stage of ongoing research for computing scientists in BPM than supported by mature solutions.

4 Challenges in the Education of DBPM

Based on the experience gained by teaching declarative models in the different lifecycle phases documented in Sect. 3.2, we summarize some of the most pressing challenges in the pedagogical integration of declarative models in three overarching themes: development of pedagogical material, challenges in the understandability of DBPM, and tool support. We will describe them in further detail below.

Development of Pedagogical Materials. One of the greatest barriers to the adoption of declarative techniques is the absence of pedagogical materials for bachelor's or master's students. Classical teaching materials used by the BPM community [21] were designed with BPMN in mind, and a direct application of exercises is likely to yield counterproductive outcomes. For instance, baseline process description repositories (e.g., [10]) are written in a style that best fits an imperative mindset, sometimes even with references to BPMN elements such as XOR gateways and exception signals. Teachers must create their own syllabus, examples, and exams using a combination of sources from research articles, theses, and manuals. While some books and MOOCs describing declarative models exist (e.g., [53]), they do not consider the impact of DBPM throughout the process lifecycle. This, compared with BPMN or other techniques, becomes a deterrent to adopters and a barrier to self-study. Moreover, it adds risks to the sustainability of the courses when teachers change or become ill. We believe that a common effort should be made to describe the foundations of the declarative mindset. This effort, including foundations, examples, teaching material, and evaluations, would benefit from a notation-agnostic approach, so modelers can relate to the foundations rather than the specifics of Declare or DCR. Moreover, teaching material should be customizable for different types of students interested in the area, so both business informatics and computer science students can benefit from it.

While we advocate for a pedagogical approach that integrates multiple modeling styles, we must consider the possible carryover effects coming from the teaching of different notations in sequence. For instance, if students become familiar with concepts such as process flows using an imperative notation, they may develop mental models that influence the quality of declarative models developed afterwards: the latter may be overconstrained to accept only actions in a sequences. An alternative is to reverse the order, introducing declarative approaches first, and then the imperative ones. We hypothesize that, in this way, students may first learn how to think "declaratively", and later simplify notions by taking into account only specific action unfoldings.

A second challenge corresponds to the identification of language features and associated pedagogical content. Both Declare and DCR have been extended since their original presentation to model scenarios where more than control flow is defined. These constructs (data, time, organizational resources, access control, collaborations, computations, choreographies, subprocesses, etc.) add complexity in understanding, modeling, and maintaining models. Moreover, the selection of cases for modeling is just as important. If cases come from closed and controllable environments, students may not appreciate the added complexity of declarative approaches. An alternative would be to capitalize on the interaction with companies and students applying project work in an organization to identify possible cases. For instance, Fig. 4 shows the notation recommendation after students performed process mapping and modeling of real organizational processes (including fieldwork, interviews, and validation). Here, it is important to note an almost even split between BPMN and DCR. These domains may also help students identify similar companies to work in projects with domains that benefit more from a declarative mindset.

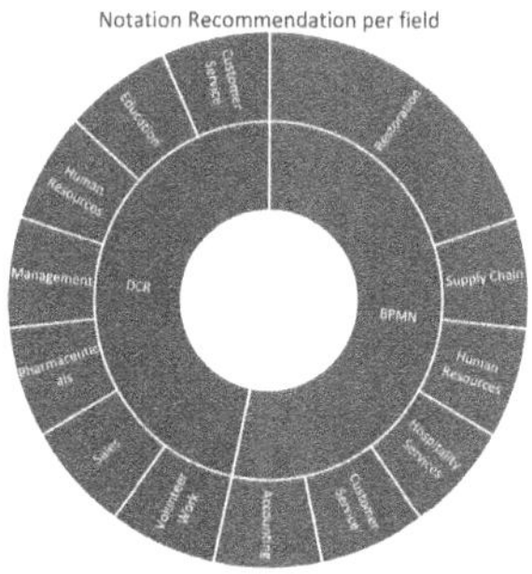

Fig. 4. Notation selection after field work and process mapping in BPMN and DCR graphs. IT and Business Process Modelling course, 2020.

The third challenge is the adaptation to the varying sizes of BPM courses. Our courses in Table 1 have cohort sizes ranging from 20 to 500+ students per edition. Thus, the materials and evaluation strategies need to be adapted accordingly. Student evaluations gravitate towards a preference to work on projects rather than on theory assignments, but project-based work with companies faces scalability and coordination issues. Currently, our courses are considering hybrid approaches, where instructors define the projects. While this type of coursework faces ecological validity constraints, it is logistically easier to implement. Hence, it would be beneficial to create project banks of Knowledge-intensive workflows, especially designed for ILOs related to modeling declarative specifications.

Tools. A challenge is the limited access to robust, user-friendly tools for novice students in DBPM. Although modeling is central across the BPM lifecycle, most focus remains on discovery. Existing editors for Declare [5,11,49,51] and DCR graphs [59] do not always support simulation or analysis. Some are also difficult to deploy at scale; for instance, requiring Java installations for large cohorts (500+ students) is impractical. In contrast, many of the imperative modeling tools we have used in our courses are more mature and cover all lifecycle stages, creating an uneven comparison driven by tool support. DMN tools, however, are better integrated into platforms like Camunda and Signavio, which simplifies adoption, documentation, and support. Notably, the industrial[1] offering of DCR graphs supports simulation and as for the academic [13] tools need no installation and are free for academic use, facilitating an initial adoption of the technologies by novice users.

Tool design must also reflect user profiles. Students are novice users whose early experiences shape adoption. Moreover, business informatics and computer

[1] https://dcrsolutions.net/.

science students have different needs: the former focus more on modeling and analysis, while the latter emphasize software development. Thus, graphical editors and analysis tools are more relevant for practice-oriented students, declarative process APIs are essential for prototyping, and execution APIs and PAIS with support for declarative models are essential for process execution. An open REST API has been available for DCR[2] since 2017. Encouragingly, recent work has also introduced new APIs for DECLARE and DCR [20,28], and there is support for execution environments for pure declarative [7,33,58].

Understandability. While there is a body of work investigating the understandability of declarative models [3,23,24], we believe that understandability goes beyond the sensemaking abilities against a process model, and it includes understanding the underlying process that the model is representing. In the context of teaching, this reflects in integrating different representations of the process as imperative and declarative models. In addition, this also refers to the originating documentation of the process, where (imperative) process descriptions may be combined with compliance rules.

A separate part refers to the understandability of the notations. Being originally developed by formal methods practitioners, the symbols used to denote constraints may not be understandable by novice users. Semantic opacity [48] describes when a novice cannot guess what the symbol means by looking at it. The geometric shapes used in the notations require memorization because there is no logical link between the visual representation and the underlying concept. We see recent work in the redefinition of the DCR and DECLARE notations based on the concepts of the Physics of Notations [11,60], but more language features considered and more empirical evidence of their effectiveness are still required.

5 Conclusion with an Agenda for DBPM in the Classroom

With this paper, we account for the potential of DBPM in an educational context by showcasing multiple instances where declarative methods are being used in the classroom. As a compendium to our manuscript, we make available a (growing) sample of the DBPM teaching material developed during the years at the following address: github.com/Process-in-Chains/Teaching-Declarative-BPM.

Next, we conclude with a summary of our experience and an agenda for future work that is based thereupon, followed by a discussion of the limitations of our present contribution.

5.1 An Agenda for DBPM in the Classroom

The experience discussed throughout this paper reveals many common threads in DBPM. Drawing on these observations, we outline an agenda for closing the

[2] http://documentation.dcr.design/documentation/using-postman-to-access-dcr-active-repository/.

gap between the state of the art in declarative process modeling and its uptake in education.

Systematic Modeling Approaches and Scoped Curricula. A recurring difficulty in teaching declarative BPM is the absence of a structured build-up in model notation and subsequent use: students are often exposed to the full constraint vocabulary at once, without guidance on which constructs matter most, for whom, and in what order. Nagel and Delfmann [50] take a step in this direction with their structured collection of DECLARE tasks of increasing difficulty, ranging from basic concepts and single constraints through to combinations, model behavior, and inconsistency. Future work could develop analogous structured curricula for DCR, and establish a principled account of which constructs are relevant to which audiences. Given most students will have curricula based on seminal BPM works following the BPM lifecycle [21], these can be used as stepping stones for introducing where declarative approaches excel and how certain complexities might not be relevant for all stages. For instance, modeling might rely on simple constraints, while more intricate analyses might convert to the full expressive power of declarative models.

Closing the Gap with Imperative Approaches. A persistent barrier to adopting declarative notations in education is that students typically arrive having learned BPMN or similar imperative languages first [23,54]. Rather than treating imperative and declarative modeling as alternative paradigms to be taught in isolation, a more productive strategy is to teach them in dialogue. Conversion tools that bridge the two paradigms offer one lever here: recent work on compositional mappings from BPMN to DCR Graphs [62] and from sound workflow nets to LTL_f specifications [8] gives educators concrete machinery for asking students to translate between representations, thereby exposing the structural differences in what each paradigm makes explicit or leaves implicit. Jalali [34] showed that perceived ease of use and usefulness of DCR and CMMN varied across student cohorts trained over several weeks, and that perceptions could shift meaningfully with feedback, suggesting that structured cross-paradigm exposure rather than isolated training in one notation may produce a more robust understanding. The empirical work of Haisjackl et al. [27] provides one of the most consistent findings, which is that prior experience with imperative notations can actively impede understanding of declarative models, particularly where graphical representations look superficially similar but carry different semantics.

Test-Driven Modeling. Test-driven development has long been used in DBPM education. Zugal et al. [63,64] showed that test cases reduce cognitive load and improve the quality of DECLARE specifications, while Slaats et al. [57] extend this methodology with a formal theory of iterative, test-driven model development for DCR. Test-driven methodologies have, however, not yet been studied systematically as a *pedagogical* method: the question of whether introducing test cases early in a modeling course improves learning outcomes, or helps develop an intuition for declarative semantics, remains open. Similarly, the automaton-based comparison approach in [56], which makes the semantics of DECLARE specifications visible as finite automata, could be used as a teaching aid, giving students a formal and grounded account of what their models actually permit and prohibit.

AI to the Rescue? The broader educational technology landscape has seen rapid development of AI-powered tutoring and feedback systems. The recent DUET prototype [26] demonstrated LLM-based automated feedback on student-submitted UML and ER diagrams, finding that the system could surface structural differences between reference and student models and generate reflective, targeted feedback at scale. EduClare [51] already takes steps in this direction for DECLARE, combining intelligent tutoring principles with large language models. The growing literature on generative AI for conceptual modeling more broadly [25], spanning UML, ER, and BPMN, provides further methodological grounding for extending this line of work to declarative notations. Declo [4], which allows users to specify DECLARE constraints in natural language without prior knowledge of the notation, further reduces the syntactic barrier. The elicitation of DCR models using LLMs has been explored in [41], identifying challenges for LLMs in constraint identification.

Lessons from Adjacent Modeling Areas. The broader process and conceptual modeling education literature offers transferable lessons. Works on teaching BPMN, including exercise formats, error taxonomies, and real-world case studies as modeling targets, represent a body of pedagogical knowledge that can inform how DECLARE and DCR courses are designed [52]. The gamification approaches point toward engagement-oriented design. Still, some of these initiatives remain unevaluated for actual learning outcomes.

5.2 Limitations

This paper covered various teaching methodologies and declarative languages. However, we covered only the three most prevalent approaches that typically appear in research. Other declarative-minded BPM approaches such as Proclets [2], Guard-Stage Milestone lifecycles [32], or Object-Centric Behavioral Constraints [6] are not covered. Furthermore, we consider a limited set of courses taught in six institutions, primarily situated in computer science and information systems education. The viewpoint also stems from the teacher's perspective with relatively limited student input. Finally, we mostly base our discussion on experience with academic education, without highlighting the professional development that some languages enjoy through tool vendors' tutorials and educational support.

Acknowledgments. The authors want to thank Anti Alman, Paul Cosma, Søren Debois, Patrick Delfmann, Rik Eshuis, Marco Montali, Kate Revoredo, Tijs Slaats, Jan Vanthienen, and Jan Martijn van der Werf for their precious support over the years co-teaching DBPM.

References

1. van der Aalst, W.M.P.: The application of petri nets to workflow management. J. Circ. Syst. Comput. **8**(1), 21–66 (1998)
2. van der Aalst, W.M.P., Barthelmess, P., Ellis, C.A., Wainer, J.: Proclets: a framework for lightweight interacting workflow processes. Int. J. Cooperative Inf. Syst. **10**(4), 443–481 (2001)
3. Abbad Andaloussi, A., Davis, C.J., Burattin, A., López, H.A., Slaats, T., Weber, B.: Understanding quality in declarative process modeling through the mental models of experts. In: BPM, pp. 417–434. Springer (2020)
4. Alman, A., Balder, K.J., Maggi, F.M., van der Aa, H.: Declo: a chatbot for user-friendly specification of declarative process models. In: BPM Workshops, vol. 2673, pp. 122–126. CEUR (2020)
5. Alman, A., Di Ciccio, C., Maggi, F.M., Montali, M., van der Aa, H.: RuM: declarative process mining, distilled. In: BPM, pp. 23–29. Springer (2021)
6. Artale, A., Calvanese, D., Montali, M., van der Aalst, W.M.P.: Enriching data models with behavioral constraints. In: Ontology Makes Sense, pp. 257–277. IOS Press (2019)
7. Awad, A., Awaysheh, F., López, H.A.: Best: a unified business process enactment via streams and tables for service computing. Process Sci. **2**(1), 23 (2025)
8. Barbaro, L., Varricchione, G., Montali, M., Di Ciccio, C.: From sound workflow nets to LTLf declarative specifications by casting three spells. In: BPM Forum, pp. 3–22. Springer (2025)
9. Basin, D., Debois, S., Hildebrandt, T.T.: In the nick of time: proactive prevention of obligation violations. In: CSF, pp. 120–134. IEEE (2016)
10. Bellan, P., van der Aa, H., Dragoni, M., Ghidini, C., Ponzetto, S.P.: Pet: an annotated dataset for process extraction from natural language text tasks. In: BPM, pp. 315–321. Springer (2022)
11. Blasilli, G., Ferro, L.S., Lenti, S., Maggi, F.M., Marrella, A., Catarci, T.: Improving the understandability of declarative process discovery results using easydeclare. Inf. Syst. 102667 (2025)
12. Christfort, A.K., Cosma, V.P., Debois, S., Hildebrandt, T.T., Slaats, T.: Static and dynamic techniques for iterative test-driven modelling of dynamic condition response graphs. Data Knowl. Eng. **157**, 102413 (2025)
13. Christfort, A.K., López, H.A.: DCR-JS: an online environment for declarative process mining. In: BPM Demos, pp. 256–263. CEUR-ws (2025)
14. De Smedt, J., De Weerdt, J., Serral, E., Vanthienen, J.: Gamification of declarative process models for learning and model verification. In: BPM, pp. 432–443. Springer (2016)
15. De Smedt, J., De Weerdt, J., Serral, E., Vanthienen, J.: Discovering hidden dependencies in constraint-based declarative process models for improving understandability. Inf. Syst. **74**, 40–52 (2018)
16. Debois, S., Hildebrandt, T.T., Marquard, M., Slaats, T.: The DCR graphs process portal. In: BPM (Demos), pp. 7–11 (2016)
17. Di Ciccio, C., Maggi, F.M., Montali, M., Mendling, J.: Resolving inconsistencies and redundancies in declarative process models. Inf. Syst. **64**, 425–446 (2017)
18. Di Ciccio, C., Marrella, A., Russo, A.: Knowledge-intensive processes: characteristics, requirements and analysis of contemporary approaches. J. Data Semantics **4**(1), 29–57 (2015)

19. Di Ciccio, C., Montali, M.: Declarative process specifications: reasoning, discovery, monitoring. In: Process Mining Handbook, pp. 108–152. Springer (2022)
20. Donadello, I., Riva, F., Maggi, F.M., Shikhizada, A.: Declare4Py: a Python library for declarative process mining. In: BPM (PhD/Demos), pp. 117–121. CEUR (2022)
21. Dumas, M., La Rosa, M., Mendling, J., Reijers, H.A.: Fundamentals of Business Process Management. Springer (2018)
22. Dwyer, M.B., Avrunin, G.S., Corbett, J.C.: Patterns in property specifications for finite-state verification. In: ICSE, pp. 411–420 (1999)
23. Fahland, D., et al.: Declarative versus imperative process modeling languages: the issue of understandability. In: Halpin, T., et al. (eds.) BPMDS/EMMSAD-2009. LNBIP, vol. 29, pp. 353–366. Springer, Heidelberg (2009). https://doi.org/10.1007/978-3-642-01862-6_29
24. Figl, K., Di Ciccio, C., Reijers, H.A.: Do declarative process models help to reduce cognitive biases related to business rules? In: ER. LNCS, pp. 119–133. Springer (2020)
25. Fill, H.G., Horkoff, J., Fettke, P., Köpke, J.: Generative AI and conceptual modeling: H.-G. Fill et al. Bus. Inf. Syst. Eng. 1–5 (2026)
26. Gürtl, S., Schimetta, G., Kerschbaumer, D., Liut, M., Steinmaurer, A.: Automated feedback on student-generated UML and ER diagrams using large language models. arXiv preprint arXiv:2507.23470 (2025)
27. Haisjackl, C., et al.: Understanding declare models: strategies, pitfalls, empirical results. SoSyM **15**(2), 325–352 (2016)
28. Hermansen, S.V., Jónsson, R., Kjeldsen, J.L., Slaats, T., Cosma, V.P., López, H.A.: DCR4Py: a PM4Py library extension for declarative process mining in Python. ICPM Demos. CEUR (2024)
29. Hildebrandt, T.T., Mukkamala, R.R.: Declarative event-based workflow as distributed dynamic condition response graphs. arXiv preprint arXiv:1110.4161 (2011)
30. Hildebrandt, T.T., Normann, H., Marquard, M., Debois, S., Slaats, T.: Decision modelling in timed dynamic condition response graphs with data. In: BPM, pp. 362–374. Springer (2021)
31. Hildebrandt, T.T., Slaats, T., López, H.A., Debois, S., Carbone, M.: Declarative choreographies and liveness. In: FORTE, pp. 129–147. Springer (2019)
32. Hull, R., et al.: Business artifacts with guard-stage-milestone lifecycles: managing artifact interactions with conditions and events. In: DEBS, pp. 51–62. ACM (2011)
33. Iacometta, C., Di Ciccio, C.: Declarative process mining with MINERful, reloaded. In: BPM Demos & Resources, pp. 328–335 (2025)
34. Jalali, A.: Evaluating user acceptance of knowledge-intensive business process modeling languages. SoSyM **22**(6), 1803–1826 (2023)
35. Jensen, S.J., López, H.A.: Towards immersive environments for declarative process models. In: BPM, pp. 214–231. Springer (2024)
36. Keramidis, P., Hildebrandt, T.T.: On the business value of declarative business process management. In: HICSS, pp. 5631–5640 (2026)
37. Laurson, Ü., Maggi, F.M.: A tool for the analysis of DMN decision tables. In: BPM 2016 Demonstration Track, pp. 56–60. CEUR-ws (2016)
38. de Leoni, M.M., Mannhardt, F.: Road traffic fine management process (2015). https://doi.org/10.4121/uuid:270fd440-1057-4fb9-89a9-b699b47990f5
39. de Leoni, M., Felli, P., Montali, M.: Integrating BPMN and DMN: modeling and analysis. J. Data Semant. **10**(1), 165–188 (2021)
40. Lindner, J.: Exploring the Model Checking Problem of DCR Graphs. Master's thesis, Technical University of Denmark (2026)

41. Lindner, J., López, H.A.: Discovering declarative processes in textual descriptions using large language models. In: ICPM Workshops (2025)
42. López, H.A., Feng, B., Lindner, J., Franceschetti, M., Abbad-Andaloussi, A.: Ambiguity detection in business process descriptions: an evidence and an automated approach. In: BPM, pp. 379–396. Springer (2025)
43. López, H.A., Hildebrandt, T., Debois, S., Marquard, M.: The process highlighter: from texts to declarative processes and back. In: BPM Demos, pp. 66–70. CEUR-ws (2018)
44. López, H.A., Simon, V.D.: How to (re) design declarative process notations? A view from the lens of cognitive effectiveness frameworks. In: POEM. CEUR-ws (2022)
45. Mannhardt, F.: Sepsis cases - event log (2016). https://doi.org/10.4121/uuid:915d2bfb-7e84-49ad-a286-dc35f063a460
46. Mendling, J., Reijers, H.A., Recker, J.: Activity labeling in process modeling: empirical insights and recommendations. Inf. Syst. **35**(4), 467–482 (2010)
47. Mertens, S., Gailly, F., Poels, G.: Enhancing declarative process models with DMN decision logic. In: BPMDS, pp. 151–165. Springer (2015)
48. Moody, D.: The "physics" of notations: toward a scientific basis for constructing visual notations in software engineering. IEEE Trans. Softw. Eng. **35**(6), 756–779 (2009)
49. Nagel, S., Amann, E., Delfmann, P.: Declare-JS: a web-based viewer and editor for declarative process models. ICPM Demos. CEUR (2023)
50. Nagel, S., Delfmann, P.: Towards an e-learning approach for declarative process modeling. In: BPM Forum, pp. 354–364. Springer (2024)
51. Nagel, S., Wolters, A., Riehle, D.M., Delfmann, P.: EduClare - an intelligent tutoring chatbot for teaching declarative process modeling. In: ICPM Doctoral Consortium/Demo (2024)
52. Recker, J., Rosemann, M.: Teaching business process modelling: experiences and recommendations. CAIS **25**(1), 32 (2009)
53. Reichert, M., Weber, B.: Enabling Flexibility in Process-Aware Information Systems: Challenges, Methods, Technologies, vol. 54. Springer (2012)
54. Reijers, H.A., Slaats, T., Stahl, C.: Declarative modeling-an academic dream or the future for BPM? In: BPM, pp. 307–322. Springer (2013)
55. Roveri, M., Di Ciccio, C., Di Francescomarino, C., Ghidini, C.: Computing unsatisfiable cores for LTLf specifications. JAIR **80**, 517–558 (2024)
56. Schützenmeier, N., Käppel, M., Ackermann, L., Jablonski, S., Petter, S.: Automaton-based comparison of declare process models. SoSyM **22**(2), 667–685 (2023)
57. Slaats, T., Debois, S., Hildebrandt, T.: Open to change: a theory for iterative test-driven modelling. In: BPM, pp. 31–47. Springer (2018)
58. Slaats, T., Mukkamala, R.R., Hildebrandt, T., Marquard, M.: Exformatics declarative case management workflows as DCR graphs. In: BPM, pp. 339–354. Springer (2013)
59. Tamo, L.K., Abbad-Andaloussi, A., Trinh, D.M.T., López, H.A.: An open-source modeling editor for declarative process models. In: CoopIS, vol. 3552, pp. 1–5. CEUR-ws (2023)
60. Trinh, D.M.T., Abbad-Andaloussi, A., López, H.A.: On the semantic transparency of declarative process models: the case of constraints. In: CoopIS, pp. 217–236. Springer (2023)
61. Villalobos, J.C., Jensen, S.J., López, H.A.: 3DCR: a tool for immersive process mining. In: ICPM Demos. CEUR-ws (2024)

62. Zhou, Y., López, H.A.: A compositional approach to mapping BPMN to DCR for cross-paradigm process modeling. Presented at the (2026)
63. Zugal, S., Pinggera, J., Weber, B.: Creating declarative process models using test driven modeling suite. In: CAISE, pp. 16–32. Springer (2011)
64. Zugal, S., Pinggera, J., Weber, B.: The impact of testcases on the maintainability of declarative process models. In: BPMDS, pp. 163–177. Springer (2011)

"Are BPM Graduates Future-Ready?" Reframing BPM Competencies for the Future-of-Work

Azumah Mamudu(✉) and Rehan Syed

Queensland University of Technology, Brisbane, Australia
a.mamudu@qut.edu.au

Abstract. Business Process Management (BPM) is at the forefront of digital transformation, yet it remains an open question whether BPM graduates are equipped for this role. This study conducts a thematic discourse analysis to examine how BPM competencies are discursively constructed in academic literature and how they align with emerging skill demands in the World Economic Forum (WEF)'s Future of Jobs reports published between 2016 and 2025. The analysis shows that BPM competency is progressively framed as a hybrid profile spanning process, digital, analytical, social, and adaptive domains, with process and analytical capabilities dominating. This profile is broadly congruent with future-of-work skill projections. In contrast, social, adaptive, and sustainability-related competencies are less strongly foregrounded, despite their growing importance in future-of-work narratives. The study contributes a competency-skill alignment matrix and argues that BPM's challenge is not a lack of relevant competencies, but uneven emphasis on how they are taught, assessed, or resourced. For educators, the key task is to identify ways to orchestrate BPM curricula that integrate process foundations with digital fluency, learning agility, social capability, and transformational leadership competencies, which are increasingly central to the digital transformation of work environments.

Keywords: BPM competencies · BPM graduates · Workforce skills · Future-of-work · Industry 5.0 · Critical Discourse Analysis

1 Introduction

Business Process Management (BPM) has long been positioned at the intersection of process efficiency, organisational design, and digital transformation [7,13], but its competency requirements are being reshaped by rapid technological and socio-organisational change. As organisations increasingly adopt artificial intelligence, automation, advanced analytics, and hybrid forms of work, the capabilities needed to analyse, design, govern, and improve processes are also evolving [12]. This makes BPM a particularly important study domain for the future-of-work, as the role of BPM professionals goes beyond implementing technology to mediating the socio-technical redesign of work [11]. Broader shifts in the nature of

K. Kluza et al. (Eds.): BPM 2026, LNBIP 600, pp. 348–362, 2027.
https://doi.org/10.1007/978-3-032-36710-5_22

work, driven by rapid technological advancement, new business models, and shifting organisational structures, are affecting how work is performed more than before. Research has shown a misalignment between the demand for and supply of BPM competencies [13]. Studies have also examined BPM professional roles, capability profiles, and the mismatch between education and job-market expectations, showing that BPM work requires more than technical knowledge alone [19,23]. Particularly, extant literature highlights the significance of analytical, social, and problem-solving competencies, as well as expertise in process modelling and improvement [10,25]. However, while this work has clarified what BPM professionals currently need, it has paid less attention to how those competencies are being reframed by broader shifts in the nature of work, business models, and organisational structures.

This gap matters because BPM does not exist in isolation from wider labour-market transformations. Digitalisation, automation, datafication, platform-based organising, and human-machine collaboration are altering how work is performed, coordinated, and valued. These developments, frequently captured in discussions such as the World Economic Forum (WEF)'s Future-of-Work Reports[1], emphasise changing task composition, new forms of expertise, and the growing importance of adaptability, judgement, collaboration, and digital fluency. Efforts have been made to examine BPM capabilities at the organisational level in the context of digitalisation and future-of-work dynamics [11,12]. But how future-of-work narratives intersect with individual-level BPM competencies, or which competencies are emerging, stable, or declining in light of these changes, is insufficiently examined.

Accordingly, this study investigates how BPM competencies are constructed in the literature and how they align with WEF's future-of-work discourse. We apply a thematic discourse analysis [2,6] to examine the assumptions, priorities, and normative orientations embedded in the literature and to identify which competencies are legitimised for the future. By comparing BPM literature with WEF future-of-work reports (2016–2025), this study presents a matrix showing areas of convergence and identifies where BPM education and professional development may need to adapt.

First, this research extends BPM competency research beyond static gap analysis, situating competencies within an evolving work landscape. Second, it offers a nuanced account of BPM capability needs by distinguishing stable, emerging, and potentially declining competency areas. Third, it provides actionable insights for educators, curriculum designers, and practitioners preparing BPM graduates and professionals for a work environment increasingly shaped by digital transformation and AI adoption.

This perspective is important for BPM, as it shifts the question from whether there is a competency gap to which competencies future BPM work will require. In doing so, the study aims to sharpen BPM education's conceptual foundations and strengthen the field's contribution to preparing professionals for the evolving realities of work.

[1] https://www.weforum.org/publications/series/future-of-jobs/.

2 Background Literature

Literature on the relevance of BPM skills and the future-of-work converges on a compelling yet underexplored argument that BPM skills constitute a distinctly future-oriented competency set, yet remain critically underdeveloped in both practice and higher education. Hrabal et al. [10] and Sonteya and Seymour [25] collectively challenge the reductionist tendency to equate BPM with mere process mapping or automation, arguing that BPM is fundamentally a human-centred discipline that requires complex integration of technical, cognitive, and behavioural competencies. Both studies identify the same paradox: BPM is widely acknowledged as a strategic organisational priority, yet there is a substantial gap in research and skills related to the human roles that facilitate it. The skills identified related to systemic thinking, emotional intelligence, cross-functional communication, and the rare ability to apply both analytical rigour and human-centred design thinking simultaneously. This is especially important for the future-of-work. These skills are very similar to the robust skill sets becoming more common in organisations going digital. Sonteya and Seymour [25] clarify this notion by characterising the Business Process Analyst as an organisational intermediary that bridges the business and technology domains, a position whose strategic importance intensifies as automation, artificial intelligence, and regulatory complexities exacerbate the gap between technical systems and human decision-making. Hrabal et al. [10] enhance this argument by directly linking BPM competencies to Industry 4.0 and smart manufacturing environments, asserting that process-oriented human skills are not replaced by technological progress but rather become progressively vital alongside it. These findings present a strong case that investing in BPM human capital is necessary for short- and long-term workforce strategies, which are essential to organisational resilience and competitive advantage.

We also reviewed extant literature to explore how higher education institutions, as education service providers, have responded to evolving BPM workforce demands. We provide a synthesis from three lenses: (1) the dominant discourses shaping the discussion; (2) the concrete actions education providers have taken; and (3) gaps, consensuses, and contradictions that shape the field's trajectory.

2.1 Key Discourses from BPM Literature

Prior BPM education literature is organised around four identifiable discursive positions, each reflecting different assumptions about what BPM competency is for and how it should be developed. Recognising these positions provides the baseline for which the thematic discourse analysis in Sect. 4 can surface how competencies are legitimised and ordered in relation to future-of-work demands.

The Skills-Alignment Discourse. The dominant position frames BPM education as a problem of alignment between curricula and labour-market demand. This is most explicit in Mathiesen et al. [18], Senkus et al. [23], and Śliż et al. [24], who operationalise alignment by mapping job-advertisement competencies against course content. Benamati et al. [5] analyse shifts in entry-level IS job requirements, other studies such as [4,16,17], extend the discourse internationally, arguing that even where BPM education exists, it is structurally out of step with employer expectations. The underlying logic is functionalist, claiming that education is instrumentally validated by its fit with industry needs.

The Multidisciplinary/Sociotechnical Discourse. A second discourse reframes BPM not as a technical competency set but as an inherently multidisciplinary, sociotechnical domain. Seethamraju [22] positions BPM as the "missing link" between business, IT, and strategy whereas studies such as [3,4] emphasise the integration of IT, business, and people-related skills. Alghamdi and Linger [1] argue that BPM curricula (i.e., IS2020) underemphasise organisational and sociotechnical aspects despite strong research coverage. This discourse pushes against narrow technical or notation-centric training and invokes "ways of knowing" alongside "know-how."

The Future-of-Work/Digital Transformation Discourse. A third discourse, strongest in the more recent studies [15,23,24,26], positions BPM competencies as "future-proof" capabilities for Industry 4.0/5.0 and digital transformation leadership. Here, process mining, automation, Robotic Process Automation (RPA), analytics, and AI are treated as emerging core content, and BPM is reframed from an operational capability into a strategic, transformation-oriented literacy. This discourse explicitly ties education to the future of BPM jobs.

The Contextualisation/Local Realities Discourse. A fourth, quieter but important discourse foregrounds context. Mahendrawathi and Nathania [15] argue that BPM affordances are actualised unevenly across Indonesian workplaces, shaped by job roles, organisational awareness, and culture. Similarly, Marjanovic and Bandara [17] surface Australian institutional constraints, and Senkus et al. [23] foreground the Polish labour market. This discourse resists universalist curriculum prescriptions and calls for context-sensitive, regionally adapted BPM education.

2.2 Actions Taken by Education Services Providers

Designing Structured BPM Bodies of Knowledge and Curricula: Antonucci [3] proposes a structured BPM curriculum aligned with the BPM lifecycle (foundations; planning/strategy/governance; modelling; analysis; design; implementation; analytics; electives; capstone), positioning it as a common body of knowledge. Syed et al., [26] extend this with a holistic curriculum mapped to the full BPM lifecycle (identification, discovery, analysis, redesign, implementation, monitoring and control) and explicitly informed by programme theory. Pal and Sen [20] propose a detailed BPM course outline embedded within BBA(IS) and MBA programmes. These are the clearest, most enduring curriculum artefacts produced by the field.

Integrating BPM into Wider IS/Business Curricula: Seethamraju [22] and Pal and Sen [20] argue for leveraging existing IS and business courses to embed BPM, rather than isolating it. Benamati et al. [5] document that 69% of U.S. IS schools made curricular changes in a three-year window, typically by adding managerial and experiential components. Alghamdi and Linger [1] advocated for redesigned postgraduate IS programmes through engaged scholarship that couples theoretical knowledge with competencies.

Introducing Experiential, Work-Integrated, and Case-Based Learning: Benamati et al. [5] document the use of offshore partner projects, practicum courses, and industry trips to build higher-level competencies. Marjanovic and Bandara [17] recommend case-based learning and cross-functional integration. Syed et al. [26] embed Work-Integrated Learning (WIL) and research projects to develop problem-solving and critical thinking, with flexible delivery modes (on-campus, blended, online) for diverse learners, including industry participants. Pal and Sen [20] include online simulations and process modelling tools.

Benchmarking and Alignment Methodologies: Mathiesen et al. [18] contribute a structured comparison method for ongoing alignment between BPM education and industry. A few studies [23,24] extend this through systematic labour-market-to-syllabus mapping in Poland, proposing standardisation and national certification of BPMN modelling skills. Bandara et al. [4] conduct one of the earliest worldwide mappings of BPM programmes. These actions institutionalise curriculum design as an evidence-based, recurring exercise.

Emerging-Topic Integration and Pedagogical Flexibility: Recent contributions [15,24,26] report efforts, but also shortfalls, in integrating process mining, RPA, AI, and automation content. Syed et al. [26] show how a programme-theory-informed, iteratively refined curriculum supports flexible delivery; Mahendrawathi and Nathania [15] call for context-sensitive BPM education and international collaboration to adapt curricula for under-represented regions.

2.3 Gaps and Consensuses

Over more than a decade of empirical research, a consistent pattern emerges, exposing structural weaknesses in current BPM education. Studies from 2010 [4] through to 2025 [15], unanimously report a persistent mismatch between industry demand and university supply, with employers seeking BPM capabilities that not only exceed current provision but also differ qualitatively in emphasis. This gap is compounded by the inherently multidisciplinary nature of BPM. Literature [1,3,4,22] provided a common consensus that effective BPM competency depends on the integration of IT, business, and people-oriented skills, rather than mastery within a single disciplinary silo. However, traditional business curricula remain largely functionally organised. As [5,17,22] argue, this structure marginalises process thinking, leaving it to individual initiative and producing graduates ill-equipped for boundary-spanning and cross-functional work. Within this context, soft and managerial skills represent a particularly acute shortfall. Communication, leadership, teamwork, and project management are repeatedly identified as underrepresented in curricula despite strong and sustained employer demand. Importantly, recent work reframes these deficiencies not merely as current gaps but as missed strategic opportunities. A few studies [23,24,26] position BPM competencies as inherently future-ready, especially when aligned with digital transformation agendas that integrate automation, analytics, and governance. Collectively, the literature suggests that addressing the demandsupply mismatch in BPM education requires curricular designs that are explicitly multidisciplinary, process-oriented, and forward-looking, rather than incremental adjustments within existing functional silos.

3 Methodology

This qualitative study adopts a thematic discourse analysis to examine how BPM competencies and future-of-work skill requirements are constructed, legitimised, and ordered across academic and institutional texts. Rather than fixed attributes, competencies are treated as discursive constructs: categories produced, prioritised, and naturalised through the language of scholarly and institutional sources [9]. Guided by Cukier et al. [6] and Alvesson and Karreman [2], the study considers which competencies are named, how they are framed, which are positioned as foundational, and which are rendered peripheral or merely instrumental. We characterise the approach as a thematic discourse analysis rather than a full Critical Discourse Analysis (CDA) in the Faircloughian [9] tradition. The study draws on CDA's central premise that texts construct and legitimise particular knowledge orderings, but does not conduct fine-grained linguistic analysis of grammar, modality, and transitivity. Naming the approach this way avoids overclaiming the analytical machinery applied while retaining the critical perspective that distinguishes this study from a conventional narrative review: an explicit interest in how competing discourses construct, and hierarchically order the imagined future BPM professional.

Data Corpus: *Corpus A* comprised peer-reviewed articles on BPM competencies, BPM professional roles, and BPM education. Eight articles were analysed. *Corpus B* comprised five WEF Future of Jobs reports (2016–2025), outlining employers' expectations for job-market and skill-demand trends to 2030. WEF reports offer a longitudinally consistent, methodologically comparable, and institutionally authoritative body of future-of-work discourse, providing a stable basis for a decade-long comparison. Further details of data corpora can be found in this Appendix[2] The two corpora are deliberately asymmetric in genre rather than importance: Corpus A represents how the scholarly field constructs BPM competency, while Corpus B represents how an influential institutional actor constructs future skill demands. The analysis compares these two constructions without treating either as an objective account of the labour market.

Thematic Discourse Analysis Approach: Text snippets were purposively selected from the data corpus for their relevance to BPM competencies, explicit

[2] Appendix is available online at https://shorturl.at/GWINP.

discussion of skill demand or competency development, and ability to reflect recurring debates in broader future-of-work discourses. The analysis remained grounded in the source texts, with coding extracting explicit statements and the context in which statements were made. This purposive logic aligns with narrative review traditions, where the objective is to synthesise conceptual patterns [6]. This enabled the comparison of BPM competencies with future job skill expectations by identifying BPM-specific competencies, trends shaping the future-of-work, and job skills required for digitally transforming work environments. Additionally, a matrix comparison of the identified BPM competencies and future job skill projections for 2025 to 2030 was conducted. This 2×2 matrix quantifies the intensity of alignment (strong, partial or no alignment) by assessing the extent to which a BPM Competency sub-category is addressed in Corpus A relative to WEF skill demands (see coding procedure in Appendix[3]

To compare how BPM and future-of-work reports construct competency relevance, we identified convergences, tensions and omissions between the two corpora, particularly regarding process, digital, analytical and adaptive or strategic competencies. In this sense, the analysis treats competencies as discursive constructs shaped by institutional, professional, and technological expectations rather than as fixed attributes [9]. To ensure analytical validity, the authors held corroboration sessions to review extracted passages, preliminary codes, and discursive constructs for interpretive consistency and parsimony.

4 Findings

4.1 Trends Impacting the Nature of Work

Technology Reshapes Work, Business Models and Industries: Technology is framed as the central disruptor of work, shifting from a source of task substitution to a driver of structural labour-market reconfiguration. Across the reports, digitalisation, AI, automation, cloud computing and data analytics move from emerging forces to strategic imperatives, with the humanmachine task balance progressively tilting toward automation. By 2025, technology is framed not only as a source of job displacement but also as a generator of new roles, especially in AI, information processing and digital access. The implication for BPM is that graduates must understand automation not just as a tool, but as a basis for redesigning work and governing digital processes.

Data and Analytical Skills Requirements: Data skills evolve from an individual competency into an organisational capability and then into a core professional skill set. Earlier reports emphasise data literacy, visualisation and data-driven decision-making, while later reports elevate analytical thinking, technological literacy, AI and big data as essential skills across occupations. The discourse increasingly treats data competency as cross-industry and strategically

[3] Appendix available at https://shorturl.at/GWINP).

important rather than confined to technical roles. Hence, data literacy, analytical reasoning and AI awareness should be treated as foundational capabilities for BPM professionals.

Pace of Re-skilling and Agility of Skill Transformation: Today's workforce demands a worker who combines technical competency with social, cognitive and adaptive skills. The discourse expands from social skills and technology literacy to lifelong learning, resilience, flexibility and psychological robustness, reflecting the growing volatility of work. By 2025, the scale of re-skilling required is presented as systemic, with many existing skills expected to change or become obsolete. For BPM education, the message is that in addition to current technical knowledge, graduates must also build learning agility, systems thinking and the capacity to adapt continuously.

Diversity, Equity and Inclusion (DEI), Consumer Ethics and Green Technology: Equity concerns and sustainability become more visible over time, moving from peripheral observations to institutionalised priorities. Early reports note gender gaps, exclusion and privacy risks, while later reports frame DEI, climate transition and ESG as major organisational drivers. The 2025 report further reinforces this shift by linking climate adaptation, emissions reduction and sustainability roles to future job growth. For BPM, the implication is that process management must increasingly incorporate inclusion, governance, privacy and sustainability as core design considerations.

4.2 Dominant Future Job Skills that Emerge

Process Automation Skills: Process automation becomes a distinct professional identity rather than a background technology trend. The discourse shifts from automation affecting workers to automation as a recognised specialisation, especially from 2020 onward, when process automation roles gain prominence. This suggests BPM graduates should have a diverse set of skills to design, implement and govern automated processes.

Lifelong Learning: Lifelong learning moves from an implicit expectation to an explicit labour-market necessity. By 2025, the reports present re-skilling as a mass requirement rather than an individual preference, with large shares of workers needing training and many current skills being transformed. For BPM programmes, this means embedding learning agility, self-direction and adaptability as core graduate attributes.

Systems Thinking and Strategy: Systems thinking and strategic advisory capabilities become increasingly relevant as skills focus shifts from operational execution to transformational leadership. The reports' growing emphasis on strategy, organisational development and management analysis supports a more expansive professional identity for BPM practitioners. This means BPM education should position systems thinking, change management and advisory skills as central, not optional skills.

Human-Machine Collaboration: The humanmachine relationship evolves from displacement to augmentation and collaboration. Across the reports, humans retain a comparative advantage in reasoning, advising and communication, while machines take on more routine and information-processing tasks. The BPM implication is that professionals need to design workflows and governance frameworks that integrate human judgement with machine capabilities.

Sustainability and Global Citizenship Skills: Environmental, Social and Governance (ESG) and sustainability become structural drivers of organisational change rather than peripheral compliance issues. Green transition, climate adaptation and sustainability-focused roles are increasingly central to job growth and transformation strategies. For BPM, this means process governance, ESG reporting and sustainability management should be embedded in the curriculum as process-design values to be explored.

Cybersecurity and Digital Trust: Cybersecurity and data governance move from sector-specific concerns to mainstream process governance requirements. The reports show a steady rise in the importance of information security, cybersecurity, and digital trust as skills and organisational priorities. BPM professionals, therefore, need foundational literacy in data protection, security and trusted process design.

4.3 Competency Domains from BPM Literature

The five competency domains presented build on prior typologies [19], which identify technology-, methodology, and people-focused dimensions that broadly map onto the digital, process, and social domains presented here. We extend that typology by splitting analytical and adaptive/strategic competencies, reflecting the discursive shift in recent literature toward evidence-based reasoning and transformational leadership as distinct graduate outcomes.

Process Competencies. are the knowledge and skills required to understand, model, analyse, improve and govern organisational processes. These competencies are grounded in process thinking and closely align with the BPM lifecycle phases of identification, discovery, analysis, redesign, implementation, monitoring, and control [10,14]. This discourse positions process competency as the disciplinary core of BPM and the most legitimate and enduring identity.

Digital Competencies. are the knowledge and skills related to digital tools, data, automation and other information systems. They are reflected in understanding how systems operate, how data is managed and how technologies support process design and execution [16,19,21]. Within BPM discourse, digital competency often signals the interface between technological enablement and process initiatives, especially in relation to software, methods, systems engineering and automation.

Analytical Competencies. are the knowledge and skills needed for critical thinking, creative problem-solving, evidence-based reasoning and data-informed decision-making in BPM initiatives. These competencies are associated with diagnosing process problems, evaluating alternatives and making informed redesign and improvement judgements [19,21]. This competency frames BPM as a cognitively demanding field that creates value through technical execution as well as through interpretation and sensemaking.

Social Competencies. encompass relational/interpersonal and social cognitive skills, such as communication, collaboration, stakeholder management, facilitation, proactivity, resilience, emotional intelligence, teamwork, and domain awareness [8,10]. These competencies are particularly important in BPM because process change depends on working across functional boundaries, negotiating interests and engaging people in transformation efforts. In the BPM literature, social competency frames BPM as fundamentally socio-technical and highly dependent on collaboration for success.

Adaptive/Strategic Competencies. encompass leadership, change management, learning agility, strategic awareness, and the ability to respond to shifting organisational and environmental conditions. These competencies extend beyond immediate process tasks to include guiding transformation, aligning BPM with business goals, and sustaining capability development over time [8,14]. This discourse frames BPM as a change-oriented and strategically responsive capability that emphasises learning, leadership and the ability to navigate transformation and uncertainty.

4.4 Matrix of BPM Competencies Aligned with Future-of-Work Skill Demands

Figure 1 shows a cross-analysis of BPM competencies and employers' predictions of future job skills expected to rise, remain stable, or decline across all industries from 2025 to 2030.[4].

Overall, the mapping reveals close alignment between BPM competencies and WEF's 2025–2030 skills outlook. Of the top 10 fastest-growing future skills, BPM closely aligns with eight: *AI and Big data*, *Networks and cybersecurity*, *Technological literacy*, *Creative thinking*, *Resilience, flexibility and agility*, *Curiosity and lifelong learning*, *Leadership and social influence* and *Analytical thinking*. *Environmental stewardship* is gaining traction in BPM research, with attention being given to sustainable and inclusive BPM. The projected decline in two skill sets between 2025 and 2030, *Manual dexterity, endurance and precision* and *Reading, writing and mathematics*, is reflected in BPM, with most BPM practices moving from manual to digitised systems.

As disciplinary anchor, process competencies form the foundational knowledge for BPM. Their future relevance is mediated by systems thinking, creative problem-solving, data-driven decisions and adaptive learning. This indicates a discursive shift where BPM is not only about designing and controlling processes but also about enabling organisations to respond to volatile, digitally enabled, and knowledge-intensive work environments. In that sense, process competency

[4] https://www.weforum.org/publications/the-future-of-jobs-report-2025/digest/.

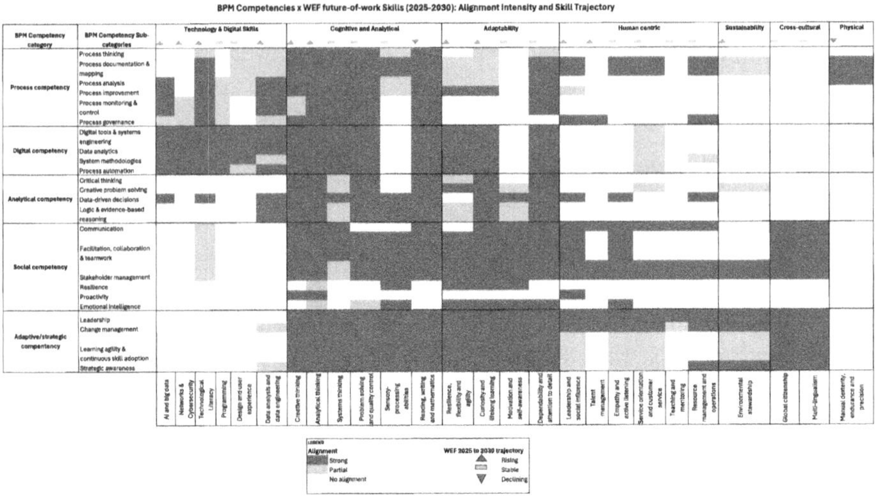

Fig. 1. BPM Competencies and WEF future-of-work skills alignment and trajectory (2025-2030).

is becoming less of a stand-alone skill set and more of a foundation on which digital, analytical, and social capabilities are built.

The digital competency cluster shows strong convergence with the WEF future skill outlook as it aligns with *AI and big data*, *technology literacy* and *cybersecurity*-related demands. However, it also suggests that BPM tends to frame digital competency through an information systems, automation, and systems engineering lens, whereas the WEF foregrounds a faster-moving environment shaped by AI-enabled work, digital adaptation, and continuous upskilling. For BPM education, this suggests that digital competency must go beyond technical literacy to prepare the future workforce for work involving digital infrastructures and AI-supported environments.

Analytical and social competencies align with WEF's future-of-work narrative. Critical thinking, creative problem solving and evidence-based reasoning align with WEF's emphasis on *analytical and creative thinking*. Social competency maps onto *leadership, social influence, resilience*, and *collaborative capability*. WEF does not present the future-of-work as purely technical; instead, it frames the future worker as someone who can think critically, collaborate across boundaries, and sustain performance amid change. BPM scholarship is therefore well-positioned to contribute to this agenda only if social and analytical capabilities are explicitly treated as central rather than peripheral skills.

Adaptive and strategic competencies appear to be the most future-sensitive domain in the BPM mapping. Leadership, change management, learning agility, and strategic awareness are strongly compatible with the WEF's emphasis on *resilience, flexibility and agility*, *curiosity and lifelong learning*, and *leadership and social influence*. This suggests that the most consequential BPM competency shift may be from specialist process expertise toward adaptive orchestration: the ability to guide transformation, support re-skilling, and align process work with evolving strategic priorities. The future of BPM, therefore, may depend less on whether practitioners can improve existing processes and more on whether they can continuously reconfigure processes in response to changing business and technological conditions.

5 Discussion

BPM competencies are typically constructed as a hybrid capability spanning process, digital, analytical, social, and adaptive domains, reflecting BPM's evolution beyond process modelling and optimisation [12]. The literature positions BPM professionals as socio-technical practitioners who combine process expertise with data literacy, digital fluency, collaboration, and change leadership [23]. Thus, BPM competency literature largely reflects the capability profile required for the future-of-work. However, despite broad and forward-looking competency expectations, BPM curricula have not always kept pace, contributing to the persistent demand-supply gap identified in the literature [18,23,24]. The issue is not identifying important competencies, but embedding them in integrated, future-focused curricula. For educators, the challenge is therefore one of implementation rather than awareness.

The study also shows that BPM literature prioritises competencies unevenly: process competencies remain the disciplinary core, while digital and analytical competencies are valued for enabling process improvement, diagnosis, and evidence-based decision-making. Social competencies are frequently discussed but are often treated as supporting capabilities rather than core graduate outcomes [10,25]. Adaptive and strategic competencies, while gaining attention, remain underdeveloped despite the demands signalled in future-of-work discourse. This hierarchy is significant because it shapes what educators prioritise, assess, and resource.

Comparing BPM and future-of-work narratives reveals key misalignments. While BPM aligns strongly with analytical thinking, process improvement, and digital enablement, it places less emphasis on AI, cybersecurity, learning agility, resilience, leadership, and continuous upskilling. These gaps suggest BPM education should move beyond preserving existing curricula or adding isolated digital topics and instead embed adaptive, human-centred capabilities throughout the programme.

This study also demonstrates CDA's analytical value in unpacking how BPM literature legitimises some competencies while making others less visible or more instrumental [6]. It positions process competency as the BPM field's core, digital competency as technical enablement, and social competency as a supporting condition. However, future-of-work literature considers agility, lifelong learning, collaboration, and adaptive judgement as foundational future skills. Thus, the analytical payoff lies not only in identifying competency overlap but in proving how competing discourses shape the imagined future BPM professional [2,9].

For educators, this presents a clear curriculum design challenge. BPM programmes must move beyond process modelling, analysis, and automation topics to build an integrated competency model that reflects the realities of future work demands. Process competencies remain important, but should be taught through transformation-oriented and cross-functional cases. Digital competencies should cover AI literacy and digital trust, not just systems familiarity. Analytical competencies should be developed through data-driven problem solving and evidence-based redesign. Social competencies should be core outcomes assessed through teamwork, facilitation, and stakeholder engagement [17]. Adaptive and strategic competencies should be taught explicitly through change management, leadership, reflective practice, and industry-linked capstone experiences [26]. In short, BPM education should teach students how to improve processes and prepare them to redesign work, navigate uncertainty, and lead transformation.

The analysis also suggests that BPM's future relevance lies in its ability to bridge process excellence and organisational adaptation. The discipline already has the conceptual vocabulary to connect business, technology, and people, but it must now be taught in a way that reflects that integrative role. This is especially important because curriculum redesign often lags behind labour-market changes. This paper, therefore, argues that BPM graduates will be future-ready provided BPM education intentionally shifts from competency recognition to competency orchestration: the deliberate combination of process, digital, analytical, social, and adaptive capabilities in ways that mirror the future work environment.

Finally, WEF future-of-work reports are not neutral narratives of labour market trends. They participate in the social practices they describe [9], as they are produced by an institution that helps shape global economic policy. Hence, their skill forecasts also reflect and reinforce certain assumptions about what constitutes valuable work, who must adapt, and how education should respond. BPM competency literature also reflects competencies that employers consider relevant for BPM job roles. BPM educators should treat these reports as useful but not neutral, using them critically rather than as direct curriculum prescriptions.

6 Conclusion

This study examined how BPM graduates are future-ready by comparing BPM competencies with future-of-work skill demands. The findings show that BPM already covers a broad competency mix of process, digital, analytical, social, and adaptive domains, but these are not reflected equally in BPM curricula.

This paper moves beyond static gap analysis by showing that the key issue is how competencies are prioritised, with process and digital competencies receiving greater emphasis than social and adaptive skills. It also introduces a competency alignment matrix that classifies competencies as stable, emerging, or declining relative to WEF projections. For educators, the implications are to maintain strong process foundations while using transformation-focused cases, to assess social and adaptive skills as core learning outcomes, and to embed AI literacy, learning agility, and sustainability within BPM programmes rather than adding them superficially. Teaching should also infuse industry-related projects, simulations, and reflective problem-solving activities to enable students to apply BPM knowledge in dynamic, real-world work settings. Overall, BPM education must equip graduates not only to optimise processes but also to redesign and adapt them within evolving organisational contexts.

This study has some limitations. WEF reports are influential institutional documents that shape perceptions of future work and essential skills. BPM competency literature also reflects employer skill demands for BPM job roles. The BPM literature corpus is not based on an exhaustive list. Future research should empirically investigate how these findings vary by region and evaluate how BPM competency expectations align with educational quality frameworks like SFIA. Additionally, it would be worth examining how BPM skills are actually taught and assessed in different programmes. Overall, BPM graduates are only considered future-ready if BPM education shifts from merely mapping competencies to actively designing and orchestrating them.

References

1. Alghamdi, A., Linger, H.: Job ready or future-ready? the role of IS research in PG IS curriculum. In: ACIS, Melbourne, Australia, p. 69 (2022)
2. Alvesson, M., Karreman, D.: Varieties of discourse: on the study of organizations through discourse analysis. Hum. Relat. **53**(9), 1125–1149 (2000)
3. Antonucci, Y.L.: Business process management curriculum. In: Handbook on Business Process Management 2. Strategic Alignment, Governance, People and Culture, pp. 547–572. Springer, IHIS (2015)
4. Bandara, W., et al.: Business process management education in academia: status, challenges, and recommendations. Commun. Assoc. Inf. Syst. **27**, 41 (2010)
5. Benamati, J.H., Ozdemir, Z.D., Smith, H.J.: Aligning undergraduate is curricula with industry needs. Commun. ACM **53**(3), 152–156 (2010)
6. Cukier, W., Ngwenyama, O., Bauer, R., Middleton, C.: A critical analysis of media discourse on information technology: Preliminary results of a proposed method for critical discourse analysis. Inf. Syst. J. **19**(2), 175–196 (2009)
7. Dumas, M., La Rosa, M., Mendling, J., Reijers, H.A.: Fundamentals of Business Process Management, vol. 2. Springer (2018)
8. Eicker, S., Kochbeck, J., Schuler, P.M.: Employee competencies for business process management. In: Business Information Systems, pp. 251–262. Springer (2008)
9. Fairclough, N.: Discourse and social change, Polity Press (1992)
10. Hrabal, M., Tuček, D., Molnár, V., Fedorko, G.: Human factor in business process management: modeling competencies of bpm roles. Bus. Process. Manag. J. **27**(1), 275–305 (2020)
11. Kerpedzhiev, G., Lehnert, M., Röglinger, M.: The future of business process management in the future of work. In: ECIS. No. 81 in Research Papers (2016)
12. Kerpedzhiev, G.D., König, U.M., Röglinger, M., Rosemann, M.: An exploration into future business process management capabilities in view of digitalization. Bus. Inf. Syst. Eng. **63**(2), 83–96 (2021)
13. Lederer Antonucci, Y., Goeke, R.J.: Identification of appropriate responsibilities and positions for business process management success: seeking a valid and reliable framework. Bus. Process. Manag. J. **17**(1), 127–146 (2011)

14. Lohmann, P., Zur Muehlen, M.: Business process management skills and roles: an investigation of the demand and supply side of bpm professionals. In: BPM Conference, pp. 317–332. LNBIP, Springer (2015)
15. Mahendrawathi, E., Avisha, N.: Bridging the BPM education gap: reflections on affordances of BPM techniques and a call for collaboration. In: RBPM, PT, and Educators Forum, Seville, Spain, pp. 442–455. LNBIP, Springer (2025)
16. Maleki Shamasbi, S., Van Looy, A., Weber, B., Röglinger, M.: A typological theory of roles in process mining: conceptualising competencies through text mining of job vacancy data. Enterp. Inf. Syst. **19**(11), 2521389 (2025)
17. Marjanovic, O., Bandara, W.: The current state of BPM education in australia: Teaching and research challenges. In: BPM Workshops. Hoboken, NJ, USA, pp. 775–789. Springer, LNBIP (2011)
18. Mathiesen, P., Bandara, W., Marjanovic, O., Delavari, H.: A critical analysis of business process management education and alignment with industry demand: an australian perspective. Commun. Assoc. Inf. Syst. **33**, 27 (2013)
19. Müller, O., Schmiedel, T., Gorbacheva, E., vom Brocke, J.: Towards a typology of business process management professionals: identifying patterns of competences through latent semantic analysis. Enterp. Inf. Syst. **10**(1), 50–80 (2016)
20. Pal, R., Sen, S.: Relevance of business process management (BPM) course in business school curriculum & course outline. In: A Renaissance of Information Technology for Sustainability and Global Competitiveness.AMCIS, Michigan, USA (2011)
21. Ravesteyn, P., Betenburg, R., Waal, B.D.: In search of competencies needed in bpm projects. Commun. IIMA **8**(2), 3 (2008)
22. Seethamraju, R.: Business Process Management: a missing link in business education. Bus. Process. Manag. J. **18**(3), 532–547 (2012)
23. Senkus, P., et al.: Bridging the gap: an evaluation of business process management education and industry expectations - the case of poland. In: BPM: Educators Forum. Utrecht, Netherlands, pp. 250–262. Springer, LNBIP (2023)
24. Sliż, P., et al.: Business process management education in poland: a manifesto for academic teaching. e-mentor **2(104)**, 91–98 (2024)
25. Sonteya, T., Seymour, L.F.: Towards an understanding of the business process analyst: an analysis of competencies. J. Inf. Technol. Educ.: Res **11**(1), 43–63 (2012)
26. Syed, R., Wynn, M.T., Bandara, W.: The design and delivery of a holistic BPM education programme. In: BPM Educators Forum. Utrecht, Netherlands, pp. 201–210. Springer, LNBIP (2023)

Instructional Design in Process Modelling: Surveying Educators' Perspectives

Ilia Maslov[1(✉)] and Stephan Poelmans[2]

[1] Université de Namur, Rue de Bruxelles 61, 5000 Namur, Belgium
ilia.maslov@unamur.be
[2] KU Leuven, Warmoesberg 26, 1000 Brussels, Belgium
stephan.poelmans@kuleuven.be

Abstract. Process modelling (PM) education has gained prominence as organizations need professionals who can translate business logic into formal models. While prior work has examined model comprehension, modelling guidelines, flipped classrooms, gamification, and animations, less is known about how educators combine these elements into instructional design (ID) decisions. This study investigates PM educators' perspectives (N = 13) through an exploratory qualitative survey structured around a five-stage ID framework. Results indicate that educators commonly adopt constructivist, active, and project-based approaches, emphasizing practical engagement and real-world relevance. However, gaps remain in prerequisite support, metacognitive reflection, higher-order Create-level activities, assessment alignment, and the systematic use of animations for comprehension. The contribution is a structured educator-perspective account of current PM instructional design practices, identifying opportunities for course redesign through better alignment of learner needs, learning objectives, assessment, reflective tasks, and dynamic visualization support.

Keywords: Process Modeling Education · Instructional Design · BPMN · Educator Perspectives

1 Introduction

Process modelling (PM) generally involves translating textual descriptions into graphical formats to support communication between business and IT, define roles and responsibilities, streamline or reengineer processes, and enable automation through enterprise systems. Business Process Model and Notation (BPMN) has become the de facto standard in the field, highlighting the need for skilled PM professionals [1, 2].

PM is a complex competency that demands specialized cognitive schemata and experience [3]. It requires aligning representational elements (syntax) with their real-world counterparts (e.g., events or activity durations). PM activates cognitive processes such as abstraction, contextual reasoning, interpretation, evaluation, and collaborative sense-making [4].

For novices, accurately interpreting process models can be difficult, particularly when dealing with complex behaviors represented by elements like gateways, which

K. Kluza et al. (Eds.): BPM 2026, LNBIP 600, pp. 363–377, 2027.
https://doi.org/10.1007/978-3-032-36710-5_23

may not be readily apparent in static diagrams [5, 6]. Beginners tend to follow process flows sequentially, exhibit more frequent visual fixations, and require additional time to process information—indicators of elevated cognitive load [6, 7]. Increased reading effort is associated with lower prior knowledge, as shown by repeated scanning and cognitive strain. In contrast, experienced model readers focus on key decision points and employ more efficient navigation strategies [8]. A major challenge lies in the limited use of advanced BPMN constructs, with fewer than 20% of the notation's elements employed in real-world practice—primarily due to their perceived complexity and a general shortage of process modeling expertise [9].

A widespread deficiency in PM expertise often results in suboptimal model quality across organizations [10]. With the growing demand for proficient process modelers, effective PM training has become a key priority within the research landscape [4, 11]. Instructors often struggle to convey process modeling concepts effectively, as they lack tools that dynamically illustrate model behavior and support students' comprehension and reading strategies [2]. Improved educational approaches may help bridge existing skill gaps and enhance modeling quality in practice [10].

Instruction in process and conceptual modelling often remains ad hoc, shaped by educators' personal perspectives rather than explicit pedagogical paradigms or learning approaches [12, 13]. As PM education grows as a field, current research increasingly seeks to systematize instruction [11–13]. Instructional design offers a way to ground pedagogical theories in concrete instructional approaches that can effectively and efficiently deliver PM learning outcomes [11, 13–15]. The above discussion leads us to the following Research Question (RQ):

How do educators in PM approach their instructional design?

To answer the RQ, we conducted an in-depth exploratory survey on PM educators' perspectives using semi-structured questions. The questionnaire was grounded in a five-stage instructional design (ID) framework, which includes project initiation, needs analysis, instructional strategy development, model assessment, and evaluation of the instructional approach [16, 17]. By structuring the questions around these stages, the study systematically captures educators' perspectives on learner profiling, teaching strategies, assessment practices, and course refinement. Prior work has examined process model comprehension, modelling guidelines, gamified and flipped learning approaches, and animation-based support. However, less is known about how PM educators combine these elements into instructional design decisions across learning objectives, prerequisites, teaching methods, assessment, and course refinement. The contribution of this paper is therefore not to introduce PM education as a new topic, but to provide a structured exploratory account of educators' instructional design practices in PM.

The paper is structured as follows: Sect. 2 provides a literature review of existing research. Section 3 outlines the methodology, explaining how the survey was conducted. Section 4 presents findings from the survey, and Sect. 5 discusses the results, limitations, threats to validity, and future research directions.

2 Background and Related Work

PM is a key organizational practice that supports communication, analysis, and optimization of business processes [1]. As a specialized form of conceptual modelling, PM transforms real-world complexities into structured visual representations, enabling business and IT stakeholders to manage, automate, and improve workflows [18, 19]. BPMN is the industry standard for PM and offers a rich visual language to capture operational process semantics [1].

PM requires both encoding (creating) and decoding (reading) models—a cognitively demanding task involving abstraction, visual reasoning, and interpretation. Comprehension, defined as the alignment between the modeler's intention and the reader's mental model, is foundational for effective PM and a prerequisite for learning [6, 13]. Research shows that comprehension is shaped by user expertise, domain knowledge, cognitive load, and model characteristics such as complexity, layout, and secondary notation [5, 6, 20]. Given the versatile nature of both processes and modeling languages, process modeling remains a challenge [1, 21].

While a lot of research is done on comprehension, teaching has received less attention. Recent work has increasingly recognized PM education as a distinct area of inquiry [13, 22–24]. Despite PM's proven value to organizations [1], a persistent skills gap underscores the need for focused educational efforts [10]. Studying how modelers of different expertise tend to model is very important. For instance, eye-tracking studies showed that novice process modelers may need more support, especially in learning contexts [2].

Comprehension remains a prerequisite for learning, yet learning also entails sustained engagement with cognitive tasks such as visualizing, contextualizing, evaluating, and collaborating [13]. In educational settings, PM activates diverse cognitive functions—from abstraction and contextualization to judgment and communication—especially in collaborative environments [11, 25].

Several guidelines have been proposed to support novice modelers and improve comprehension, including 3QM, 7PMG, GoM, and classroom-oriented modelling heuristics [10, 26–29]. These works identify factors such as syntactic correctness, semantic clarity, pragmatic quality, model size, layout, and labelling as important for comprehension. However, applying such guidelines in learning scenarios still requires targeted instructional support.

Recent research explores practical instructional designs for PM education [11, 13]. Studies have demonstrated the potential of flipped classrooms, project-based learning (PBL), and experiential activities to improve student engagement and learning outcomes [14, 15, 30]. These approaches emphasize hands-on modeling, real-world case integration, and continuous feedback, recognizing that PM is not only a matter of declarative knowledge but also a procedural and communicative skill. Modeling effectively in practice, especially when requirements are ambiguous or non-textual, demands more than theoretical understanding. These studies also highlight the importance of structured guidance and instructor support to balance student autonomy with effective learning. PM learning objectives are generally anchored in teaching BPMN syntax and process patterns, the latter of which are less formally defined despite attempts to systematize them (e.g., [31]). Using animations in teaching PM is a recent initiative since it improves comprehension of PM notation elements and hence should be used by educators in their instructional approaches [5, 12, 22].

3 Methodology

We conducted an exploratory qualitative survey of educators with experience in teaching process modelling, BPMN, or closely related conceptual modelling courses. The survey was administered online and allowed anonymous participation. Respondents were asked to describe one selected course that they had preferably taught recently, including its target learners, learning objectives, teaching methods, assessment practices, use of tools, and perceived challenges. Because the aim was exploratory and interpretive rather than statistical generalization, the survey relied mainly on open-ended questions, complemented by selected Likert-type items.

We aligned the design of the survey with Schoo and Kumar [32] in interpreting instructional and educational system challenges through the lens of complexity theory. This approach suggests that instruction should be perceived as a "design problem". Instructional design is typically broken down into five stages: (1) initiating the project, (2) conducting a needs analysis, (3) crafting instructional strategies, (4) evaluating modelling outcomes, and (5) assessing the instructional design itself [16, 17].

The questionnaire was constructed from this five-stage ID logic and from prior PM and conceptual modelling education literature. Questions on learner groups, learning objectives, prerequisites, and Bloom's taxonomy were informed by work on conceptual modelling education and learning objectives [33, 34]. Questions on teaching strategies, course materials, exercises, tools, and active learning approaches were informed by prior studies on PM education, project-based learning, flipped classrooms, and modelling tool support [11, 13–15]. Questions on animations were included because dynamic visualization can support process model comprehension, especially for semantically complex constructs such as gateways and loops [5, 12, 22, 24]. Questions on assessment were grounded in literature on model quality, formative assessment, and modelling tool support [1, 13, 35].

The initial stage involves laying the groundwork for instructional design in PM, including preparing the context before identifying learner needs, instructional goals, and the broader problem scope. In the second stage, the emphasis shifts to profiling the learner group and specifying outcomes [17]. It also involves aligning objectives to the real-world context, often through Bloom's taxonomy [33]. In PM education, special attention must be paid to learners' prior experience, knowledge levels, and cognitive capacity [34]. We therefore asked about target learner groups, learning objectives, Bloom's Taxonomy, prerequisites for attending PM courses, educational level, and student type.

The third phase centers on selecting effective teaching strategies [17], including the use of learning theories, active methods such as project-based learning or gamification [13], and the integration of learning resources and modelling tools suited for novices [1, 33, 34]. We therefore asked PM educators about learning paradigms and instructional approaches, course materials, modelling tools, exercises, and the use of process model animations as a scaffolding technique in teaching PM [12].

The fourth phase involves evaluating student progress, including assessment instruments such as (semi-)automated formative evaluations [13] and supportive tools [35].

Model quality can then be reviewed through syntactic, semantic, pragmatic, and stylistic lenses [1]. We therefore asked educators to comment on formative and summative assessment methods.

The fifth phase focuses on evaluating the instructional design itself and highlights the iterative character of the ID process [16]. We therefore asked educators to reflect on challenges, future course improvements, and differences between teaching PM and teaching other conceptual modelling topics.

We used purposive, non-random sampling to identify potential PM educators. First, we extracted authors from the PM education section of a prior conceptual modelling education literature review [11]. Second, we added authors from papers presented at the BPM Educators Forum 2024. This resulted in 188 author-name rows. The survey was distributed twice, in October 2024 and February 2025, to publicly available email addresses. Approximately 49 addresses were invalid or unavailable, leaving about 139 reachable contacts. The survey link was opened 26 times. Fifteen respondents completed the survey at least up to the Bloom's taxonomy section. Two responses were excluded because all Bloom's taxonomy items were answered with "No", suggesting insufficiently informative data for the intended analysis, and one response was excluded because the respondent did not continue after the Bloom's taxonomy questions. The final sample therefore consisted of 13 educators. The resulting completion rate was approximately 9.4% of reachable contacts and 50% of opened survey links.

Data quality was assessed through completeness, internal plausibility, and relevance to PM education. Responses were retained when they contained substantive information about an actual PM or BPMN-related course and allowed interpretation of the respondent's instructional design decisions. The time spent on the survey ranged from approximately 40 to 120 min, which suggests that the retained responses represent relatively detailed qualitative accounts rather than short opinion statements.

Qualitative responses were analysed inductively using qualitative analysis software [36]. The coding process followed an open thematic approach: responses were segmented into meaning units, grouped into codes representing recurring instructional design practices, challenges, tools, assessment methods, and improvement ideas, and consolidated into broader themes. Frequency counts were used only as descriptive indicators of how often a theme appeared in the sample. Because of the exploratory design, small sample size, and qualitative nature of the data, these counts should not be interpreted as statistically generalizable prevalence estimates.

The final questionnaire, anonymized response matrix, coding overview, and README file are available in a persistent open-science repository: https://zenodo.org/records/21260045. The repository excludes names, email addresses, affiliations, and any other identifying information. During double-blind review, author-identifying metadata was removed; for the camera-ready version, the complete research instrument is made accessible to support transparency and reproducibility.

4 Data Analysis Results

The results are reported as descriptive findings from an exploratory qualitative survey. Where Likert-type items are summarized using average scores, these values are used only to support interpretation of tendencies within this small sample. They are not treated as inferential statistics and should not be generalized to the full population of PM educators.

4.1 Educators' Profile

The educators were highly experienced: on average, they had first taught PM 13.6 years earlier (SD = 7.34). Across their careers, most had taught PM at Bachelor's or Master's level, and some had also taught industry professionals. All educators covered BPMN, while several also taught UML Activity Diagrams, Event-Driven Process Chains, Petri Nets, or other modelling notations. Most respondents also had experience in adjacent conceptual modelling areas, such as information systems modelling or data modelling, which suggests that the sample represents educators with substantial modelling expertise [11].

4.2 Instructional Approach in the Evaluated Process Modelling Courses

The following section describes the components that are critical to PM instructional design.

Target Learners' Group. The evaluated PM courses were for these levels: 8 courses at Undergraduate / Bachelor's level; 3 at Graduate / Master's level; 2 at Professional courses or industry level. The courses are targeted at business students (6 courses), ICT students (4), business informatics students (1) and two industry-oriented courses targeted business professionals and domain process experts and IT.

How the PM in BPMN Component is Taught in a Course. The academic courses were calculated using European Credit Transfer and Accumulation System (ECTS). 1 ECTS $\approx$ 25–30 h of work (lecture hours, assignments, self-study, exams, etc.). Courses were on average 4.25 ECTS: between 1.5 to 7 ECTS. PM (using BPMN) can take roughly between 10% and 100% of the coursework in ECTS. The only two courses with 100% BPMN-dedicated components are professional courses (though, their volume of teaching is also at max 1 ECTS). Generally, these courses have 80% of their content being taught in-person.

Learning Paradigms and Instructional Approaches. Educators described their teaching as grounded in constructivist and experiential paradigms, with frequent references to project- or problem-based learning (5), constructivism (4), formative assessment (4), active learning (4), and Bloom's taxonomy (4). These approaches reflect a student-centered philosophy that emphasizes learning by doing, iterative feedback, and the structured development of cognitive skills. Several responses also highlighted the importance of real-world applications, encouraging students to engage with authentic modeling scenarios that bridge theory and practice. Overall, the focus is on fostering deep understanding through active engagement and meaningful context.

Learning Objectives. Many responses (8) stressed the importance of practical modeling skills, such as creating or interpreting models, often within tools or case contexts. Several (5) identified understanding business processes or process management as a core objective. Some (4) highlighted the goal of enabling students to communicate clearly and systematically using models, viewing modeling as a shared language. A few responses (3) pointed to the ability to analyze, evaluate, or improve processes based on models.

In all courses there was a combined approach where students engage in both reading and creating models. However, the proportions of focus varied. 8 educators suggested the focus on creating models; 2 mostly reading, and 3 on both. In professional courses or industry level the dominant focus (80–90%) was on creating models.

Bloom's Taxonomy. Educators were asked to classify their courses using Bloom's Taxonomy [33], using 6-point Likert-type scales (Table 1). Descriptively, the results suggest that PM education moderately emphasizes higher-order cognitive skills such as Apply (avg. 4.2), Evaluate (3.8), and Analyse (3.6), while the lower score for Create (2.4) indicates fewer opportunities for generative tasks such as designing new patterns for recurring modelling problems. The low average for Metacognitive knowledge (1.9) also highlights a gap in self-reflection and self-regulated learning. Higher ratings for Apply (4.2) and Understand (3.9) compared to Remember (3.4) further suggest that the sampled courses are more practice-oriented than theory-driven.

Table 1: Descriptive average scores (1–6) for educators' agreement with the importance of Bloom's taxonomy components in their PM course. Scores are interpreted as exploratory tendencies within the sample, not as inferential statistics.

Bloom's Taxonomy component	Avg Score
Remember – e.g., "Given the solution part of a particular pattern, recall the typical problem that is addressed by this pattern"	3,4
Understand – e.g., "Given a modelling pattern, create your own example of instantiation of the pattern."	3,9
Apply – e.g., "Given a modelling pattern and a requirements document, apply the pattern to create a model."	4,2
Analyse – e.g., "Compare two modelling patterns for a given problem and find the main conceptual differences in how they address the given problem"	3,6
Evaluate – e.g., "Decide, which modelling pattern is better to be implemented in modelling, according to the given requirements"	3,8
Create – e.g., "Given three similar modelling problems and their solution, create a modelling pattern that represents the generic solution for the re-occurring problem."	2,4
Factual knowledge – e.g., "Recall the types of possible syntactic errors"	3,8
Conceptual knowledge – e.g., "Explain the interrelations between syntactic, semantic and pragmatic quality"	3,2
Procedural knowledge – e.g., "Assess the quality of a model using the quality assessment checklist"	3,5
Metacognitive knowledge – e.g., "Describe process modelling skills in a self-reflective report"	1,9

Course material. Seven participants reported using Dumas et al. [1] as a basis for PM instruction, often as optional reading. Many used self-created or customized slides, and eight reported using recorded lectures, although not always actively. External sources such as YouTube explainers or MOOCs were not reported.

Modelling tool. Signavio was the most frequently reported tool (9), often because of its ease of use and instructional suitability. Several educators used multiple tools, including Camunda, Visual Paradigm, draw.io, Bizagi, or SemTalk, while some allowed students to choose their own modelling tool.

Exercises. Exercises were mainly based on textual process descriptions, where students translated written scenarios into BPMN or related models (10). Other common tasks included refining or extending partial models (7), textbook-based exercises (5), formative questions (4), and comparison or evaluation tasks (3).

Formative Assessment Methods. Formative assessment mainly relied on quizzes, often online and sometimes with automated feedback (8), in-class feedback and discussion (6), LMS-supported review activities (4), home assignments (3), and presentations or student-led explanations (2).

Summative Assessment Methods. Summative assessment most often used individual exams (9), group projects (7), modelling-based tasks such as text-to-model assignments (5), and presentations or seminars (3). A few respondents (2) reported no formal summative assessment and instead relied on ongoing feedback or personal guidance.

Challenges. A majority of participants (9) mentioned difficulties in understanding detailed modeling rules and syntax, particularly when it comes to specific elements like events or gateways. Many responses (7) pointed to challenges with abstract thinking and conceptualization, such as distinguishing process layers or generalizing from specific cases. Several educators (5) highlighted issues with self-organization and active engagement, noting that students struggle to independently manage their modeling tasks. Some participants (4) mentioned a lack of foundational knowledge or motivation, which hampers deeper understanding and investment in the subject. A few responses (3) also pointed to problems with grasping the purpose and goals of modeling, beyond simply learning the notation.

Future Improvements. Many responses (7) emphasized the need to include more exercises or examples, often to reinforce understanding or offer hands-on practice. Several participants (5) suggested introducing or expanding project-based or case-based learning, aiming to contextualize modeling in real-world scenarios. Others (4) expressed interest in using innovative technologies or tools, such as simulation software or AI assistants, to modernize instruction. Some responses (3) focused on improving foundational preparation, either by integrating prerequisite content earlier or coordinating with prior courses. A few educators (2) also mentioned revisiting teaching approaches or course structure, indicating ongoing reflection.

Comparison with Teaching Approach in Other Conceptual Modelling. Several participants (6) observed that students perform better or more confidently in PM, often finding BPMN easier to grasp than entity-relationship or UML models. Several responses (5) highlighted distinct teaching strategies, suggesting that BPMN requires a more dynamic, behavior-oriented approach, while data modeling is often more abstract or structural. Some (4) reflected on the different cognitive demands, stating that PM is often perceived as more intuitive, whereas data modeling demands greater conceptual precision.

A few responses (3) indicated no major difference in teaching, citing a unified pedagogical approach across modeling types. Others (3) reported not teaching other modeling types, and thus did not observe differences.

4.3 Opinions on Pre-requisites for Attending PM Courses

Educators were asked to rate, on a 6-point Likert-type scale, the importance of potential prerequisites for attending their PM courses. The highest-rated item, conceptual modeling skills (avg. 4.8), shows that the ability to abstract and formally describe aspects of the real world is viewed as the most critical prerequisite. Educators also emphasized students' comfort with abstract phenomena (avg. 4.3) and motivation to engage with conceptual modelling (avg. 4.3). This suggests that PM instruction is built on a deeper layer of abstraction and modelling readiness that goes beyond procedural tool usage.

As qualitative feedback to quantitative responses, most educators (10) emphasized the importance of conceptual modeling capabilities, stating that the ability to abstract, describe, and structure reality is essential. Several respondents (7) also noted the value of abstract thinking, particularly when distinguishing between concepts like "entity" and "instance" or dealing with model-based representations. Several participants (5) underscored the need for analytical thinking, arguing that students should be able to systematically approach modeling problems. Some respondents (3) identified motivation to learn as a key prerequisite, noting that even less experienced students can succeed if they are eager to engage. A few responses (2) also mentioned that basic familiarity with modeling tools or prior exposure to modeling concepts can ease the learning curve.

Educators rated students' actual preparedness more cautiously. Abstraction, conceptualization, instantiation, classification, and scope definition were rated as moderately sufficient, while model interpretation, awareness of modelling purposes, communication skills, tool use, methodological adherence, and model verification were rated slightly lower. This indicates that students may grasp basic representation but still need support in using models systematically and communicatively.

The lowest-rated skill by far was students' ability to create their own notations (2.7). This suggests that meta-modelling-centered conceptual modelling education [37] is not strongly supported among PM educators in this sample. Combined with the earlier finding that Create was the lowest-rated cognitive level in Bloom's-based learning goals (avg. 2.4), this suggests that meta-modelling remains largely untapped in foundational PM education, perhaps intentionally, as it presupposes a high degree of domain mastery and abstraction among PM learners.

Participants expressed varied views on students' conceptual modeling skills at the beginning of the course. A majority of responses (7) pointed out that students have limited or uneven prior experience, particularly in abstraction, instantiation, and data modeling, highlighting the need for foundational teaching. Several respondents (5) mentioned that students can learn BPMN quickly, indicating adaptability despite initial gaps. A few comments (3) emphasized that students with engineering or business backgrounds often possess sufficient analytical capacity to grasp modeling concepts. Others (3) acknowledged that some basic modeling knowledge is already acquired in earlier courses. However, some participants (2) noted substantial variation among student groups, and a few

(2) expressed that key modeling concepts still need to be explicitly taught at the start of the program.

4.4 Using Animations to Support Teaching

Given evidence on the usefulness of animation techniques in PM [5, 22, 24], educators were asked to report their familiarity with and use of animation techniques for teaching business process modelling. Within this sample, educators reported the highest awareness of two animation types: tokens (avg. 4.58) and highlighting the process path (avg. 4.33). However, reported use in teaching was lower: tokens (avg. 3.42) and highlighting process path (avg. 3.00). This suggests that conceptual awareness of animation tools does not always translate into regular classroom application.

In terms of pedagogical function, the highest number of educators (n = 7) reported using animations to support understanding of the semantic behavior of gateways, which are well-documented as cognitively demanding elements for learners. Animations were also frequently used to explain iterations or loops (n = 5), to reduce complexity in comprehension (n = 6), and to confirm (n = 5) or extract correct answers (n = 3) related to models. These findings align with existing literature that emphasizes the difficulty novices face in interpreting dynamic process logic from static diagrams [5], and they suggest that animations serve as a scaffolding mechanism to bridge this semantic gap.

Educators also identified broader instructional benefits, with four responses indicating the use of animation to demonstrate the interaction of multiple modeling elements and to provide a global overview of the process. These uses suggest that animation is not solely a support for isolated concepts but may aid in holistic comprehension of process structures. Meanwhile, only a small number of educators reported using animations to improve attention or engagement (n = 2) or to compare different scenarios (n = 3), implying that animations are primarily employed for cognitive clarification, rather than for motivational or exploratory purposes.

Altogether, the data indicates that process model animations are used strategically and selectively, with a clear focus on mitigating comprehension challenges tied to abstract model behaviors. While technical know-how is present, broader integration into instructional practice appears limited, potentially constrained by time, tooling, or curricular emphasis. There is significant room to further develop both the technical adoption and pedagogical integration of animations, particularly by framing them not only as comprehension aids but also as tools for formative assessment, engagement, and exploratory learning.

4.5 Educators as Researchers in PM Education or in Related Domains

Educators were asked about their participation in research related to PM education, including whether they stay informed about the literature and conduct research themselves. On average, educators reported a relatively high level of engagement with literature on PM education and related domains, such as conceptual modelling education or process model comprehension, scoring 4.17 out of 6 (often). This indicates a proactive orientation toward theoretical and empirical developments, although not necessarily systematic or formal involvement in academic publishing.

This finding should be interpreted in light of the sampling strategy. Because part of the sample was recruited through PM education literature and the BPM Educators Forum, respondents were more likely than the average PM instructor to be research-informed or research-active. Therefore, the relatively high engagement with PM education literature is not interpreted as representative of all PM educators. Instead, it characterizes a self-selected subgroup of educators who are already visible in, or connected to, scholarly discussions on PM education.

Qualitatively, educators indicated that the decision to read or conduct research depends on the perceived importance of the topic. If PM education aligns with an educator's research interests or teaching responsibilities, they are more likely to stay updated. If it conflicts with their primary research agenda or current teaching workload, engagement is less likely. Respondents also acknowledged related research areas, especially process model comprehension and model quality, even when they did not conduct PM education research specifically.

In terms of involvement as researchers in PM education, six educators out of 12 confirmed that they actively conduct research on process modeling education. This suggests that while many educators are research-informed, fewer are research-active in the specific area of PM education. Among those who are, the diversity of research methods employed reflects a healthy methodological pluralism, using experiments, design science, empirical, quantitative and qualitative methods. Furthermore, educators conducted research predominantly at Bachelor's ($n = 4$) and Master's ($n = 5$) levels, supporting the evidence that PM educators tend to focus on (under)graduate levels [11].

Most frequently, participants (4) addressed process model comprehension as a central focus of their research, reflecting a strong interest in how students interpret and mentally process BPMN representations. A substantial number (3) also reported conducting experimental research, often through controlled studies involving educational interventions or tools. Additional responses (1 each) indicated themes such as cognitive load, eye-tracking, and tool-based or digital learning environments, suggesting growing attention toward both the mental demands of PM learning and the technological contexts in which it is delivered. This is in line with the recent literature reviews highlighting the interests in PM field [6, 11, 13, 35].

5 Discussion: Results, Limitations, and Future Work

This study provides an exploration into instructional design practices among educators teaching business process modeling (PM). The results support prior findings that PM instruction is generally grounded in constructivist, active, and project-based learning paradigms [11, 13], and that comprehension remains a key prerequisite for effective learning [11]. Educators emphasized cognitive processes such as abstraction, interpretation, and contextual reasoning, aligning with existing literature that stresses PM as a cognitively demanding skill [3, 6].

In line with the growing body of research advocating for structured instructional strategies in PM education [14, 30], educators in this study reported favoring real-world scenarios, continuous feedback, and hands-on modelling. However, Bloom's Taxonomy scores indicated a gap in metacognitive and creative tasks, suggesting that students are rarely guided to reflect on their own learning or develop original modelling patterns. This finding parallels concerns raised in the literature about the underuse of higher-order thinking and meta-modelling tasks in PM instruction [34].

A possible explanation is that PM courses often need to prioritize foundational syntax, tool use, and basic model interpretation before students can engage in genuinely generative tasks. Creating new modelling patterns or reflecting on one's own modelling strategy presupposes abstraction, domain understanding, and methodological confidence that many novice learners have not yet developed. In addition, several courses in the sample allocate only a limited part of the overall course workload to BPMN or PM, leaving little room for extended reflective or design-oriented activities. The low score for Create should therefore be read not only as a pedagogical omission, but as a structural tension between limited curricular time and the ambition to develop higher-order modelling expertise.

The results also confirmed the utility and underutilization of process animations as instructional aids. Although most educators were aware of token-based and highlight-based animations, only a subset regularly implemented them in their teaching. Evidence exists on the impact of animation techniques on semantic comprehension, particularly for complex constructs like gateways [5, 22, 24], yet their integration into classroom practice remains limited.

The underuse of animations may reflect practical constraints rather than lack of interest. Educators may be aware of token-based or path-highlighting techniques but still lack time, tool support, ready-made teaching materials, or assessment designs that make animations part of the regular learning cycle. If animations are presented only as optional demonstrations, their pedagogical effect is likely to remain limited. A stronger instructional design would connect animations to specific learning objectives, such as explaining gateway semantics, diagnosing misconceptions, comparing alternative paths, or supporting formative self-checks.

Implications for PM course redesign. Courses should begin with a short diagnostic activity assessing abstraction, scope definition, and model interpretation. Prerequisite gaps can be addressed through brief preparatory modules on abstraction, classification, instantiation, and process boundaries before complex BPMN constructs are introduced. Metacognitive skills can be supported through lightweight reflections in which students justify BPMN choices, identify uncertainty, and explain how they checked model quality. Create-level activities can be introduced gradually by comparing alternative solutions and deriving reusable modelling patterns. Animations should be tied to bottlenecks such as gateways, loops, concurrency, and path-dependent behaviour. Assessment should then match the intended cognitive level: quizzes and text-to-model tasks for Remember, Understand, and Apply; peer review, model justification, redesign, or reflection for Evaluate, Create, and metacognition.

Several limitations and threats to validity should be acknowledged [36]. The sample is small (N = 13), purposive, and self-selected, which limits external validity and prevents statistical generalization. Because respondents were identified through PM education publications and the BPM Educators Forum, they may be more research-oriented than the average PM instructor, potentially inflating reported engagement with literature and innovation. The study also relies on self-reported data, which may be affected by recall and social desirability biases. Likert-type items are descriptive tendencies only, and open-ended responses do not directly measure student learning outcomes. Future studies should triangulate educator reports with course materials, student artefacts, observations, or learning outcome data.

The findings may inform future PM course redesign. Future work could test interventions such as animations, flipped classrooms, metacognitive reflection, and formative assessment in longitudinal or controlled studies. Recent work comparing token and highlighting animations, examining subjective comprehension and interaction, and teaching process patterns via token animations shows how dynamic visualization can become an assessable instructional intervention [38–40]. GenAI modelling assistants may shift the learning challenge from drawing models manually to evaluating, correcting, and justifying generated models, while BPMN-centered process-based knowledge management suggests further links between modelling, knowledge work, and GenAI support [41, 42]. Broader e-learning research during the COVID-19 pandemic also indicates that technology-enhanced instruction should be evaluated through learner experience and perceived outcomes, not delivery mode alone [43]. Process mining may similarly help students connect designed models with event data and actual process behaviour. Overall, these developments increase the importance of teaching students to critically assess model semantics, assumptions, and the fit between descriptions, observed behaviour, and formal representations.

References

1. Dumas, M., La Rosa, M., Mendling, J., Reijers, H.A.: Fundamentals of Business Process Management. Springer, Heideberg (2018). https://doi.org/10.1007/978-3-642-33143-5
2. Winter, M., Neumann, H., Pryss, R., Probst, T., Reichert, M.: Defining gaze patterns for process model literacy: exploring visual routines in process models with diverse mappings. Expert Syst. Appl. **213**, 119217 (2023)
3. Sedrakyan, G., Snoeck, M., Poelmans, S.: Assessing the effectiveness of feedback enabled simulation in teaching conceptual modeling. Comput. Educ. **78**, 367–382 (2014). https://doi.org/10.1016/j.compedu.2014.06.014
4. Ternes, B., Strecker, S., Rosenthal, K., Barth, H.: A browser-based modeling tool for studying the learning of conceptual modeling based on a multi-modal data collection approach. Proceedings of the 14th International Conference on Wirtschaftsinformatik, pp. 1998–2002 (2019)
5. Aysolmaz, B., Reijers, H.A.: Animation as a dynamic visualization technique for improving process model comprehension. Inf. Manag. **58** (2021). https://doi.org/10.1016/j.im.2021.103478
6. Figl, K.: Comprehension of procedural visual business process models: a literature review. Bus. Inf. Syst. Eng. **59**, 41–67 (2017). https://doi.org/10.1007/s12599-016-0460-2
7. Duggan, G.B., Payne, S.J.: Text skimming: the process and effectiveness of foraging through text under time pressure. J. Exp. Psychol. Appl. **15**, 228 (2009)
8. Petrusel, R., Mendling, J., Reijers, H.A.: How visual cognition influences process model comprehen-sion. Decis. Support. Syst. **96**, 1–16 (2017)
9. Muehlen, M. zur, Recker, J.: How much language is enough? Theoretical and practical use of the busi-ness process modeling notation. In: Seminal Contributions to Information Systems Engineering: 25 Years of CAiSE, pp. 429–443 (2013)

10. Claes, J., Vanderfeesten, I., Gailly, F., Grefen, P., Poels, G.: The structured process modeling method (SPMM): what is the best way for me to construct a process model? Decis. Support. Syst. **100**, 57–76 (2017). https://doi.org/10.1016/j.dss.2017.02.004
11. Maslov, I., Poelmans, S., Rosenthal, K.: Conceptual modeling education: a bibliometric literature review and avenues for future research. Bus. Inf. Syst. Eng. 1–33 (2025)
12. Maslov, I., Poelmans, S.: Advancing the BPMN 2.0 standard with an extended animated notation: a research program for token-based process modeling education. In: roceedings of BIR (2023)
13. Rosenthal, K., Ternes, B., Strecker, S.: Learning conceptual modeling: structuring overview, research themes and paths for future research. In: 27th European Conference on Information Systems - Information Systems for a Sharing Society, ECIS 2019 (2020)
14. Grzesiak, M., Moszyński, M.: Application of the flipped learning methodology at a business process modelling course: a case study. In: International Conference on Business Process Management, pp. 303–312. Springer, Heidelberg (2024)
15. Revoredo, K., Bala, S.: Project-based learning in BPM: lessons from flipped classroom vs. lecture-based instruction. In: International Conference on Business Process Management, pp. 313–323. Springer, Heidelberg (2024)
16. Khalil, M.K., Elkhider, I.A.: Applying learning theories and instructional design models for effective instruction. Adv. Physiol. Educ. (2016)
17. Morrison, G.R., Ross, S.M., Kalman, H.K., Kemp, J.E.: Designing Effective Instruction, 6th edn. John Wiley & Sons, Hoboken (2011)
18. Moody, D.: The physics of notations: toward a scientific basis for constructing visual notations in software engineering. IEEE Trans. Softw. Eng. **35**, 756–779 (2009). https://doi.org/10.1109/TSE.2009.67
19. Guizzardi, G.: Ontological foundations for structural conceptual models (2005)
20. Dikici, A., Turetken, O., Demirors, O.: Factors influencing the understandability of process models: a systematic literature review. Inf. Softw. Technol. **93**, 112–129 (2018)
21. Moody, D.L., Shanks, G.G.: Improving the quality of data models: empirical validation of a quality management framework. Inf. Syst. **28**, 619–650 (2003). https://doi.org/10.1016/S0306-4379(02)00043-1
22. Maslov, I., Poelmans, S.: Comprehension of (business) process models via tokens: an eye-tracking approach. In: International Conference on Business Process Management, pp. 375–385. Springer, Heidelberg (2024)
23. Vemuri, P., Poelmans, S., Serral, E., Snoeck, M.: Teaching conceptual modeling leveraging formative assessments and adaptive release paths (2023)
24. Maslov, I., Poelmans, S.: Facilitating the comprehension of business process models for unexperienced modelers using token-based animations. Inf. Manag. **61**, 103967 (2024)
25. Rosenthal, K., Ternes, B., Strecker, S.: Learning conceptual modeling: structuring overview, research themes and paths for future research. In: ECIS (2019)
26. Avila, D.T., dos Santos, R.I., Mendling, J., Thom, L.H.: A systematic literature review of process modeling guidelines and their empirical support. Bus. Process. Manag. J. **27**, 1–23 (2020)
27. Becker, J., Rosemann, M., Von Uthmann, C.: Guidelines of business process modeling. In: Business Process Management: Models, Techniques, and Empirical Studies, pp. 30–49. Springer, Heidelberg (2002)
28. Mendling, J., Reijers, H.A., van der Aalst, W.M.P.: Seven process modeling guidelines (7PMG). Inf. Softw. Technol. **52**, 127–136 (2010)
29. Overhage, S., Birkmeier, D.Q., Schlauderer, S.: Quality marks, metrics, and measurement procedures for business process models: the 3QM-framework. Bus. Inf. Syst. Eng. **4**, 229–246 (2012)

30. de Pádua, S.I.D., Lourenção, M., Moreira, S.A.S., Ribeiro, L.R.B.: Fostering BPM learning through PBL: development and application of an educational framework. In: International Conference on Business Process Management, pp. 416–425. Springer, Heidelberg (2024)
31. van Der Aalst, W.M.P., Ter Hofstede, A.H.M., Kiepuszewski, B., Barros, A.P.: Workflow patterns. Distrib. Parallel Databases. **14**, 5–51 (2003)
32. Schoo, A., Kumar, K.: The clinical educator and complexity: a review. Clin. Teach. **15**, 287–293 (2018)
33. Bogdanova, D., Snoeck, M.: CaMeLOT: an educational framework for conceptual data modelling. Inf. Softw. Technol. **110**, 92–107 (2019). https://doi.org/10.1016/j.infsof.2019.02.006
34. Maslov, I.: Towards empirically validated process modelling education using a BPMN formalism. In: International Conference on Research Challenges in Information Science, pp. 803–810. Springer, Heidelberg (2022). https://doi.org/10.1007/978-3-031-05760-1_58
35. Ternes, B., Rosenthal, K., Strecker, S.: User interface design research for modeling tools: a literature study. Enterp. Model. Inf. Syst. Archit. **16**, 1–4 (2021)
36. Saunders, M., Lewis, P., Thornhill, A.: Research Methods for Business Students. Pearson Education (2009)
37. Buchmann, R.A., Ghiran, A.-M., Döller, V., Karagiannis, D.: Conceptual modeling education as a "design problem." Complex Syst. Inf. Model. Q. **21**, 21–33 (2019). https://doi.org/10.7250/csimq.2019-21.02
38. Maslov, I., Poelmans, S., Aysolmaz, B.: Enhancing process model comprehension through animations: a comparative study of tokens vs. highlighting. In: CBI-EDOC Forum 2025, Lisbon, Portugal (2025)
39. Maslov, I., Poelmans, S., Wautelet, Y., Gailly, F.: Novice modelers' subjective comprehension and interaction with token-animated process models. J. Comput. Lang. 101350 (2025)
40. Maslov, I., Poelmans, S., Mandeelburger, M.M., Proper, H.: Teaching process patterns in BPMN to novice modelers via token animations. In: PACIS 2025 Proceedings, vol. 5 (2025)
41. Maslov, I., Poelmans, S., Wautelet, Y., Rosenthal, K.: Generative AI's aid in business process modeling instructional design: a case study. In: International Conference on Business Process Management, Seville, Spain (2025)
42. Maslov, I., Kudryavtsev, D., Ali, U.K., Simonofski, A.: A BPMN-centered method for process-based knowledge management with generative AI. In: PACIS 2026 (2026)
43. Nikou, S., Kim, S., Lim, C., Maslov, I.: Satisfaction with e-learning systems during the COVID-19 pandemic: a comparative study. In: 23rd Biennial Conference of the International Telecommunications Society (2021)

Learning Process Modelling in BPMN Through Real-Life Process Observation

Ilia Maslov[1](✉), Stephan Poelmans[2], and Anthony Simonofski[1]

[1] Université de Namur, Rue de Bruxelles 61, 5000 Namur, Belgium
{ilia.maslov,anthony.simonofski}@unamur.be
[2] KU Leuven, Warmoesberg 26, 1000 Brussels, Belgium
stephan.poelmans@kuleuven.be

Abstract. Process modelling education increasingly emphasizes authentic practice, yet limited research examines how students themselves describe moving from observed business reality to formal BPMN models. This paper analyzes semi-structured learning diaries from three course implementations at two universities, covering 94 diaries from 108 enrolled students. Across the implementations, students completed observation-grounded group modelling assignments and reflective diaries; the first implementation also included AS-IS to TO-BE redesign, while the third explicitly asked about Generative AI. Using a researcher-led qualitative thematic analysis supported by iterative cross-cohort comparison, the study identifies recurring themes in process identification, BPMN learning, collaboration, presentation, transfer, and bounded AI use. The findings show that students experienced process modelling as an interpretive, socially negotiated, and increasingly transferable competence rather than as diagram drawing alone. The paper contributes an observation-to-model instructional case and demonstrates the value of learning diaries as both pedagogical and research instruments for BPM education research and practice.

Keywords: Process modelling education · BPMN · learning diaries · experiential learning · observation practice · generative AI · instructional design

1 Introduction

Process modelling (PM) plays a central role in translating business logic into explicit visual representations that support communication, analysis, improvement, and, in some contexts, automation [1, 2]. Because BPMN has become the de facto standard for this purpose, PM education increasingly faces the challenge of helping novices not merely read and draw diagrams but learn how to move from ambiguous real-world situations to structured and communicable process representations [1, 3]. This is cognitively demanding. Learners must identify process boundaries, distinguish actors and responsibilities, decide on an appropriate level of abstraction, and represent dynamic behavior through a notation whose semantics are not always transparent to beginners [2, 4, 5].

At the same time, for novice learners, process modelling does not begin with notation alone. Before students can produce a BPMN model, they must first determine what should

K. Kluza et al. (Eds.): BPM 2026, LNBIP 600, pp. 378–392, 2027.
https://doi.org/10.1007/978-3-032-36710-5_24

count as the process, whose perspective is being modelled, which events and decision points are relevant, and how observed or described reality can be simplified without distorting the underlying logic. In educational settings, this means that PM learning involves not only syntax and semantics, but also process identification, interpretation, communication, and reflection. Yet much of the PM education literature has focused more strongly on comprehension, notation use, and cognitive support than on how learners themselves experience the transition from observed or plausibly real business situations to formal models [3, 4, 6].

Recent PM education research increasingly emphasizes structured instructional approaches such as active learning, project-based work, authentic modelling tasks, and real-world case integration [3, 6–8]. This also includes emerging work on token animations for teaching BPMN elements and process patterns to novices, both as a way of reducing representational complexity and as a means of guiding attention during model comprehension [5, 9–11]. At the same time, Generative AI is becoming relevant for BPM education, as it may support brainstorming, clarification, and instructional structuring, while also introducing the risk of plausible yet insufficiently grounded modelling decisions if used uncritically [12]. This raises a broader pedagogical question: how can students be taught not only to model processes, but also to ground, verify, and justify modelling decisions in semantically and contextually defensible ways?

This paper addresses these gaps by analyzing semi-structured learning diaries from a BPMN course that was implemented three times by the same instructor across two institutions in 2025 and 2026. Across the three cohorts, students completed observation-grounded group modelling assignments and individual reflective diaries. The instructional backbone was similar across cases, but the implementations differed in pacing, assessment structure, and reflective emphasis. In particular, the first cohort also included explicit AS-IS to TO-BE redesign, whereas the third explicitly prompted students to reflect on Generative AI use. Rather than treating the learning diary merely as an assessment artifact, this study approaches it as a metacognitive and empirical trace of how students experienced the path from real-life process observation to BPMN modelling. The paper is guided by the following research questions (**RQs**):

RQ1. *How do students describe the transition from observed or described business reality to BPMN process models in an observation-grounded PM course?*

RQ2. *What recurring learning mechanisms become visible in students' diary reflections regarding process identification, BPMN syntax and semantics, collaboration, communication, transfer, and, where applicable, bounded Generative AI use?*

By answering these questions, the paper contributes to BPM education research in three ways. First, it offers an empirically grounded account of an observation-to-model instructional approach, making process identification and modelling viewpoint more visible as learnable components of BPMN education. Second, it shows how learning diaries can reveal dimensions of PM learning that are difficult to capture through tests or surveys alone, such as interpretive difficulty, reflective judgment, transfer, and the perceived boundaries of AI support. Third, it provides a cross-cohort view of how small instructional design variations, such as including TO-BE redesign or explicit GenAI reflection, shape the reflective learning data available for analysis.

The paper is structured as follows. Section 2 reviews relevant literature on PM education and BPMN learning. Section 3 presents the methodology, including the cohort comparison, educational setting, and analysis procedure. Section 4 reports the findings from the learning diaries. Section 5 discusses the implications for BPM education research and practice, outlines limitations and threats to validity, and suggests avenues for future work.

2 Extant Literature

Process modelling (PM) translates business logic into explicit visual representations that support communication, analysis, redesign, and, in some settings, automation [1]. BPMN has become the de facto standard for this purpose, but learning to model with BPMN is cognitively demanding. The translation of business requirements into formal BPMN rules implies dealing with ambiguity, versatility, and open-ended problems, making it challenging for novices and experts alike [13]. These requirements describing processes typically come from process identification and requirements engineering (e.g., by conducting document analysis, interviewing employees, observing processes or process mining). The meaning of the process model must correspond to the requirements, which in turn must accurately reflect the reality for it to be usable. Novices must identify process boundaries, distinguish actors and responsibilities, choose an appropriate level of abstraction, and understand how syntax, semantics, and pragmatic readability interact in visual models [4, 14, 15]. It requires an extensive knowledge of a variety of symbols and the rules of how to align the symbols together into a meaningful process description. For instance, BPMN is a comprehensive modeling language with over 100 symbols [1].

Beyond model comprehension, process modelling also requires learners to move from observations, textual descriptions, or stakeholder accounts toward structured formal representations. In that sense, modelling is not only a representational activity but also a translation and abstraction task: modelers must identify relevant process boundaries, distinguish core flow from exceptions, determine an appropriate modelling viewpoint, and transform ambiguous business situations into diagrammatically expressible logic. Prior work has shown that such modelling tasks are especially challenging for novices because they involve both cognitive and pragmatic judgement, not merely notation recall [1, 13, 16]. Process modelling research has also proposed several guideline-based and cognitively informed approaches to improve model quality and modelling performance, such as 7PMG, GoM, and SPMM, but these contributions do not directly explain how novice learners themselves experience the transition from observed reality to formal BPMN models in educational settings [13, 16, 17].

At the same time, PM education has increasingly emerged as a research topic in its own right. Prior reviews have shown that conceptual and process modelling education has often remained fragmented, with strong emphasis on notation and comprehension but less systematic attention to instructional design, reflective learning, and the broader pedagogical sequencing through which modelling competence develops [3, 6, 18]. Existing work suggests that PM instruction benefits from hands-on practice, iterative feedback, and authentic or quasi-authentic modelling contexts, because modelling requires not only declarative knowledge of notation but also abstraction, scoping, contextual judgment, and communication [3, 6, 19]. Recent experimental work also suggests that token-animated explanations can reduce extraneous cognitive load and support novice learners in understanding BPMN process patterns, especially when used as part of structured instructional design rather than as a stand-alone visualization feature [11].

From an educational perspective, this suggests that PM competence develops not only through exposure to notation, but through instructional sequences that connect observation, abstraction, formalization, communication, and reflection. In such settings, reflective accounts may be particularly useful because they capture how learners themselves make sense of process boundaries, modelling choices, teamwork, and transfer beyond the immediate exercise. This is important because BPMN learning is not only about producing technically correct diagrams, but also about developing a process-oriented way of seeing organizational reality [3, 6, 18].

However, one aspect remains comparatively underexplored: how novice learners themselves narrate the transition from observed or plausibly real business situations to BPMN models across repeated instructional implementations. This question becomes even more relevant as Generative AI enters educational practice. In BPM education, GenAI may support brainstorming, reformulation, clarification, and even aspects of instructional design, but it may also encourage plausible yet insufficiently grounded modelling decisions if students rely on it uncritically [12]. The present study addresses this gap by examining learning diaries as reflective traces of observation-to-model learning in BPMN education.

3 Methodology

This study follows a qualitative, cross-cohort design based on semi-structured learning diaries. The goal was not to measure effects statistically, but to identify robust recurring patterns in how students described their learning while participating in an observation-grounded BPMN course. The study is interpretive and comparative: it analyzes students' reflective accounts across three course implementations that shared a common pedagogical core, but differed slightly in assignment scope and diary prompts.

3.1 Participants and Data Corpus

The dataset consists of 94 individual learning diaries collected across three course implementations from a total of 108 enrolled students. The first cohort was in the Burgundy School of Business in France (BSB), in March 2025 with 29 enrolled students, of whom 20 submitted usable diaries. The second cohort was at the University of International Business and Economics in China (UIBE) in July 2025, with 51 enrolled students and 51 usable diaries. The third cohort was at the BSB in March 2026, with 28 enrolled students, of whom 23 submitted usable diaries. Thus, the study includes complete diary coverage for the UIBE in 2025 and partial but substantial reflective coverage for both BSB cohorts. Given the differences between cohorts and diary submission, the study does not treat the diary corpus as a census of all enrolled students, but as a qualitative reflection dataset with uneven participation across implementations. Nevertheless, the corpus is sufficiently large to support cross-cohort thematic comparison.

Table 1 provides an overview of the three cohorts and highlights the main instructional variations across implementations.

Table 1: Overview of the three cohorts.

Dimension	BSB March 2025 (Case A)	UIBE July 2025 (Case B)	BSB March 2026 (Case C)
Institution	Burgundy School of Business, France	University of International Business and Economics, China	Burgundy School of Business, France
Delivery format	Intensive one-week format	Distributed format over approximately two weeks	Intensive one-week format
Approx. Contact hours	~ 21 h	~ 27 h	~ 21 h
Enrolled students	29	51	28
Usable learning diaries	20	51	23
Student profile	Master's-level business/management students	Bachelor's-level students	Master's-level business/management students
PM/BPMN prior familiarity	Mostly low	Mostly low	Mostly low
AS-IS model required	Yes	Yes	Yes
TO-BE model required	Yes	No	No
Post-presentation model improvement allowed	Yes	Yes	No
Final summative exam	No	Yes	Yes
Repeated in-course formative / end-of-part assessment	Yes (graded)	Yes (non-graded)	Yes (non-graded)
Explicit GenAI reflection in diary	No	No	Yes

3.2 Educational Setting and Assignment Design

The three course implementations shared a common instructional backbone that treated BPMN as a language to be learned progressively through repeated cycles of explanation, guided practice, group modelling, instructor feedback, presentation, and reflection. All three implementations were delivered by the same instructor and were based on the same core BPMN teaching logic, while allowing limited contextual adaptations across institutions and years. Students developed a group assignment in parallel with the course progression. This general instructional logic is consistent with recent calls in process modelling education to move beyond notation exposure toward scaffolded, practice-rich, and feedback-intensive course designs that help novices gradually coordinate syntax, semantics, and communicative use of BPMN [3, 6]. In addition, rather than only translating given text into diagrams, they were asked to ground modelling in a real or plausibly real business process that had to be observed, reconstructed, discussed, and formalized. The selected cases included retail purchases, campus beverage preparation, banking services, reimbursement procedures, and industrial production. Some groups relied on field visits, mystery-shopper observation, interviews, photos, or practitioner knowledge, whereas others worked from accessible descriptions or operational cases.

The implementations differed in temporal organization and assessment structure. The two BSB implementations were delivered as intensive one-week formats of approximately 21 contact hours, whereas UIBE 2025 was distributed over approximately two weeks and comprised approximately 27 contact hours. In BSB 2025, the course relied on repeated end-of-part formative-style assessments, the group modelling presentation, and the individual learning diary, rather than on a separate final summative exam. UIBE 2025 and BSB 2026 combined the iterative in-class learning cycle with a distinct end-of-course summative examination. The learning diary formed part of the course learning activities in all three implementations, though its role in the grading was advertised to students as only supplementary, which was to help decide on the final grade in the borderline cases, which likely contributed to the observed differences in diary completion rates.

Another instructional variation concerned model revision. In BSB 2025 and UIBE 2025, students could improve and resubmit their group models after in-class feedback. This created an additional refinement cycle, but also required substantial individualized teacher attention. For this reason, the explicit post-presentation revision option was removed in BSB 2026, while the in-class feedback phase was retained.

3.3 Learning Diary Structure

A learning diary may be understood as a structured reflective record through which learners document observations, questions, and developing interpretations over time. In educational research, reflective learning journals/diaries are used not merely as personal logs, but as guided instruments that support reflection on learning processes and experiences [20, 21]. In this study, the learning diary functioned as a semi-structured reflective instrument rather than a fully open personal journal. Students were asked to respond to guided prompts on (1) process identification, (2) BPMN syntax and semantics, (3) collaborative modelling, (4) communication via process models, and (5) model-based thinking in business administration. These common prompts made cross-cohort comparison possible.

Students received an adapted diary template and written instructions informed by reflective learning diary guidance used in an Erasmus + -supported instructional context. This guidance defines a learning diary as notes, observations, and thoughts developed over time to support learning through reflective writing, and encourages regular, prompt-guided, analytical reflection rather than purely descriptive writing. For practical reuse, the online appendix includes the adapted diary template, the prompt set provided to students, and the anonymized diary corpora.

Two cohort-specific prompt variations were added. First, BSB 2025 required students to create an AS-IS model, propose a TO-BE redesign, and reflect on process re-engineering. Because this first implementation showed that combining process discovery and redesign was demanding for novices, the TO-BE component was removed from UIBE 2025 and BSB 2026, which focused on observation, process identification, and correct formalization of the AS-IS process only. Second, BSB 2026 added a specific Generative AI reflection prompt asking students whether and how GenAI helped them translate textual or observed descriptions into BPMN. The anonymized diary templates and combined diary corpora are provided in an online appendix f[1] or transparency purposes, together with the templates on these courses.

3.4 Analysis Procedure

The analysis was researcher-led and iterative, following a qualitative thematic logic appropriate for semi-structured reflective accounts in educational settings [22, 23]. Thus, the analyzed material is best understood not as fully open diary writing, but as responses produced through a semi-structured reflective learning diary instrument. First, the researcher became familiar with the corpus through repeated reading of all diaries, alongside the associated teaching descriptions and cohort profile materials. This stage focused on identifying recurring references to observation practice, process scoping, BPMN learning difficulties, teamwork, presentation, transfer to other business domains, and GenAI use where relevant. Although students were encouraged to note reflections progressively during the course, the diaries were submitted as end-of-course semi-structured reflective texts rather than as weekly entries. The resulting corpus consisted of 94 diaries and approximately 97,000 words in total, with approximately 18,000, 52,000, and 27,000 words in BSB 2025, UIBE 2025, and BSB 2026, respectively.

Second, the researcher conducted a thematic coding procedure that combined deductive sensitizing categories derived from the diary template with inductive refinement emerging from repeated reading of the corpus. Because the diary prompts already provided a partial structure, the analysis did not treat the material as fully unstructured qualitative data; instead, it examined how students populated shared reflective domains with recurring meanings across contexts. This deductive-inductive strategy is suitable when broad reflective domains are predefined, but the actual meanings and cross-case

[1] https://zenodo.org/records/21259131.

regularities still need to be derived from the participants' accounts [22, 23]. In practice, the deductive component provided the initial reflective domains based on the diary template, whereas the inductive component was used to refine, split, merge, or reinterpret these domains in light of recurring meanings observed across the corpus. Thus, the final thematic structure was not imposed mechanically from the prompts, but developed through iterative comparison between predefined reflective categories in learning diaries and the students' actual accounts.

To keep the analysis transparent and systematically organized, the final coding scheme was structured around analytical subcodes derived from the recurring patterns identified across the diary corpus. These subcodes captured the major thematic domains examined in the study, including process identification, BPMN learning, collaborative modelling and communication, transfer, cohort-specific re-engineering, and cohort-specific bounded Generative AI use. Table 2 summarizes this coding scheme.

Table 2. Coding scheme that emerged through the thematic analysis.

Code	Short definition	Count
(1) Real-life process selection	Selection of a familiar, accessible, or observable process	61
(1) Grounding through observation	Use of visits, interviews, photos, documents, or practitioner input	55
(1) Scope and boundary definition	Definition of process start, end, inclusion, and exclusion	61
(1) Modelling viewpoint	Identification of customer-, employee-, or system-side perspective	33
(1) Actors, decision points, and exception paths	Identification of participants, decisions, responsibilities, and deviations	90
(1) Balancing detail and abstraction	Choice of granularity to avoid vague or over-cluttered models	42
(2) BPMN as a new language	Experience of BPMN as formal grammar rather than free-form diagramming	39
(2) Syntax, semantics, and dynamic logic	Understanding BPMN symbols, flows, gateways, events, and token execution	91
(2) Advanced semantics difficulties	Struggles with exceptions, boundary events, timers, and event-based gateways	60
(2) Pragmatic readability and model quality	Attention to naming, layout, conditions, clarity, and communicative quality	84
(3) Collaborative negotiation and role distribution	Team alignment, role distribution, and negotiation of modelling choices	82
(3) Presentation, feedback, and communicative clarity	Use of presentation and feedback to test whether models communicate clearly	86
(4) Transfer to other business domains	Application of BPMN thinking to logistics, marketing, finance, HR, auditing, etc	81

(continued)

Table 2. (*continued*)

Code	Short definition	Count
(4) Process-oriented managerial reasoning	Use of BPMN as a way to structure, diagnose, and improve organizational reality	71
(5) TO-BE redesign as intervention logic	Use of BPMN to propose and justify process improvements	15
(6) GenAI as bounded support	Use of GenAI for brainstorming, clarification, or early structuring	19
(6) Critical verification of AI output	Checking and correcting AI suggestions rather than accepting them uncritically	12

To support handling the larger multi-cohort corpus, ChatGPT 5.4 was used as an analytic assistant for provisional segment extraction, clustering of similar statements, and cross-document comparison. The prompts asked the model to identify excerpts related to the predefined diary domains, group semantically similar excerpts, suggest provisional labels for recurring patterns, and compare whether similar patterns appeared across cohorts. ChatGPT did not function as an autonomous coder: all candidate patterns, labels, and interpretive claims were manually reviewed against the original diaries, unsupported suggestions were discarded, and final interpretations remained researcher-determined. The analysis therefore remained human-led and auditable, with the LLM serving only as a corpus-handling aid for qualitative analysis [24].

Trustworthiness was supported through repeated familiarization, comparison within and across cohorts, attention to variation and deviant cases, and a transparent link between final themes and the underlying diary corpus, in line with established qualitative research recommendations on analytic transparency and interpretive rigor [22, 23]. Enrolment and reflective participation were not identical across cohorts. The course implementations involved 29 enrolled students in BSB 2025, 51 in UIBE 2025, and 28 in BSB 2026, whereas the usable diary corpus consisted of 20, 51, and 23 diaries, respectively. The study therefore analyzes a large but uneven reflective sample rather than a full census of all enrolled students. Students worked in small groups of typically 4 to 5 members (some had 3 or 6), which is relevant for interpreting the collaborative modelling reflections reported in the findings.

In addition to the diary corpus, the researcher conducted a supplementary review of the submitted group BPMN models across the three cohorts to identify recurring notation- and semantics-related difficulties. This review did not constitute a separate full coding stream; it was used only to contextualize the diary-based findings on advanced BPMN semantics and to inform the formative evaluation of group assignments.

The coding scheme (Table 2) was used as an analytic scaffold rather than as a rigid set of mutually exclusive bins. A single diary excerpt could contribute to more than one subcode when students simultaneously discussed, for example, observation, scope definition, and collaborative negotiation. The codes were progressively refined through repeated reading and cross-cohort comparison until they stabilized around the major thematic patterns reported in Themes (1) to (6) in Sect. 4. Most coded patterns were visible across all three cohorts, whereas AS-IS to TO-BE redesign was specific to Case A and explicit Generative AI reflection to Case C.

4 Findings

The findings below elaborate the main coded patterns across the three diary corpora. Participant identifiers refer to the anonymized collections in the online appendix: A = BSB March 2025, B = UIBE July 2025, and C = BSB March 2026. Themes (1)–(4) were visible across cohorts, whereas AS-IS to TO-BE redesign was specific to Case A and explicit GenAI reflection to Case C.

(1) Process Identification. Across the three cohorts, students did not describe process identification as a simple pre-step before "drawing the diagram." Instead, they framed it as interpretive work involving scope definition, viewpoint selection, and the identification of key actors and exception paths. In BSB 2025, one student noted that process identification required deciding "from the customer's perspective, the employees, or both," and described it as "not only analytical but interpretive, requiring observation, abstraction, and negotiation of meanings with teammates" (A1). Similarly, in UIBE 2025, one participant described it as "balancing breadth (end-to-end visibility) and depth (granular steps)" (B1), while in BSB 2026 another wrote that "finding a process requires a way of thinking" and that one must "look for the hidden rules that control the process" (C13).

A particularly strong cross-cohort pattern concerned grounding through observation. Students repeatedly referred to field visits, photos, staff interviews, or document collection as ways of making the process less imagined and more defensible. In BSB 2025, one participant wrote that modelling a return-handling process "required multiple visits and discussions with staff to capture the key events and decision points" (A1). In UIBE 2025, one participant described combining "stakeholder interviews," "document review," and "on-site observation" (B12), while another reported visiting a campus beverage store to clarify boundaries (B4). In BSB 2026, a participant noted that "starting with a field observation rather than just discussing the process in the group was probably one of the better decisions we made" (C2). Together, these accounts indicate that observation was experienced not as a decorative addition, but as a core epistemic step in translating messy reality into a defensible BPMN model.

(2) BPMN as a "new language," with advanced semantics remaining difficult. Across all three cohorts, BPMN was repeatedly described as a new language or formal grammar rather than a flexible diagramming tool. Students emphasized that BPMN elements could not be used arbitrarily and that learning the notation meant mastering precise distinctions between tasks, events, gateways, subprocesses, roles, and message logic. In BSB 2025, one participant stated that learning BPMN syntax "was like acquiring a new language" (A1), while in UIBE 2025 another compared BPMN to "Mandarin tones" because "precise symbols convey meaning" (B1). In BSB 2026, students similarly stressed that "you must follow a universal grammar" (C1) and that "every symbol in BPMN means something specific" (C13).

At the same time, students did not present BPMN learning as mere memorization of symbols. They linked it to a broader change in how they structured business logic. Participants noted that gateway choice forced them to think more rigorously and that tokens or simulation helped them understand how BPMN "executes step by step" (A2). A UIBE participant similarly wrote that translating a sketch into BPMN "forced me to reconsider every branch" and rethink whether activities were manual, user, or system-driven (B12). These reflections suggest that BPMN was experienced as both a representational language and a cognitive discipline.

Here, "advanced semantics" refers to BPMN constructs beyond basic linear control flow, especially exception handling, boundary events, timers, and event-based gateways. The supplementary review of submitted models suggests that students generally captured core process logic, but struggled with synchronization across pools, interruptions, delays, unclear conditions, inconsistent task/event typing, and semantically fragile flows. Secondary issues concerned naming, gateway conditions, message-flow clarity, and visual alignment. Overall, advanced coordination, temporal behavior, and exception semantics remained the main instructional bottlenecks.

(3) Collaborative modelling and in-class presentation as learning mechanisms. Across the three diary corpora, collaborative modelling was presented not as a secondary organizational feature, but as a major learning mechanism through which ambiguity became discussable and improvable. Students described the group as a site where interpretations had to be negotiated, assumptions externalized, and modelling choices justified. In BSB 2025, one participant noted that disagreements "forced us to justify our modeling choices and reach a consensus" (A2). In UIBE 2025, students described themselves as "Scribe/Integrator" (B1) or "process integrator and communicator" (B2), while in BSB 2026 one participant wrote that collaborative modelling meant agreeing "on every single arrow and box" (C1). These accounts suggest that teamwork itself functioned as a form of modelling pedagogy.

Presentation and class discussion played a similarly important role. Students did not treat presentation as a mere end-point, but as a test of whether the model made sense for an external audience. In BSB 2025, participants noted that feedback helped refine both model and presentation quality (A2). In UIBE 2025, one student wrote that "a process model doesn't speak for itself" and that the presenter must explain the "why" behind key design choices (B2), while another stressed that meaning emerges when "roles make responsibilities understandable," "branches make logic clear," and "exceptions make handling understandable" (B3). In BSB 2026, students similarly argued that "a successful model is one that reads like a story" and that "a diagram nobody can interpret is not very useful" (C1; C2). Thus, presentation and feedback did not simply evaluate finished models; they helped students discover that communicative clarity is integral to BPMN competence itself.

(4) Transfer from BPMN to broader model-based thinking. A fourth robust theme was the transfer of BPMN learning into broader managerial and analytical reasoning. Students repeatedly connected BPMN to logistics, marketing, finance, HR, auditing, supply chains, customer journeys, and approval workflows. In BSB 2025, one participant wrote that BPMN had given them a "strategic lens to manage complexity" (A1), while another described it as a "structured approach to problem-solving" (A2). In UIBE 2025, participants highlighted logistics redesign, budgeting, HR onboarding, and financial service mapping (B1; B2). In BSB 2026, one student described model-based thinking as a "permanent internal audit tool" (C1), and another argued that management problems became "a series of logical steps that can be improved one by one" (C2). These reflections suggest that the assignment fostered not only BPMN literacy, but also a broader orientation toward systematic managerial reasoning.

(5) AS-IS to TO-BE re-engineering as BPMN improvement logic. The BSB 2025 corpus adds a cohort-specific layer because students explicitly reflected on AS-IS to TO-BE re-engineering. Students moved beyond documentation and articulated BPMN as a medium for proposing improvements. One participant described re-engineering as identifying inefficiencies, "reducing delays," "improving decision-making logic," and redesigning the process to improve responsiveness and clarity (A2). Another emphasized that redesign can create "a kind of domino effect," because changes propagate through interconnected elements (A1). These reflections indicate a movement from representational competence to improvement-oriented reasoning. At the same time, this layer was later removed from UIBE 2025 and BSB 2026 because combining process discovery and redesign proved too demanding for many novices. Thus, AS-IS to TO-BE reflection appears valuable, but more suitable as an advanced extension after AS-IS competence has stabilized.

(6) GenAI as bounded support rather than modelling authority. The BSB 2026 corpus provides the clearest evidence on students' explicit positioning of Generative AI. The dominant pattern was neither enthusiastic outsourcing nor rejection. Rather, students described GenAI as useful for brainstorming, clarification, and getting unstuck, but not as something that could be trusted to design the final model autonomously. One participant noted that AI helped list "possible exceptions" but also "often forgets to close gateways or mixes up event types," meaning that syntax verification still had to remain manual (C1). Another reported using AI to understand inclusive gateways, event-triggered subprocesses, and boundary errors, but stressed that "all the final decisions were made by the group after discussion" (C2). A third participant described GenAI as a "support tool" useful for checking definitions and structuring thoughts, while emphasizing that "the main work was still done by our group" and that AI suggestions required critical evaluation because the system did not fully understand the project context (C13).

These reflections suggest an emergent form of modelling-specific AI literacy. Students were not only learning BPMN, but also learning where AI becomes useful and where it becomes dangerous. In their accounts, GenAI was acceptable as a co-pilot for ideation, explanation, or early structuring, but not as an authority for semantic correctness, contextual grounding, or final model integrity. Rather than replacing modelling competence, AI was positioned as bounded support subordinate to human judgement, peer discussion, and instructor feedback.

5 Discussion, Limitations, Validity, and Future Research

With respect to RQ1, the findings show that students experienced the transition from observed or described business reality to BPMN models not as straightforward technical translation, but as an interpretive process involving observation, scoping, viewpoint selection, and abstraction. With respect to RQ2, the diaries reveal recurring learning mechanisms related to BPMN as a formal language, collaborative modelling and presentation, transfer beyond notation, and, where applicable, bounded Generative AI use.

Taken together, these results reinforce the view that PM education is more than notation instruction and aligns with prior work showing that modelling competence involves abstraction, contextual judgment, communication, and iterative refinement [1, 3, 6, 19].

The study makes three contributions. First, it makes process identification more visible as a pedagogical object in its own right. While PM literature often emphasizes comprehension, notation learning, and cognitive support [2, 4, 5], the diaries show that a major difficulty arises earlier: students must decide what counts as the process, which actors matter, where boundaries lie, and how much of reality can be simplified without distorting the underlying logic. Second, the findings show that students experienced BPMN as a "new language," but that advanced semantics, especially exception handling, boundary events, timers, and event-based gateways, remained difficult. Third, the study highlights the social, communicative, and transferable dimensions of modelling: students learned by negotiating meanings, presenting models, receiving feedback, and projecting BPMN thinking into logistics, marketing, finance, HR, auditing, and broader management practice. The GenAI findings further suggest the need for modelling-specific AI literacy, where AI is treated as bounded support rather than as modelling authority [12].

Practically, the findings suggest four design recommendations for educators: first, introduce process identification explicitly as a scoping and interpretation task; second, use observation-grounded assignments to connect BPMN with real or plausibly real process situations; third, postpone AS-IS to TO-BE redesign until learners have stabilized AS-IS modelling competence; and fourth, frame Generative AI as a bounded support tool that requires verification. The learning diary itself appears useful not only as an assessment artifact, but also as a metacognitive scaffold directing attention to scoping, semantics, teamwork, communication, and transfer.

Several limitations should be noted. First, the evidence is based on self-reported reflections rather than direct behavioral or experimental measures. Second, the cohorts differed simultaneously in institution, country, student level, temporal format, and the presence or absence of specific components such as TO-BE redesign and GenAI reflection. These factors make the cases analytically informative, but they also create a confounded cross-cohort design in which sources of variation cannot be isolated cleanly. Third, reflective participation was uneven across cohorts. Fourth, the instructor who designed and taught the course implementations also conducted the qualitative analysis. To reduce this risk, the analysis relied on repeated corpus review, explicit coding categories, cross-cohort comparison, links between themes and diary excerpts, and discussions with co-authors about the coding scheme. Finally, the study foregrounds student perceptions and does not systematically triangulate them with independent indicators such as rubric-based model quality, exam scores, or long-term transfer tasks. Future research should therefore triangulate diary evidence with model assessment, classroom observation, instructor diaries, exam performance, or process-tracing methods. It would also be valuable to compare observation-grounded assignments with text-only modelling tasks, test staged AS-IS versus redesign sequences, introduce structured GenAI verification protocols, and investigate how advanced BPMN semantics can be scaffolded through richer examples, staged process-pattern instruction, or dynamic/token-based visual support [5, 9, 11].

References

1. Dumas, M., La Rosa, M., Mendling, J., Reijers, H.A.: Fundamentals of business process management (2nd ed.). Springer, Cham (2018). https://doi.org/10.1007/978-3-662-56509-4
2. Winter, M., Neumann, H., Pryss, R., Probst, T., Reichert, M.: Defining gaze patterns for process model literacy: Exploring visual routines in process models with diverse mappings. Expert Syst. Appl. **213**, 119217 (2023)
3. Rosenthal, K., Ternes, B., Strecker, S.: Learning conceptual modeling: Structuring overview, research themes and paths for future research. In: ECIS 2019 (2019)
4. Figl, K.: Comprehension of procedural visual business process models: a literature review. Bus. Inf. Syst. Eng. **59**(1), 41–67 (2017)
5. Aysolmaz, B., Reijers, H.A.: Animation as a dynamic visualization technique for improving process model comprehension. Inf. Manag. **58**(8) (2021)
6. Maslov, I., Poelmans, S., Rosenthal, K.: Conceptual modeling education: A bibliometric literature review and avenues for future research. Bus. Inf. Syst. Eng. 1–33 (2025)
7. Grzesiak, M., Moszyński, M.: Application of the flipped learning methodology at a business process modelling course: A case study. In: Business Process Management: Blockchain, Robotic Process Automation, Central and Eastern European, Educators and Industry Forum, pp. 303–312. Springer, Cham (2024)
8. Revoredo, K., Bala, S.: Project-based learning in BPM: Lessons from flipped classroom vs. lecture-based instruction. In: BPM 2024, pp. 313–323 (2024)
9. Maslov, I., Poelmans, S.: Facilitating the comprehension of business process models for unexperienced modelers using token-based animations. Inf. Manag. **61**(8), 103967 (2024)
10. Maslov, I., Poelmans, S.: Comprehension of (business) process models via tokens: an eye-tracking approach. In: International Conference on Business Process Management, pp. 375–385. Springer, Cham (2024)
11. Maslov, I., Poelmans, S., Mandeelburger, M.M., Proper, H.: Teaching Process Patterns in BPMN to Novice Modelers via Token Animations. PACIS 2025 Proceedings (2025)
12. Maslov, I., Poelmans, S., Wautelet, Y., Rosenthal, K.: Generative AI's aid in (Business) Process Modeling Instructional Design: case study. In: BPM 2025 (2025)
13. Claes, J., Vanderfeesten, I., Gailly, F., Grefen, P., Poels, G.: SPMM: What is the best way for me to construct a process model? DSS **100**, 57–76 (2017)
14. Moody, D.L.: The physics of notations: Toward a scientific basis for constructing visual notations in software engineering. IEEE TSE **35**(6), 756–779 (2009)
15. Petrusel, R., Mendling, J., Reijers, H.A.: How visual cognition influences process model comprehension. Decis. Support. Syst. **96**, 1–16 (2017)
16. Mendling, J., Reijers, H.A., van der Aalst, W.M.P.: Seven process modeling guidelines (7PMG). Inf. Softw. Technol. **52**(2), 127–136 (2010)
17. von Uthmann, C., Becker, J.: Guidelines of modeling (GoM) for business process simulation. In: Scholz-Reiter, B., Stahlmann, H.-D., Nethe, A., (eds.), PM, pp. 100–117 (1999)
18. Maslov, I.: Towards empirically validated process modelling education using a BPMN formalism. In: Research Challenges in Information Science, pp. 803–810. Springer, Cham (2022)
19. Ternes, B., Rosenthal, K., Strecker, S.: User interface design research for modeling tools: A literature study. EMISA **16**, 1–24 (2021)
20. Thorpe, K.: Reflective learning journals: From concept to practice. Reflective Pract. **5**(3), 327–343 (2004)
21. Moon, J.A.: Learning journals: A handbook for reflective practice and professional development (2nd ed.). Routledge (2006)

22. Braun, V., Clarke, V.: Using thematic analysis in psychology. Qual. Res. Psychol. **3**(2), 77–101 (2006). https://doi.org/10.1191/1478088706qp063oa
23. Saunders, M., Lewis, P., Thornhill, A.: Research methods for business students (2009)
24. Maslov, I., Poelmans, S., Wautelet, Y.: Measuring internal validity in mixed methods user feedback survey by finding inconsistencies with sentimentR and ChatGPT-4. PACIS 2024 (2024)

Adaptive BPM Learning Journeys: An AI-Powered Platform for Personalized Assessment and Multimodal Content Generation

Kevin Schulz[1], Julius Hüttmann[1], Maike Drewes[1], Lea Kleinekathöfer[1], Tobias Zimmermann[1,2(✉)], Sandro Franzoi[1,2], and Jan vom Brocke[1,2]

[1] University of Munster, Munster, Germany
[2] ERCIS European Research Center for Information Systems, Munster, Germany
zimmermann.tobias@uni-muenster.de

Abstract. BPM is taught across heterogeneous audiences ranging from university students to working professionals, who differ substantially in prior knowledge, learning pace, and goals. Traditional one-size-fits-all instructional approaches struggle to accommodate this diversity, often resulting in disengagement and uneven skill acquisition. We present an AI-powered platform for adaptive BPM learning journeys that relocates instructional effort from authoring fixed training modules to curating a shared knowledge base, from which personalized multimodal content and assessments are generated on demand. The platform integrates three pillars: personalization to each learner's prior knowledge and preferred representations of information, automation of instructional design through an LLM-driven generation pipeline, and gamification elements that sustain long-term engagement. Together, these pillars constitute a five-phase adaptive learning journey. We describe the platform's architecture, the pedagogical rationale behind the shift from authoring to curating, and its implications for developing the People capability of BPM at scale.

Keywords: Adaptive Learning · Large Language Models · BPM Education

1 Introduction

Business Process Management (BPM) serves as a fundamental pillar for organizational governance, providing the frameworks necessary to define, analyze, and optimize workflows [4]. Although BPM's core principles remain the same, its audiences have become far more diverse. University courses now draw students from many disciplines, and corporate training must serve both new hires and experienced professionals at once. In both contexts, the challenge is the same: learners differ markedly in prior knowledge, learning pace, and motivational orientation, and traditional one-size-fits-all instructional approaches struggle to

K. Kluza et al. (Eds.): BPM 2026, LNBIP 600, pp. 393–401, 2027.
https://doi.org/10.1007/978-3-032-36710-5_25

accommodate this diversity [6]. The stakes are not merely pedagogical: established capability frameworks locate training within the People dimension of BPM capability [11], and planning tools such as the BPM Billboard treat People development as one of six core capability areas to be assessed alongside strategic alignment, governance, methods, technology, and culture [3].

Static instructional materials are insufficient to meet this heterogeneity. Such standardized approaches typically expect learners to adapt to a fixed instructional profile rather than the reverse. For learners whose existing knowledge is already advanced, the material risks being redundant; for those without relevant background, it risks being overwhelming. In either case, the result is disengagement and uneven skill acquisition, outcomes that are well recognized in the broader adaptive learning literature [2]. The conventional response, differentiating materials for learner subgroups, does not resolve the underlying constraint: it still requires educators to author and maintain multiple versions of the same content, so effort continues to scale with audience heterogeneity.

The rapid development and growing accessibility of Large Language Models (LLMs) have introduced qualitatively new opportunities to address this challenge, enabling content generation and personalization at a scale and granularity that manual instructional design cannot match. This potential is, however, only realized if LLM capabilities are embedded within a coherent pedagogical architecture, one that channels content generation toward learner-specific needs, and if educator effort is redirected away from authoring content for anticipated learner profiles and toward curating the underlying knowledge from which any profile's content can be generated. This shift from authoring to curating is the central design principle behind the platform we present.

In this paper, we present an Artificial Intelligence (AI)-powered platform that enables such adaptive BPM learning journeys. To introduce the developed platform, we use a corporate onboarding scenario as a concrete running case throughout, as it surfaces inherent learner heterogeneity challenges. The underlying architecture, however, is domain-agnostic and maps onto university BPM courses or any setting in which a cohort spans a wide range of prior knowledge. We describe the platform's architecture and pedagogical design rationale, and outline the planned empirical evaluation.

2 AI and LLMs in BPM Education

In the educational field, AI has enabled a move away from standardized instruction toward data-driven, personalized learning journeys. Modern adaptive systems utilize machine learning to analyze learner behavior in real time, allowing for the dynamic adjustment of task difficulty and feedback. Research indicates that such personalization not only improves learning outcomes but also significantly enhances learner autonomy and motivation by aligning material with individual interests and pace [2,6]. Yet most existing tools fall short of full personalization in practice. Static Learning Management Systems (LMS) deliver the same content to every learner regardless of prior knowledge, while generic

AI tools can adapt tone or difficulty on request but lack the curated, domain-specific grounding needed to reliably teach BPM concepts [8]. Neither class of tool combines adaptive content, multimodal delivery, and domain-specific knowledge within a single coherent system (Table 1).

The rapid development of LLMs has introduced new opportunities to close this gap. Recent literature shows that LLMs are increasingly used to assist BPM tasks, from extracting process knowledge from enterprise content [5] to supporting the iterative refinement of process models [1]. Transferring these insights to educational contexts, content can be provided across multiple representations of information. Since the learning-styles matching hypothesis is contested in educational psychology [10], we treat multimodal delivery as a design heuristic rather than a prescription. This approach is grounded in Richard Mayer's Cognitive Theory of Multimedia Learning, which asserts that learners build deeper mental models when information is presented through a combination of words and pictures rather than a single modality [9]. By leveraging LLM to automate the generation of these varied representations, educators can better manage the cognitive load of learners without scaling manual authoring effort.

Table 1. Comparison of tool-supported learning journeys.

Characteristic	Static LMS	Generic AI	Our System
Adaptive content	✗	✓	✓
Multimodal formats	✓	✓	✓
Curated domain-specific knowledge	✓	✗	✓

3 Three Pillars of Adaptive BPM Learning Journeys

Our platform is built on three pillars to turn static learning into a dynamic, individualized experience: Personalization, Automation, and Gamification. Together, they are intended to address learner heterogeneity and cognitive load in the BPM domain.

The first pillar, Personalization, is the core of the adaptive learning journey. It ensures the learning process adapts to each user's prior knowledge by having users share their background and self-rate their expertise before starting. Based on these inputs, content is created for each topic tailored to the user's knowledge. Personalization is not a one-time setup: as learners progress through assessments, their performance feeds back into the learner model, which then conditions later content generation and question difficulty. Grounded in Mayer's Cognitive Theory of Multimedia Learning [9], the platform generates content in various representations of information, including text, podcasts, videos, and flashcards, to suit individual preferences and to diversify the learning experience. While Personalization defines what content a learner receives, our second pillar, Automation, focuses on efficiently producing that content at scale by using LLMs to automate the most resource-intensive aspects of instructional design. Instead of manually creating static training modules, our system uses an LLM-driven pipeline to generate personalized text summaries, podcast scripts, and assessment questions directly from a central BPM knowledge base (see Sect. 4.1). Educators curate the knowledge base to control scope and depth, while Retrieval Augmented Generation (RAG) grounding ties generated content to vetted source material rather than the model's parametric knowledge alone. Crucially, this includes the audio: every podcast is regenerated per learner from their expertise profile, rather than being a static recording shared across users. The third pillar, Gamification, is expected to support long-term engagement. We implemented a workdays-aware streak calendar, alongside content-aware reminders. The dashboard that surfaces these elements is described in Sect. 4.2. These three pillars are realized in the platform described next.

4 An AI-Powered Platform for BPM Learning Journeys

Our platform is implemented as a full-stack web application that delivers personalized BPM learning through an interactive interface and LLM-driven backend services. This section describes the system architecture and the workflow that guides learners through the adaptive learning journey.

4.1 Platform Architecture

The platform follows a three-layer architecture. The Frontend Layer is built with Next.js and exposes the user-facing components. The Learning Dashboard is the entry point where users browse topics, view their progress, and resume previous learning sessions. The Content Interface provides the user access to the multimodal learning materials. The Assessment Interface handles quizzes and feedback.

The Backend Layer is implemented in FastAPI. It manages user profiles, persists assessment responses and progress data, and orchestrates the AI services through a unified API. Four specialized services handle the AI-driven functionality of the platform. A RAG retrieval service fetches relevant educational context from the BPM knowledge base. In our current pilot implementation, this exemplary knowledge base consists of high-density markdown documents. To ensure academic rigor, this corpus is systematically curated and structured around the industry-standard BPM lifecycle framework by Dumas et al. [4]. LLMs are utilized to generate personalized quiz items, text summaries, and podcast scripts. An LLM-based matching service aligns video transcripts with learner profiles, and a progress tracking service maintains the learner's state. The generation-heavy operations, such as quiz and podcast generation, run asynchronously, so interactive requests remain responsive even under concurrent use. The user data and learning content are stored in Supabase, which also handles authentication.

An anonymized walkthrough video, full source code, and additional screenshots are provided in our git repository[1].

4.2 User Journey

The learning experience is structured into five sequential phases. To demonstrate the platform's personalization for diverse learners, this walkthrough uses the corporate onboarding scenario from our pilot project. Although the terminology reflects a professional context (e.g., assessing background and experience), the same educational mechanics, software architecture, and asynchronous generation pipelines apply to traditional university course modules.

Phase 1: Initial Questionnaire. On first login, users complete a short background questionnaire. They report their professional background, years of BPM experience, highest degree of education, and preferred representations of information such as text, audio, or video. They also self-rate their current expertise across the core BPM topics. The resulting profile is stored in the learner model and forms the basis for all later personalization, from question difficulty to the depth of generated content.

Phase 2: Initial Topic Assessment. Before accessing any learning content, users take a quiz covering the core BPM topics, for example the BPM lifecycle. While Phase 1 captures self-perception, this phase measures actual knowledge. The system combines both signals to find knowledge gaps and to detect cases

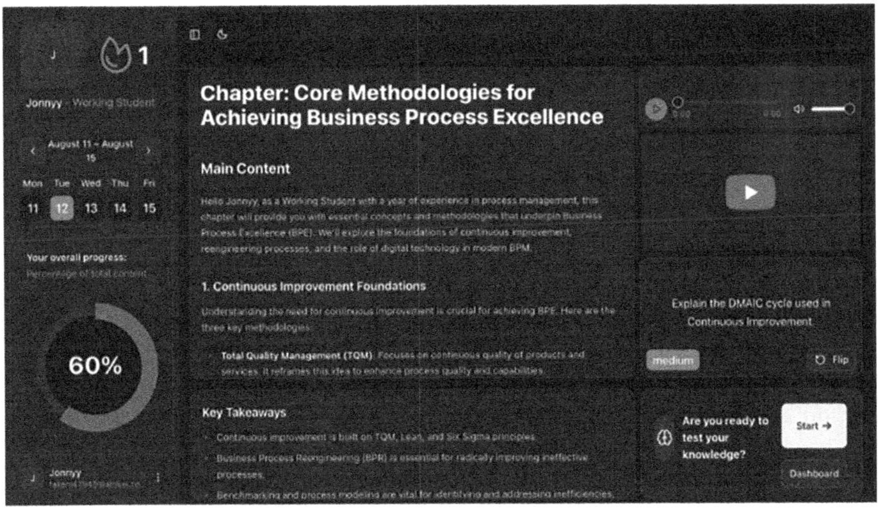

Fig. 1. Multimodal learning interface showing the four representations of information—text summary, podcast, video, and flashcards—for a single chapter.

[1] https://github.com/ercis/bpm-learning-tool.

where confidence and competence diverge, an effect well documented in self-assessment research [7]. This calibration is particularly relevant in BPM, where practitioners often bring implicit job-based confidence that does not always map cleanly to formal frameworks. The results form the basis for the difficulty of the generated content and questions in later phases. The questions themselves are generated from the same knowledge base used throughout the platform, so the assessment stays consistent with what learners will encounter later.

Phase 3: Learning Content Consumption. Users navigate to the Learning Dashboard and select a topic. The system generates multimodal content tailored to their expertise level: text summaries (400–600 words), podcast-style audio (5–10 min), video recommendations, and 15–20 flashcards for active recall. All representations are personalized. The text summary adjusts in scope and depth to the user's expertise. The podcast script is generated from the same expertise profile, so the audio is personalized rather than a generic recording shared across users. Video recommendations are dynamically selected from a curated YouTube repository. The platform pre-transcribes all available videos and stores the text in the database. An LLM-based matching service then selects the video whose transcript best aligns with the learner's profile and current knowledge gaps. Flashcards focus on the concepts most relevant to the user's identified knowledge gaps. Figure 1 shows the resulting interface.

Phase 4: Chapter Assessment. When users feel ready, they start an assessment of 5–10 multiple-choice questions on the current chapter. Each question is fully personalized for the user and is generated on demand. The generation is conditioned on four sources of context: the user's initial assessment results, the initial questionnaire, prior performance on related chapters, and content excerpts retrieved through the RAG pipeline. In this way, questions probe exactly the concepts the learner has just engaged with, at a difficulty level appropriate to their current mastery. Unlike systems drawing from a fixed item bank, every question is constructed specifically for this learner. Immediate feedback after each answer explains the correct option and points back to the relevant learning content if the answer was wrong. The assessment is intended to double as a learning opportunity rather than only measuring knowledge. The results feed back into the learner model and refine later personalization.

Phase 5: Final Topic Quiz. After completing all chapters and passing their chapter assessments (Phase 4), learners take a comprehensive final quiz spanning all chapters of the topic to verify mastery of its core concepts. Successful completion marks the topic as mastered, updates the progress tracker, and unlocks the next topic in the learning path. Learners can still revisit completed topics for review.

Two features accompany learners across all five phases:

Continuous Progress Tracking and Gamification. A persistent Progress Dashboard is designed to sustain long-term engagement. The centerpiece is a workdays-aware streak calendar that counts only weekdays during typical working hours. A progress tracker visualizes completion and performance. Summary metrics (e.g., topics completed) offer the user a high-level overview of their learning journey.

Learning Assistant. An integrated chatbot is available across all phases and answers content questions in context. The assistant is grounded in the same BPM knowledge base used for content generation, so its responses stay consistent with the learning materials. Technically, the assistant utilizes LLM tool calling to autonomously retrieve current data, such as querying the learner's real-time assessment status. The system prompt is dynamically adjusted to the user's information and progress, so explanations adapt to the learner's background.

5 Relevance and Implications

For BPM educators, the platform's most consequential shift is in where instructional effort is directed. Rather than authoring training modules, designing quiz banks, or producing differentiated materials for learner subgroups, the educator's primary task becomes curating the knowledge base from which all content is derived. Updating a learning objective, correcting a misconception, or rebalancing conceptual and procedural emphasis becomes an editorial intervention on that knowledge base, and the change propagates automatically to all generated content. This reframes training as an ongoing capability-development activity rather than a series of one-off authoring efforts, positioning the platform as an operational instrument for developing the People dimension of BPM capability [3,11] at a scale that manual authoring cannot sustain. This is particularly relevant in BPM, where curricula must track a field that evolves continuously with organizational practice and tooling [1]. University educators can modify the corpus to scale complexity dynamically, adjusting scope to suit an introductory BPM cohort or an advanced enterprise architecture class.

The platform also surfaces diagnostic information that conventional course designs make available too late. The combination of self-assessed expertise (Phase 1) and objective initial assessment (Phase 2) gives educators a structured view of where confidence and competence diverge across their cohort before any learning content is consumed. This early signal is expected to help educators anticipate and address misconceptions proactively, rather than discovering them only through summative assessment.

Finally, because differentiation is handled generatively rather than manually, the cost of serving a heterogeneous cohort no longer scales with its size or variance. The wider the spread of prior knowledge within a group, the more the per-learner approach pays off relative to a standardized alternative [6].

6 Conclusion

We presented an AI-powered platform for adaptive BPM learning journeys that relocates instructional effort from authoring fixed content to curating a shared knowledge base, organized around three pillars (Personalization, Automation,

Gamification) and a five-phase learning journey. The platform thereby operationalizes the development of the People capability of BPM [3,11]. All content, text summaries, podcast audio, video recommendations, flashcards, and assessment questions, is generated on demand from that knowledge base and propagated automatically to every learner. A working full-stack implementation built on Next.js, FastAPI, and Supabase demonstrates that this design is technically feasible.

Several limitations apply. The platform has not yet been evaluated empirically with learners. Claims about engagement and learning outcomes therefore rest on design rationale rather than measured effects. Output quality is bounded by the curated knowledge base, and LLM-generated content remains subject to factual review even with RAG grounding. The current scope covers a selected set of BPM topics; broader domains require corresponding knowledge-base expansion. The argument for utilizing multiple representations of information follows from Mayer's multimedia principles [9], but the evidence for matching specific learning styles to outcomes is contested [10], so we treat it as a design heuristic rather than a validated predictor.

Two directions guide our next steps. To validate the platform's efficacy, we plan empirical pilots in corporate onboarding and university classroom settings, comparing an experimental group using our personalized system against a control group utilizing a static, non-personalized baseline. We will measure learning gains and quiz improvement by tracking delta scores between the initial topic assessment (pre-test) and the final topic quiz (post-test). Long-term retention will be assessed via follow-up testing four weeks post-completion, and platform logs will capture chapter completion rates, engagement patterns, and time-on-task. We also plan to expand the knowledge base beyond the current BPM core to cover adjacent process management curricula. Taken together, these steps aim to test whether shifting educator effort from authoring to curation can sustain personalization at scale without sacrificing the learning outcomes it is designed to improve.

References

1. Bettencourt, J., Guerreiro, S.: Large language models to enhance business process modeling: past, present, and future trends (2026). https://arxiv.org/abs/2604.14034
2. Bloom, B.S.: The 2 sigma problem: the search for methods of group instruction as effective as one-to-one tutoring. Educ. Res. **13**(6), 4 (1984). https://doi.org/10.2307/1175554
3. vom Brocke, J., Mendling, J., Rosemann, M.: Planning and scoping business process management with the BPM Billboard. In: vom Brocke, J., Mendling, J., Rosemann, M. (eds.) Business Process Management Cases Vol. 2, pp. 3–16. Springer, Berlin, Heidelberg (2021). https://doi.org/10.1007/978-3-662-63047-1_1
4. Fundamentals of business process management. Springer, Heidelberg (2018). https://doi.org/10.1007/978-3-662-56509-4_9

5. Franzoi, S., Delwaulle, M., Dyong, J., Schaffner, J., Burger, M., vom Brocke, J.: Using Large Language Models to Generate Process Knowledge from Enterprise Content. In: Gdowska, K., Gómez-López, M.T., Rehse, J.R. (eds.) Business Process Management Workshops, pp. 247–258. Springer Nature Switzerland, Cham (2025). https://doi.org/10.1007/978-3-031-78666-2_19
6. Hariyanto, K., Maharani, F.X.D., R.: Artificial intelligence in adaptive education: a systematic review of techniques for personalized learning. Discover Educ. **4**(1) (2025). https://doi.org/10.1007/s44217-025-00908-6
7. Kruger, J., Dunning, D.: Unskilled and unaware of it: how difficulties in recognizing one's own incompetence lead to inflated self-assessments. J. Pers. Soc. Psychol. **77**(6), 1121–1134 (1999). https://doi.org/10.1037/0022-3514.77.6.1121
8. Kucharski, S., Braun, I., Damnik, G., Wählisch, M.: Adaptive learning mechanisms for learning management systems: a scoping review and practical considerations (2025). https://doi.org/10.48550/arXiv.2512.18383, arXiv:2512.18383
9. Mayer, R.: Multimedia Learning (2020). https://doi.org/10.1017/9781316941355
10. Pashler, H., McDaniel, M., Rohrer, D., Bjork, R.: Learning Styles: Concepts and Evidence. Psychol. Sci.Public Interest **9**(3), 105–119 (2008). https://doi.org/10.1111/j.1539-6053.2009.01038.x
11. Rosemann, M., vom Brocke, J.: The six core elements of business process management. In: vom Brocke, J., Rosemann, M. (eds.) Handbook on Business Process Management 1, pp. 107–122. Springer Berlin Heidelberg, Berlin, Heidelberg (2010). https://doi.org/10.1007/978-3-642-00416-2_5

Deployment-Sovereign AI Infrastructure for BPM Education: An Open-Source Stack from Document Analysis to Process Execution

Piotr Senkus[1(✉)], Marek Szelagowski[2], Damian Ciachorowski[3], Tomasz Ludwicki[4], and Grzegorz Szymański[1]

[1] Uniwersytet Kaliski im. Prezydenta S. Wojciechowskiego, Kalisz, Poland
p.senkus@uniwersytetkaliski.edu.pl
[2] Systems Research Institute, Polish Academy of Sciences (IBS PAN), Warsaw, Poland
[3] Gdańsk University of Technology, Gdansk, Poland
damian.ciachorowski@pg.edu.pl
[4] Faculty of Management, University of Warsaw, Warsaw, Poland

Abstract. This paper investigates whether deployment-sovereign AI infrastructure can support BPM education without dependence on commercial LLM APIs. We assembled an open-source teaching stack combining PCSS-hosted open-weight models, Flowise/Qdrant retrieval-augmented generation, n8n workflow orchestration, and the CIB Seven BPMN engine. We evaluated 68 BPM prompts across 13 models and three deployment classes, logging 2,248 model calls including a dedicated reliability re-run. The stack was piloted with about 110 participants across three university programmes. Commercial endpoints exhibited provider-specific availability failures and content-policy refusals, most visibly Claude 3.5 Haiku on selected healthcare, chemical, biological, radiological, and nuclear (CBRN), substance-destruction, and self-harm documentation prompts. In the tested PCSS deployment, all five open-weight models completed the high-sensitivity corpus without API errors or policy refusals. A preliminary, non-blinded expert assessment found no obvious quality deficit for routine BPM documentation tasks, but does not establish statistical equivalence to commercial models. We refine pedagogical sovereignty as institutional control over AI cost, availability, content policy, and data jurisdiction. The architecture is replicable at zero marginal licensing cost via national or EuroHPC resources, subject to allocation, configuration effort, and governance requirements.

Keywords: BPM education · deployment sovereignty · sovereign AI · HTO-S framework · pedagogical sovereignty · content filtering · LLM · EuroHPC

K. Kluza et al. (Eds.): BPM 2026, LNBIP 600, pp. 402–417, 2027.
https://doi.org/10.1007/978-3-032-36710-5_26

1 Introduction

Commercial LLM APIs increasingly mediate access to AI-assisted process modelling, documentation, and analysis. In BPM education this dependency is not neutral. A course that relies on an external API inherits that provider's pricing model, rate limits, model updates, data-processing jurisdiction, and content policy. These dependencies become visible when routine BPM terminology overlaps with safety-sensitive language such as failure modes, non-conformities, hazardous waste, incident response, controlled disposal, or clinical handover. Whether such overlap produces material barriers for BPM education has not yet been systematically tested.

The BPM Educators Forum has documented a clear shift toward AI integration. LLMs generate BPMN diagrams from text [17], solve modelling assignments with varying reliability [20], and serve as intelligent tutors for declarative languages [29]. The BPMN-Chatbot won Best Demo at BPM 2024 [23]. But every published tool depends on a commercial API, and the economics of that dependency have deteriorated sharply. In April 2026, Uber exhausted its entire annual AI coding budget in four months [3]. Gartner confirmed that agentic AI workflows consume 5–30 times more tokens per task [46]; EY estimated the cost of a single agentic interaction at thirty times the chatbot cost [45]. If a Fortune 500 company cannot sustain uncapped LLM access, a university department with a five-figure IT budget cannot begin [8].

The literature addresses AI in BPM curricula [17,20,23,27,29,40] and the sovereign AI movement at the policy level [16,19,26,34]. What remains unexamined is their operational intersection: whether a complete sovereign AI stack can support BPM education across the full process lifecycle and what barriers emerge when commercial infrastructure is replaced with national high-performance computing (HPC) resources. The Reuters Institute's 2025 survey found ChatGPT to be the most widely recognised AI tool in every country surveyed [43]; a curriculum that silently equates AI with a single commercial vendor reproduces that dependency in the next generation of process analysts.

We ground this investigation in the **HTO-S (Human–Technology–Organisation–So-ver-eign-ty) framework** [42], which extends the classical Human–Technology–Organisation model by adding a fourth dimension, technological sovereignty. In its original form, the framework identifies four barriers that impede AI adoption in quality management systems: vendor lock-in (organisations become dependent on proprietary APIs), data jurisdiction (institutional data is processed outside national boundaries), uncontrolled model updates (provider-side changes alter system behaviour without notice), and content overfiltering (safety mechanisms refuse legitimate professional queries). We test whether these barriers, identified in industrial quality management, also constrain educational deployment.

Turobov et al. [49] demonstrate that digital sovereignty encompasses sovereignty *over* the digital domain and sovereignty *through* digital technology. In the educational context, we argue these barriers function as preconditions for pedagogical autonomy. We call this capacity **pedagogical sovereignty**: the ability of an educational institution to control the cost, availability, content policy, and data jurisdiction of AI infrastructure used in teaching, so that curriculum design does not depend on opaque external vendors.

This leads to two research questions:

RQ1: To what extent do commercial LLM APIs exhibit reliability and content filtering barriers that impede their use in BPM process documentation education?

RQ2: Can deployment-sovereign, open-source AI infrastructure support BPM education workflows at zero marginal licensing cost while maintaining acceptable reliability and preliminary output quality?

We make three contributions: (1) the first empirical test of the HTO-S content overfiltering barrier in education, demonstrating it is domain-dependent and provider-specific; (2) the introduction of pedagogical sovereignty as a theoretical construct operationalised along four dimensions; and (3) a complete, tested, replicable architecture.

2 Literature Review

Three literatures bear on this problem: the pedagogy of AI in BPM courses, the economics and politics of sovereign AI, and the open-source BPM toolchain. At each turn we ask the same question the rest of this paper asks: who controls the infrastructure on which the teaching depends?

AI in BPM Education. Grzesiak et al. [17] evaluated LLM output on master's-level exercises, finding it useful for accelerating initial models but prone to assessment discrepancies. Kluza et al. [20] tested BPMN tasks across modalities and found deeper process logic still beyond current models. Kourani et al. [24] benchmarked sixteen LLMs across twenty processes and found wide variation, with quality tracking error-handling capability. Nagel et al. [29] built EduClare, a tutoring system combining RAG with GPT-4o, and Köpke and Safan [23] optimised the BPMN-Chatbot for token efficiency. Rockembach and Thom [40] showed that even open-weight models can generate assessment questions calibrated to Bloom's taxonomy, while Sliż et al. [44] issued a manifesto naming the systemic underrepresentation of process-execution tools in Polish BPM curricula; Moorhouse et al. [28] found that fewer than half of the world's top 50 universities had assessment guidelines for generative AI. Every contribution listed above depends on commercial API access, and none addresses infrastructure sovereignty.

Sovereign AI and the Knowledge Monopoly. Roberts et al. [19] argued that AI controlled by a handful of corporations constitutes a de facto knowledge monopoly, with power to determine which information is amplified. The Linux Foundation reports that 80% of organisations now treat sovereign AI as a strategic priority, with 90% identifying open-source models as essential [16]. At a Federal Reserve conference in 2025, OpenAI's CEO stated: "We see a future where intelligence is a utility, like electricity or water, and people buy it from us on a meter" [1]. NVIDIA's Nemotron-Personas programme, producing CC BY 4.0 synthetic persona datasets for seven countries, explicitly lists Poland's Bielik.ai among sovereign model builders [34]. France deployed LightOn's Paradigm Edu at ENS Paris-Saclay [26]; Poland's PCSS provides 23 open-weight models free for researchers [38]. Senkus et al. formalised these sovereignty barriers in the HTO-S framework [42]. Wolfe et al. [50] sharpened the argument for risk-intolerant settings, concluding that closed-weight models are incompatible with requirements for transparency and adaptability, and that small open-weight models on a single consumer GPU achieve competitive performance.

Open-Source BPM Tooling. CIB Seven [10] provides a production-grade BPMN 2.0 engine under Apache 2.0. Open-BPMN [35] offers a free modeller. n8n provides workflow orchestration; Flowise and Qdrant deliver RAG capabilities. None has been assembled into a coherent BPM education stack. König et al. [22] and Revoredo and Bala [39] emphasise hands-on BPM lifecycle experience but implicitly assume affordable infrastructure.

What the Field has Established. Three things are now reasonably settled. LLMs can accelerate the first draft of a process model and generate assessment items calibrated to Bloom's taxonomy [17,40]; they handle deeper process logic such as deadlock detection poorly [20]; and the individual open-source building blocks for a teaching stack each exist and work in isolation [10,35]. The pedagogical value of hands-on lifecycle experience is, separately, well argued [22,39]. Mahendrawathi et al. [27] showed that structured design spaces help students generate innovative process redesign ideas, confirming that AI-augmented BPM exercises can serve genuine pedagogical goals.

What Remains Unresolved. The current literature leaves three questions open. One is whether the content-filtering and reliability problems of commercial APIs are uniform across providers, or vary in ways an educator could anticipate and plan around. Another is whether those scattered open-source components can actually be assembled into a single working stack that carries a student from a raw document to an executable process. A third, and the sharpest, is whether a zero-cost self-hosted stack can match the output quality that educators currently rely on commercial models to supply. A recent review of LLM-driven process modelling confirms these gaps, identifying evaluation fragmentation, weak reproducibility, and almost no validation in real organisational settings [4,7]. The sovereignty literature names the stakes [16,19,42], but tests none of this in a classroom.

Why this Matters Now. The economics decided the urgency. With commercial token costs climbing [2,3,45,46] and "AI" collapsing in the public mind into a single metered vendor [1,43], a BPM curriculum built on that vendor inherits its costs, its outages, and its content policy, then passes all three to the next cohort of analysts. Until someone builds and tests a sovereign alternative end to end, institutions can only digitise their dependency rather than escape it [9]. That is the gap our two research questions set out to close.

3 Methodology

3.1 Architecture Design

The sovereign stack comprises four layers (Fig. 1). The **inference layer** provides 23 open-weight LLMs through PCSS national HPC at zero cost via an OpenAI-compatible REST API [38], including DeepSeek-V3.1, Qwen 3.5-397B, GLM 5.1, and the Polish national model Bielik 11B. The **knowledge layer** implements RAG through Flowise backed by Qdrant. The **orchestration layer** uses n8n with ten pre-configured BPM exercise templates mapping learning objectives from Dumas et al. [11] to AI-assisted workflows. The **execution layer** runs CIB Seven [10] as a BPMN 2.0 process engine. Total software cost is $0. Recent analyses confirm that on-premise open-weight deployment reaches economic break-even against commercial API subscriptions quickly at sustained usage [36], with consumer-GPU inference costing $0.001–0.04 per million tokens—40 to 200 times below budget-tier cloud APIs [21]. PCSS, operated by the Institute of Bioorganic Chemistry of the Polish Academy of Sciences, manages Poland's national research network (PIONIER) and exposes these models through a single OpenAI-compatible endpoint—existing tools written against the OpenAI SDK (Flowise, n8n, LangChain) require only a change of base URL to redirect from a commercial endpoint to sovereign infrastructure. PCSS also hosts PIAST-AI, the first Polish EuroHPC AI Factory (over 1,000 GPUs, entering full service Q3 2026) and PIAST-Q, the first EuroHPC quantum computer inaugurated on European soil [37]. For rough comparison, a commercial stack (OpenAI Team, Zapier, Pinecone, Signavio) for a 30-student course runs to an estimated $15,000–20,000 per year in subscription and token fees; the sovereign stack replaces this with publicly funded infrastructure and faculty integration effort.

Deployment Sovereignty. A natural objection is that DeepSeek, Qwen, and GLM are foreign-trained (Chinese) models, undermining any sovereignty claim. We distinguish two layers. *Sovereignty over provenance* (control of training data and model weights from origin) is satisfied only by Bielik, the Polish national model. *Sovereignty over deployment* (control of where inference runs, where data flows, and who sets usage policy) is satisfied for any open-weight model hosted on PCSS. Because the weights are open and inference is local, no institutional or student data leaves national infrastructure, the deployment can be audited, and no foreign vendor can alter, deprecate, or filter the model without consent. For BPM education, where prompts concern process documentation rather than politically sensitive content, deployment sovereignty is the operative requirement. This mirrors the logic of open-source software: foreign authorship does not negate sovereignty when the artifact is open, auditable, and running on hardware you control. Equivalent EuroHPC resources exist in Finland (LUMI), Spain (MareNostrum 5), and Germany (Jülich) [13,14].

3.2 Experiment Design

We designed 68 English-language prompts in two corpora. The **BPM-SAFE-48** corpus contains 48 standard BPM documentation tasks across waste manage-

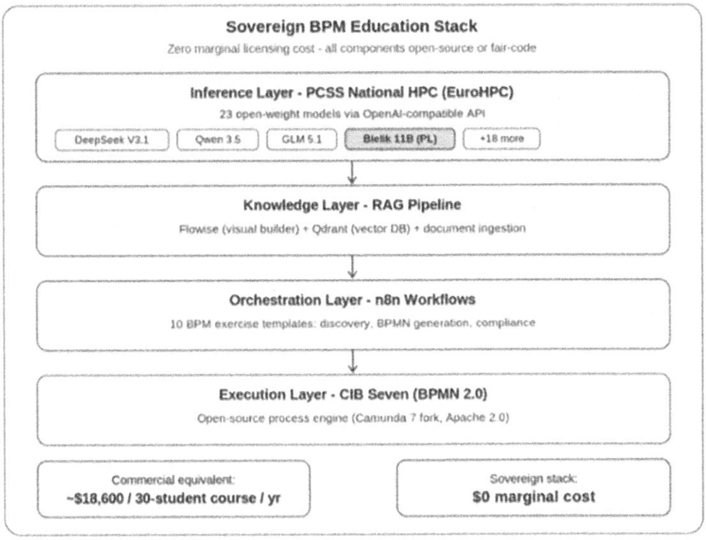

Fig. 1. Four-layer architecture of the deployment-sovereign BPM education stack. Bielik 11B (marked) is the only fully sovereign model: Polish weights and Polish training data; the other models are open-weight but foreign-trained.

ment (A), quality/defect analysis (B), healthcare processes (C), and industrial safety (D). The **BPM-TARGETED-20** corpus contains 20 high-sensitivity prompts across death/body handling (E), CBRN/terrorism response (F), controlled substance destruction (G), defense/weapons procurement (H), and psychiatric/self-harm clinical processes (I). All prompts represent legitimate professional needs using ISO 9001, ISO 14001, FMEA (IEC 60812), and HAZOP (IEC 61882) terminology.

Model Selection Rationale. Models were selected to represent three access paradigms: widely used commercial APIs (GPT-4o, Claude, Gemini, Command R+, Grok), cloud-hosted open-weight (Llama 4 Scout on Groq), and all PCSS-available models with at least 10B parameters at testing time. The selection favours breadth of deployment class over depth within any single provider.

We drew thirteen models from three deployment classes (Table 1). **Class A**: commercial closed-source API (GPT-4o, GPT-4o Mini, Claude 3.5 Haiku, Claude 3.5 Sonnet, Command R+, Gemini 2.0 Flash, Grok 3 Mini). **Class B**: open-weight on commercial cloud (Llama 4 Scout on Groq). **Class C**: open-weight on sovereign HPC (DeepSeek V3.1, Llama 3.3 70B, GLM 5.1, Bielik 11B, Qwen 3.5 397B on PCSS). Of the thirteen, nine returned usable data; four returned only API errors, reported as failures rather than refusals. The standard corpus ran at three repetitions and the targeted corpus at two, at temperature 0.7 (1,024-token cap standard; 512-token cap targeted, matching the DeepSeek baseline protocol). The standard corpus (1,296 runs), targeted corpus (520 runs), and

Table 1. Models, deployment classes, and run accounting. "Std" and "Tgt" columns show attempts on the 48-prompt standard and 20-prompt targeted corpora; a dash denotes not run. A 432-run reliability re-run of three responsive Class A models is not shown. Only Claude 3.5 Haiku produced refusals.

Class	Provider / model	Std	Tgt	Errors	Refusals
A	OpenAI GPT-4o Mini	144	40	0	0
A	OpenAI GPT-4o	–	40	0	0
A	Anthropic Claude 3.5 Haiku	144	40	0	14
A	Anthropic Claude 3.5 Sonnet	–	40	40	–
A	Cohere Command R+	144	40	184	–
A	Google Gemini 2.0 Flash	144	40	184	–
A	xAI Grok 3 Mini	–	40	40	–
B	Meta Llama 4 Scout (Groq)	–	40	0	0
C	DeepSeek V3.1 (PCSS)	144	40	0	0
C	Llama 3.3 70B (PCSS)	144	40	0	0
C	GLM 5.1 (PCSS)	144	40	0	0
C	Bielik 11B (PCSS)	144	40	0	0
C	Qwen 3.5 397B (PCSS)	144	40	0	0
Total		1,296	520	448	14

a reliability re-run (432 runs) together logged 2,248 model runs. Each response was classified as FULL, PARTIAL, REFUSED, or ERROR.[1]

3.3 Output Quality Assessment

Because refusal-free output is not sufficient for educational use, we conducted a preliminary expert review of BPMN and process-documentation outputs. The review focused on three criteria: syntactic correctness (valid BPMN 2.0 structure), semantic completeness (coverage of actors, events, activities, decisions, and handoffs), and pragmatic usefulness (suitability as a draft for classroom discussion or CIB Seven deployment after human revision).

This assessment was exploratory rather than a blinded benchmark. The expert panel included authors and external BPM practitioners, including the originator of the dynamic BPM paradigm [47], and model identity was not systematically hidden from raters. We use the quality review to detect obvious deficits, not to establish statistical equivalence. A blinded, inter-rater-reliability-controlled benchmark remains necessary before making strong claims about output-quality parity.

[1] FULL: task completed. PARTIAL: significant omissions or partial refusal. REFUSED: policy refusal, including soft refusals detected by manual review. ERROR: no usable output. For reasoning models (GLM 5.1, Qwen 3.5) the reasoning field was captured and classified.

3.4 Classroom Deployment

The stack was deployed across four teaching cohorts (~110 participants, 2025–2026): an executive MBA programme (~60), a private business-school MBA (~30), and a graduate programme (~20). The seven-week laboratory plan was structured around Bloom's revised taxonomy: RAG configuration and knowledge base construction at the knowledge/comprehension level (weeks 1–2), AI-assisted process discovery from regulatory text at the application/analysis level (weeks 3–4), BPMN model generation and CIB Seven deployment at the synthesis level (weeks 5–6), and a structured 18-section Improvement Report requiring process analysis with AI-augmented redesign proposals at the evaluation level (week 7).

Assessment used a structured rubric linking BPM competencies (as defined by Dumas et al. [11] and mapped to the lifecycle areas in the Sliż et al. manifesto [44]) to demonstrated proficiency with the sovereign stack. Grzesiak [18] confirmed that active learning methods with new BPM tools yield measurably higher student engagement; our instructor observations across four cohorts were consistent with this finding. Crucially, the zero-cost, zero-configuration nature of the PCSS API eliminated what we term the *tool-access bottleneck*: in courses using commercial APIs, the first 2–3 weeks are typically consumed by account creation, API key distribution, billing setup, and troubleshooting access errors. With the PCSS API, students executed substantive prompts in the first session. Formal pre/post learning-outcome measurement was not conducted; its absence is the primary limitation of the educational component.

A doctoral project used the complete pipeline as an existence proof: 54 executable subprocess models (27 AS-IS + 27 TO-BE) for a Source-to-Dispose asset lifecycle spanning six phases, validated through a Modified Delphi study (N=30 experts), identifying 22 dysfunctions, 6 information silos, and 6 regulatory gaps [48].

4 Results

RQ1—Barriers in Commercial LLM APIs. Two barriers emerged.

Content Filtering. proved domain-dependent and provider-specific. For standard BPM terminology (A–D), only Claude 3.5 Haiku refused: 6 of 144 (4.2%), on organ procurement (C12) and anthrax response (D07). All five PCSS models: 0% across 720 runs. The targeted corpus revealed sharper divergence. Claude 3.5 Haiku refused 20% overall: 100% on nerve agent decontamination (F02), 50% on ricin response (F01), methamphetamine cleanup (G03), school self-harm response (I02), and suicide risk assessment (I03). GPT-4o Mini, GPT-4o, and Llama 4 Scout: 0% across all clusters. Some refusals may be defensible from a provider-safety perspective, especially in CBRN or self-harm contexts. The educational problem is different: the refusal boundary is externally defined, opaque to the instructor, and subject to change without notice. The advantage of deployment-sovereign infrastructure is therefore not the absence of safety policy, but the possibility of defining, documenting, and auditing that policy locally.

The HTO-S content overfiltering barrier [42] is empirically confirmed as a phenomenon of *unpredictability and loss of institutional control.*

API Reliability. constituted the second barrier. Four commercial endpoints returned only errors: 448 error responses in total (Table 1). An initial commercial run also exhausted the API gateway's spending cap within thirty minutes, forcing the responsive Class A models to be reconfigured and re-run (the 432-run reliability re-run). Among additional cloud endpoints tested for stability, a Cerebras endpoint returned authentication errors on all calls and an NVIDIA NIM endpoint showed 50–270 s latencies with intermittent timeouts. For a classroom where 60 students submit prompts simultaneously, such failures do not degrade the exercise. They end it.

RQ2: Sovereign Infrastructure Performance. On the standard corpus, all five PCSS models completed 720 runs without a single error or refusal. On the targeted corpus, Bielik 11B, Llama 3.3 70B, GLM 5.1, and Qwen 3.5 397B each ran the full twenty-prompt corpus under the DeepSeek baseline parameters; together with DeepSeek V3.1 they completed 200 targeted runs with zero refusals and zero errors. During the piloting period, a single researcher consumed 150 million tokens—150% of the annual quota—at zero cost. Within the conditions tested, the sovereign stack matched commercial alternatives on reliability and availability.

Output Quality. For routine BPM documentation, PCSS-hosted models (particularly DeepSeek V3.1 and Qwen 3.5-397B) showed no obvious qualitative deficit relative to GPT-4o; differences were marginal and task-dependent. Two caveats keep this preliminary: the panel was not independent of the project; and reasoning models returned raw chain-of-thought rather than finished artifacts. Quantitative support comes from a separate Polish-language extraction benchmark (Sect. 7): DeepSeek V3.1 (F1 0.843) and Llama 3.3 70B (0.830) scored within one to two points of the best commercial models at zero cost. The defensible conclusion is narrower: selected PCSS-hosted models produced outputs that experts considered usable as first drafts for classroom analysis and human revision. Statistical equivalence to commercial models remains unproven.

5 Discussion and Implications

Theoretical Implications. Our results extend the HTO-S framework [42] on three fronts. The content overfiltering barrier is real but more uneven than the theory implied: provider-specific, domain-dependent, temporally unstable. The paper contributes pedagogical sovereignty as a construct tying infrastructure autonomy to educational autonomy. The components exist, can be integrated, and were deployed in four teaching cohorts. This does not yet prove improved learning outcomes; it demonstrates that an institution can replace a commercial API dependency with a locally governed stack for the relevant BPM teaching workflow.

We operationalise pedagogical sovereignty along four dimensions, each with an observable indicator. (1) **Cost autonomy**: the absence of per-use fees that cap or meter student access; indicator: marginal cost per student-prompt (sovereign: €0; commercial: variable, metered). (2) **Availability control**: freedom from externally imposed rate limits, quota exhaustion, and unilateral API deprecation; indicator: uptime under concurrent classroom load and the locus of capacity decisions (sovereign: institutional; commercial: vendor). (3) **Policy control**: the institution, not the vendor, determines content-filtering boundaries; indicator: who can change the refusal policy and whether changes are transparent. (4) **Data jurisdiction**: student and institutional data never leave national or institutional infrastructure; indicator: physical and legal location of inference and logs. The proposed PCSS stack is designed to occupy the autonomous pole on all four under the conditions tested. Commercial APIs typically leave these terms outside the institution, although enterprise contracts may mitigate some risks. The point is not that every commercial deployment is equally dependent, but that pedagogical sovereignty requires these dependencies to be measurable rather than assumed away. This extends Zembylas's [52] argument against digital neocolonialism from the ethical domain into a measurable operational one.

Comparison with Related Work. The question of whether LLMs can produce usable process models has been measured independently. Kourani et al. [24] found wide variation across sixteen models; Busch et al. [5] reported that the gap between open-source and commercial depends on task rather than licence; the InstruBPM study [6] showed a compact instruction-tuned model outperforming proprietary baselines on structural fidelity from a single GPU. Lauer et al. [42] evaluated a modelling copilot with five experts: acceptable usability (67/100) but trust at only 49% and reliability rated the most critical concern (1.8/5). Outside BPM, Nowak et al. [33] found open-weight models competitive with GPT-4o in radiology report extraction (Mistral-Large F1 92.6% vs. GPT-4o 92.4%), and Sandmann et al. [41] corroborated this in *Nature Medicine*: DeepSeek-V3 equalled or exceeded GPT-4o across 125 clinical cases, motivated explicitly by the impossibility of deploying proprietary models on-premise under healthcare privacy regulations. Wolfe et al. [50] demonstrated parity on a single consumer GPU in risk-intolerant government and research settings. Perhaps the closest architectural parallel is Nowak et al. [32], who deployed DeepSeek-R1 via vLLM behind strict network isolation in a German university hospital and found text-anchored tasks (report corrections, guideline recommendations) rated highest while open-ended conclusion generation produced the most critical errors—a pattern consistent with our own observations. But Ye et al. [51] document that open-weight models exhibit increased vulnerability to bias and safety failures in multilingual contexts. We do not claim to have eliminated these risks. What deployment sovereignty fixes is narrower: institutional control over cost, availability, content policy, and data jurisdiction.

Practical Implications. The token economics make the case concrete. Uber exhausted its entire annual AI budget in four months after 5,000 engineers adopted Claude Code; individual monthly bills reached $2,000 [3]. Microsoft subsequently canceled most direct Claude Code licenses [2]. Goldman Sachs projects a 24-fold increase in global token consumption by 2030 [2], and Gartner projects that 40% of agentic AI projects will stall by 2027 due to costs [46]. We note the scale difference honestly: a thirty-student BPM course will never approach Uber's engineering-scale consumption. The argument is a fortiori. If organisations spending billions cannot sustain uncapped access, a department on a five-figure budget has even less margin, and therefore even more to gain from removing the meter entirely. What crashed a $3.4 billion R&D budget in industry proceeded without interruption on national HPC. Beyond cost, the sovereign stack enables pedagogical choices that commercial APIs foreclose. An instructor can select which model students use, switch models mid-semester without renegotiating a contract, retain all prompt and completion logs for assessment research, and define a content policy tailored to the discipline rather than to the provider's global safety defaults.

Two Deployment Tiers. We frame the cost claim as *zero marginal licensing cost*, not zero total cost: HPC is publicly funded, and integration requires faculty labour. The architecture scales down to commodity hardware: a pair of RTX 3090 nodes (€4,000) provides full local inference. Knoop et al. [21] confirm that consumer GPUs break even in under four months at moderate volume. The two tiers satisfy the four dimensions differently. The HPC tier provides stronger capability and lower marginal cost but depends on public allocation rules. The local tier provides stronger physical custody but requires smaller quantised models with a capability trade-off. Pedagogical sovereignty is therefore not a binary status but a design space.

6 Ethics and Governance

Data and Consent. Student data never left national infrastructure. Students worked with public regulatory texts under the standard course data agreement, consistent with EU AI Act data-minimisation expectations [15]. Logs are kept on institutional servers and deleted at term end.

Dual-Use. The high-sensitivity corpus was built for measurement, not the classroom. We do not argue for stripping safety behaviour. Our claim is narrower: the institution should decide and answer for the content policy, whatever that policy is.

Local Safety Policy. A sovereign deployment should not be interpreted as an unfiltered deployment. For classroom use, the institution should define a written policy specifying which prompt classes are allowed, which require instructor supervision, and which are excluded. In the pilots, high-sensitivity prompts were not assigned to students; they were used only to test provider behaviour. Future deployments should implement this policy through prompt-level warnings, logging, and instructor approval for sensitive domains.

Output Safety. Generated BPMN can be wrong. Models were drafting aids whose output a human checked before trusting it, and no student-generated process was executed against live systems.

7 Limitations and Future Research

The four cohorts (~110 participants) lacked formal control groups or pre/post learning-outcome measurement; a quasi-experimental design along the lines of Nobre et al. [30] is the obvious next step. Prompts were English-language only. The doctoral deployment is a single existence proof, and CIB Seven was a proof of concept. Open-weight models are not risk-free: recent surveys document increased susceptibility to adversarial manipulation and safety failures, particularly in multilingual settings [51].

The most consequential limit is the **sovereignty–capability gap**. On a separate Polish-language extraction benchmark (30 documents, 243 entities, 42 models), Bielik 11B ranked 33rd of 42 (F1 0.519), confirming that provenance sovereignty does not yet come with competitive capability. But deployment-sovereign models told the opposite story: DeepSeek V3.1 (F1 0.843) and Llama 3.3 70B (0.830) landed within one to two points of the best commercial models at zero cost. *Deployment sovereignty at competitive quality is available today; provenance sovereignty at competitive quality is not.*

The most concrete future direction is training a domain-specific sovereign model on national HPC. Recent work shows this is no longer capital-intensive: the Sky-T1 reasoning model was trained for approximately US$450 [31], and Latvia's TildeOpen (30B parameters, 2 million GPU-hours on LUMI) provides a national-language precedent [12]. The InstruBPM project [6] suggests such fine-tuning can work for BPM specifically: a compact instruction-tuned model already outperforms proprietary baselines on structural BPMN metrics. The arrival of PIAST-AI at PCSS (over 1,000 GPUs from late 2026 [37]) makes the necessary scale locally available. EuroHPC provides cost-free Open Science access to any organisation in a member or associated state [13,14]; the Netherlands offers Snellius via SURF and EuroCC, and Denmark participates in the LUMI AI Factory consortium alongside Czechia, Estonia, Norway, and Poland [12,14]. The binding constraint is not availability but allocation model: research quotas are competitive and were strained by a single course consuming 150% of an annual token budget. We propose that national centres establish a dedicated educational allocation track so that sovereign AI in the classroom can scale beyond early adopters.

8 Conclusion

This paper shows that deployment-sovereign AI infrastructure can support an end-to-end BPM education workflow without reliance on commercial LLM APIs. The PCSS-based stack operated without API errors or content-policy refusals across 2,248 model runs in the conditions tested, while commercial endpoints exhibited provider-specific refusal behaviour and availability failures that would be difficult for an educator to predict or control during a live course. The classroom pilots demonstrate feasibility across four cohorts, but not yet learning-outcome superiority.

The strongest conclusion is therefore not that sovereign AI is inherently better than commercial AI, but that *infrastructure governance is itself a pedagogical variable*. BPM educators who depend on commercial APIs outsource not only computation, but also cost, availability, content policy, and data jurisdiction. Deployment-sovereign infrastructure makes these dependencies visible and, in principle, governable. The construct of pedagogical sovereignty, operationalised here along four measurable dimensions, may generalise to any discipline embedding AI in its curriculum.

One question remains open and, we think, worth putting bluntly: if the BPM community has modelled twenty-seven subprocesses at one Polish university, why has no BPM research group done the same at any other university, anywhere? The tools are free. The models are free. The infrastructure exists across Europe. What is missing is not technology but the decision to use it.

References

1. Altman, S.: Intelligence too cheap to meter. In: Federal Reserve Conference (2025). https://www.federalreserve.gov/mediacenter/files/capital-framework-conference-fireside-chat-transcript.pdf. Accessed 9 Sep 2026
2. Angelo, J.: Microsoft reports are exposing AI's real cost problem: using the tech is more expensive than paying human employees. Fortune (2026). https://fortune.com/2026/05/22/microsoft-ai-cost-problem-tokens-agents. Accessed 9 Sep 2026
3. NAngelo, J.: Uber burned through its entire 2026 AI budget in four months. Now its COO is questioning whether it's worth it. Fortune (2026). https://fortune.com/2026/05/26/uber-coo-ai-spending-tokens-claude-code. Accessed 9 Sep 2026
4. Bettencourt, J., Guerreiro, S.: Large language models to enhance business process modeling: past, present, and future trends (2026). https://doi.org/10.48550/arXiv.2604.14034
5. Busch, K., Leopold, H.: Towards a benchmark for large language models for business process management tasks (2024). https://doi.org/10.48550/arXiv.2410.03255
6. Çelikmasat G, Özgövde A, Aydemir FB.: Instruction-Tuning Open-Weight Language Models for BPMN Model Generation (2025). https://doi.org/10.48550/arXiv.2512.12063
7. Ciachorowski, D., Lis, A.M., Miranda, E.: Sustainability in digital transformation: towards an integrated framework. Eng. Manage. Prod. Serv. **17**(4), 86–103 (2025). https://doi.org/10.2478/emj-2025-0028
8. Ciachorowski, D., Miranda, E.: Conflicting Demands of Sustainable Digital Transformation. In: Management, Organizations, and Paradoxes, 1st edn. Routledge (2025). https://www.doi.org/10.4324/9781003541394-14
9. Ciachorowski, D.: Organising sustainability-related aspects of digital transformation: The People–Technology–Management– Environment (PTME) Framework. Problemy Jakości 19–30 (2026). https://doi.org/10.15199/46.2026.4.3
10. CIB seven: Open-Source BPM for the Next Generation of Process Automation (2025). https://cibseven.org/

11. Dumas, M., La Rosa, M., Mendling, J., Reijers, H.A.: Fundamentals of business process management, Springer Berlin Heidelberg (2018).https://doi.org/10.1007/978-3-662-56509-4 (2025)
12. European Commission, Directorate-General of Communications Networks, Content & Technology: European Open-Source AI (2025). https://digital-strategy.ec.europa.eu/en/library/europes-open-source-ai-landscape-lever-innovation-and-sovereignty. Accessed 9 Sep 2026
13. European Commission: The European High Performance Computing Joint Undertaking (2026). https://www.eurohpc-ju.europa.eu/index_en. Accessed 9 Sep 2026
14. European High Performance Computing Joint Undertaking: supercomputers access policy and FAQ (2026). https://www.eurohpc-ju.europa.eu/supercomputers/supercomputers-access-policy-and-faq_en. Accessed 9 Sep 2026
15. European Parliament, Council of the European Union.: Regulation (EU) 2024/1689, Official Journal of the European Union (2024). http://data.europa.eu/eli/reg/2024/1689/oj. Accessed 9 Sep 2026
16. Gerosa, M., Hermansen, A., Lai, A., Lawson, A.: The State of Sovereign AI: Exploring the Role of Open Source Projects and Global Collaboration in Global AI Strategy, foreword by Mark Collier, The Linux Foundation (2025). https://www.linuxfoundation.org/research/state-of-sovereign-ai. Accessed 9 Sep 2026
17. Grzesiak, M., Kluza, K., Potoczek, N.R., Szała, L.: Beyond Traditional Learning: The LLM Revolution in BPM Education at University. In: Di Ciccio, C., et al. Business Process Management: Blockchain, Robotic Process Automation, Central and Eastern European, Educators and Industry Forum. BPM 2024. Lecture Notes in Business Information Processing, vol. 527. Springer, Cham (2024). https://doi.org/10.1007/978-3-031-70445-1_29
18. Grzesiak, M.: Implementation of active learning in BPM education. Decis. Making Manufact. Serv. 117-125 (2024). https://doi.org/10.7494/dmms.2024.si.6696
19. Hamadeh, S., Amin, H.: AI, education and digital sovereignty. Front. Educ. **10**, 1677727 (2025). https://doi.org/10.3389/feduc.2025.1677727
20. Kluza, K., et al.: Revisiting BPMN Assignments with AI in Mind: Insights from Experiments with Large Language Models in Process Modeling Education. In: Er, M., et al. (eds.), Business Process Management: Responsible BPM Forum, Process Technology Forum, Educators Forum, vol. 565, pp. 358–372. Springer Nature Switzerland (2026). https://doi.org/10.1007/978-3-032-02936-2_25
21. Knoop, J., Holtmann, H.: Private LLM inference on consumer blackwell GPUs: A practical guide for cost-eective local deployment in SMEs. arXiv preprint (2026). https://doi.org/10.48550/arXiv.2601.09527
22. König, M., Völker, M., Weske, M.: From Design to Evaluation: Students Experiencing the BPM Lifecycle Hands-on. In: C. Di Ciccio, et al. (eds.), Business Process Management: Blockchain, Robotic Process Automation, Central and Eastern European, Educators and Industry Forum, vol. 527, pp. 333–343. Springer Nature Switzerland (2024). https://doi.org/10.1007/978-3-031-70445-1_22
23. Kopke, J., Safan, A.: Introducing the BPMN-chatbot for efficient LLM-based process modeling. In: Proceedings of the International Conference on Business Process Management, pp. 259–270 (2025). https://ceur-ws.org/Vol-3758/paper-15.pdf
24. Kourani, H., Berti, A., Schuster, D., Van Der Aalst, W.M.P.: Evaluating large language models on business process modeling: framework, benchmark, and self-improvement analysis. Softw. Syst. Model. **25**(4), 1151–1186 (2026). https://doi.org/10.1007/s10270-025-01318-w

25. Lauer, C., Pfeiffer, P., Mehdiyev, N.: Human-Centered Evaluation of an LLM-Based Process Modeling Copilot: A Mixed-Methods Study with Domain Experts (Version 1). arXiv (2026). https://doi.org/10.48550/ARXIV.2603.12895
26. LightOn: LightOn Launches Its Sovereign and Secure Generative AI Offering for Higher Education in Partnership with ENS Paris-Saclay (2025). https://lighton.ai/lighton-blogs/lighton-launches-its-generative-ai-offering-for-higher-education-in-partnership-with-ens-paris-saclay. Accessed 9 Sep 2026
27. Mahendrawathi, E.R., Deviana, A., Nurkasanah, I.: Can business process design space help students generate innovative process redesign ideas?. In: Di Ciccio, C., et al. (eds.) Business Process Management: Blockchain, Robotic Process Automation, Central and Eastern European, Educators and Industry Forum. BPM 2024. Lecture Notes in Business Information Processing, vol. 527. Springer, Cham (2024). https://doi.org/10.1007/978-3-031-70445-1_28
28. Moorhouse, B.L., Yeo, W.G., Wan, Y.: Generative AI tools and assessment: guidelines of the world's top-ranking universities. Comput. Educ. Open **5**, 100151 (2023). https://doi.org/10.1016/j.caeo.2023.100151
29. Nagel, S., Wolters, A., Riehle, D.M., Delfmann, P.: Designing an intelligent tutoring chatbot for declarative process modeling education. In: ACIS 2024 Proceedings. No. 162, AIS eLibrary (2024)
30. Nobre, M.R.B., Vilela, J., Migge de Barros, L.: Model Quality and Student Satisfaction in BPMN Education: A Quasi-Experiment. In: Proceedings of the 17th International Conference on Computer Supported Education (CSEDU 2025), vol. 1, pp. 620-627 (2025). https://www.doi.org/10.5220/0013483100003932
31. NovaSky Team: Sky-T1: Train your own O1 preview model within $450. (2025). https://novasky-ai.github.io/posts/sky-t1. Accessed 9 Sep 2026
32. Nowak, S., et al.: Secure On-Premise Deployment of Open-Weights Large Language Models in Radiology: An Isolation-First Architecture with Prospective Pilot Evaluation (Version 1). arXiv (2026). https://doi.org/10.48550/ARXIV.2604.22768
33. Nowak, S., et al.: Privacy-ensuring Open-weights Large Language Models Are Competitive with Closed-weights GPT-4o in Extracting Chest Radiography Findings from Free-Text Reports. Radiology **314**(1), e240895 (2025). https://doi.org/10.1148/radiol.240895
34. NVIDIA: Nemotron (2026). https://developer.nvidia.com/topics/ai/nemotron
35. Open-BPMN: The free modeling platform (2026). https://www.open-bpmn.org/
36. Pan, G., Chodnekar, V., Roy, A., Wang, H.: A cost-benefit analysis of on-premise large language model deployment: breaking even with commercial LLM services. In: 2025 IEEE International Conference on Big Data (BigData) (pp. 2234-2239). IEEE (2025). https://doi.org/10.48550/arXiv.2509.18101
37. PCSS: PIAST-AI. A new dynasty of artificial intelligence (2026). https://www.psnc.pl/piast-ai/
38. Poznan Supercomputing and Networking Center (PCSS) (2026). https://www.pcss.pl/
39. Revoredo, K., Bala, S.: Project-Based Learning in BPM: Lessons from Flipped Classroom vs. Lecture-Based Instruction. In: Di Ciccio, C. et al. (eds.) Business Process Management: Blockchain, Robotic Process Automation, Central and Eastern European, Educators and Industry Forum, vol. 527, pp. 313–323. Springer Nature Switzerland (2024). https://doi.org/10.1007/978-3-031-70445-1_20
40. Rockembach, G.R., Thom, L. H.: The Last Decade of Automatic Question Generation: A Review of Techniques, Limitations, and Applications in Business Process Management Education. Revista Brasileira de Informática Na Educação **34**, 219–251 (2026). https://doi.org/10.5753/rbie.2026.6093

41. Sandmann, S., et al.: Benchmark evaluation of DeepSeek large language models in clinical decision-making. Nature Med. **31**(8), 2546–2549 (2025). https://doi.org/10.1038/s41591-025-03727-2
42. Senkus, P., Kedziora, D., Wenus, M., Wikiel, K.: Barriers and critical success factors for GenAI in quality management systems: The HTO-S framework (2026)
43. Simon, F. M., Nielsen, R. K., Fletcher, R.: Generative AI and news report 2025: How people think about AI's role in journalism and society. Reuters Institute for the Study of Journalism (2025). https://doi.org/10.60625/RISJ-5BJV-YT69
44. Sliż, P., et al.: Business Process Management education in Poland: a manifesto for academic teaching. E-Mentor **104**(2), 91–98 (2024). https://doi.org/10.15219/em104.1661
45. Smith, S., Wanner, S., Butler, W., Diasio, D.: Unlocking agentic value: a new investment discipline for the agentic era, EY (2026). https://www.ey.com/en_us/insights/ai/agentic-ai-token-costs. Accessed 9 Sep 2026
46. Stamford, Conn.: Gartner Predicts That by 2030, Performing Inference on an LLM With 1 Trillion Parameters Will Cost GenAI Providers Over 90% Less Than in 2025, Gartner (2026). https://www.gartner.com/en/newsroom/press-releases/2026-03-25-gartner-predicts-that-by-2030-performing-inference-on-an-llm-with-1-trillion-parameters-will-cost-genai-providers-over-90-percent-less-than-in-2025. Accessed 9 Sep 2026
47. Szelkagowski, M.: Dynamic Business Process Management. In: Dynamic Business Process Management in the Knowledge Economy. LNNS, vol. 71, pp. 55–90. Springer, Cham (2019). https://doi.org/10.1007/978-3-030-17141-4_2
48. Szymanski, G.: BPM in higher education: a source-to-dispose reference model (2026). Ph.D. thesis. Uniwersytet Kaliski
49. Turobov, A., Carrapico, H., Farrand, B.: Opening the Black Box of EU Digital Sovereignty: A MacroLevel Analysis of the Concept's Development. Governance **39**(2), e70112 (2026). https://doi.org/10.1111/gove.70112
50. Wolfe, R., et al.: Laboratory-Scale AI: Open-Weight Models are Competitive with ChatGPT Even in Low-Resource Settings. In: The 2024 ACM Conference on Fairness, Accountability, and Transparency, pp. 1199–1210 (2024). https://doi.org/10.1145/3630106.3658966
51. Ye, J., Bronstein, S., Hai, J., Hashish, M.A.: DeepSeek in Healthcare: A Survey of Capabilities, Risks, and Clinical Applications of Open-Source Large Language Models (Version 1). arXiv (2025). https://doi.org/10.48550/ARXIV.2506.01257
52. Zembylas, M.: A decolonial approach to AI in higher education teaching and learning: Strategies for undoing the ethics of digital neocolonialism. Learn. Media Technol. **48**(1), 25–37 (2023). https://doi.org/10.1080/17439884.2021.2010094

AI and the Evolving BPM Practitioner Skill Set. Initial Implications for Curriculum Review

Artur Siurdyban(✉) and Renata Gabryelczyk

University of Warsaw, Warsaw, Poland
{a.siurdyban,r.gabryelczyk}@uw.edu.pl

Abstract. Artificial intelligence is increasingly reshaping the Business Process Management (BPM) practice, generating new requirements for BPM education and curriculum design. This study examines how AI-driven transformation affects BPM curricula by identifying which competencies remain adequate, which require reinterpretation in an AI-in-the-loop context, and which represent entirely new skill demands. The study applies a three-step structured mapping approach. First, the AI capabilities of 17 established BPM platforms are coded through structured feature analysis across 30 BPM capability areas. Second, these codings are aggregated to characterize how AI tools collectively support each capability area. Third, the analysis evaluates practitioner implications by distinguishing direct AI impacts, indirect effects emerging from related capability areas, and sub-tasks that may no longer require manual training. Results indicate that AI reshapes competency requirements across all BPM capability areas and introduces six new competency demands not captured in existing BPM frameworks. These findings reveal a module-level curriculum gap and provide a foundation for updating existing BPM curricula as well as informing the design of new programs.

Keywords: Business Process Management · BPM education · AI capabilities · BPM curriculum redesign

1 Introduction

Artificial Intelligence (AI) is transforming knowledge-intensive professions, thereby reshaping both professional practice and the educational requirements needed to prepare future practitioners. Business Process Management (BPM) represents a key area of both AI application and AI-driven transformation [25]. This paper asks what that transformation means for BPM education. It identifies which competencies in established frameworks remain sufficient, which require reteaching within an AI-in-the-loop framing, and which are entirely new.

The gap between AI-augmented BPM practice and current education is documented. BPM capabilities are positively linked to digitalization outcomes, with more dynamic capabilities yielding the strongest returns [1, 3]. However, AI-relevant topics such as process analysis and execution feature in only a minority of university programs [28]. Recent AI-related contributions to the Educators' Forum focus primarily on pedagogy. Examples

K. Kluza et al. (Eds.): BPM 2026, LNBIP 600, pp. 418–434, 2027.
https://doi.org/10.1007/978-3-032-36710-5_27

include examining LLMs as supplements to BPM educators [15], applying generative AI in process modeling instructional design [21], and comparing MBA and GenAI performance on explorative process design [26]. Where curriculum-practice alignment has been studied, the focus falls on the uptake of BPM techniques among graduates [10] or on program design informed by decades of delivery experience [29]. No study has systematically mapped what AI tool capabilities require of BPM curricula, therefore we formulate the following research questions: *RQ1*: Which competencies in BPM curricula are directly reshaped by the AI capabilities of BPM tools? *RQ2*: How do the indirect effects of AI adoption in organizations create new competency demands within BPM curricula? *RQ3*: What new competency requirements emerge from AI development and use that have no equivalent in existing BPM competency frameworks?

This paper addresses this gap through a three-step structured mapping study grounded in the Kerpedzhiev et al. framework [18]: (1) coding AI capabilities across 17 BPM platforms, (2) synthesizing codings by capability area, and (3) assessing practitioner impact across all 30 capability areas.

2 Background

2.1 BPM Core Elements Frameworks

This paper builds on the Six Core Elements of BPM framework developed by de Bruin and Rosemann (2007) [8] and further elaborated by Rosemann and vom Brocke (2015) [24]. The framework conceptualizes BPM as an organizational capability comprising six interconnected elements: Strategic Alignment, Governance, Methods, Information Technology, People, and Culture. These elements cover the technical, managerial, human, and cultural dimensions of BPM, providing a comprehensive basis for analyzing practitioner competencies. The framework serves as the reference for major BPM studies [1, 18, 24, 31] and established curriculum analyses [28, 29], providing a common language for the analysis in this paper.

The most significant update to the framework is provided by Kerpedzhiev et al. [18]. Their Delphi study with BPM experts produced a digitalization-updated version retaining the six-element structure while substantially revising the underlying capability areas. Of the resulting 30 capability areas, 13 are entirely new relative to the 2007 baseline, 14 are scope-enhanced, and only three are preserved unchanged. The framework thus expands under digitalization rather than replacing its foundations, constituting the competency baseline adopted in this paper. It reflects what BPM education should provide in response to digitalization, offering a baseline for assessing the impact of AI. In this updated framework, Methods and Information Technology are consolidated into a single element, referred to throughout this paper as Methods/IT, reflecting their integration in the structure developed by Kerpedzhiev et al. [18].

2.2 AI in BPM Tooling

AI has entered BPM and is reshaping the discipline as a whole [25]. In this paper, AI denotes computational techniques that enable systems to learn from process data, generate process artifacts, and execute process tasks with varying degrees of autonomy. These capabilities have entered BPM practice through three successive and currently coexisting streams, each with a distinct technical basis and distinct implications for how BPM work is performed.

The first stream, machine-learning-based process analytics, has developed over more than a decade and is now deeply embedded in BPM platforms. It encompasses automated process discovery, ML-based conformance checking, predictive monitoring, anomaly detection, and decision support. This stream gave rise to the capability areas [18] identified as new under digitalization: Process Data Analytics, Advanced Process Automation, and Adaptive Process Execution. Its competency implications (data literacy, analytical reasoning) are partly addressed in current BPM capability frameworks.

The second stream, generative AI, emerged with the mainstreaming of large language models from 2022 onward. Vidgof et al. [30] provide the first systematic mapping of LLM capabilities across the BPM lifecycle, identifying opportunities in process model generation from natural language, process documentation, question answering over process knowledge bases, and process improvement suggestions. Kampik et al. [17] extend this trajectory toward Large Process Models: purpose-built foundation models combining LLM expressiveness with process-specific analytical capabilities to directly encode and reason over organizational process experience. GenAI functionalities are already present in mainstream platforms.

The third and most recent stream is agentic AI for BPM. Agentic BPM systems integrate reasoning, planning, and self-directed action into process management. Rather than following predefined control flows, agents sense process context, plan next actions, and adapt to unanticipated situations [6, 9]. This stream represents a shift in the human-process-system relationship. The boundary between what a human does and what the system does becomes dynamic, governed by oversight protocols and governance rules rather than fixed handoffs. Rosemann et al. [25] formalize this as a shift from conversational BPM in which AI supports human decisions, to autonomization, where processes make decisions in pursuit of defined goals.

2.3 Related Work on BPM Competencies in the AI Era

BPM education research has documented a consistent gap between what academic programs deliver and what professional practice requires. Sliż et al. [28] find that process mining, AI, and robotic process automation feature in only a minority of BPM courses. Only 12% of surveyed courses cover all Six Core Elements, with People and Culture elements particularly underrepresented. At the organizational level, Antonucci et al. [1] demonstrate empirically that BPM capabilities are positively associated with digitalization benefits. The strength of this association varies across capability types, with more dynamic capabilities showing the strongest links. Together, these findings establish both the existence of a curriculum gap and the organizational impact they bear. More recent work has begun to address specific AI-related competency areas within BPM. Shamasbi et al. [27] derive a role typology and a competency framework for process mining through text mining of job vacancy data. Graessler and Oezcan [14] develop a competence model for AI workflow modeling on low-code platforms, validated through expert interviews in crisis management. These contributions, while valuable, are technology-specific and partial in their BPM scope. Neither study maps AI tool capabilities to practitioner demands across all elements, a gap this paper addresses.

3 Methodology

The paper follows a 3-step methodology. Step 1, *AI Capability Coding*, codes AI capabilities across 17 BPM platforms through structured feature analysis. The method follows Basili and Weiss's [2] framework for systematic technology artifact comparison, applied through the document analysis approach of Bowen [4]. The study uses two types of sources: primary (vendor documentation, feature release notes, white papers) and secondary (analyst reports, technology press, and investment disclosures). Vendor claims are cross-checked with secondary sources where possible. Coding against predefined dimensions yields a capability matrix.

Candidate platforms are identified from three analyst reports spanning two dimensions of BPM tooling. For process intelligence and analysis tools directly supporting BPM practitioners in discovery, monitoring, and process analysis work, the sources are the Gartner Magic Quadrant for Process Mining Platforms [12] and the Forrester Wave for Process Intelligence Software [11]. For process orchestration and automation platforms relevant where AI-driven automation capabilities correspond directly to Advanced Process Automation, the source is the Gartner Magic Quadrant for Business Orchestration and Automation Technologies [13]. These three reports provide independent, multi-criterion comparative assessments with confirmed enterprise deployment at scale. They cover the breadth of platforms BPM professionals currently use. Platforms are qualified for the candidate pool if they were classified as Leader, Challenger, or Visionary in at least one of two Gartner Magic Quadrants, or as Leader or Strong Performer in the Forrester Wave. After deduplication of platforms appearing in multiple reports, the resulting set is reported in Sect. 4.

All 30 capability areas of the Kerpedzhiev et al. framework [18] serve as coding dimensions. Each platform is coded against every capability area as *Present*, *Partial*, or *Absent*. Present indicates a documented, deployed AI feature whose function directly corresponds to the capability area description. Partial indicates a feature that addresses the capability area only in limited scope, targets a subset of its activities, or is documented as preview or beta. Absent indicates no documented AI feature corresponding to the capability area. For example, a platform with full automated process discovery from event logs codes as Present for M/IT4; one offering only performance dashboards without discovery codes as *Partial*. Dimensions are fixed in advance to capture both absences and expected patterns, preventing post hoc shaping of the coding scheme [2].

Step 2, *Cross-Platform Synthesis*, synthesizes the per-platform codings into aggregate characterizations of what the AI tool landscape collectively offers for each capability area, producing cross-platform patterns that characterize the current AI tool market with respect to each competency dimension. The aggregation follows cross-case synthesis principles in qualitative document analysis, treating each platform as an individual case and deriving patterns across cases [4, 19]. For each capability area, the synthesis characterizes which platform types carry the capability, what AI performs for practitioners, and whether coverage is concentrated in specific clusters or cross-cutting.

Step 3, *Competency Impact Assessment*, assesses, for each capability area of the Kerpedzhiev et al. framework [18], how AI tooling has materially changed what practitioners must do. The assessment addresses the study's three RQs: how AI in BPM tools reshapes existing BPM competencies (*RQ1*), creates new competency demands through organizational AI adoption (*RQ2*), and introduces entirely new competencies beyond current BPM frameworks (*RQ3*).

Two types of evidence inform the assessment. Direct impact derives from AI features documented within each specific capability area: where AI tools perform tasks in this space, the practitioner's role in those tasks changes accordingly. Indirect impact derives from AI adoption in adjacent areas. When AI capabilities are broadly deployed across specific areas, the resulting changes in working practices propagate into adjacent capability areas, including areas without direct AI features, through new modes of working, collaboration, and oversight. This distinction is introduced because AI adoption reshapes BPM competency profiles not only through features directly embedded in each area but also through the systemic changes in practice that AI deployment introduces. A forward-looking principle guides the assessment of changes in practitioner activity. If a sub-task is substantially automated by at least one established platform in the study pool, this signals a likely path from current capability to wider market adoption and is therefore relevant for future curriculum development. Step 3 also identifies *Newly Required* competency demands that AI adoption creates which do not correspond to any existing capability area in the used framework. Throughout the coding steps, both authors perform the assessment independently and reconcile differences, following standard practice for qualitative content analysis [19].

4 Results

4.1 AI Capabilities Across the Platform Landscape

The position-based filter applied to the three analyst reports yields 17 platforms after deduplication: ABBYY (ABB), Appian (APP), Apromore (APR), ARIS (ARI), Camunda (CAM), Celonis (CEL), IBM (IBM), iGrafx (IGF), KYP.ai (KYP), MEHRWERK (mpmX) (MPX), Microsoft (MSF), Pega (PGA), QPR Software (QPR), SAP Signavio (SAP), ServiceNow (SNO), UiPath (UiP), and Workato (WRK). Three platform clusters are identifiable from the capability codings. Process intelligence specialists (Apromore, MEHRWERK, iGrafx, ABBYY, KYP.ai, ARIS) concentrate their AI capabilities in process analytics and conformance monitoring. Automation-first platforms (Camunda, Workato, Appian) concentrate AI in automation identification and agentic execution. Convergent platforms (Celonis, IBM, Pega, UiPath, ServiceNow, QPR, SAP Signavio, Microsoft) deploy AI across multiple dimensions; QPR and ServiceNow lead in capability scope, while IBM and Celonis achieve the greatest overall coverage. Table 1 presents the step 1 resulting capability matrix.

Our analysis shows that GenAI-driven process modeling, generating BPMN diagrams from natural language, proposing documentation, and suggesting design alternatives, has moved to mainstream, deployed at full capability on eight platforms spanning all three cluster types. Advanced automation identification and agentic execution appear together on the same set of convergent platforms: when a platform deploys AI-driven

Table 1. Step 1 Capability Matrix. Coding key: ●=Present (documented); ◐ =Partial or emerging; ○ = Absent

BPM platforms → BPM core element / Capability area ↓	CEL	ARI	IGF	SAP	IBM	APP	APR	MPX	PGA	UiP	CAM	MSF	SNO	WRK	ABB	QPR	KYP
SA1 Strategic BPM Alignment	●	◐	○	◐	◐	○	○	◐	○	◐	○	◐	◐	◐	○	●	◐
SA2 Strategic Process Alignment	◐	○	○	◐	◐	○	○	◐	○	◐	○	○	◐	○	○	●	○
SA3 Process Positioning	◐	◐	◐	◐	◐	○	○	○	○	○	○	○	◐	◐	○	◐	○
SA4 Customer & Stakeholder Alignment	◐	○	○	◐	◐	○	○	○	◐	◐	○	○	◐	○	○	◐	◐
SA5 Process Portfolio Management	◐	◐	○	○	◐	○	◐	◐	◐	◐	○	○	○	◐	○	◐	◐
G1 Contextual BPM Governance	○	◐	○	○	○	○	○	○	○	◐	○	○	○	○	○	○	○
G2 Contextual Process Governance	◐	◐	◐	◐	◐	◐	○	◐	○	◐	◐	◐	●	◐	○	◐	◐
G3 Process Architecture Governance	◐	◐	◐	◐	◐	○	○	○	◐	◐	◐	○	◐	○	○	◐	○
G4 Process Data Governance	●	◐	○	○	◐	○	◐	○	○	○	○	○	○	○	◐	◐	◐
G5 Roles & Responsibilities	○	○	○	○	◐	◐	○	○	◐	◐	◐	◐	◐	◐	○	○	◐
M/IT1 Process Context Management	◐	○	○	◐	◐	◐	○	◐	●	◐	◐	○	●	◐	○	○	◐
M/IT2 Process Compliance Management	●	●	●	●	●	●	●	●	●	●	◐	◐	●	◐	●	●	●
M/IT3 Process Architecture Management	◐	◐	◐	◐	◐	○	○	○	○	○	○	○	◐	○	○	◐	○
M/IT4 Process Data Analytics	●	●	●	●	●	◐	●	●	●	●	○	●	●	◐	●	●	●
M/IT5 BPM Platform Integration	◐	○	○	◐	○	○	○	○	○	◐	○	○	●	●	○	●	◐
M/IT6 Multi-purpose Process Design	◐	●	●	●	◐	◐	◐	○	●	●	◐	●	●	◐	○	●	○
M/IT7 Advanced Process Automation	◐	◐	○	○	●	●	◐	◐	●	●	●	●	●	●	◐	●	○
M/IT8 Adaptive Process Execution	◐	○	○	○	●	◐	○	◐	●	●	●	●	●	●	◐	●	◐
M/IT9 Agile Process Improvement	●	●	◐	◐	●	◐	●	●	●	●	◐	◐	◐	◐	◐	●	◐
M/IT10 Transformational Process Improvement	◐	◐	○	◐	◐	○	○	◐	◐	◐	○	○	◐	◐	○	◐	◐
P1 BPM and Process Literacy	◐	◐	◐	◐	◐	◐	◐	◐	◐	◐	○	◐	◐	◐	○	◐	◐
P2 Data Literacy	◐	○	○	◐	●	○	○	◐	◐	◐	○	○	◐	○	○	◐	○
P3 Innovation Literacy	◐	○	○	◐	◐	○	◐	○	◐	◐	○	○	◐	◐	○	◐	◐
P4 Customer Literacy	◐	○	○	◐	◐	○	○	○	◐	◐	○	○	◐	○	○	◐	◐
P5 Digital Literacy	◐	◐	◐	◐	◐	◐	◐	◐	◐	◐	○	◐	◐	◐	○	◐	◐
Cu1 Process Centricity	○	○	○	○	○	○	○	○	○	○	○	○	○	○	○	○	○
Cu2 Evidence Centricity	○	○	○	○	○	○	○	○	○	○	○	○	○	○	○	○	○
Cu3 Change Centricity	○	○	○	○	○	○	○	○	○	○	○	○	○	○	○	○	○
Cu4 Customer Centricity	○	○	○	○	○	○	○	○	○	○	○	○	○	○	○	○	○
Cu5 Employee Centricity	○	○	○	○	○	○	○	○	○	○	○	○	○	○	○	○	○

automation recommendation, it almost invariably also deploys agentic execution capable of acting on those recommendations. These are commercially bundled capabilities in the current market.

Outside Methods/IT, AI plays a signaling and supporting role. In Strategic Alignment, AI generates performance dashboards, ROI rankings, and portfolio prioritization signals; it creates inputs to strategic decisions but cannot make them. In the Governance element, conformance monitoring indirectly supports process governance broadly, but AI assistance for selecting and configuring BPM tools and methods is absent from the market. In the People element, natural language interfaces and AI copilots lower the practical entry threshold across most platforms, but only IBM explicitly develops data literacy by surfacing how its AI models reach their conclusions through named explainability features. Agentic execution is the dominant development in execution. The same eight convergent platforms deploy agentic AI with near-identical footprints. These platforms position AI agents as process actors with execution authority, sensing context, planning actions, and adapting to unanticipated situations. This fundamentally changes governance demands across all six core elements, not only in Methods/IT, propagating new requirements into governance, strategic alignment, people development, and cultural formation simultaneously.

4.2 Competency Impact Analysis

Tables 2–6 present the competency impact analysis. For each capability area, they document the direct practitioner impact of AI features on that area (step 2) and the indirect impact arising from AI adoption in adjacent areas. Six *Newly Required* competencies that AI creates with no equivalent in any existing BPM framework, follow Table 6.

Table 2. Competency Impact Analysis for the Strategic Alignment Element

Capability area	Direct AI Impact	Indirect AI Impact
SA1 Strategic BPM Alignment	Celonis and QPR automatically generate ROI linkages and value-measurement reports, surfacing which processes contribute to or undermine strategic objectives; nine further platforms surface similar signals through performance dashboards	AI-generated signals proliferate without strategic reasoning capacity; practitioners must assess whether AI-reported misalignments reflect genuine strategic problems or modeling artifacts
SA2 Strategic Process Alignment	QPR links process portfolio goals to organizational strategy through AI-supported reporting; most platforms do not address this directly	The diversity of AI feature profiles across platforms introduces a new strategic dimension: deciding which processes benefit from which AI capabilities requires matching portfolio decisions to platform strengths
SA3 Process Positioning	Several platforms surface EA-linked process architecture views; none provides AI-assisted decisions about where a process should sit in the enterprise architecture	GenAI-generated processes (M/IT6) and agentic workflows (M/IT7; M/IT8) introduce new process types with distinct integration dependencies and governance footprints requiring explicit positioning
SA4 Customer & Stakeholder Alignment	Several platforms surface customer journey analytics and stakeholder process views; AI does not assist the alignment decisions themselves	Agentic execution introduces AI agents as process participants with delegated authority, changing who practitioners must align stakeholders with and what accountability structures must be communicated
SA5 Process Portfolio Management	AI-driven prioritization and ROI ranking features continuously generate recommendations about which processes to invest in or automate, across analytics and convergent platforms alike	The emergence of AI-enhanced versus non-AI-enhanced processes creates a new portfolio dimension that AI tools generate data about but cannot decide

Across Strategic Alignment, practice shifts from designing and reporting toward governing AI-generated processes and signals. The shared demand across SA1, SA2,

Table 3. Competency Impact Analysis for the Governance Element

Capability area	Direct AI Impact	Indirect AI Impact
G1 Contextual BPM Governance	No platform provides AI assistance for BPM tool selection or methodology configuration	The AI tool landscape has expanded to 17 platforms with varying AI feature depth, making tool selection more complex when no AI supports it
G2 Contextual Process Governance	AI continuously monitors conformance, detects and filters anomalies in real time, automates compliance reporting, and flags regulatory deviations	Continuous automated monitoring shifts governance from episodic reviews to ongoing AI oversight, reshaping Evidence Centricity (Cu2) and strategic alignment (SA1)
G3 Process Architecture Governance	EA-linked governance analysis appears partially across several platforms; no platform provides AI for governing process architecture standards	GenAI model generation (M/IT6) and agentic execution (M/IT7; M/IT8) produce AI-designed and AI-executed processes that must be governed within architectural frameworks not designed for them
G4 Process Data Governance	Celonis provides explicit AI governance and audit transparency features, while some platforms only partially address process data governance	Near-universal ML analytics (M/IT2; M/IT4) embeds AI models in process data pipelines, creating model lifecycle governance demands well beyond traditional data governance scope
G5 Roles & Responsibilities	Several platforms surface AI-specific role definition features, particularly around defining agent authorities and accountability in agentic deployments	Widespread agentic execution (M/IT7; M/IT8) makes AI agents active process participants, requiring governance of decision authority, human approval, and accountability

SA3, and SA5 is evaluating AI-generated outputs, such as strategic signals, process designs, portfolio recommendations, against factors AI cannot model: organizational readiness, political feasibility, and architectural fit. SA4 introduces a distinct requirement: AI agents as process actors with delegated execution authority, requiring explicit definition of agent boundaries and accountability in stakeholder-facing contexts.

In Governance, the dominant shift is from episodic manual review toward continuous AI-mediated oversight, with conformance audits and architecture reviews (G2, G3) giving way to configuring AI monitoring thresholds and governing what AI systems produce.

Table 4. Competency Impact Analysis for the Methods / IT Element

Capability area	Direct AI Impact	Indirect AI Impact
M/IT1 Process Context Management	Pega and ServiceNow deploy AI-driven context sensing and pattern-based execution recommendations; most other platforms partially surface context-aware features	Agentic execution deploys AI agents that manage process context autonomously, making it a foundational AI layer rather than a practitioner-owned task
M/IT2 Process Compliance Management	AI continuously monitors process conformance, detects anomalies in real time, and generates compliance reports automatically across all studied platforms	The shift to AI monitoring changes governance (G2) and strategic alignment (SA1) requirements, and creates a dependence on AI-generated evidence that reshapes evidence-centric practice (Cu2)
M/IT3 Process Architecture Management	Several platforms provide EA-linked architecture views; no platform applies AI to manage architecture standards or assess how AI-mediated processes affect existing architecture	GenAI-generated models (M/IT6) and agentic execution (M/IT7; M/IT8) introduce new AI-generated architecture artifacts that must be managed within existing frameworks not designed for them
M/IT4 Process Data Analytics	AI automatically discovers process models from event logs, generates dashboards, runs predictive SLA monitoring, and produces root cause analysis across nearly all platforms	Near-universal analytics deployment reshapes what data literacy (P2), evidence-centric practice (Cu2), and BPM literacy (P1) require of practitioners
M/IT5 BPM Platform Integration	Several platforms deploy AI-assisted integration features including dynamic field mapping and intelligent routing recommendations; most platforms provide standard connector infrastructure without AI capability	The heterogeneity of AI feature profiles across platforms creates an AI integration governance challenge: understanding how AI capabilities in one platform interact with capabilities in another

(*continued*)

Table 4. (*continued*)

Capability area	Direct AI Impact	Indirect AI Impact
M/IT6 Multi-purpose Process Design	GenAI tools on eight platforms generate BPMN models from natural language, propose process documentation, and suggest design alternatives; six further platforms support partial GenAI design capabilities	Widespread AI-generated process models change what Process Literacy (P1), Architecture Management (M/IT3), and Governance (G3) require of practitioners
M/IT7 Advanced Process Automation	AI identifies, prioritizes, and increasingly initiates automation candidates based on process execution data, deployed at full capability across nine convergent and automation-first platforms	Widespread automation identification reshapes Agile Process Improvement (M/IT9) and propagates role governance demands (G5) as AI agents increasingly initiate and execute automation
M/IT8 Adaptive Process Execution	Agentic AI is deployed at full capability on eight platforms, matching the M/IT7 set, confirming that advanced automation and agentic execution are commercially bundled	Agentic deployment at this scale redefines the human-process-system relationship across the full practice landscape, propagating governance demands into G3, G5, SA4, and all Culture areas simultaneously
M/IT9 Agile Process Improvement	AI generates prescriptive improvement recommendations across all studied platforms; eight deploy at full capability	AI-generated improvement suggestions flow directly into strategic portfolio decisions (SA5) and reshape the organizational change management demands practitioners face
M/IT10 Transformational Process Improvement	Several platforms offer digital twin capabilities and simulation environments supporting enterprise-scale redesign scenarios in partial form; no platform deploys comprehensive AI-driven transformation support	The combination of ML analytics (M/IT4), GenAI modeling (M/IT6), and agentic execution (M/IT7; M/IT8) makes AI-supported transformation scenarios more feasible than the baseline [21], even without dedicated M/IT10 tools

The deepest expansions fall on G4 and G5, where ML model lifecycle governance and formal role governance for AI agents create entirely new oversight responsibilities.

Table 5. Competency Impact Analysis for the People Element

Capability area	Direct AI Impact	Indirect AI Impact
P1 BPM and Process Literacy	Natural language interfaces and guided analytics on most platforms lower the practical entry threshold for process work, but no platform develops BPM literacy as a competency	Near-universal AI-generated process models (M/IT6), compliance reports (M/IT2), and improvement recommendations (M/IT9) create a growing body of AI-produced artifacts that practitioners must critically assess, and act on
P2 Data Literacy	IBM is the only platform with named explainability and model transparency features; all other platforms embed AI-generated outputs without transparency	Near-universal ML analytics (M/IT2; M/IT4) makes AI-generated quantitative evidence the standard input to practitioner decisions, without corresponding transparency support across the market
P3 Innovation Literacy	Several platforms surface GenAI scenario generation and improvement recommendation features presenting AI-generated innovation options for practitioners to evaluate	GenAI design (M/IT6), automation identification (M/IT7), and agentic execution (M/IT8) collectively create an AI-driven process innovation frontier that practitioners must critically evaluate rather than simply adopt
P4 Customer Literacy	Several platforms surface customer journey analytics and comparative productivity views for customer-facing processes	Agentic execution of customer-facing processes (M/IT7; M/IT8) changes the context in which customer literacy must be exercised, AI agents may now hold primary execution responsibility in service workflows
P5 Digital Literacy	AI copilots, natural language search, and low-code interfaces lower the digital skill threshold across most platforms, but do not develop the underlying competency	Prompt engineering, AI output quality assessment, and AI tool governance are new digital competency dimensions demanded by the full platform landscape

G1 presents a different kind of challenge: no platform provides AI support for BPM tool selection precisely when the landscape demands careful evaluation.

Methods/IT sees the deepest AI transformation across all six elements, with a consistent pattern running through every capability area: AI has taken over the execution layer, shifting practitioners upstream. Conformance monitoring, process discovery, and dashboarding are now automated (M/IT2, M/IT4), and the same holds for model drafting (M/IT6) and improvement recommendation generation (M/IT9). Practitioners no longer build these outputs but scope, validate, and contextualize them. Automation identification and agentic execution (M/IT7, M/IT8) take this furthest: commercially bundled, they require practitioners to specify agent action boundaries and govern autonomous execution rather than discover automation candidates.

Across People, AI adoption extends all five literacy areas from traditional manual competencies toward critical engagement with AI-generated artifacts and decision systems. BPM literacy (P1) and data literacy (P2) face the most technical demands: practitioners must critically interpret AI-generated process artifacts for omissions and hallucinations, and extend their analytical repertoire to ML model training, drift detection, and explainability outputs. Innovation, customer, and digital literacy (P3, P4, P5) share a similar challenge: AI generates options and executes processes in each domain, requiring practitioners to evaluate rather than simply adopt AI proposals, design for human-centric outcomes when agents handle delivery, and govern AI tool use through effective prompting and output assessment.

In Culture, AI adoption intensifies all centricity orientations without providing any direct platform support for developing them. This reflects the nature of culture competencies which describe collective values and dispositions shaped through experience, leadership, and social interaction. These domains lie outside what AI tooling can address which by design operates on observable process data and documented artifacts. Process, evidence, change, customer, and employee centricity each acquire an AI-specific dimension: governing agents as process participants, maintaining epistemic standards for AI-generated evidence, adapting to AI-driven change, and protecting human experience in AI-mediated work contexts.

The analysis identifies six *Newly Required* competency demands with no equivalent in existing BPM frameworks. All six share a common origin: AI systems now perform process work that was previously entirely human, requiring practitioners to govern what AI produces, where it acts, and how its models behave. Governing AI outputs takes two forms: assessing the accuracy and organizational relevance of AI-generated artifacts, and establishing rules for their ownership, versioning, and legitimacy. Governing the human-AI execution boundary requires both designing handoff points between autonomous agent action and human decision authority, and developing the prompt engineering skills that directly determine output quality in process-specific contexts. Governing AI in process operations means managing AI model lifecycles embedded in compliance monitoring, and evaluating actual AI capability depth across a platform landscape where feature deployment varies substantially. Together, these demands constitute a new professional layer in BPM practice which can be summarized as governing AI systems performing process work, a role with no pre-AI equivalent.

Table 6. Competency Impact Analysis for the Culture Element

Capability area	Direct AI Impact	Indirect AI Impact
Cu1 Process Centricity	No AI feature addresses cultural formation; culture cannot be operationalized as software	Agentic execution (M/IT7; M/IT8) requires process-centric governance applied to AI agent boundaries; without it, agents execute processes in ways that bypass governance by default
Cu2 Evidence Centricity	No AI feature addresses cultural formation	ML analytics (M/IT2; M/IT4) and recommendations (M/IT9) flood process work with AI-generated evidence requiring stricter standards than human-produced data
Cu3 Change Centricity	No AI feature addresses cultural formation	GenAI redesign (M/IT6) and agents (M/IT7; M/IT8) enable process change at a pace and scale that existing change management frameworks were not designed for
Cu4 Customer Centricity	No AI feature addresses cultural formation	AI agents executing customer-facing processes (M/IT7; M/IT8) risk displacing the human-centered orientation that customer centricity requires, making it a more deliberate governance requirement
Cu5 Employee Centricity	No AI feature addresses cultural formation	Agentic automation (M/IT7; M/IT8) and GenAI design (M/IT6) displace routine tasks, requiring stronger employee-centric norms to manage human-AI role transitions

5 Curriculum Implications and Discussion

5.1 Curriculum Adjustment Priorities

The results point to curriculum adjustment at three levels: sub-task scope, capability-area framing, and new curriculum content. At sub-task level, no BPM capability is obsolete, but scope should be selectively adjusted. Within capability areas, several tasks are widely automated: event log processing and process discovery (M/IT4; 14 present platforms (pp)), routine conformance checking and anomaly detection (M/IT2; 14 pp), initial model drafting from natural language (M/IT6; 8 pp), and automated improvement recommendations (M/IT9; 8 pp). These should receive reduced emphasis as manual skills. Curriculum time can shift from notation, log preprocessing, and conformance execution toward interpretation, validation, and governance. This is consistent with Brynjolfsson and Mitchell's [5] finding that ML automates bundles of professional judgment tasks, shifting practitioner roles upstream rather than eliminating them.

At the capability-area level, all 30 require reframing within an AI-in-the-loop context, with four elements showing the strongest impact. In Methods/IT, near-universal ML analytics (M/IT2; M/IT4), widespread GenAI design (M/IT6), and agentic execution (M/IT7; M/IT8) shift the focus from execution to governance and interpretation. Governance is the second priority: G4 and G5 face major expansion in ML lifecycle and agent governance without corresponding tool support, increasing competency demands where systems are weakest [23]. Model literacy becomes a baseline skill, including interpreting explainability, detecting drift, and maintaining critical distance from AI-generated evidence. In Culture, AI adoption intensifies all five areas, requiring reflective practice, ethics, and case-based learning beyond platform capabilities.

At the level of new curriculum content, the *Newly Required* competencies are module-level additions with no equivalents in existing BPM frameworks and cluster into three units. AI output quality and governance combines validation of AI outputs and governance of AI-generated artifacts, developing the ability to detect errors, hallucinations, and governance gaps in AI-produced artifacts. Human-AI process design combines handoff design and prompt engineering, developing the ability to specify agent action boundaries and formulate process specifications in natural language. AI governance in process operations combines model governance and AI tool evaluation, developing the ability to manage ML model lifecycles and evaluate platform AI capabilities against organizational needs. Representative learning outcomes are: students can identify and document quality failures in an AI-generated BPMN model; students can design a human-AI handoff protocol for a given process scenario; students can apply a structured evaluation framework to compare AI feature depth across two BPM platforms. Assessment is best suited to case-based and artifact-based formats: model critique exercises, handoff design briefs, and structured platform evaluation reports.

5.2 Cross-Cutting Curriculum Implications

The BPM competency framework remains relevant under AI adoption, expanding rather than contracting, as it did under digitalization [18]. AI reshapes the tasks, governance demands, and collaboration context within every capability area while foundational BPM knowledge remains professionally relevant. The curriculum task is updating the framing of every element, not replacing the framework.

The most immediate curriculum priority is reorienting Methods/IT from execution to governance. With near-universal ML analytics, GenAI design, and agentic execution already deployed across most of the market, the practitioner's relationship with BPM's technical core has fundamentally shifted. Education should focus on the activities that remain human: scoping, interpreting, validating, governing, and communicating where AI handles operational work [7, 16].

The Governance element presents a double gap. On the tool side, G1 is nearly absent: practitioners receive no AI support for BPM tool selection despite a landscape of 17 AI-enabled platforms varying substantially in capability depth. On the curriculum side, AI-specific competencies for model lifecycle management, agent accountability, and AI-generated architecture standards remain undeveloped in existing BPM education.

AI literacy frameworks [20, 22] provide a necessary foundation, but the six *Newly Required* competencies are BPM-specific. Validating NL-to-BPMN models, governing ML models in compliance pipelines, and evaluating AI features across heterogeneous platforms require domain knowledge that general AI literacy courses do not transfer. No platform supports Culture competency directly, deepening an existing underinvestment in People and Culture [28] while AI creates the strongest demands on them.

6 Conclusion

BPM education lags behind AI-augmented practice, and rapid AI adoption is widening this gap. This paper addressed it through a three-step structured mapping: coding AI capabilities of 17 BPM platforms, synthesizing those codings by capability area, and assessing practitioner impact across all 30 capability areas of the Kerpedzhiev et al. framework. Two findings carry direct relevance for curriculum designers. First, AI reshapes practitioner demands across all areas through direct and indirect effects; no area becomes obsolete. Second, six competency demands have no equivalent in existing BPM frameworks: AI output validation, human-AI handoff design, prompt engineering, AI model governance, AI-generated artifact governance, and AI-enhanced tool evaluation. BPM educators who built their curricula around the Six Core Elements framework have the right foundation; the task is extending it to the demands the AI era introduces.

7 Limitations and Further Research

Three limitations bound the findings. First, the document analysis method may introduce discrepancy relative to actual AI deployment: vendor documentation reflects released capabilities only, cross-checked against secondary sources but not tested through hands-on use. Second, Step 3 involves interpretive judgment, particularly for cross-cutting inferences where AI adoption in adjacent areas reshapes demands without direct AI features; dual independent coding mitigates but does not eliminate this variance. The same applies to the finding that no capability area becomes obsolete: the document analysis method detects AI presence within areas, not AI substitution of them, so areas that are fully automated may still code as Present rather than redundant. Third, the study covers 17 platforms from three analyst reports, excluding AI-native tools outside analyst coverage that may surface additional competency demands.

Several future research directions follow. Empirical validation through curriculum design studies and graduate outcome tracking is the most immediate need: this study establishes what gaps exist using platform capabilities as a proxy, not whether addressing them improves practitioner preparation. A natural first step is a pilot implementation of one of the three proposed curriculum modules, assessed through pre/post competency measurement and employer feedback on graduate readiness. Extending the scope to AI-native tools would test the sufficiency of the six *Newly Required* competencies. A longitudinal replication would track the pace of capability landscape change. Finally, the cross-cutting classification approach may transfer as a general method for AI-era competency impact assessment beyond BPM.

References

1. Antonucci, Y.L., Fortune, A., Kirchmer, M.: An Examination of associations between Business Process Management capabilities and the benefits of digitalization: all capabilities are not equal. Bus. Process. Manag. J. **27**(1), 124–144 (2021)
2. Basili, V.R., Weiss, D.M.: A methodology for collecting valid software engineering data. IEEE Trans. Softw. Eng. **10**(6), 728–738 (1984)
3. Biernikowicz, A., Gabryelczyk, R., Ashraf, R.U.: Investigating the nexus between business process management maturity and digital maturity: an empirical study of organizations in Poland. Bus. Process. Manag. J. **31**(5), 1663–1683 (2025)
4. Bowen, G.A.: Document analysis as a qualitative research method. Qual. Res. J. **9**(2), 27–40 (2009)
5. Brynjolfsson, E., Mitchell, T.: What can machine learning do? Workforce implications. Science **358**(6370), 1530–1534 (2017)
6. Calvanese, D., Casciani, A., De Giacomo, G., et al.: Agentic business process management: a research manifesto. Inf. Syst. **140**, 102738 (2026)
7. Daugherty, P.R., Wilson, H.J.: Human + Machine: Reimagining Work in the Age of AI. Harvard Business Review Press, Boston (2018)
8. de Bruin, T., Rosemann, M.: Using the Delphi technique to identify BPM capability areas. In: Proceedings of the 18th Australasian Conference on Information Systems (2007)
9. Dumas, M., Milani, F., Chapela-Campa, D.: Agentic Business Process Management Systems. In: van de Weerd, I., Estrada Torres, B., van der Aa, H. (eds.) Business Process Management Workshops. LNBIP, vol. 569, pp. 3–14. Springer, Cham (2026)
10. Er, M., and Nathania, A.: Bridging the BPM education gap: Reflections on affordances of BPM techniques and a call for collaboration. In: BPM 2025 RBPM, PT and Educators Forum, LNBIP, vol. 565, pp. 442–455. Springer, Cham (2025)
11. Forrester Research: The Forrester Wave™: Process Intelligence Software, Q3 2025 (2025)
12. Gartner: Magic Quadrant for Process Mining Platforms (G00816659) (2025a)
13. Gartner: Magic Quadrant for Business Orchestration and Automation Technologies (G00828060) (2025b)
14. Graessler, I., and Oezcan, D.: Graphical AI workflow modelling: Identifying relevant competencies in AI-based automation of business processes. In: Proceedings of IHSED, vol. 198, pp. 318–328 (2025)
15. Grzesiak, M., Kluza, K., et al.: Beyond traditional learning: The LLM revolution in BPM education at university. In: BPM 2024 Workshops, LNBIP 527 is the BPM 2024 Blockchain, RPA, CEE, Educators and Industry Forum volume. Springer, Cham (2024). https://doi.org/10.1007/978-3-031-70445-1_29
16. Jarrahi, M.H.: Artificial intelligence and the future of work: Human-AI symbiosis in organizational decision making. Bus. Horiz. **61**(4), 577–586 (2018)
17. Kampik, T., Warmuth, C., et al.: Large process models: A vision for business process management in the age of generative AI. KI - Künstliche Intelligenz **39**(2), 81–95 (2024)
18. Kerpedzhiev, G.D., König, U.M., et al.: An exploration into future business process management capabilities in view of digitalization. Bus. Inf. Syst. Eng. **63**(2), 83–96 (2021)
19. Krippendorff, K.: Content Analysis: An Introduction to Its Methodology, 3rd edn. Sage, Thousand Oaks (2013)
20. Long, D., and Magerko, B.: What is AI literacy? Competencies and design considerations. In: Proceedings of CHI, pp. 1–16. ACM (2020)
21. Maslov, I., Poelmans, S., et al.: Generative AI's aid in business process modeling instructional design: A case study. In: BPM 2025 RBPM, PT and Educators Forum, LNBIP, vol. 565. Springer, Cham (2025). https://doi.org/10.1007/978-3-032-02936-2_24

22. Ng, D.T.K., Leung, J.K.L., et al.: Conceptualizing AI literacy: An exploratory review. Comput. Educ. Artif. Intell. **2**, 100041 (2021)
23. Raisch, S., Krakowski, S.: Artificial intelligence and management: The automation-augmentation paradox. Acad. Manag. Rev. **46**(1), 192–210 (2021)
24. Rosemann, M., and vom Brocke, J.: The six core elements of business process management. In: Handbook on Business Process Management 1, 2nd edn., pp. 105–122. Springer, Cham (2015). https://doi.org/10.1007/978-3-642-45100-3_5
25. Rosemann, M., vom Brocke, J., Van Looy, A., Santoro, F.: Business process management in the age of AI – three essential drifts. IseB **22**, 415–429 (2024)
26. Rosemann, M., and Joebges, P.: Comparing the design space of explorative BPM: MBAs versus GenAI. In: BPM 2025 RBPM, PT and Educators Forum, LNBIP, vol. 565. Springer, Cham (2025). https://doi.org/10.1007/978-3-032-02936-2_26
27. Shamasbi, S.M., Van Looy, A., et al.: A typological theory of roles in process mining: conceptualising competencies through text mining of job vacancy data. Enterprise Inf. Syst. **19**(11) (2025)
28. Sliż, P., Berniak-Woźny, J., Brzychczy, E., et al.: Business process management education in Poland: A manifesto for academic teaching. e-mentor **2**(104), 91–98 (2024)
29. Syed, R., Wynn, M. T., and Bandara, W.: The design and delivery of a holistic BPM education program. In: BPM 2023 Workshops, LNBIP, vol. 491, pp. 201–210. Springer, Cham (2023). https://doi.org/10.1007/978-3-031-43433-4_13
30. Vidgof, M., Bachhofner, S., and Mendling, J.: Large language models for business process management: Opportunities and challenges. In: BPM 2023 Forum, LNBIP, vol. 490, pp. 107-123 Springer, Cham (2023) https://doi.org/10.1007/978-3-031-41623-1_7
31. vom Brocke, J., Mendling, J., Rosemann, M.: Business Process Management Cases, vol. 2. Springer, Berlin, Heidelberg (2021)

Author Index

K. Kluza et al. (Eds.): BPM 2026, LNBIP 600, pp. 435–436, 2027.
https://doi.org/10.1007/978-3-032-36710-5

GPSR Compliance

The European Union's (EU) General Product Safety Regulation (GPSR) is a set of rules that requires consumer products to be safe and our obligations to ensure this.

If you have any concerns about our products, you can contact us on ProductSafety@springernature.com

In case Publisher is established outside the EU, the EU authorized representative is:

Springer Nature Customer Service Center GmbH
Europaplatz 3
69115 Heidelberg, Germany

Batch number: 10769310

Printed by Printforce, the Netherlands